Energy Resources
through Photochemistry and Catalysis

Energy Resources through Photochemistry and Catalysis

Edited by

Michael Grätzel
Institut de Chimie Physique
École Polytechnique Fédérale
Lausanne, Switzerland

1983

ACADEMIC PRESS
A Subsidiary of Harcourt Brace Jovanovich, Publishers
New York London
Paris San Diego San Francisco São Paulo Sydney Tokyo Toronto

ACADEMIC PRESS, INC.
111 Fifth Avenue, New York, New York 10003

United Kingdom Edition published by
ACADEMIC PRESS, INC. (LONDON) LTD.
24/28 Oval Road, London NW1 7DX

Library of Congress Cataloging in Publication Data

Main entry under title:

Energy resources through photochemistry and catalysis.

Includes index.
1. Renewable energy sources. 2. Photochemistry.
3. Catalysis. I. Grätzel, Michael.
TJ163.2.E486 1983 621.47 83-9983
ISBN 0-12-295720-2

PRINTED IN THE UNITED STATES OF AMERICA

83 84 85 86 9 8 7 6 5 4 3 2 1

Contents

Contributors

Numbers in parentheses indicate the pages on which the authors' contributions begin.

VINCENZO BALZANI (1), Istituto Chimico "G. Ciamician" dell'Università *and* Istituto di Fotochimica e Radiazioni d'Alta Energia del Consiglio Nazionale delle Ricerche, Bologna, Italy

ARTHUR J. FRANK (467), Solar Energy Research Institute, Golden, Colorado 80401

AKIRA FUJISHIMA (359), Department of Synthetic Chemistry, Faculty of Engineering, University of Tokyo, Tokyo 113, Japan

MICHAEL GRÄTZEL (71), Institut de Chimie Physique, École Polytechnique Fédérale, CH-1015 Lausanne, Switzerland

M. HALMANN (507), Isotope Department, The Weizmann Institute of Science, Rehovot, Israel

ANTHONY HARRIMAN (163), Davy Faraday Research Laboratory, The Royal Institution, London W1X 4BS, England

ADAM HELLER (385), Bell Laboratories, Murray Hill, New Jersey 07974

GARY HODES (421), Department of Plastics Research, The Weizmann Institute of Science, Rehovot, Israel, and Ormat Turbines Ltd., Yavneh, Israel

KENICHI HONDA (359), Department of Synthetic Chemistry, Faculty of Engineering, University of Tokyo, Tokyo 113, Japan

PIERRE P. INFELTA (49), Institut de Chimie Physique, École Polytechnique Fédérale, CH-1015 Lausanne, Switzerland

K. KALYANASUNDARAM (217), Institut de Chimie Physique, École Polytechnique Fédérale, CH-1015 Lausanne, Switzerland

T. KAWAI[1] (331), Institute for Molecular Science, Okazaki 444, Japan

J. KIWI (297), Institut de Chimie Physique, École Polytechnique Fédérale, CH-1015 Lausanne, Switzerland

GEORGE McLENDON (99), Department of Chemistry, The University of Rochester, Rochester, New York 14627

V. N. PARMON (123), Institute of Catalysis, Novosibirsk 90, U.S.S.R.

EZIO PELIZZETTI (261), Istituto di Chimica Analitica, Università di Torino, Torino, Italy

T. SAKATA (331), Institute for Molecular Science, Okazaki 444, Japan

FRANCO SCANDOLA (1), Istituto Chimico dell'Università *and* Centro di Fotochimica del Consiglio Nazionale delle Ricerche, Ferrara, Italy

A. E. SHILOV (535), Institute of Chemical Physics of the U.S.S.R. Academy of Sciences, Chernogolovka 142432, U.S.S.R.

MARIO VISCA (261), Centro Ricerche SIBIT (Montedison), Spinetta Marengo, Italy

TADASHI WATANABE (359), Department of Synthetic Chemistry, Faculty of Engineering, University of Tokyo, Tokyo 113, Japan

K. I. ZAMERAEV (123), Institute of Catalysis, Novosibirsk 90, U.S.S.R.

[1]Present address: Institute of Scientific and Industrial Research, Osaka University, Osaka 565, Japan.

Preface

The development of new energy resources constitutes a very active and challenging area of modern-day research. Confronted with dwindling supplies of fossil reserves, scientists face the task of conceiving alternative ways of energy conversion. New energy carriers need to be generated for use in future economic systems. This requires a vast research effort covering such diverse fields as solid-state physics, chemistry, and biology. Such an interdisciplinary venture has to be undertaken now when there appears to be still plentiful coal and petrol, in order to acquire the fundamental scientific knowledge on which mankind can rely and build once the conventional resources are exhausted.

While research in the area of new energy-conversion devices has been carried out for a relatively short time only, there has been an explosion of information on this subject. To cover all the significant discoveries in this field would be beyond the scope of a single book. Therefore, we decided to concentrate on topics of general importance where rapid progress has been achieved over the past few years and where a comprehensive documentation of the state of the art is needed. Thus, catalysis intervenes in most chemical transformations where energy is converted or needs to be saved. Catalysis of redox reactions and their application to the photocleavage of water, reduction of carbon dioxide, and fixation of nitrogen therefore constitute the central themes of the present book.

In photochemical or photoelectrochemical conversion systems, these catalytic events are linked to light-energy-harvesting and charge-separation processes. Including a discussion of some fundamental aspects of these phenomena appeared to us as being useful, in particular since many concepts that helped in the design of these devices and the understanding of their operation were developed only recently. This concerns, for example, light-induced redox reactions, reaction dynamics in organized assemblies such as micelles, colloidal metals, or semiconductors, and strategies for molecular engineering of artificial photosynthetic devices. Furthermore, the principles of electrochemical conversion of light energy via semiconductor electrodes or semiconducting particles are treated.

To deal with all these points in an encyclopedic manner would be a tantalizing experience for a single author. Fortunately, outstanding scientists from all over the world agreed to participate in this effort and address the important issues in individual contributions. Their expertise, acquired through extensive and excellent research in the particular areas of the field covered by the book, is thus made available to a wide readership. I am most grateful to these authors for their enthusiastic participation in the work which has made this venture successful.

1 Light-Induced and Thermal Electron-Transfer Reactions

Vincenzo Balzani

Istituto Chimico "G. Ciamician" dell'Università
and *Istituto di Fotochimica e Radiazioni d'Alta Energia*
del Consiglio Nazionale delle Ricerche
Bologna, Italy

Franco Scandola

Istituto Chimico dell'Università
and *Centro di Fotochimica del Consiglio Nazionale delle Ricerche*
Ferrara, Italy

ISBN 0-12-295720-2

I. Introduction

In principle, conversion of solar energy into chemical energy can be obtained by means of any thermodynamically uphill reaction produced by visible-light excitation (*1–4*). In practice, however, the conversion and storage of solar energy into a real energy resource requires the transformation of an abundant and low-cost raw material into a fuel (i.e., into a highly energetic chemical species that can be stored and transported). Simple economical, ecological, and energetic considerations show that water, carbon dioxide, and nitrogen are the most attractive raw materials that can be used as feedstocks of solar reactors, and that hydrogen, methane, methanol, and ammonia are among the most valuable fuels that one would like to obtain (*5, 6*). Thus it is not surprising that most of the current activity in the field of solar photochemistry is devoted to the four processes shown in Table I. Such processes, as well as the natural photosynthetic processes (*7*) are based on electron-transfer reactions. The elucidation of the factors that govern electron-transfer reactions (*8–16*) is fundamental for any progress in the field of photochemical conversion of solar energy.

A photochemical conversion system based on a redox process must involve a light-induced electron-transfer reaction. As we shall better see in Section VI,C, when a molecule absorbs a photon of suitable energy an electronically excited state is obtained that is a better oxidant and reductant than the ground state. An electron-transfer reaction between such an excited state and a suitable reaction partner may convert a fraction of the

TABLE I

Some Fuel-Forming Reactions Starting from Abundant and Low-Cost Materials[a]

	ΔG (kJ mol^{-1})[b]	n[c]	$E(V)$[d]
$H_2O(l) \xrightarrow{h\nu} H_2(g) + \frac{1}{2}O_2(g)$	237	2	1.23
$CO_2(g) + 2H_2O(l) \xrightarrow{h\nu} CH_3OH(l) + \frac{3}{2}O_2(g)$	703	6	1.21
$CO_2(g) + 2H_2O(l) \xrightarrow{h\nu} CH_4(g) + 2O_2(g)$	818	8	1.06
$N_2(g) + 3H_2O(l) \xrightarrow{h\nu} 2NH_3(g) + \frac{3}{2}O_2(g)$	678	6	1.17

[a] From Bolton and Hall (*6*).
[b] Free-energy change.
[c] Number of electrons transferred.
[d] Potential energy stored per electron transferred.

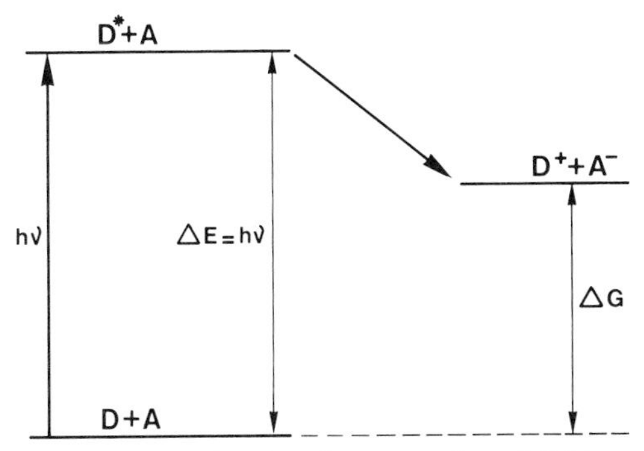

Fig. 1. Schematic energy diagram showing the conversion of light energy into chemical energy. D, Donor; A, acceptor.

absorbed light into chemical energy (Fig. 1). Usually, the raw material that we would like to convert into fuel (Table I) cannot be electronically excited by solar radiation. A typical example is that of water, the electronic absorption spectrum of which does not overlap the emission spectrum of the sun. In such a case the process must be mediated by a suitable chemical species called a *photosensitizer* (P) (*17*) (Fig. 2). Electron-transfer reactions converting raw materials into fuels are usually very slow because they involve the transfer of more than one electron (Table I). It follows that the excited state of the photosensitizer would usually undergo deactivation before reacting with the raw material. Therefore, a *relay* (R) species is usually needed that must first undergo a fast electron-transfer reaction with the excited state of the photosensitizer and then induce a thermal electron-transfer process that transforms the raw material into fuel (*18, 19*). The latter process is again slow because of its multielectron nature, whereas the competing back electron-transfer reac-

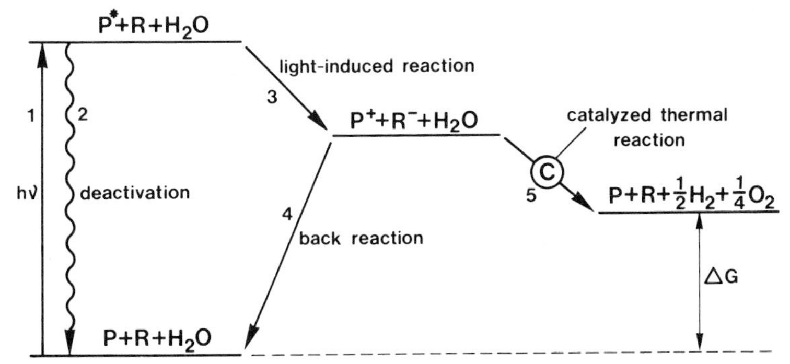

Fig. 2. Schematic energy diagram showing the photochemical conversion of a raw material (exemplified by H_2O) into fuel mediated by a photosensitizer (P) and a relay (R).

tion is very fast (Fig. 2). Thus a homogeneous or heterogeneous catalyst (C) is usually needed to speed up the thermal reaction leading to fuel (*18, 19*).

In conclusion, any artificial system for photochemical conversion and storage of solar energy consists of a light-induced electron-transfer reaction, which is followed by a sequence of thermal electron-transfer reactions that may involve homogeneous and/or heterogeneous catalysts. The efficiency of the system will depend on the relative rates of the various electron-transfer steps. In this chapter, the state of the art of the theories of electron-transfer reactions will be presented, and the role played by the various factors in governing the reaction rates will be discussed. We shall deal explicitly with homogeneous reactions, but many concepts that will be illustrated can also be applied to heterogeneous processes, such as those involving electrodes or heterogeneous catalysts.

II. Kinetic Formulation

A bimolecular electron-transfer reaction originating from a weak interaction[1] between a donor and an acceptor [Eq. (1)] can be discussed in terms of elementary steps as shown in the scheme

$$D + A \xrightarrow{k_{exp}} D^+ + A^- \tag{1}$$

$$D + A \underset{k_{-d}}{\overset{k_d}{\rightleftarrows}} D\cdots A \underset{k_{-e}}{\overset{k_e}{\rightleftarrows}} D^+\cdots A^- \xrightarrow{k'_{-d}} D^+ + A^- \tag{2}$$

where the electronic states of D and A are left unspecified; k_d, k_{-d}, and k'_{-d} are diffusion or dissociation rate constants; and k_e and k_{-e} the (unimolecular) rate constants for the forward and back electron transfer in the encounter. In the inorganic literature $D\cdots A$ and $D^+\cdots A^-$ are often called precursor and successor complex, respectively (*12, 20*). Note that the $D^+\cdots A^-$ successor complex in some systems can disappear via additional channels not shown in Eq. (2); in such cases the following kinetic treatment retains its validity, provided that k'_{-d} is substituted by the sum of the rate constants of the various channels. One such channel may be the rapid dissociation of D^+ and/or A^- into fragments. For excited-state electron-transfer reactions an important way for the disappearance of

[1] Classical cases of weak-interaction electron-transfer processes are outer-sphere electron-transfer reactions of transition-metal complexes (*10*). The problem of the magnitude of the interaction in electron-transfer processes will be discussed in later sections.

$D^+ \cdots A^-$ may be the back electron-transfer reaction to give the ground-state reactants, as we shall see in Section VI,B.

A simple steady-state treatment shows that the overall rate constant for the disappearance of the reactants [k_{exp}, Eq. (1)] can be expressed as a function of the rate constants of the various steps of Eq. (2):

$$k_{exp} = \frac{k_d}{1 + (k_{-d}/k_e) + (k_{-d}k_{-e}/k'_{-d}k_e)}. \quad (3)$$

In Eq. (3), $k_{-e}/k_e = \exp(\Delta G/RT)$, where ΔG is the free-energy change of the electron-transfer step. This quantity is given by (*11*)

$$\Delta G = \Delta G_{exp} - w_r + w_p = E(D^+/D) - E(A/A^-) - w_r + w_p, \quad (4)$$

where ΔG_{exp} is the free-energy change of reaction (1), which can be obtained from the standard redox potentials, and w_r and w_p are the work terms related to the formation of the precursor and successor complexes, respectively (see the following text).

For slow electron-transfer processes, that is, when $k_{-e} \ll k'_{-d}$ and $k_e \ll k_{-d}$, Eq. (2) reduces to the preequilibrium formulation (*12*)

$$k_{exp} = \frac{k_d k_e}{k_{-d}} = K_o k_e, \quad (5)$$

where K_o is the formation constant of the precursor complex. Note that in the preceding treatment the true electron-transfer step is a unimolecular reaction. The often used alternative approach (*8*) of considering an electron-transfer reaction as a bimolecular process whose rate constant is a function of the collision frequency between D and A (see Section VII,A) is less flexible for a classical treatment and less appropriate for a quantum mechanical treatment.

The diffusion and dissociation rate constants k_d and k_{-d} in Eqs. (3) and (5) are usually obtained from the Debye (*21*) and Eigen (*22*) equations

$$k_d^{(\mu=0)} = \frac{8RT}{3000\eta} \frac{b}{r} \frac{1}{e^{b/r} - 1} \quad (6)$$

and

$$k_{-d} = \frac{2kT}{\pi r^3 \eta} \frac{b/r}{1 - e^{-b/r}}, \quad (7)$$

where μ, η, and r are the ionic strength, viscosity, and encounter distance, and the term b is given by

$$b = Z_D Z_A e^2/\varepsilon kT, \quad (8)$$

where e is the electron charge, ε the dielectric constant, and Z_D and Z_A are the electronic charges of the two reactants. The ionic strength dependence of k_d is given by

$$\log k_d^{\mu} = \log k_d^{(\mu=0)} + \frac{1.02 Z_D Z_A \sqrt{\mu}}{1 + \sqrt{\mu}}. \tag{9}$$

Of course, k'_{-d} can also be obtained from Eq. (7) by using the appropriate quantities of the $D^+ \cdots A^-$ encounter.

According to the preceding mechanism [Eq. (2)], the key step of the electron-transfer process is the conversion of the precursor into the successor complex during the encounter. As we shall see in the next sections, this step can be dealt with using a classical or a quantum mechanical approach.

III. Classical Approach to Electron-Transfer Reactions

A. *Classical Adiabatic Model*

In a classical approach, the rate constant of the electron-transfer step of Eq. (2) that converts the precursor into the successor complex can be expressed by an equation that is formally the same as that used in the absolute reaction-rate theory. According to this theory (*23*), the rate constant for a unimolecular reaction is given by

$$k_e = k^\circ_e \exp(-\Delta G^\ddagger / RT), \tag{10}$$

where k°_e is a frequency factor and $\Delta G^\ddagger$ the standard free energy of activation. According to the absolute reaction-rate theory, the frequency factor is given by

$$k^\circ_e = \frac{\kappa k T}{h}, \tag{11}$$

where κ is the so-called transmission coefficient and kT/h a universal frequency factor ($\sim 6 \times 10^{12}$ sec^{-1} at 25°C) for the passage over the barrier.

To understand the meaning of k°_e and $\Delta G^\ddagger$ for electron-transfer reactions, it is convenient to discuss the problem on the basis of potential-energy diagrams (*8, 9, 12, 16*). We first consider the zero-order potential-energy surfaces of the initial and final states of the system as a function of the nuclear configuration (including inner molecular vibrations and outer solvation sphere). A cross section of these surfaces is shown in Fig. 3.

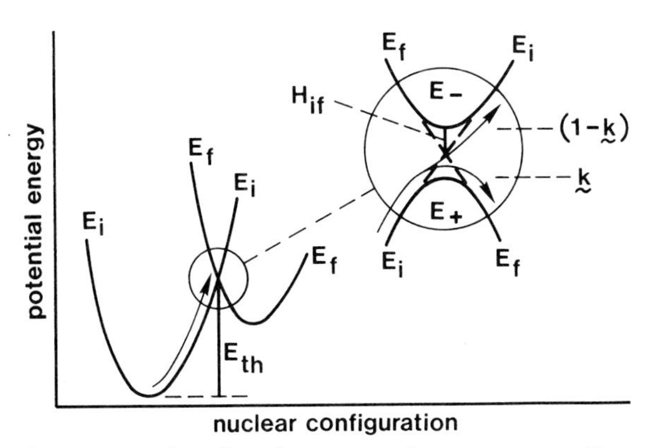

Fig. 3. Schematic representation of an electron-transfer process according to the classical approach. E_i and E_f are the zero-order potential-energy surfaces of the initial and final states.

The minima of the two curves correspond to the stable nuclear configurations of the reactants and products. The electron-transfer step must obey the Franck–Condon principle, which states that the nuclear positions and nuclear velocities remain essentially unchanged during the electronic transition. Thus the only possibility for electron transfer to occur is near the crossing point of the two curves. This requires an adjustment of the inner vibrational coordinates of the reactants and of the outer solvation spheres to some nonequilibrium configuration *prior* to electron transfer. The energy separation E_{th} between the minimum of the initial state and the crossing point where the reaction can occur is called reorganization energy, and it is usually taken to be equal to the reorganization free energy $\Delta G^{\ddagger}$ that appears in Eq. (10), on the assumption (*24, 25*) that the entropy contribution related to the reorganization is negligible. Strictly speaking, it should be noted (*15*) that $\Delta G^{\ddagger}$ of the electron-transfer reactions does not have the meaning of an activation energy because it does not refer to a transition state but is related to a nonequilibrium nuclear configuration of the reactants that has to be reached to obey the Franck–Condon principle. For the same reason, the frequency factor of an electron transfer reaction is not given by Eq. (11) but by

$$k^{\circ}_{e} = \kappa \nu_{N}, \tag{12}$$

where the universal frequency kT/h has been replaced by an effective frequency for nuclear motion ν_N, which is defined by the parabolic curves in Fig. 3 and can be expressed as a function of the inner and outer vibrational frequencies (*12, 25*).

Once the suitable nuclear configuration has been reached, whether or not electron transfer occurs is a matter of electronic factors. If there is no electronic interaction between the zero-order initial and final states, the surfaces will intersect, ignoring each other, and the reaction cannot occur. If there is an electronic interaction, the situation in the intersection region can no longer be described by the separate zero-order functions. Instead, it is necessary to form linear combinations of these wave functions that correspond to new first-order surfaces that avoid each other (Fig. 3) (*8, 9, 12, 16*). The separation between the two first-order surfaces is equal to $2H_{if}$, where H_{if} is the interaction energy (electronic coupling matrix element). In the classical theories (like the Marcus theory) (*8, 26–28*) the interaction energy is assumed to be small enough to be neglected in calculating E_{th} but large enough so that when the initial system reaches the intersection region the probability of the reactants being converted into products is unity. This is the so-called *adiabatic* assumption, according to which the system always remains on the lower potential-energy surface on passing through the intersection region. This is equivalent to saying that the transmission coefficient κ is unity. We shall see subsequently that for a more general treatment this assumption must be removed and the possibility considered that a system arriving from the initial state may continue along the original zero-order curve and then return without undergoing reaction. This so-called *nonadiabatic* behavior may be taken into account by a lower than unity transmission coefficient. We shall come back to this problem in Section III,B.

Thus in a classical adiabatic approach the rate constant of the electron-transfer reaction converting the precursor into the successor complex is given by

$$k_e = \nu_N \exp(-\Delta G^{\ddagger}/RT). \tag{13}$$

To discuss the activation-energy term in more detail, let us consider a so-called exchange reaction, that is, a symmetrical ($\Delta G = 0$) electron-transfer reaction

$$\text{Red}\cdots\text{Ox} \longrightarrow \text{Ox}\cdots\text{Red} \tag{14}$$

where Red and Ox are the reduced and oxidized forms of the same molecule, respectively. A classical case of this kind of reaction is that between $Fe(H_2O)_6^{2+}$ and $Fe(H_2O)_6^{3+}$ [Eq. (15)], which has been the object of sev-

$$Fe(H_2O)_6^{2+}\cdots Fe(H_2O)_6^{3+} \longrightarrow Fe(H_2O)_6^{3+}\cdots Fe(H_2O)_6^{2+} \tag{15}$$

eral experimental and theoretical investigations (*9, 16, 24, 25*). The reorganization free energy is made up by two parts, one related to the internal vibrations of the two reactants (inner-shell contribution) and the other to the solvent repolarization (outer-shell contribution):

$$\Delta G^{\ddagger} = \Delta G^{\ddagger}_{in} + \Delta G^{\ddagger}_{out}. \tag{16}$$

We first consider the inner-shell contribution, which is essentially related to the different metal–ligand bond lengths and frequencies in the oxidized and reduced forms of the molecule. If it is assumed that the vibrations of the coordination shells are harmonic (at least to the intersection region), the inner-shell reorganization energy can be easily estimated (*8, 12*). Calling a_{red} and a_{ox} the metal–ligand equilibrium distances and f_{red} and f_{ox} the metal–ligand breathing force constants of the reactants, the metal–ligand distance at the intersection point is given by

$$a = \frac{f_{red}a_{red} + f_{ox}a_{ox}}{f_{red} + f_{ox}}, \tag{17}$$

and the potential energy required to adjust the six metal–ligand distances in Red and Ox to the common distance a of the intersection point is given by

$$E_{in} = \frac{3f_{red}f_{ox}(a_{red} - a_{ox})^2}{f_{red} + f_{ox}}. \tag{18}$$

If the vibrational partition functions of the reactants do not change in going to the intersection region (which is a good approximation for weakly interacting molecules), $\Delta G^{\ddagger}_{in}$ is equal to E_{in}, and thus Eq. (18) gives the inner-shell contribution to the free activation energy. For reaction (15), $(a_{red} - a_{ox}) = 0.14$ Å and the breathing frequencies $\bar{\nu}_{red}$ and $\bar{\nu}_{ox}$ (which are related to the force constants f_{red} and f_{ox} by $f = 4\pi^2\nu^2\mu$, where μ is the mass of a single water molecule) are 390 and 490 cm^{-1}, respectively. The calculated inner-shell contribution to the free activation energy results to be 8.4 kcal mol^{-1} (*24*).

We next consider the energy required to reorganize the solvent sphere outside the coordination shell of the reactants. The solvent can be treated as a dielectric continuum having electronic, atomic, and orientation polarization. In the initial state of the exchange reaction the polarization of the medium is in equilibrium with the charges of the reactants. Because the solvent electrons move very rapidly, the electronic polarization of the medium can change in phase with the transferring electron. In contrast, the atomic and orientation polarization relaxes too slowly to change in phase with the transferring electron (Franck–Condon principle). Thus the atomic and orientation polarization must be adjusted to a suitable non-equilibrium value prior to electron transfer. It can be shown (*8, 20, 26–28*) that in the approximation of the continuum model for the solvent, the free energy needed for this reorganization is given by

$$\Delta G^{\ddagger}_{out} = (\Delta e)^2\left(\frac{1}{2a_{red}} + \frac{1}{2a_{ox}} - \frac{1}{r}\right)\left(\frac{1}{\varepsilon_{op}} - \frac{1}{\varepsilon_s}\right), \tag{19}$$

where Δe is the change in the charge of each reactant, a_{red} and a_{ox} are the previously seen equilibrium bond distances of Red and Ox, r is the distance between the centers of the two reactants in the encounter complex (generally taken to be equal to the close contact distance $a_{red} + a_{ox}$), ε_{op} the optical dielectric constant, and ε_s the static dielectric constant of the solvent. In the case of reaction (15) in aqueous solution, the value of $\Delta G^{\ddagger}_{out}$ calculated from Eq. (19) is 6.4 kcal mol^{-1} (*24*).

For an exchange reaction [Eq. (14) or (15)] the standard free-energy change ΔG is zero, and the free activation energy is only related to the horizontal displacement and the curvature of the two curves representing the initial and final states of the system (Fig. 4a). For the more general case of cross-electron-transfer reactions [i.e., electron-transfer reactions

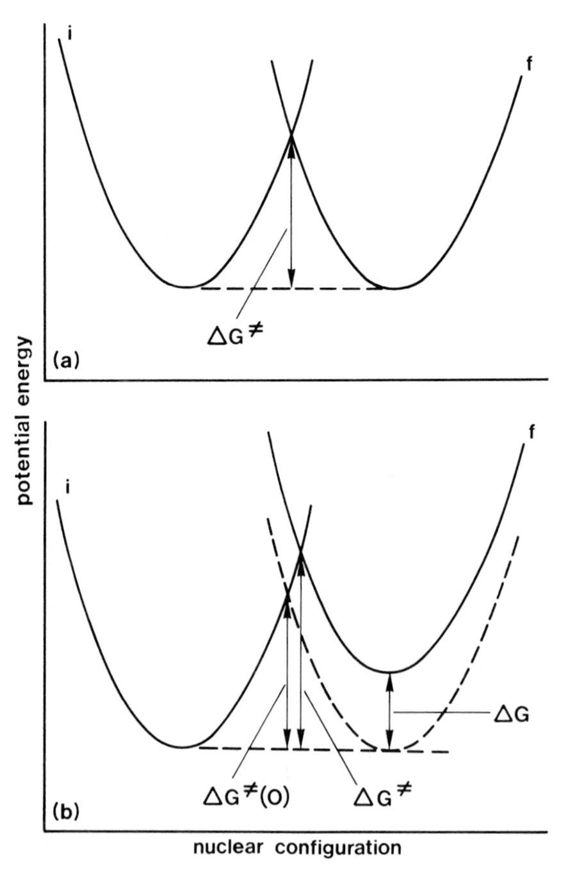

Fig. 4. Schematic representation of the energy surfaces of the initial and final states for (a) exchange and (b) cross-electron-transfer reactions. i, Initial; f, final.

$$D\cdots A \longrightarrow D^{+}\cdots A^{-} \tag{20}$$

$$Fe(H_2O)_6^{2+}\cdots Ru(NH_3)_6^{3+} \longrightarrow Fe(H_2O)_6^{3+}\cdots Ru(NH_3)_6^{2+} \tag{21}$$

having net chemical change, Eqs. (20) and (21)], it is clear that the free activation energy depends not only on the horizontal displacement and curvature of the two curves, but also on the standard free-energy change ΔG that represents the vertical displacement of the minima of the two curves under the usual assumption of a negligible entropy change (Fig. 4b). It can be shown (*8, 12*) that if the initial and final states can be described by the same harmonic function, the free activation energy of the cross reaction (20) is given by

$$\Delta G^{\ddagger}_{DA} = \Delta G^{\ddagger}(0)_{DA}\{1 + [\Delta G_{DA}/4\ \Delta G^{\ddagger}(0)_{DA}]\}^2, \tag{22}$$

where $\Delta G^{\ddagger}(0)_{DA}$ is the so-called intrinsic reorganizational parameter (i.e., the free energy of activation for $\Delta G = 0$), which is only related to the horizontal displacement and curvature of the two curves and is given by

$$\Delta G^{\ddagger}(0)_{DA} = (\Delta G^{\ddagger}_{D} + \Delta G^{\ddagger}_{A})/2, \tag{23}$$

where $\Delta G^{\ddagger}_{D}$ and $\Delta G^{\ddagger}_{A}$ are the previously discussed [Eq. (16)] intrinsic barriers for the exchange reactions:

$$D\cdots D^{+} \longrightarrow D^{+}\cdots D \tag{24}$$

$$A\cdots A^{-} \longrightarrow A^{-}\cdots A \tag{25}$$

Equation (22) is the famous Marcus quadratic equation, which has been extensively used since 1960 to rationalize and predict the rate of electron-transfer processes (see Section VII,A). When Eq. (22) is explicitly used in Eq. (13), one obtains the plot shown by curve a in Fig. 5 for $\ln(k_e/\nu_N)$ versus ΔG_{DA}. As one can see, the rate constant has its maximum for $\Delta G_{DA} = -4\ \Delta G^{\ddagger}(0)_{DA}$ and should drastically *decrease* with increasing exergonicity when $\Delta G_{DA} < -4\ \Delta G^{\ddagger}(0)_{DA}$ (Marcus inverted region).

Relationships such as Eq. (22), where the free energy of activation is expressed as a function of the net free-energy change and of some intrinsic parameter, are very common in chemical kinetics and are called free-energy relationships (FERs) (*29*). For electron-transfer reactions, other *empirical* FERs have been proposed and used in the chemical literature. They are the Polanyi linear equation (*30, 30a*)

$$\Delta G^{\ddagger} = \alpha\ \Delta G + \beta, \tag{26}$$

the Rehm–Weller equation (*31, 31a*)

$$\Delta G^{\ddagger} = (\Delta G/2) + \{(\Delta G/2)^2 + [\Delta G^{\ddagger}(0)]^2\}^{1/2}, \tag{27}$$

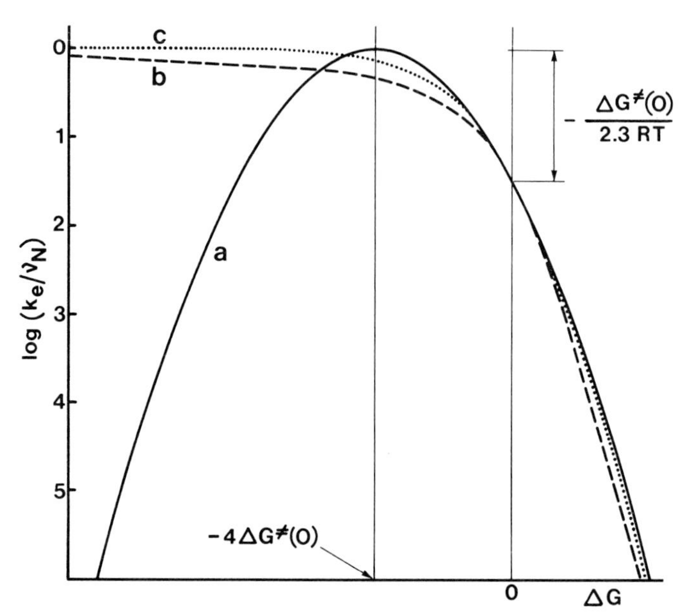

Fig. 5. Plot of $\log(k_e/\nu_N)$ versus ΔG according to Eq. (13) and (a) the Marcus quadratic [Eq. (22)], (b) the Rehm–Weller [Eq. (27)], and (c) the Marcus–Agmon–Levine hyperbolic [Eq. (28)] free-energy relationships.

and the hyperbolic equation derived first by Marcus (*32*) for atom-transfer reactions and formulated later empirically by Agmon and Levine (*33, 34*):

$$\Delta G^{\ddagger} = \Delta G + \frac{\Delta G^{\ddagger}(0)}{\ln 2} \ln\left\{1 + \exp\left[-\frac{\Delta G \ln 2}{\Delta G^{\ddagger}(0)}\right]\right\}. \tag{28}$$

In Eq. (26) α (usually $0 < \alpha < 1$) and β are empirical parameters; β is conceptually similar to, but not always equal to, $\Delta G^{\ddagger}(0)$. It has been shown (*35, 35a, 36*) that the linear FERs of Eq. (26) can be viewed as tangents of the curves corresponding to nonlinear FERs. Thus they can be considered to be approximations of the other FERs over narrow ranges and will no longer be discussed. When Eq. (27) or (28) is explicitly used in Eq. (13), one obtains the plots shown by curves b or c in Fig. 5. As one can see, according to these empirical FERs the rate constant tends asymptotically to its maximum value with increasing exergonicity.

More generally, one can see that Eqs. (22), (27), and (28) behave very much the same around $\Delta G = 0$, but they differ for very negative ΔG values, as discussed previously, and for very positive ΔG values, where Eqs. (27) and (28) give a linear instead of a quadratic dependence of $\ln(k_e/\nu_N)$ on ΔG. It should also be noted that the shapes of the curves depend on

$\Delta G^{\ddagger}(0)$. The curve corresponding to Eq. (22) becomes broader and broader as $\Delta G^{\ddagger}(0)$ increases, and at the same time its maximum moves toward more negative ΔG values. For the curves corresponding to Eqs. (27) and (28) the nonasymptotic region becomes more and more extended as $\Delta G^{\ddagger}(0)$ increases. These problems will be discussed in more detail in Section VII,B.

B. Extension of the Classical Model

As we have previously seen, in the adiabatic model it is assumed that once the system arrives at the intersection region of the zero-order surfaces (Fig. 3), the probability that the reactants are converted into products is unity. This however, may not be the case when the interaction between the two zero-order curves is small owing to spin, symmetry, or overlap problems. In such cases the reaction is called *nonadiabatic*.

Within the classic approach, nonadiabaticity can formally be taken into account by using the transmission coefficient κ of the absolute reaction-rate theory,

$$k_e = \kappa \nu_N \exp(-\Delta G^{\ddagger}/RT), \tag{29}$$

and relating it to the interaction energy between the two zero-order surfaces (Fig. 3) by means of the semiclassical Landau–Zener theory (*15, 25, 37*). According to this theory, the probability of the reactants being converted into products per single passage through the intersection region is given by

$$P_{if} = 1 - \exp[-(4\pi^2 H_{if}^2/hv|s_i - s_f|)], \tag{30}$$

where H_{if} is the previously seen interaction energy at the crossing point, v the average velocity with which the system moves through the intersection region, and s_i and s_f are the derivatives of the zero-order surfaces in the intersection region. The velocity v is generally put equal to $(2RT/\pi\mu)^{1/2}$, where μ is the reduced mass of the system, which in the case of a transition-metal complex is usually the mass of a single ligand. The electronic transmission coefficient, which expresses the overall probability that the system will undergo a transition from the initial to the final zero-order surfaces on passing over the potential energy barrier (Fig. 3), is given by

$$\kappa = 2P_{if}/(1 + P_{if}). \tag{31}$$

From Eqs. (30) and (31) it is evident that when H_{if} is large (strong coupling between the initial and final states), P_{if} and κ are unity (adiabatic behavior); when H_{if} is small (weak coupling), P_{if} and κ are also small (nonadiabatic behavior). Equation (30) can be further elaborated for the case in which $|s_{\mathrm{i}}| = |s_{\mathrm{f}}|$, which is valid for an exchange reaction (Fig. 4a). In such a case $|s_{\mathrm{i}}| = |s_{\mathrm{f}}| = (2f_{\mathrm{i}}E_{\mathrm{th}})^{1/2}$, where $f_{\mathrm{i}} = f_{\mathrm{f}}$ are the force constants of the parabolic curves of Fig. 4a. Recalling that $f = 4\pi^2\nu_{\mathrm{N}}^2\mu$ and $E_{\mathrm{th}} = \Delta G^{\ddagger}$, Eq. (30) yields

$$P_{\mathrm{if}} = 1 - \exp\left[-\left(\frac{H_{\mathrm{if}}^2}{h\nu_{\mathrm{N}}}\right)\left(\frac{\pi^3}{\Delta G^{\ddagger}RT}\right)^{1/2}\right]. \tag{32}$$

In the weak-interaction case this equation can be expanded and substituted into Eq. (31), yielding

$$\kappa = \frac{2H_{\mathrm{if}}^2}{h\nu_{\mathrm{N}}}\left(\frac{\pi^3}{\Delta G^{\ddagger}RT}\right)^{1/2}, \tag{33}$$

which taken together with Eqs. (16) and (29) then gives

$$k_{\mathrm{e}} = \frac{H_{\mathrm{if}}^2}{h}\left[\frac{\pi^3}{(\Delta G_{\mathrm{in}}^{\ddagger} + \Delta G_{\mathrm{out}}^{\ddagger})RT}\right]^{1/2} \exp[-(\Delta G_{\mathrm{in}}^{\ddagger} + \Delta G_{\mathrm{out}}^{\ddagger})/RT], \tag{34}$$

which is a classical expression for the rate constant for a nonadiabatic electron-exchange reaction (*15, 25*).

An important difference between the adiabatic and nonadiabatic classical expressions must be emphasized. In the adiabatic expression [Eq. (13)] the preexponential factor (i.e., the maximum value for the rate constant) is a nuclear frequency, and thus the electron-transfer process resembles an ordinary chemical reaction that is consummated by a nuclear motion. In contrast, in the nonadiabatic expression [Eq. (34)] the preexponential term does not contain a nuclear frequency but is essentially an electron frequency.

Aside from the adiabaticity problem, a central assumption of the classical model is that reaction can only occur when the system goes over the barrier represented by E_{th} (Fig. 3). This would result in an Arrhenius-type dependence of the rate constant on temperature. It is well known, however, that a chemical reaction may have an activation energy at higher temperatures but may become essentially temperature independent at low temperatures. This is explained by assuming that the reaction proceeds via thermal excitation at the higher temperatures and via nuclear tunneling from the potential-energy surface of the reactants to that of the products at low temperatures. In this view, Eq. (29) represents the high-temperature limit of a more general equation that should contain a nuclear-tunneling factor. This problem has been discussed (*25*) in the case of exchange reactions. It has been shown that the tunneling correction is not important at room temperature unless the difference between the equilibrium configurations of the two oxidation states is extremely large. As we shall see in the next section, the tunneling problem finds its natural explanation within the frame of the quantum mechanical approach.

IV. Quantum Mechanical Approach to Electron-Transfer Reactions

Quantum mechanical models for homogeneous electron-transfer reactions have been elaborated by various authors (*14, 15, 38–43*). All of these models consider the electron-transfer process as a radiationless transition between zero-order vibronic states of the total system consisting of the donor molecule, the acceptor molecule, and the solvent. The rate constant for the electron-transfer process is then obtained following perturbation theory as a thermally averaged transition probability between the initial and final manifolds of vibronic states. The basic features of the model (*15, 40*) can be schematized as follows.

Consider the "supermolecule" consisting of the donor D, the acceptor A, and the polar solvent S, at a fixed intermolecular D—A distance. The total Hamiltonian of the system can be written in the equivalent alternative forms:

$$\mathcal{H} = \mathcal{H}^{e}_{(DA)} + T_N + V_{(eA)}, \tag{35}$$

$$\mathcal{H} = \mathcal{H}^{e}_{(D^+A^-)} + T_N + V_{(eD^+)}. \tag{36}$$

In Eqs. (35) and (36) T_N is the sum of the nuclear kinetic energy operators for the whole system, and $\mathcal{H}^{e}_{(DA)}$ and $\mathcal{H}^{e}_{(D^+A^-)}$ are the "electronic" Hamiltonians of the total system with the electron localized on the donor or on the acceptor, respectively; that is,

$$\mathcal{H}^{e}_{(DA)} = T_e + \mathcal{H}_{D^+} + \mathcal{H}_A + \mathcal{H}_S + V_{(D^+A)} + V_{(D^+S)} + V_{(AS)} + V_{(eD^+)} + V_{(eS)}, \tag{37}$$

$$\mathcal{H}^{e}_{(D^+A^-)} = T_e + \mathcal{H}_{D^+} + \mathcal{H}_A + \mathcal{H}_S + V_{(D^+A)} + V_{(D^+S)} + V_{(AS)} + V_{(eA)} + V_{(eS)}, \tag{38}$$

where T_e is the kinetic energy of the transferred electron; $\mathcal{H}_{D^+}$, $\mathcal{H}_A$, and $\mathcal{H}_S$ are the electronic Hamiltonians of the bare donor, the acceptor, and the solvent; $V_{(D^+A)}$ is the electrostatic interaction between the bare donor and the acceptor; $V_{(D^+S)}$ and $V_{(AS)}$ are the interactions that couple the ionic charges of bare donor and acceptor with the solvent; and $V_{(eS)}$ is the interaction between the transferred electron and the solvent. In Eqs. (35)–

(38), $V_{(eD^+)}$ and $V_{(eA)}$ are the interactions between the transferred electron and the bare donor and the acceptor, respectively. According to the adiabatic Born–Oppenheimer approximation, one can define a complete orthonormal set of zero-order "initial" electronic states of the system (i.e., the ground and excited states of the DA pair) as solutions of

$$\mathcal{H}^{e}_{(DA)}\Psi_{(DA)i}(r, Q) = U_{(DA)i}(Q)\Psi_{(DA)i}(r, Q), \qquad (39)$$

where r and Q denote electron and nuclear coordinates, respectively, and subscript i identifies the initial state. $U_{(DA)i}(Q)$ are the nuclear potential-energy surfaces of the zero-order initial states. Similarly, the zero-order final electronic states and potential-energy surfaces can be obtained by solving

$$\mathcal{H}^{e}_{(D^+A^-)}\Psi_{(D^+A^-)j}(r, Q) = U_{(D^+A^-)j}(Q)\Psi_{(D^+A^-)j}(r, Q). \qquad (40)$$

To follow the time evolution of the system during the electron-transfer process, it is useful to consider the total, time-dependent wave function of the system as an expansion in terms of both sets $\Psi_{(DA)i}$ and $\Psi_{(D^+A^-)j}$ of the initial and final zero-order states. To a usually valid first approximation, the expansion can be limited to two zero-order states, which in the following will simply be called $\Psi_{(DA)}$ and $\Psi_{(D^+A^-)}$ [with potential-energy surfaces $U_{(DA)}$ and $U_{(D^+A^-)}$]:

$$\Psi(r, Q, t) = \chi_{(DA)}(Q, t)\Psi_{(DA)} + \chi_{(D^+A^-)}(Q, t)\Psi_{(D^+A^-)}, \qquad (41)$$

where the χs are time-dependent nuclear expansion coefficients. Inserting this wave function into the time-dependent Schrödinger equation

$$i\hbar \frac{\partial\Psi(r, Q, t)}{\partial t} = \mathcal{H}\Psi(r, Q, t) \qquad (42)$$

that contains the total Hamiltonian [Eq. (35) or (36)] results in a pair of coupled equations (*15, 40*) for the expansion coefficients $\chi_{(DA)}$ and $\chi_{(D^+A^-)}$. The next step is to ignore the Born–Oppenheimer breakdown operator (nonadiabaticity operator) and to define the zero-order vibrational wave functions $\chi^{\circ}_{(DA)v}$ and $\chi^{\circ}_{(D^+A^-)w}$ of the initial and final electronic states through the eigenvalue equations

$$\Big[T_N + \varepsilon_{(DA)}(Q) + \langle\Psi_{(DA)}|V_{(eA)}|\Psi_{(DA)}\rangle$$

$$- \frac{S}{1 - S^2}\langle\Psi_{(D^+A^-)}|V_{(eA)}|\Psi_{(DA)}\rangle\Big]\chi^{\circ}_{(DA)v} = E^{\circ}_{(DA)v}\chi^{\circ}_{(DA)v} \qquad (43)$$

and

$$\left[T_N + \varepsilon_{(D^+A^-)}(Q) + \langle\Psi_{(D^+A^-)}|V_{(eD^+)}|\Psi_{(D^+A^-)}\rangle - \frac{S}{1-S^2}\langle\Psi_{(DA)}|V_{(eD^+)}|\Psi_{(D^+A^-)}\rangle\right]\chi^\circ_{(D^+A^-)w} = E^\circ_{(D^+A^-)w}\chi^\circ_{(D^+A^-)w}, \quad (44)$$

in which S is the electronic overlap integral and $E^\circ_{(DA)v}$ and $E^\circ_{(D^+A^-)w}$ are the energies of the zero-order vibronic states $|(DA)v\rangle \equiv \Psi_{(DA)}\chi^\circ_{(DA)v}$ and $|(D^+A^-)w\rangle \equiv \Psi_{(D^+A^-)}\chi^\circ_{(D^+A^-)w}$. Expanding the general time-dependent nuclear factors $\chi_{(DA)}$ and $\chi_{(D^+A^-)}$ in terms of these zero-order vibronic levels, and using second-order perturbation theory, one obtains a "golden rule" expression for the transition probability:

$$W_{(DA)v} = (2\pi/h)|H_{if}|^2 \sum_w |\langle\chi^\circ_{(DA)v}|\chi^\circ_{(D^+A^-)w}\rangle|^2 \times \delta[E^\circ_{(DA)v} - E^\circ_{(D^+A^-)w}]. \quad (45)$$

In Eq. (45), $W_{(DA)v}$ is the transition probability from an initial vibronic level $|(DA)v\rangle$ to the manifold of final levels $\{|(D^+A^-)w\rangle\}$, and δ is the Dirac delta function that ensures energy conservation. The Condon approximation has been used in Eq. (45), allowing the separation of the electronic matrix element H_{if} from the vibrational overlap term. The electronic matrix element is an "effective" perturbation of the form given by

$$H_{if} = \langle\Psi_{(DA)}|V_{(eA)}|\Psi_{(D^+A^-)}\rangle - \frac{S}{1-S^2}\langle\Psi_{(D^+A^-)}|V_{(eA)}|\Psi_{(D^+A^-)}\rangle, \quad (46)$$

which couples the initial and final states of the process.

In the limit of weak electronic interaction between D and A, the energy spacing between adjacent vibronic levels in DA is much larger than the width $\Gamma_{(DA)v} = \hbar W_{(DA)v}$ of each level, so that each vibronic level of DA has an independent exponential decay.[2] In this limit (kinetically called the *nonadiabatic limit*), the overall probability of transition from the initial manifold $\{|(DA)v\rangle\}$ to the final one $\{|(D^+A^-)w\rangle\}$, that is, the unimolecular rate constant for the electron-transfer process, can be obtained by thermally averaging $W_{(DA)v}$, that is,

$$k_e = Z^{-1}\sum_v \exp[-E^\circ_{(DA)v}/RT]\,W_{(DA)v}, \quad (47)$$

[2] An additional condition for the decay of each vibronic level of DA is that the levels of D^+A^- constitute a dense, quasi-continuum manifold. These two conditions, taken literally for the forward electron-transfer reaction, would preclude the occurrence of the reverse electron-transfer reaction.

where

$$Z = \sum_{v} \exp[-E^\circ_{(DA)v}/RT]. \tag{48}$$

The complete expression for the unimolecular rate constant of an electron-transfer step can, therefore, be written as

$$k_e = k_{el} k_{nucl}, \tag{49}$$

where

$$k_{el} = |H_{if}|^2/h \tag{50}$$

and

$$k_{nucl} = (4\pi^2/Z) \sum_{v} \sum_{w} \exp[-E^\circ_{(DA)v}/RT] \langle \chi^\circ_{(DA)v} | \chi^\circ_{(D^+A^-)w} \rangle \times \delta[E^\circ_{(DA)v} - E^\circ_{(D^+A^-)w}]. \tag{51}$$

Equations (49)–(51) provide a general expression for the unimolecular rate constant of electron-transfer processes that satisfy the following conditions: (i) only two electronic states (one for the reactant and one for the product) need to be considered, (ii) the density of final levels is large, and (iii) the transition probability is small (a nonadiabatic process).

The electronic term k_{el} of the unimolecular electron-transfer rate constant, which in this model is assumed to be small, depends crucially on the extent of donor–acceptor orbital overlap, so that its value maximizes at essentially contact distances. Its accurate calculation, however, is probably still beyond the capability of quantum chemistry because it involves the behavior of the tails of the electronic wave functions. Estimates based on band-structure calculations in organic solids suggest values of the order of 1 cm^{-1} (*40*), whereas a calculation performed on the $Fe(H_2O)_6^{2+}$–$Fe(H_2O)_6^{3+}$ couple gives ~10 cm^{-1} for the electronic matrix element of this system (*24*). On general grounds, situations that may cause anomalously low electronic factors include: (i) the participation of "inner" orbitals in the electron-transfer process (e.g., the *f* orbitals in rare-earth ions), (ii) steric hindrance to close contact between the chromophores involved in the electron-transfer process, and (iii) spin factors (e.g., low- and high-spin transitions in transition-metal complexes).

The nuclear part k_{nucl} of the electron-transfer rate constant can be conveniently treated by separating the nuclear motions of the system into two categories: (i) discrete, high-frequency modes, characterized by nuclear coordinates Q_c and vibrational frequencies $\nu_c \gg kT/h$ (quantum modes), and (ii) low-frequency modes, characterized by normal modes q_k and the corresponding frequencies $\nu_k \ll kT/h$ (classical modes). Typical modes of type (i) are the C—C and C—H vibrations in organic molecules and metal–ligand vibrations in coordination compounds. The solvent reorganizational motions, in contrast, can be considered as type (ii) modes. If the equilibrium nuclear configurations and frequencies of the quantum modes are identical in the initial and final states, the nuclear term of the rate constant is determined by the classical modes only. It can be shown (*38, 39, 39a, 39b*) that in such a case Eq. (51) becomes

$$k_{\text{nucl}} = 2(\pi^3/RTE_r)^{1/2} \exp -[(\Delta E + E_r)^2/4E_rRT)], \tag{52}$$

where E_r is the solvent reorganizational energy and ΔE the difference between the minimum of the zero-order potential energy surface $U_{(D^+A^-)}$ and that of $U_{(DA)}$. For an isoergic reaction ($\Delta E = 0$), this result is equivalent to that obtained using the classical model and allowing for nonadiabatic behavior [Eq. (34)], where $\Delta G^\ddagger_{\text{in}} + \Delta G^\ddagger_{\text{out}} = E_r/4$. Thus the use of the classical model (corrected for nonadiabaticity) corresponds to considering the inner-reorganizational energy as being made up of low-frequency classical modes.

If the equilibrium configurations and frequencies of the quantum modes change in going from the initial to the final state, the nuclear term of the rate constant is determined by both classical and quantum modes. The vibrational wave functions and energies can be split [Eqs. (53) and (54)] into contributions by D, A, D^+, A^-, and solvent:

$$\chi^\circ_{(DA)v} = \chi^\circ_{Dl}\chi^\circ_{Am}\chi^\circ_{Sn}, \tag{53a}$$

$$\chi^\circ_{(D^+A^-)w} = \chi^\circ_{D^+p}\chi^\circ_{A^-t}\chi^\circ_{Su}, \tag{53b}$$

$$E^\circ_{(DA)v} = E^\circ_{Dl} + E^\circ_{Am} + E^\circ_{Sn} = \varepsilon_{Dl} + \varepsilon_{Am} + \varepsilon_{Sn} - \Delta E, \tag{54a}$$

$$E^\circ_{(D^+A^-)w} = E^\circ_{D^+p} + E^\circ_{A^-t} + E^\circ_{Su} = \varepsilon_{D^+p} + \varepsilon_{A^-t} + \varepsilon_{Su}, \tag{54b}$$

where l, m, n, and p, t, u define the initial and final vibrational states, respectively, of donor, acceptor and solvent; ε represents vibrational energy; and ΔE is the energy difference between the minima of the $U_{(D^+A^-)}$ and $U_{(DA)}$ potential-energy surfaces. Under this assumption Eq. (51) can be reduced to

$$\begin{aligned} k_{\text{nucl}} = {} & 2(\pi^3/E_rRT)^{1/2}(Z_DZ_A)^{-1} \\ & \times \sum_l \sum_m \sum_p \sum_t \exp[-(\varepsilon_{Dl} + \varepsilon_{Am})/RT] \\ & \qquad \times \exp -[(\Delta E + E_r + \varepsilon_{D^+p} + \varepsilon_{A^-t} \\ & \qquad - \varepsilon_{Dl} - \varepsilon_{Am})^2/4E_rRT] \\ & \qquad \times |\langle\chi^\circ_{Dl}|\chi^\circ_{D^+p}\rangle|^2 |\langle\chi^\circ_{Am}|\chi^\circ_{A^-t}\rangle|^2, \end{aligned} \tag{55}$$

where $Z_D = \Sigma_l \exp -(\varepsilon_{Dl}/RT)$ and $Z_A = \Sigma_m \exp -(\varepsilon_{Am}/RT)$. Equation (55) provides a general picture of the effect of classical and quantum modes on the nuclear part of the electron-transfer rate constant. In the high-temperature limit, Eq. (55) reduces to Eq. (52).

It is interesting to examine the behavior predicted by Eq. (55) for the dependence of log k_{nucl} on ΔE because this dependence can be regarded as a quantum mechanical counterpart of a free-energy relationship (Sections III,A and VII,B). If the quantum modes have identical equilibrium configurations and frequencies in DA and D^+A^-, Eq. (52) can be used, leading to a parabolic plot of log k_{nucl} versus ΔE such as that shown in Fig. 6a. The maximum of the parabola is located at $\Delta E = -E_r$, and log k_{nucl} at $\Delta E = 0$ is lower than the maximum value by $E_r/4$. This behavior is clearly reminiscent of that predicted classically using the transition-state theory and the Marcus FER [Eq. (22)]. If the quantum modes have displaced equilibrium configurations and/or different frequencies in D^+A^- with respect to DA, Eq. (55) should be used. The calculations in this case involve more or less standard techniques to evaluate the Franck–Condon factors in terms of displacements and frequency shifts (*15, 40, 44*). Although calculations on practical systems are hardly feasible without using severe approximations, sample calculations can be performed to show the role of the quantum modes in the log k_{nucl} versus ΔE plots (*41*). The shape of a plot obtained using the same solvent reorganizational energy as in the para-

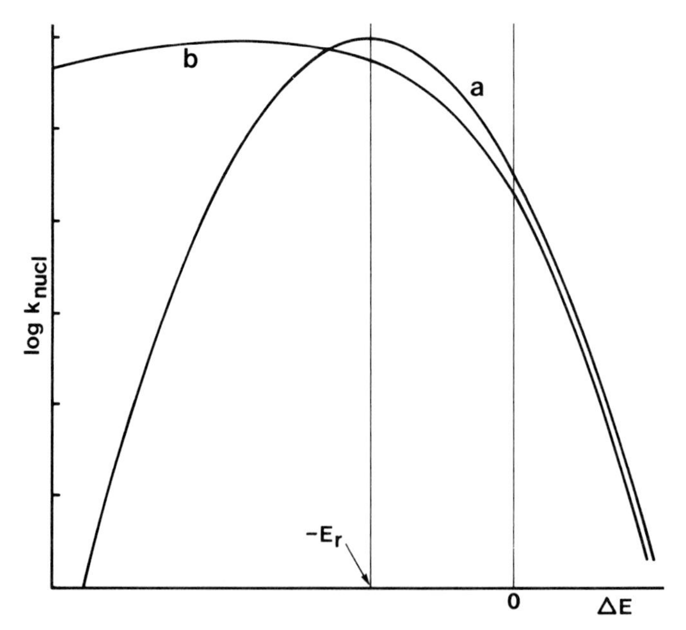

Fig. 6. Schematic plot of log k_{nucl} versus ΔE for (a) distorted classical modes and undistorted quantum modes [Eq. (52)] and (b) distorted classical and quantum modes (see text).

bolic case, considering a single quantum mode having a displaced equilibrium configuration (with constant frequency), is shown schematically in Fig. 6b. It can be seen that the log k_{nucl} versus ΔE plot is asymmetric and, with respect to the parabolic case, has its maximum displaced toward negative ΔE, with a lower value at $\Delta E = 0$ and a lower slope for large, negative ΔE. It can be shown that this asymmetric behavior becomes more pronounced as (i) the displacement in the equilibrium configuration of the quantum modes becomes larger and (ii) the number of displaced quantum modes becomes larger. It can also be shown that asymmetry, though not involving a shift in the maximum, is also brought about by changes in the frequency of the quantum modes in going from reactants to products. In conclusion, the quantum model predicts that if a sufficient number of highly distorted quantum modes is present in the donor–acceptor redox system, a relatively flat (or only slightly "inverted") behavior of log k_{nucl} versus ΔE should be observed in the exoergic region.

V. Comparison between the Classical and Quantum Mechanical Models

From a theoretical point of view, the full quantum mechanical treatment of electron-transfer reactions is vastly superior to the classical one. However, the classical model, particularly in the Marcus formulation (*8, 26–28*) has enjoyed a wide popularity over the years, even after the appearance of the quantum models. This is undoubtedly because of (i) its conceptually simple basis, (ii) its ability to rationalize and correlate rates of cross reactions and exchange reactions (Section VII), and (iii) the ease with which the intrinsic barriers can be calculated from molecular and medium parameters. In contrast, the practical use of quantum models may often be made difficult or even prohibited by computational difficulties or unavailability of relevant molecular parameters. A comparison of the classical and quantum mechanical models may be worthwhile to stress similarities and differences and to arrive at a possible complementary use of the two theories. Critical discussions of quantum mechanical versus classical models have been given by various authors (*14, 15, 25, 44–46*).

Looking at the classical expression for the unimolecular electron-transfer rate constant [Eq. (13)], one can notice that, owing to the adiabaticity assumption, no electronic factor is present, and both the activation term and the maximum rate (preexponential factor) depend only on the nuclear factor. If, however, allowance is made for nonadiabaticity, the classic rate constant [Eq. (34)] contains an electronic factor in the preexponential term, much in the same way as does the quantum mechanical expression [Eq. (49)]. There is still, however, a substantial difference in the two expressions. Whereas the term representing the solvent rearrangement (which can safely be treated in the high-temperature limit) has the same form in the quantum mechanical and the classical models, the term accounting for the rearrangement of the nuclear coordinates has a sharply different form in the two models, that is, it is a simple additional term in the classic free energy of activation, whereas it amounts to a thermally averaged Franck–Condon overlap sum in the quantum mechanical expression. It has been shown (*25*) that for isoergonic reactions ($\Delta G \simeq \Delta E = 0$) the quantum mechanical expression can be recast within the classical formalism, provided that a temperature-dependent internal intrinsic barrier $\Delta G^{\ddagger}_{\text{in}}(T)$ is used, as defined by

$$
\begin{aligned}
\exp -[\Delta G^{\ddagger}_{\text{out}} + \Delta G^{\ddagger}_{\text{in}}(T)]/RT
&= [(E_{\text{out}} + E_{\text{in}})/E_{\text{out}}]^{1/2}(Z_{\text{D}}Z_{\text{A}})^{-1} \\
&\quad \times \sum_{l}\sum_{m}\sum_{p}\sum_{t} \exp -[(\varepsilon_{\text{D}l} + \varepsilon_{\text{A}m})/RT] \\
&\quad \times \exp -[(E_{\text{out}} + \varepsilon_{\text{D}^{+}p} + \varepsilon_{\text{A}^{-}t} - \varepsilon_{\text{D}l} - \varepsilon_{\text{A}m})^{2}/4E_{\text{out}}RT] \\
&\quad \times |\langle \chi^{\circ}_{\text{D}l}|\chi^{\circ}_{\text{D}^{+}p}\rangle|^{2}|\langle \chi^{\circ}_{\text{A}m}|\chi^{\circ}_{\text{A}^{-}t}\rangle|^{2}.
\end{aligned}
\tag{56}
$$

The quantitative consequences of using the classical temperature-independent $\Delta G^{\ddagger}_{\text{in}}$ instead of $\Delta G^{\ddagger}_{\text{in}}(T)$ have also been discussed, and it has been concluded (*25*) that in typical cases the exchange rate constants calculated at room temperature are not very sensitive to this choice. Thus, unless one is interested in detailed temperature dependence of electron-transfer rate constants, the classical model is reasonably adequate to discuss isoergonic (or nearly so) reactions.

Where the largest differences in the predictions of the classical and quantum mechanical models show up is in the highly exergonic region of electron-transfer processes. In such a region [$\Delta G < -4\,\Delta G^{\ddagger}(0)$] the classical Marcus theory predicts the famous "inverted" behavior, corresponding to a symmetric parabolic increase in the activation free energy with increasing exergonicity (Fig. 5). The quantum mechanical model, in contrast, by virtue of the presence of accepting distorted quantum modes, implies a much shallower decrease of reaction rates (which is equivalent to a much smaller increase in activation free energy) on the highly exoergic side (Fig. 6). The extension of the "flat" log k_{exp} versus ΔG behavior in the exoergonic region is, according to the quantum mechanical model, a function of the number of distorted quantum modes and of the degree of their distortion. Although in most practical cases it is difficult to evaluate how far the "flat" region will extend, the failure to observe inverted behavior in real systems (Section VII) indicates that such quantum effects play an important role.

On a purely operational basis, it seems that a reasonable way to include the properties of the quantum mechanical models over wide ΔG ranges is to use the classical theory (corrected for nonadiabaticity) coupled with one of the empirical FER [Eqs. (27) and (28)], which, owing to their aympototic behavior ($\Delta G^{\ddagger} \rightarrow 0$ for $\Delta G \rightarrow -\infty$), allow for sustained rates in the highly exergonic region. This approach is admittedly incorrect in principle because both the classical and quantum models predict a decrease in the rate constant for more or less exergonic reactions. Nevertheless, it may be useful as a frame for comparison and rationalization of the available data as well as for predicting the behavior of systems not yet investigated.

A point concerning the classical nonadiabatic model and the quantum mechanical model that could be stressed is that both of them consider the electronic factor as not being explicitly dependent on the energetics of the reaction. This point has been tackled (*49*) on the basis of an electron-tunneling model considering simple binding potentials for the electron on D and A. The prediction of this model is that of increasing electronic matrix elements with increasing driving force of the reaction. This would introduce a new compensation mechanism playing against the "inversion" in the behavior of the rate constant in the highly exergonic region, on the basis of a more or less gradual increase in the adiabaticity of the electron-transfer reactions in that region.

VI. Peculiar Features of Electronically Excited States as Reactants in Electron-Transfer Processes

An electronically excited state is obtained when a molecule absorbs a photon of suitable energy, for example,

$$D + h\nu \longrightarrow D^* \tag{57}$$

Each electronically excited state is virtually a new chemical species with its own chemical and physical properties. The approaches illustrated in the previous sections apply regardless of the electronic states of reactants and products, and thus they can also be used in dealing with excited-state electron-transfer reactions. In doing this, however, the following features of electronically excited states have to be taken into consideration: (i) the excited states are transient species whose lifetimes may range from 10^{-12} to 10^{-3} s in fluid solution; (ii) an excited state has always a higher electronic affinity and a lower ionization potential compared with the ground-state molecule, and thus it is both a better oxidant and a better reductant than the ground state; (iii) the equilibrium nuclear geometry of the excited state may be more or less different from that of the ground state; and (iv) formation of an excited state usually involves population of high-energy, expanded molecular orbitals.

A. *Excited-State Lifetime*

In fluid solution, an excited-state electron-transfer reaction [Eq. (58)] occurs via an encounter between the excited state and the reaction partner (which is usually called the *quencher*). The encounter rate is determined by the diffusion rate constant k_d, which for the usual solvents is of the order of $10^{10}\ M^{-1}\ s^{-1}$. Simple kinetic considerations show that even when high quencher concentrations are used, only those excited states that have lifetimes longer than 10^{-10}–10^{-9} s can be involved in encounters with the quencher. This usually precludes the participation to bimolecular processes to all the excited states but the lowest one of any multiplicity. An important consequence of this kinetic constraint is that the excited states involved in electron-transfer processes in fluid solution are thermally equilibrated species, so that their reactions can be dealt with using thermodynamic and kinetic arguments as in the case of any other chemical reaction (*48*). For those organized systems in which electron transfer may occur from levels that are not thermally equilibrated, the classical approach loses any validity, and the quantum mechanical approach must be drastically modified (*50*).

$$D^* + A \underset{k_{-d}}{\overset{k_d}{\rightleftharpoons}} D^*\cdots A \xrightarrow{k_e} D^+\cdots A^- \tag{58}$$

B. *Kinetic Scheme*

For an excited-state reaction [Eq. (59)], the kinetic scheme of Eq. (2) must be replaced by that shown in Fig. 7, where τ is the excited-state lifetime, and the rate constants concerning the ground-state electron-transfer steps have now been labeled with g for the sake of clarity. The bimolecular electron-transfer rate constant k_{exp}, which can now be obtained as a Stern–Volmer quenching constant, is given by

$$D^* + A \xrightarrow{k_{exp}} D^+ + A^- \tag{59}$$

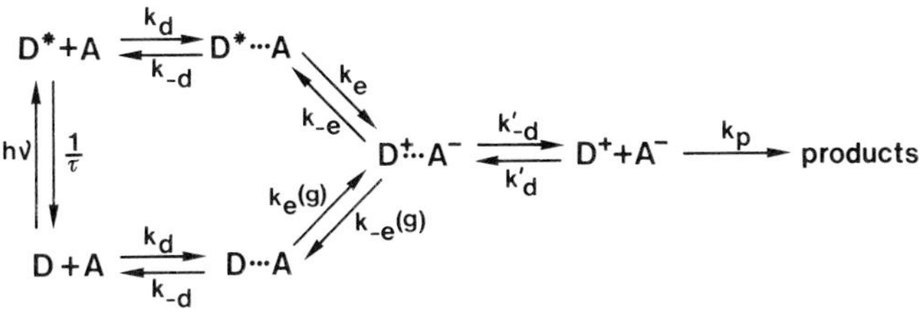

$$
\begin{array}{ccccccc}
D^{*}+A & \underset{k_{-d}}{\overset{k_d}{\rightleftharpoons}} & D^{*}\cdots A & & & & \\
h\nu \uparrow \downarrow \frac{1}{\tau} & & k_e \searrow \nwarrow k_{-e} & & & & \\
 & & & D^{+}\cdots A^{-} & \underset{k'_{d}}{\overset{k'_{-d}}{\rightleftharpoons}} & D^{+}+A^{-} & \xrightarrow{k_p} \text{products} \\
 & & k_e(g) \nearrow \swarrow k_{-e}(g) & & & & \\
D+A & \underset{k_{-d}}{\overset{k_d}{\rightleftharpoons}} & D\cdots A & & & &
\end{array}
$$

Fig. 7. Kinetic scheme for an excited-state electron-transfer reaction. For details, see text.

$$k_{\mathrm{exp}} = \frac{k_{\mathrm{d}}}{1 + (k_{-\mathrm{d}}/k_{\mathrm{e}}) + (k_{-\mathrm{d}}k_{-\mathrm{e}}/k_{\mathrm{x}}k_{\mathrm{e}})}, \tag{60}$$

which is the same as Eq. (3) except that $k'_{-\mathrm{d}}$ has now been replaced by k_{x}, where $k_{\mathrm{x}} \simeq k_{-\mathrm{e}}(\mathrm{g})$ when $k_{-\mathrm{e}}(\mathrm{g}) \gg k'_{-\mathrm{d}}$ or when k'_{d} is the only path through which D^+ and A^- can disappear, and $k_{\mathrm{x}} \simeq k'_{-\mathrm{d}}$ when $k'_{-\mathrm{d}} \gg k_{-\mathrm{e}}(\mathrm{g})$ and some other reaction, k_{p}, rapidly consumes D^+ and/or A^-.

The back electron-transfer reaction $k_{-\mathrm{e}}(\mathrm{g})$ plays an important role in determining the yield of the excited-state electron-transfer process because it causes a short circuit that voids the effect of photoexcitation (Section IX).

C. *Free-Energy Changes*

In the previous sections we have seen that an important factor in determining the rate constant of electron-transfer reactions is the free-energy change ΔG of the electron-transfer step. For an excited-state electron-transfer reaction, ΔG can be obtained as follows *(51)*. The excited-state reactant of reaction (61) possesses an extra energy with respect to the ground-state reactant of reaction (62). If both the excited molecule and

$$D^{*}\cdots A \underset{k_{-\mathrm{e}}}{\overset{k_{\mathrm{e}}}{\rightleftharpoons}} D^{+}\cdots A^{-} \tag{61}$$

$$D\cdots A \underset{k_{-\mathrm{e}}}{\overset{k_{\mathrm{e}}}{\rightleftharpoons}} D^{+}\cdots A^{-} \tag{62}$$

the ground-state molecule lie in their zero vibrational levels, the extra energy content of D^* is the zero–zero spectroscopic energy E_{oo}. Passing to thermodynamic quantities, if the vibrational partition functions of the two states are not very different, the enthalpy difference between excited- and ground-state molecules is practically equal to E_{oo}, and the entropy difference is negligible. Thus the excited states have an extra free-energy content that is approximately equal to E_{oo}. This means that to a first approximation the reduction and oxidation potentials of the excited-state molecules are given by

$$E°(D^*/D^-) = E°(D/D^-) + E_{oo}(D^*) \tag{63}$$

and

$$E°(D^*/D^+) = E°(D/D^+) + E_{oo}(D^*), \tag{64}$$

where $E°(D/D^-)$ and $E°(D/D^+)$ are the reduction and oxidation potentials of the ground-state molecules and $E_{oo}(D^*)$ is the one-electron potential corresponding to the zero–zero spectroscopic energy of the excited state. Thus the excited state is both a better oxidant and a better reductant than the ground state, and the free-energy change of reaction (61) is given by the free-energy change of reaction (62) [Eq. (4)] minus a free-energy term equal to $E_{oo}(D^*)$, that is,

$$\Delta G = E°(D^+/D) - E°(A/A^-) - E_{oo}(D^*) - w_r + w_p . \tag{65}$$

D. *Intrinsic Barrier*

An excited state may differ from the corresponding ground state in size, shape, and dipole moment. It follows that the reorganizational intrinsic barrier will generally be different for ground and excited states of the same molecule. As the change in shape, size, and solvation of an excited state with respect to the ground state is reflected in the Stokes shift between absorption and emission maxima, it can be expected that the difference in the reorganizational barrier between ground- and excited-state reactions is also related to the Stokes shift. Unfortunately, such a relation is difficult to find out except in trivial cases or unless severe approximations are used (*51*). Furthermore, the Stokes shift is often unknown. In several cases, however, it is not difficult to predict whether the intrinsic barrier of the excited-state reaction will be smaller or larger than that of the ground-state reaction. For example, for d^6 low-spin octahedral complexes the intrinsic barrier for the reduction of the lowest metal-centered excited state [Eq. (66)] is expected to be lower than that for

$$ML_6(t_{2g}{}^5e_g{}^1)^{*n+} + Red \longrightarrow ML_6(t_{2g}{}^6e_g{}^1)^{(n-1)+} + Ox \tag{66}$$

ground-state reduction [Eq. (67)]. Both the reduced complex $ML_6{}^{(n-1)+}$

$$ML_6(t_{2g}{}^6e_g{}^0)^{n+} + Red \longrightarrow ML_6(t_{2g}{}^6e_g{}^1)^{(n-1)+} + Ox \tag{67}$$

and the excited complex $ML_6{}^{n+*}$ are expected to be similarly distorted with respect to the regular octahedral structure of $ML_6{}^{n+}$ because of the presence of an electron in the e_g antibonding orbitals. Similarly, the oxidation of the same d^6 low-spin $ML_6{}^{n+}$ complex is expected to involve a smaller intrinsic barrier in the ground state than in the excited state.

E. Transmission Coefficient

As we have seen in Sections III,B and IV, the rate constant of an electron-transfer reaction in the nonadiabatic regime is proportional to the square of the electronic coupling matrix element H_{if} [Eq. (46)], which depends on the overlap between the donor and acceptor wave functions. Excited-state electron-transfer reactions generally involve more expanded molecular orbitals than the corresponding ground-state reactions, and thus they are expected to have larger transmission coefficients (i.e., a more adiabatic character) (*48*).

These effects may be particularly important for transition-metal complexes, in which some types of excited states involve very expanded molecular orbitals. Consider, for example, the reduction of $Ru(bipy)_3^{3+}$ to give either the ground-state $Ru(bipy)_3^{2+}$ or the lowest excited-state $Ru(bipy)_3^{2+*}$:

$$Ru(bipy)_3^{3+} + Red \longrightarrow Ru(bipy)_3^{2+} + Ox \tag{68}$$

$$Ru(bipy)_3^{3+} + Red \longrightarrow Ru(bipy)_3^{2+*} + Ox \tag{69}$$

In Eq. (68) the transferred electron has to enter the t_{2g} metal orbitals, whereas in Eq. (69) the electron goes into the π^* orbitals of the bipy ligand. The latter orbitals will certainly give a better overlap with the Red orbitals, and thus reaction (69) is expected to have a better electronic factor than reaction (68). As another example, consider the ground and the lowest excited state of Eu^{2+}. Reduction of an Ox species by the ground state involves the transfer of an f electron, whereas reduction by the lowest excited state (which is obtained by $f \rightarrow d$ excitation) involves the transfer of a much less shielded d electron. As a consequence, the excited-state reaction is again expected to exhibit a better transmission coefficient.

The consequences of the successive involvement of ground and excited states on the shape of the log k_{exp} versus ΔG plots will be discussed in Section VIII,C.

VII. Correlations of Rate Constants

Two kinds of relations are frequently used in the chemical literature in an attempt to rationalize the available results and to predict the behavior of systems not yet studied. Such approaches are more or less related to the theories discussed previously and are also related to each other. For historical reasons and for the sake of clarity we shall discuss these relations separately.

A. The Marcus Cross Relation

Let us consider a unimolecular cross reaction [Eq. (20)] and the corresponding unimolecular exchange reactions [Eqs. (24) and (25)]. Assuming that (i) the rate constant of each reaction can be expressed by a classical adiabatic equation [Eq. (13)], (ii) the preexponential term is the same for the three reactions, (iii) the intrinsic barrier of the cross reaction is given by the average of the intrinsic barriers of the two exchange reactions [Eq. (23)], and (iv) the free activation energy of the cross reaction is given by the Marcus quadratic equation [Eq. (22)],

$$\begin{aligned} k_{\mathrm{DA}} &= (k_{\mathrm{D}} k_{\mathrm{A}} K_{\mathrm{DA}} f)^{1/2}, \\ \log f &= (\log K_{\mathrm{DA}})^2 / [4 \log(k_{\mathrm{D}} k_{\mathrm{A}} / \nu^2_{\mathrm{N}})], \end{aligned} \tag{70}$$

is obtained, where K_{DA} is the equilibrium constant for the cross reaction. Equation (70) is the so-called Marcus cross relation. The equation originally derived by Marcus (*8, 26, 27*) was really

$$\begin{aligned} k_{\mathrm{exp}} &= (k'_{\mathrm{D}} k'_{\mathrm{A}} K'_{\mathrm{DA}} f')^{1/2}, \\ \log f' &= (\log K'_{\mathrm{DA}})^2 / [4 \log(k'_{\mathrm{D}} k'_{\mathrm{A}} / Z^2)], \end{aligned} \tag{71}$$

where k_{exp} is the bimolecular rate constant for the cross reaction [Eq. (1)], k'_{D} and k'_{A} are the bimolecular rate constants for the corresponding exchange reactions, K'_{DA} is the equilibrium constant of the cross reaction and Z a collision frequency, generally taken as $10^{11}\ M^{-1}\ \mathrm{s}^{-1}$. Equation (71) can be derived from Eq. (70) with the further assumptions that the work terms for the cross and exchange reactions are the same and that the reagents may be treated as spherical, structureless reactants.

Equation (71) is very popular among inorganic and bioinorganic chemists, who regularly use it to calculate rates of reactions that are otherwise difficult to measure (*16, 52, 53*). This equation is also used in reverse, to calculate $\log K'_{\mathrm{DA}}$ from rate data and hence to obtain otherwise unknown redox potentials (*54, 55*). The Marcus cross relation is indeed very useful, but one should not forget that it is based on several assumptions. Conspicuous failures of the cross relation occur for some systems, showing that in such cases at least one of the assumptions made is not obeyed. For example, values ranging from 10^{-3} to $10^{-10}\ M^{-1}\ \mathrm{s}^{-1}$ have been obtained for the exchange of the Eu^{3+}/Eu^{2+} couple, using five different cross reactions involving Eu^{2+} as a reductant (*56*). Similar problems are found for other redox couples, such as V^{3+}/V^{2+} (*56*). Corrections for differences in the work terms or adiabaticity can be incorporated in Eq. (71) (*56, 57*), and the consequences of removing some other assumption have also been examined (*57, 58*). Although for europium ions the failure of Eq. (71) is most probably caused by nonadiabaticity (*47*), the "non-Marcus" behavior of other couples is not easy to understand.

When f' in Eq. (71) is ~ 1 (i.e., when log K'_{DA} and/or $k'_D k'_A$ is sufficiently small), Eq. (71) may be further simplified to give

$$k_{exp} = (k'_D k'_A K'_{DA})^{1/2}. \tag{72}$$

Thus if a series of related reactions with $f' \simeq 1$ is studied as a function of driving force, a plot of log k_{exp} versus log K'_{DA} should be linear, with slope 0.5 and intercept 0.5 log $k'_D k'_A$. These linear plots are extensively used (and sometimes misused) in the chemical literature. We shall come back to this problem in the next section.

B. Free-Energy Relationships

Let us first define a *series of homogeneous reactions* (*47, 48*). Consider, for example, the electron-transfer reactions between the same reductant and a series of structurally related oxidants (A_1, A_2, ..., A_n) that have variable redox potential but the same size, shape, electric charge, and electronic structure:

$$\begin{aligned} D + A_1 &\longrightarrow D^+ + A_1^- \\ D + A_2 &\longrightarrow D^+ + A_2^- \\ &\vdots \\ D + A_n &\longrightarrow D^+ + A_n^- \end{aligned} \tag{73}$$

It can be assumed that throughout this homogeneous series of reactions the parameters k_d, k_{-d}, k'_{-d} in Eqs. (3) and (60), K_o in Eq. (50), ν_N in Eqs. (13) and (29), $\Delta G^{\ddagger}(0)_{DA}$ in Eq. (23), κ in Eq. (29), k_x in Eq. (60), and Z in Eq. (71) are constant. Under these assumptions, k_{exp} obtained from Eq. (3), (60), (71), or (72) is only a function of the free-energy change, that is, of the oxidation potentials of the members of the homogeneous series A_n. More generally, a homogeneous series of reactions can involve different reductants and/or oxidants. Procedures are also available to account for a small degree of nonhomogeneity of some of the preceding parameters (*47*).

Consider now the general equation for k_{exp} [Eq. (60)], which includes Eqs. (3) and (5) as special cases. The ratio k_{-e}/k_e is equal to $\exp(\Delta G/RT)$, and, using the classical nonadiabatic expression [Eq. (29)] for k_e, Eq. (60) can be written as

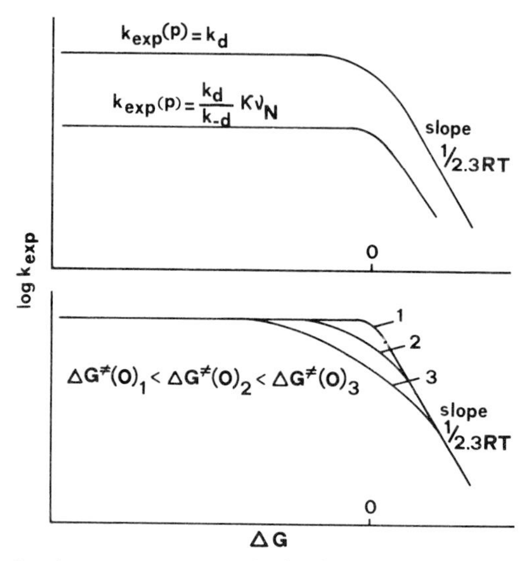

Fig. 8. Influences of various parameters on the log k_{exp} versus ΔG plots. From Balzani and Scandola (*48*).

$$k_{exp} = \frac{k_d}{1 + [k_{-d}/\kappa\nu_N \exp(-\Delta G^{\ddagger}/RT)] + (k_{-d}/k_x)\exp(\Delta G/RT)}. \quad (74)$$

As we have seen in Section III,A, $\Delta G^{\ddagger}$ can be expressed as a function of ΔG, so that in a homogeneous series k_{exp} is only a function of the free-energy change. The log k_{exp} versus ΔG plot so obtained will be different depending on the specific FER that is chosen.

When the Marcus quadratic equation [Eq. (22)] is used, the log k_{exp} versus ΔG plot exhibits the peculiar inverted behavior (Fig. 5). Such a behavior, however, is not experimentally observed, as will be discussed in Section VIII,A.

When Eq. (27) or (28) is used, the plots obtained are of the type shown in Fig. 8 (*35, 35a, 36, 47, 48*). In these plots we may distinguish the following regions: (i) a plateau region for sufficiently exergonic reactions, (ii) an Arrhenius-type linear region (limiting slope $1/2.3RT$) for sufficiently endergonic reactions, and (iii) an intermediate region (centered at $\Delta G = 0$) in which log k_{exp} increases in a complex but monotonic way as ΔG decreases. Simple mathematical considerations show that the plateau value $k_{exp}(\text{p})$ (i.e., k_{exp} for $\Delta G \rightarrow -\infty$) is equal to

$$k_{exp}(\text{p}) = k_d\kappa\nu_N/(\kappa\nu_N + k_{-d}), \quad (75)$$

and thus it does not depend on $\Delta G^{\ddagger}(0)$. Equation (75) reduces to

$$k_{exp}(\text{p}) \simeq k_d \quad (76)$$

or

$$k_{\mathrm{exp}}(\mathrm{p}) \simeq (k_{\mathrm{d}}/k_{-\mathrm{d}})\kappa\nu_{\mathrm{N}} = K_{\mathrm{o}}\kappa\nu_{\mathrm{N}}, \tag{77}$$

depending on whether $\kappa\nu_{\mathrm{N}}$ is much larger or much smaller than $k_{-\mathrm{d}}$. Thus a lower than diffusion-controlled value of $k_{\mathrm{exp}}(\mathrm{p})$ is related to a low value of the transmission coefficient κ (nonadiabatic reaction). For large and positive values of ΔG [the preceding point (ii)], k_{exp} is given by

$$k_{\mathrm{exp}}(\mathrm{p}) = \frac{k_{\mathrm{d}}\kappa\nu_{\mathrm{N}}k_{\mathrm{x}}}{k_{-\mathrm{d}}(k_{\mathrm{x}} + \kappa\nu_{\mathrm{N}})}\exp(-\Delta G/RT), \tag{78}$$

and thus in this case also there is no dependence on $\Delta G^{\ddagger}(0)$. In contrast, $\Delta G^{\ddagger}(0)$ strongly affects the values of the curve in the intermediate ΔG region. As is shown in Fig. 8, for very small values of $\Delta G^{\ddagger}(0)$ the intermediate region is almost unnoticeable, and the connection between the plateau and the Arrhenius straight line takes place (mediated by diffusion) in a very narrow ΔG range. As $\Delta G^{\ddagger}(0)$ increases, the nonlinear region increasingly broadens, and for this reason the variation of log k_{exp} over increasingly broader ΔG ranges can be approximated by straight lines, that is, by tangents to the curve. For example (*36*), when $\Delta G^{\ddagger}(0) = 15$–$20$ kcal mol^{-1}, the curve can be approximated by a tangent for four or five units of log k_{exp} values. It can be shown that, using Eq. (28) as the FER, the slope of the tangent is given by

$$\gamma = 1/2.3RT\{1 + \exp[-(\ln 2)\,\Delta G/\Delta G^{\ddagger}(0)]\}, \tag{79}$$

and thus must be in the range $0 > \gamma > -(1/2.3RT)$, with $\gamma = -0.5/2.3RT$ at $\Delta G = 0$, $\gamma < -0.5/2.3RT$ for positive ΔG and $\gamma > -0.5/2.3RT$ for negative ΔG. These tangents are exactly the straight lines that can be obtained from Eq. (74) using the linear FER of Eq. (26). Thus the experimental values of the slope α and intercept β can be related to ΔG and $\Delta G^{\ddagger}(0)$ by (*32, 33, 36*)

$$\alpha = 1/\{1 + \exp[-(\ln 2)\,\Delta G/\Delta G^{\ddagger}(0)]\} \tag{80}$$

and

$$\beta = \Delta G/\{1 + \exp[(\ln 2)\,\Delta G/\Delta G^{\ddagger}(0)]\} + \frac{\Delta G^{\ddagger}(0)}{2}\ln\{1 + \exp[-(\ln 2)\,\Delta G/\Delta G^{\ddagger}(0)]\}. \tag{81}$$

Note that

$$\alpha = -2.3RT\gamma \tag{82}$$

and that β is equal to $\Delta G^{\ddagger}(0)$ only for the tangent at $\Delta G = 0$, being lower than $\Delta G^{\ddagger}(0)$ for all the other tangents. Note also that the Marcus equation [Eq. (72)] represents the tangent for $\Delta G = 0$.

When $\Delta G^{\ddagger}(0)$ is very large (a necessary condition for "linear" behavior over a large ΔG range in the intermediate region), k_{exp} values sufficiently high to be experimentally measurable can only be obtained in the exergonic region, where α is expected to be lower than 0.5 (i.e., $\gamma > -0.5/2.3RT$). Another point should be emphasized. The slope of log k_{exp} versus ΔG (or ΔG-related quantities) has wrongly been taken as an indication of the degree of charge transferred at the reaction "transition state" (*59*). It is important to note, however, that α values between 0 and 1 are easily understood by the preceding model, which is based on reversible complete electron transfer (*36*).

A number of electron-transfer reactions have been found to obey (*31, 31a, 47, 60–70*) Eqs. (74) and (28) [or (27)], and by best-fitting procedures it is possible to evaluate important parameters such as κ, $\Delta G^{\ddagger}(0)$, and redox potentials for ground- and excited-state reactions. Free-energy relationships for irreversible electron-transfer processes have also been discussed (*36*).

It should be noted that this FER treatment of electron-transfer reactions is based on several assumptions related to the definition of a homogeneous series of reactions. An important advantage of the present FER treatment is that the reactions are allowed to be nonadiabatic. However, the assumption that the degree of nonadiabaticity is the same throughout any allegedly "homogeneous" series of reactions may hardly be obeyed. Even if no explicit dependence of κ on ΔG is considered (see, however, Section V), the electronic factor of nonadiabatic processes is expected to be very sensitive to changes in several molecular parameters. Small structural changes may modify the effective distance of approach. Changes in the spatial distribution of the electronic charge on the reactants will strongly affect the magnitude of the electronic interaction. Furthermore, even small differences in polarizability may cause substantial changes in magnitude of the electronic interaction, especially when highly charged species are involved. All these factors are expected to introduce considerable scattering in log k_{exp} versus ΔG plots for nonadiabatic reactions.

It should also be considered that when the rate constants of a homogeneous series of reactions tend to be slower than the diffusion-controlled limit because of nonadiabaticity (Fig. 8), other, electronically more efficient, channels may become important at higher exoergonicity (*48*). For example, the formation of an excited-state product A^{-*} may become thermodynamically allowed. If such a reaction has a more favorable electronic factor, owing to the transfer of the electron to an outer orbital of the acceptor, its rate constant, as schematized in Fig. 9, will reach a higher plateau. In this case, a stepwise behavior is expected for the rate constant of the overall reaction leading to the disappearance of the reac-

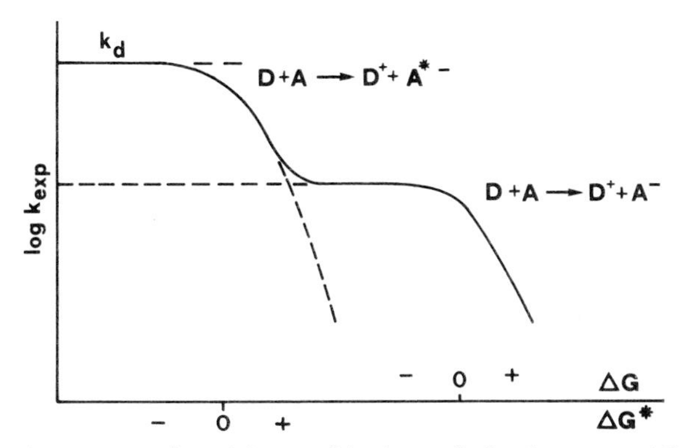

Fig. 9. Schematic representation of the possible shape of a log k_{exp} versus ΔG plot when the reaction leading to ground-state products is strongly nonadiabatic and a more adiabatic reaction (leading to an excited product) becomes thermodynamically possible at more negative ΔG values. From Balzani and Scandola (*48*).

tants, although the observation of distinct steps is possible only if the reaction leading to A^{-*} becomes important when that leading to A^- is close to saturation and if the electronic factors (κ) of the two reactions are sufficiently different.

VIII. Discussion of Selected Experimental Results

As shown in Sections III, IV, and V three basic ingredients contribute to characterize an outer-sphere electron-transfer rate constant: (i) the electronic matrix element, (ii) the Marcus reorganizational energy of the classical polar solvent, and (iii) configurational changes and frequency shifts in the nuclear coordinates of donor and acceptor. The theories described previously assign different roles to these factors.

The electronic matrix element is, according to both models, the fundamental (but not the only) part of the preexponential factor of the electron-transfer rate constant. Its value is qualitatively related to the *adiabatic* or *nonadiabatic* nature of the reaction (and, in the second case, to the degree of nonadiabaticity), although the quantitative aspects of the relationship are far from established.

The solvent reorganizational energy is assigned by both models to the exponential part of the rate constant. Its quantitative evaluation is perhaps the only universally accepted part of the computational problem of the electron-transfer rate constant, and it can be conveniently done using the Marcus–Hush formalism [Eq. (19)]. This factor constitutes a general, never negligible, contribution to the *intrinsic barrier* of electron-transfer reactions.

The configurational changes and frequency shifts in the internal donor and acceptor coordinates are simply included by the classical model as an additional contribution into the exponential term of the rate constant. In the quantum mechanical model, they contribute to the preexponential term, where they appear as a thermally averaged Franck–Condon factor. In a semiclassical model (*25*), they are included into the exponential factor as a temperature-dependent inner-shell contribution. If the reactions considered belong to the slightly endergonic or slightly exergonic region, these internal configuration and frequency changes can reasonably be considered (except for detailed temperature-dependence arguments) as an inner contribution to the *intrinsic barrier* of the electron-transfer reaction. If, in contrast, the reactions studied are in the highly exergonic (classically "inverted") region, these changes should be more specifically considered as providing quantum (nuclear tunneling) effects, which have the role of enhancing the rate-constant values with respect to the classically predicted ones. These factors, therefore, are likely to manifest themselves as deviations from the classical behavior in the so-called *inverted region*.

In this section we shall examine some experimental results in light of the theoretical treatments seen before. In doing this we shall not make an exhaustive review of the available data, but rather we shall try to show the usefulness and limitations of the various theoretical approaches.

A. The Inverted Region

One of the most remarkable predictions of the classical Marcus theory [Eq. (22)] is that for very exergonic reactions the rate constant should *decrease* with *increasing* driving force. The ΔG range in which such a behavior is predicted to occur [$\Delta G < -4\,\Delta G^{\ddagger}(0)$, see Eq. (22) and Fig. 5] is usually called the "inverted region." This rather peculiar feature of the theory has attracted a great deal of attention. Its experimental confirmation, in fact, would constitute a very specific test of the model. It should also be noted that several important practical consequences could arise from this effect, in particular, from the point of view of solar-energy conversion. Actually, in this field one of the crucial problems is that of slowing down the very exergonic, energy-wasting, back electron-transfer reactions between primary photoredox products (see Section IX).

Experimentally, it has been difficult to obtain definite evidence for or against the predicted inversion by studying thermal electron-transfer reactions [Eq. (1)] because the ΔG values of such reactions are usually not sufficiently negative. However, the use of electronically excited states as reactants (Fig. 7) has made it possible to investigate the inverted region in some detail (*31, 31a, 61, 63, 64, 66, 68*). From this point of view, photochemical electron-transfer reactions have thermodynamic and kinetic advantages over the corresponding thermal ones: (i) owing to the high oxidizing and reducing power of the excited states, very negative ΔG values can easily be achieved either in the excited-state forward electron-transfer reaction or in the thermal back electron-transfer reaction; (ii) the use of Stern–Volmer techniques for the forward reaction and of flash-photolysis relaxation methods for the back reaction allows very fast (up to diffusion-controlled) reaction rates to be measured.

A point that should be appropriately emphasized here is that, according to Eqs. (74) and (76), diffusional control could often mask to some extent the inverted behavior predicted by the Marcus model. This masking occurs provided that the preexponential factor of the electron-transfer step, $\kappa\nu_N$, is higher than the dissociation rate constant k_{-d} (Section VII,B). The result of this situation is a diffusional plateau ($k_{exp} \simeq k_d$) for a ΔG range that depends on the magnitude of the $\kappa\nu_N/k_{-d}$ ratio and on the $\Delta G^{\ddagger}(0)$ value. As an example, if $\kappa\nu_N = 6 \times 10^{12}\ s^{-1}$, $\Delta G^{\ddagger}(0) = 3$ kcal mol^{-1}, and $k_{-d} = 8 \times 10^9\ s^{-1}$, the diffusional plateau ($k_{exp} \simeq k_d = 10^{10}\ M^{-1}\ s^{-1}$) is calculated [Eqs. (28) and (74)] to extend down to $\Delta G = -25$ kcal mol^{-1}. Thus experiments at considerably more negative ΔG values than the conventional $-4\ \Delta G^{\ddagger}(0)$ inversion point could be necessary in many instances in order to have a chance to detect the inverted behavior.

Although single experiments in the highly exergonic region may be of some value and have been used to discuss this problem, it seems that the most reliable way to investigate the inverted region consists in the study of homogeneous series of reactions (Section VII,B) over wide ΔG ranges. As to the characteristics of such homogeneous series, adiabatic reactions with low intrinsic barriers should be the best choice. Low barriers will place the onset of the inverted region at reasonably low exergonicities. On the other hand, because the electronic factors are expected to be sensitive to small changes in the nature of the reactants (Section VII,B), nonadiabaticity is unlikely to remain appreciably constant throughout any series of reactions.

A number of experimental studies have been carried out along these lines since the early 1970s. The prototype study was that reported by Rehm and Weller (*31, 31a*) on the electron-transfer quenching of aromatic singlets by organic electron donors and acceptors. By using fluorescence-quenching measurements, they found a log k_{exp} versus ΔG plot of the type shown in Fig. 8 that had a diffusional plateau without any indication of inverted behavior down to $\Delta G \simeq -60$ kcal mol^{-1}. Given the intrinsic barrier of their systems [$\Delta G^{\ddagger}(0) = 2.4$ kcal mol^{-1} as the best fitting value], the inverted behavior (allowing for the maximum leveling effect of diffusion control) should have been observed at $\Delta G < -25$ kcal mol^{-1}. It was this negative result that led Rehm and Weller to propose their popular FER [Eq. (27)] as an empirical substitute for the Marcus FER. It should be noted, however, that, as stressed by Rehm and Weller (*31, 31a*), some of these "highly exergonic" results are open to question. In fact, several of these reactions generate radical-ion products that, judging from their spectra, have very low-lying electronically excited states. Thus, in many of these reactions, electronically excited products could be initially formed in the electron-transfer quenching process, with much lower exergonicity than would be assumed on the basis of known ground-state redox potentials. This could be one of the reasons for the failure to observe the inverted behavior in this study.

A number of other investigations have been carried out along these lines, and except for some "vestiges" over a narrow ΔG range (*71*), no definite evidence for an inverted region has been obtained whatsoever in homogeneous solution.[3] However, in most cases alternative paths, especially the formation of electronically excited products, cannot be ruled out (*45, 74*). A study (*48, 75*), that seems to be as far as possible free from this type of objection is that performed on a homogeneous series of reactions between excited or ground-state metal polypyridine complexes and aromatic electron acceptors or their radical-cations [Eqs. (83) and (84)].

$$M^{*n+} + Q \longrightarrow M^{(n-1)+} + Q^{+} \quad (83)$$

$$M^{(n-1)+} + Q^{+} \longrightarrow M^{n+} + Q \quad (84)$$

In this study, $Cr(bpy)_3^{3+}$, $Rh(phen)_3^{3+}$, and $Ir(Me_2phen)_2Cl_2^{+}$ were used as M^{n+}, and various methoxybenzenes and aromatic amines were used as Q. The most exergonic reactions of the series were the back electron-transfer reactions [Eq. (84)] involving those quenchers that, being difficult to oxidize, only have high-energy excited states. For these reactions, excited-state product formation could be easily ruled out on energetic grounds. In spite of that, these reactions did not give any substantial indication of inverted behavior down to $\Delta G \simeq -60$ kcal mol^{-1} (Fig. 10). Because the intrinsic barriers of these systems are estimated as ~3 kcal mol^{-1}, the diffusional plateau extends down to $\Delta G \simeq -20\ \Delta G^{\ddagger}(0)$. Of course, other reaction paths can always be invoked to justify the failure to observe the

[3] Electron-transfer reactions in rigid matrix at low temperature may (*72*) or may not (*73*) exhibit an inverted behavior.

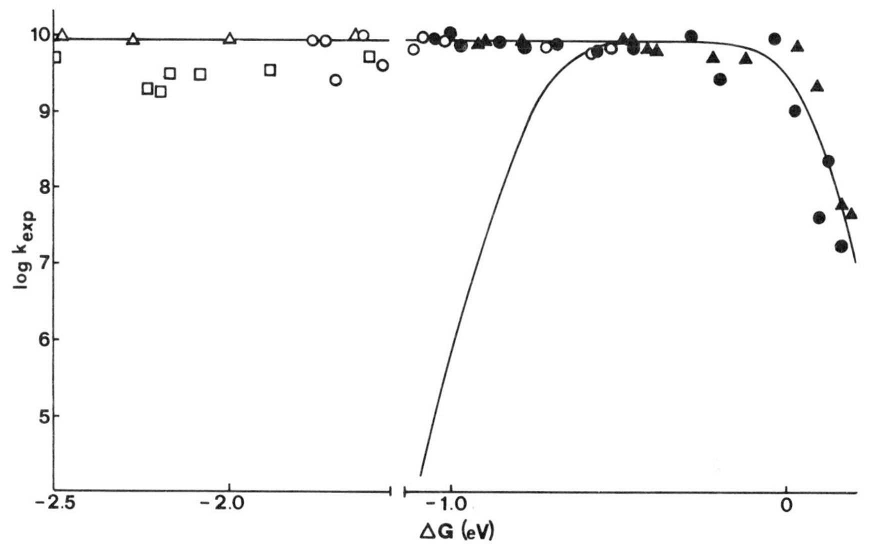

Fig. 10. Plot of log k_{exp} versus ΔG for the reaction between excited states of transition-metal complexes and aromatic amines or methoxybenzenes (●, ▲) and the back electron-transfer reactions between the reduced complex and the oxidized quencher (○, △, □): ●, ○, $Cr(bpy)_3^{3+}$; □, $Rh(phen)_3^{3+}$; ▲, △, $Ir(5,6\text{-}Me_2phen)_2Cl_2^+$. The downward-sloping line at about −1.0 eV corresponds to the Marcus "inverted" region. From Balzani and Scandola (*48*).

inverted behavior. In particular, reversible hydrogen-atom–proton transfer, although not very plausible in these systems, cannot be safely ruled out.

Altogether, the perseverance with which the inverted behavior escapes experimental detection suggests that (i) quantum effects (nuclear tunneling) are generally very efficient in enhancing rates in the classical "inverted" region; (ii) new, more efficient reaction channels, not necessarily involving excited states of the products, may take off at high exergonicities (e.g., increased adiabaticity, strong charge-transfer interactions, atom transfer, etc.). These considerations justify the operational use of empirical FERs, such as those given by Eqs. (27) and (28) as substitutes for the Marcus FER [Eq. (22)] in the exergonic region.

B. Intrinsic Barriers

The intrinsic barrier may receive contributions from changes in solvent polarization and changes in internal molecular vibrations. The former contribution [Eq. (19)] depends on the radii of the two reactants, the distance between the centers of the two reactants, and the dielectric constant of the medium. The latter contribution [Eq. (18)] depends crucially on the electronic structures of the two redox forms of the species involved. For example, reduction of $Ru(NH_3)_6^{3+}$ (t_{2g}^5 electronic configuration) involves the transfer of an electron into the t_{2g} orbitals, which are practically nonbonding. This causes only a very small difference (0.040 Å) between the Ru(II)—NH_3 and Ru(III)—NH_3 bond distances (*76*), with a consequent very small inner-sphere contribution to the intrinsic barrier (*24*). In contrast, reduction of $Co(NH_3)_6^{3+}$ (t_{2g}^6 electronic configuration) implies the transfer of an electron into the strongly antibonding e_g orbitals (with reorganization to the high-spin $t_{2g}^5e_g^2$ electronic configuration), and thus it is expected to involve large structural changes. The difference between the Co(III)—NH_3 and Co(II)—NH_3 bond distances (0.178 Å) (*76*) is in agreement with these expectations, and it is consistent with a large inner-sphere contribution to the intrinsic barrier (*77*). For some redox couples the inner- and outer-sphere contributions to the intrinsic barrier have been calculated on the basis of the treatment given in Section III, and the overall ΔG values so obtained are in relative agreement with the experimental values (*24*).

For a number of other couples the intrinsic barrier has been evaluated using the Marcus cross equation [Eq. (71)] given in Section VII,A. [For a discussion see (*78*).] As we have mentioned, such an equation must be used with care, taking into account that practical systems may not obey the assumptions made on deriving it.

Another way to evaluate unknown intrinsic barriers is that of using the log k_{exp} versus ΔG plots discussed in Section VII,B. From a qualitative point of view, a large intrinsic barrier is reflected in a broad nonlinear region connecting the asymptotic straight lines that are found for very positive and very negative ΔG values. Figure 11 shows such plots for the electron-transfer reactions of $Ru(NH_3)_6^{n+}$ and $Fe(H_2O)_6^{n+}$ ($n = 2, 3$) with the same homogeneous series of reactants (*47*). As one can see, for the $Ru(NH_3)_6^{n+}$ reactions the rate constant approaches the diffusion-controlled limit for slightly negative ΔG values, and the plot shows a narrow nonlinear region. This behavior is that expected for adiabatic or nearly adiabatic reactions having a small intrinsic barrier. For $Fe(H_2O)_6^{n+}$ reactions, the nonlinear part of the plot is considerably wider, indicating a larger intrinsic barrier. Quantitatively, these plots can be analyzed using suitable parameters for the diffusion and dissociation rate constants, the frequency factor of the electron-transfer step, and the intrinsic barrier of the reaction partners. From the best-fitting curve a fair evaluation of the intrinsic barrier for the species of interest can thus be obtained. The values of 10.6 and 17.4 kcal mol^{-1} from the curves shown in Fig. 11 obtained (*47*) for the $Ru(NH_3)_6^{3+/2+}$ and $Fe(H_2O)_6^{3+/2+}$ couples, respec-

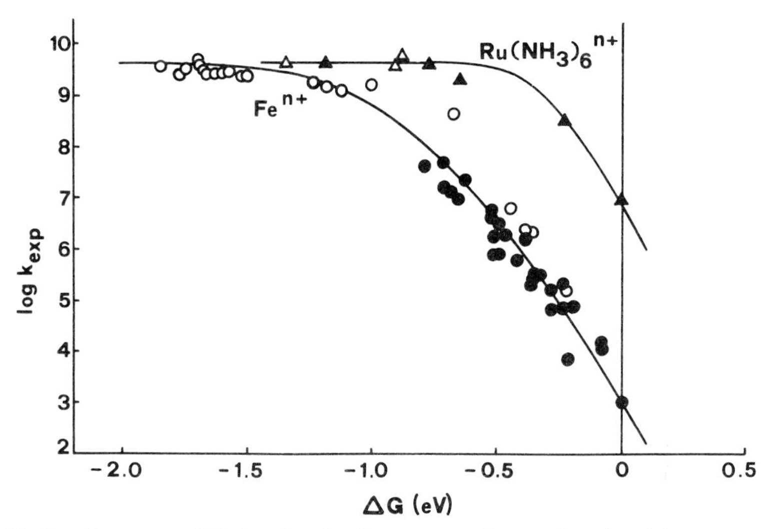

Fig. 11. Log k_{exp} versus ΔG plots for the electron-transfer reactions involving $Ru(NH_3)_6^{2+}$ (▲), $Ru(NH_3)_6^{3+}$ (△), $Fe(H_2O)_6^{2+}$ (●), and $Fe(H_2O)_6^{3+}$ (○). For details see Balzani *et al.* (*47*), from which this figure has been adapted. Copyright 1981 American Chemical Society.

tively, are in fair agreement with those measured experimentally (12.5 and 16.6 kcal mol^{-1}) or evaluated theoretically (11.8 and 18.4 kcal mol^{-1}) (*24*).

This method based on the log k_{exp} versus ΔG plots has also been used to evaluate the intrinsic barriers of the reactions involving reduction of the $Cr(bpy)_3^{3+*}$, $Ru(bpy)_3^{2+*}$, and $Ir(5,6\text{-}Me_2phen)_2Cl_2^{+*}$ excited states (*64*). The values obtained were very small, practically equal to those of the ground-state exchange reactions. This is in agreement with the small Stokes-shift values, which indicate that the excited states are very similar in size, shape, and solvation to the ground state.

When the electron-transfer reaction is accompanied by very large nuclear rearrangements (e.g., dissociation of the reaction products into fragments), the log k_{exp} versus ΔG plots assume almost linear shapes in the intermediate region and the asymptotic limits are only reached at extremely negative and extremely positive ΔG values. The treatment of these peculiar cases, including irreversible electron-transfer processes, has been discussed (*36*).

C. Adiabatic and Nonadiabatic Processes

Most biological electron-transfer processes are thought to be nonadiabatic and are generally discussed in the frame of the quantum mechanical approach (*14*). In contrast, adiabaticity has long been considered the rule for electron-transfer reactions in homogeneous solution (*8, 9, 12, 16, 52, 53*). However, suggestions or indications of nonadiabatic behavior have been discussed for several specific reactions (*12, 24, 56, 79–84*).

Establishing whether an electron-transfer reaction is adiabatic or nonadiabatic is a very difficult task from both an experimental and a theoretical point of view. The rate constant of a reaction usually depends on temperature according to the Arrhenius equation

$$k = A \exp(-E_a/RT); \tag{85}$$

the values of the preexponential factor A and the activation energy E_a can be obtained from a study of the temperature dependence of the reaction rate. Comparing Eq. (85) with the following formulation of Eq. (29),

$$k_e = \kappa \nu_N \exp(\Delta S^{\ddagger}/R) \exp(-\Delta H^{\ddagger}/RT), \tag{86}$$

one can see that the preexponential factor A reflects not only the transmission coefficient but also the effective nuclear frequency and the activation entropy. Thus when the last two quantities are not exactly known one cannot get a firm evaluation of the electronic transmission coefficient. For bimolecular reactions the situation is even worse because the frequency factor also contains the entropy changes associated with forming the precursor complex [i.e., with K_0 in Eq. (5)]. These changes may be very large for charged reactants and are difficult to evaluate. For activationless reactions [Eq. (75)], nonadiabaticity is expected to be more or less hidden (depending on ν_N and k_{-d}) because of the diffusional control of the reaction rate. Finally, it should be considered that there may be some compensation between entropic and enthalpic components of $\Delta G^{\ddagger}$ in the sense that electron transfer can occur in a large number of different nuclear configurations and that with higher activation energies a system can usually find nuclear configurations characterized by better transmission coefficients.

The nonadiabaticity problem in intramolecular electron-transfer reactions has been thoroughly discussed in the case of binuclear complexes of the type $(NH_3)_5Co(III)L—L'Ru(II)(NH_3)_4H_2O$, where L—L′ represents a bifunctional bridging group that fixes the positions of the reacting metal ions (*80, 82*). With suitable choices of the L and L′ groups it is possible to change the electronic coupling between the metal ions and to study in a systematic way the effect of these changes on the rates of reaction. When the metal ions bound to the bifunctional ligand are Ru(III) and Ru(II) and when the electronic coupling between the centers is strong enough, light absorption is observed in the near-infrared region, which corresponds to the transfer of an electron from the reducing to the oxidizing metal ion (intervalence band). The characteristics (energy, bandwidth, and intensity) of such a band yield information concerning the intrinsic barrier of the thermal electron-transfer reaction and the coupling between the two metal centers. In the nonadiabatic regime it is expected that the rates of electron transfer correlate with the intensities of the intervalence bands because both quantities depend on the electronic coupling. Such a relation has been found for some complexes, but this approach is still in an exploratory stage (*80, 82*).

Another approach used in searching for nonadiabaticity effects is that of studying electron-exchange reactions in series of complexes in which the metal ions and the nature of the ligands are kept constant but the ligands are made more bulky by adding saturated hydrocarbon groups (*79, 81*). The outer-sphere contribution to the activation energy is expected to decrease with increasing molecular size [Eq. (9)]; therefore, the reaction rate is expected to increase if the reactions are adiabatic. Conversely, a decrease in the reaction rate can be taken as an indication of a decreasing electron-transfer probability that more than compensates for the decrease in the activation energy. Evidence for the latter effect has been found for $Mn(CNR)_6^{2+/+}$ complexes (*79*) and for polypyridine Fe(III)–Fe(II) complexes (*81*).

The low rate of electron-transfer reactions involving aqueous ions has attracted the attention of several authors. Different explanations have been advanced, including anharmonicity effects (*56, 57*), very large work terms (*85*), and nonadiabaticity (*47, 56, 84*). A calculation of the electronic coupling matrix element for the $Fe(H_2O)_6^{3+/2+}$ exchange suggests that this reaction is appreciably nonadiabatic (*24, 25*).

Among the aqueous ions, the $Eu_{aq}^{3+/2+}$ couple was long suspected to behave nonadiabatically because the 4*f* orbitals involved in the electron transfer are strongly shielded by the outer 5*s* and 5*p* orbitals (*80*). The literature data for the reactions of Eu^{3+} and Eu^{2+} with a homogeneous family of partners have been collected, and the corresponding log k_{exp} versus ΔG plots have been drawn (Fig. 12) and compared with similar plots concerning $Fe(H_2O)_6^{3+}$, $Fe(H_2O)_6^{2+}$, $Ru(NH_3)_6^{3+}$, and $Ru(NH_3)_6^{2+}$ (*47*). The peculiar features of the Eu_{aq}^{3+} and Eu_{aq}^{2+} plots strongly contrast with the regular plots obtained for the other species and suggest that Eu_{aq}^{3+} and Eu_{aq}^{2+} behave nonadiabatically, at least up to moderately negative ΔG values. A theoretical estimate of the transmission coefficient based on spectroscopic information yields $\kappa < 10^{-5}$. At larger and negative ΔG values, more efficient but different channels become available for Eu_{aq}^{2+} oxidation or Eu_{aq}^{3+} reduction, which are assigned to paths involving different charge-transfer intermediates (*47*).

An important cause of nonadiabaticity may be the spin-forbidden nature of the electron-transfer step. Reactions involving interconversion of

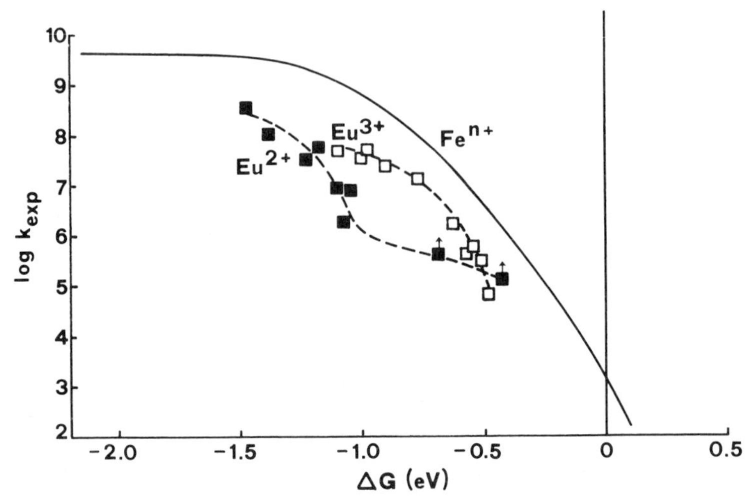

Fig. 12. Log k_{exp} versus ΔG plots for electron-transfer reactions involving Eu_{aq}^{2+} (■) and Eu_{aq}^{3+} (□). The dashed curves have been drawn only with the aim of connecting the points within the Eu_{aq}^{2+} and the Eu_{aq}^{3+} series of reactions. The solid curve is that for $Fe(H_2O)_6^{n+}$ reactions shown in Fig. 11. For details see Balzani *et al.* (*47*), from which this figure has been adapted. Copyright 1981 American Chemical Society.

low-spin Co(III) complexes (t_{2g}^6 electronic configuration) and high-spin Co(II) complexes ($t_g^5e_g^2$ configuration) are expected to suffer this prohibition. The very slow exchange between $Co(NH_3)_6^{3+}$ and $Co(NH_3)_6^{2+}$ has been critically compared with that of $Ru(NH_3)_6^{3+}$ and $Ru(NH_3)_6^{2+}$ in the frame of a quantum mechanical approach (*77, 83*). The results obtained show that nonadiabaticity plays an important role ($\kappa \simeq 10^{-4}$) in slowing down this reaction.

IX. Conclusions

After having discussed the theoretical approaches to electron-transfer kinetics and having compared the predictions of these approaches with the available results, we can try to make some comments about the role of electron-transfer kinetics in solar energy conversion systems.

As mentioned in Section I and widely discussed in other chapters of this book, conversion and storage of solar energy into chemical energy are based on a sequence of consecutive and parallel electron-transfer reactions. In a typical case (Fig. 2), such a sequence consists of (i) the excited-state electron-transfer reaction (process 3 in Fig. 2), which causes the *conversion* of part of the spectroscopic energy (i.e., of the absorbed light) into chemical energy; (ii) the thermal back electron-transfer reaction (process 4 in Fig. 2), which dissipates the converted energy; and (iii) the thermal reactions of the products of the excited-state electron-transfer reaction with the substrate (process 5 in Fig. 2), which may lead to the *storage* of the converted energy. Efficient conversion of solar energy into chemical energy requires a high efficiency in both the conversion and storage steps.

To discuss this problem, it is convenient to use the kinetic scheme shown in Fig. 7, which is related to that of Fig. 2 in this way: (i) D and A stand for P and R; (ii) a distinction is made between solvent-caged and separated primary products; (iii) k_p is to be considered as the predominant pseudo-first-order rate constant of reaction between one of the free primary products and a substrate. On the basis of Fig. 7, the overall efficiency of the conversion–storage process can be given as the product of three fundamental efficiencies:

$$\eta_{\text{tot}} = \eta_{\text{el.t.}} \eta_{\text{c.g.}} \eta_{\text{p}} . \tag{87}$$

In this equation $\eta_{\text{el.t.}}$ is the efficiency with which the excited state of the sensitizer is converted to the primary, solvent-caged, electron-transfer products; $\eta_{\text{c.g.}}$ is the efficiency with which the primary caged products escape cage recombination to give the separated primary products; and η_{p} is the efficiency with which the separated products undergo the useful reaction leading to the storable products (fuel). In the usual case of an exergonic excited-state electron-transfer reaction ($k_e \gg k_{-e}$), assuming that the fastest useful reaction (k_p) involves the reduced acceptor, the three efficiencies have the following expressions in terms of single-step rate constants:

$$\eta_{\text{el.t.}} = \frac{k_{\text{exp}}[\text{A}]}{k_{\text{exp}}[\text{A}] + (1/\tau)}, \tag{88}$$

where k_{exp} is given by Eq. (73), [A] is the concentration of the quencher, and τ the sensitizer excited-state lifetime;

$$\eta_{\text{c.g.}} = \frac{k'_{-\text{d}}}{k'_{-\text{d}} + k_{-\text{e}}(\text{g})} \tag{89}$$

and

$$\eta_{\text{p}} = \frac{k_p}{k'_{\text{d}}[\text{D}^+](1 - \eta_{\text{c.g.}}) + k_p}, \tag{90}$$

where $[\text{D}^+]$ represents the concentration of the primary oxidized product of the sensitizer.

As far as the values of $\eta_{el.t.}$ are concerned, it should be noted that k_{exp} is a second-order reaction rate constant between the excited state and a suitable redox quencher. When this reaction behaves adiabatically and does not involve large molecular rearrangements, its rate constant is of the order of 10^8 to 10^9 M^{-1} s^{-1} for $\Delta G = 0$ and reaches the diffusion-controlled limit for slightly negative ΔG values. This happens, for example, in the electron-transfer quenching of $Ru(bipy)_3^{2+}$ (or other polypyridine complexes) by quenchers such as methylviologen (*69*)

$$Ru(bipy)_3^{2+*} + MV^{2+} \longrightarrow Ru(bipy)_3^{3+} + MV^{+} \tag{91}$$

where $\Delta G = -0.41$ eV and $k_{exp} = 1.9 \times 10^9$ M^{-1} s^{-1}. When the reaction behaves nonadiabatically and/or has a large intrinsic barrier, the rate constant may be low even for large and negative ΔG values, thus compromising $\eta_{el.t.}$. This is the case, for example, of Eu^{3+} as a quencher (*86*)

$$Ru(bipy)_3^{2+*} + Eu_{aq}^{3+} \longrightarrow Ru(bipy)_3^{3+} + Eu_{aq}^{2+} \tag{92}$$

where $\Delta G = -0.46$ eV and $k_{exp} \leq 0.8 \times 10^5$ M^{-1} s^{-1}. It should also be noted that even diffusion-controlled rate constants are useless when the excited-state lifetime is too short. For example, Eq. (88) shows that when $\tau = 10^{-9}$ s, even if the rate constant is diffusion controlled (10^{10} M^{-1} s^{-1}) and the quencher concentration high (10^{-1} M), 50% of the excited states deactivate before being able to react with the quencher.

The efficiency of cage escape [$\eta_{c.g.}$, Eq.(89)] may in many cases be a crucially limiting factor for the overall efficiency. For the purpose of solar-energy conversion, the primary solvent-caged products of the excited-state electron-transfer reaction must have a high energetic content. This also means that the dissipative cage recombination reaction $k_{-e}(g)$ is strongly exergonic. According to the Marcus model, strongly exergonic reactions should be very slow because they should lie in the inverted region. If this were true, it should not be difficult (*87*) for the diffusional process k'_{-d} to compete with the dissipative cage recombination reaction $k_{-e}(g)$. Unfortunately, the inverted behavior has never been observed in homogeneous systems (Section VIII,A). To obtain a more efficient cage escape it is perhaps easier to increase k'_{-d} than to decrease $k_{-e}(g)$ (*88*). The dissociation constant can be increased by using reactants having larger electric charges of the same sign (*89*), whereas the only way to impose a kinetic control on the highly exergonic back electron-transfer reaction is to introduce a degree of nonadiabaticity (*88*). This, however, should be done without compromising the rate of the excited-state electron-transfer reaction k_e involved in k_{exp} and thus in $\eta_{el.t.}$. The ideal case would be that of a sensitizer that behaves in a strongly nonadiabatic way in the ground-state electron-transfer reactions and adiabatically in the excited-state reactions. This would imply a strong radial displacement of an electron on photoexcitation, as in the case of the $f \rightarrow d$ transition in Eu^{2+} and Ce^{3+} (which, however, do not have other properties that are required for a photosensitizer), or as could be the case for metal-to-ligand charge-transfer transitions in appropriate molecular structures. The finite yield (~25%) (*90, 91*) of cage escape observed for $Ru(bipy)_3^{3+} \cdot MV^{2+}$ implies a lower-than-diffusion rate constant for the very exergonic back electron-transfer reaction (*92*)

$$Ru(bipy)_3^{3+} + MV^{+} \longrightarrow Ru(bipy)_3^{2+} + MV^{2+} \tag{93}$$

($\Delta G = -1.70$ eV and $k = 4 \times 10^9\ M^{-1}\ s^{-1}$); this is probably due to some degree of nonadiabaticity imposed by the relatively small overlap between the donor orbital of MV^+ and the metal-centered t_{2g} acceptor orbital of $Ru(bipy)_3^{3+}$.

Even when the cage-escape efficiency is almost 100%, the dissipative back electron-transfer reaction can nontheless predominate, giving rise to a very low η_p if the useful reactions are too slow $\{k_p < k'_d[D^+](1 - \eta_{c.g.})\}$. The useful reaction involves, in principle, one of the primary products (e.g., A^-) and a substrate (e.g., H_2O). In practice, such multielectron processes only occur via an appropriate redox catalyst with which a primary product can react at a diffusion-controlled rate. For example, hydrogen evolution from an aqueous solution of MV^+ is conveniently mediated by a very fast (*93*) reaction between MV^+ and colloidal platinum (*18, 19*). It follows that the efficiency of energy storage is actually governed by the competition between the back electron-transfer reaction, the pseudo-first-order rate constant $\{k'_d[D^+](1 - \eta_{c.g.})\}$ of which depends on primary-product concentration (and thus on absorbed light intensity) and the reaction between the primary redox products and an appropriate catalyst, the pseudo-first-order rate constant of which depends on the type of catalyst and size and concentration of catalyst particles. Thus the light intensity and the concentration of the catalyst particles are other crucial parameters in the storage of the converted energy (*94*).

Acknowledgments

The authors thank Professor Giorgio Orlandi for helpful discussions. This work was supported by the Progetto Finalizzato Chimica Fine e Secondaria of the Italian National Research Council and the Commission of the European Communities (Contracts ESD-025-I and ESD-026-I).

References

1. Bolton, J. R. (ed.), "Solar Power and Fuels." Academic Press, New York, 1977.
2. Claesson, S., and Enström, L. (eds.), "Solar Energy-Photochemical Conversion and Storage." Natl. Swedish Board Energy Source Development, Stockholm, 1977.
3. Connolly, J. S. (ed.), "Photochemical Conversion and Storage of Solar Energy." Academic Press, New York, 1981.
4. Wrighton, M. S., *Chem. Eng. News.* **57,** 29 (1979).
5. Bolton, J. R., *Science* (*Washington, D.C.*) **202,** 705 (1978).
6. Bolton, J. R., and Hall, D. O., *Annu. Rev. Energy* **4,** 353 (1979).
7. Hall, D. O., and Rao, K. K., "Photosynthesis," 2nd ed. Arnold, London, 1976.
8. Marcus, R. A., *Annu. Rev. Phys. Chem.* **15,** 155 (1964).
9. Reynolds, W. R., and Lumry, R. W., "Mechanisms of Electron Transfer." Ronald Press, New York, 1966.
10. Taube, H., "Electron Transfer Reactions of Complex Ions in Solution." Academic Press, New York, 1970.
11. Sutin, N., *Acc. Chem. Res.* **1,** 225 (1968).
12. Sutin, N., *in* "Inorganic Biochemistry" (G. L. Eichorn, ed.), p. 611. Elsevier, New York, 1973.
13. Gerischer, H., and Katz, J. J. (eds.), "Light Induced Charge Separation in Biology and Chemistry." Verlag Chemie, Weinheim, 1979.
14. Chance, B., De Vault, D. C., Frauenfelder, H., Marcus, R. A., Schrieffer, J. R., and Sutin, N. (eds.), "Tunneling in Biological Systems." Academic Press, New York, 1979.
15. Ulstrup, J., "Charge Transfer in Condensed Media." Springer-Verlag, Berlin, 1979.
16. Cannon, R. D., "Electron Transfer Reactions." Butterworth, London, 1980.
17. Balzani, V., Moggi, L., Manfrin, M. F., Bolletta, F., and Gleria, M., *Science* (*Washington, D.C.*) **189,** 852 (1975).
18. Grätzel, M., Photo-induced water splitting in heterogeneous solution. *In* "Photochemical Conversion and Storage of Solar Energy" (J. S. Connolly, ed.), pp. 131–160. Academic Press, New York, 1981.
19. Lehn, J. M., Photoinduced generation of hydrogen and oxygen from water. *In* "Photochemical Conversion and Storage of Solar Energy" (J. S. Connolly, ed.), pp. 161–200. Academic Press, New York, 1981.
20. Hush, N. S., *Trans. Faraday Soc.* **57,** 557 (1961).
21. Debye, P., *Trans. Electrochem. Soc.* **82,** 265 (1942).
22. Eigen, M. Z., *Z. Phys. Chem.* (*Wiesbaden*) **1,** 176 (1954).
23. Glasstone, S., Laidler, J. K., and Eyring, H., "The Theory of Rate Processes." McGraw-Hill, New York, 1941.
24. Sutin, N., Electron transfer reactions of metal complexes in solution. *In* "Tunneling in Biological Systems" (B. Chance, D. C. De Vault, H. Frauenfelder, R. A. Marcus, J. R. Schrieffer, and N. Sutin, eds.), pp. 201–224. Academic Press, New York, 1979.
25. Brunschwig, B. S., Logan, J., Newton, M. D., and Sutin, N., *J. Am. Chem. Soc.* **102,** 5798 (1980).
26. Marcus, R. A., *Discuss. Faraday Soc.* **29,** 21 (1960).
27. Marcus, R. A., *J. Chem. Phys.* **43,** 679 (1965).
28. Marcus, R. A., *Electrochim. Acta* **13,** 995 (1968).
29. Clark, E. D., and Wayne, R. P., *in* "Comprehensive Chemical Kinetics" (C. H. Bamford, and C. F. H. Tipper, eds.), p. 365. Elsevier, Amsterdam, 1969.
30. Evans, M. D., and Polany, M., *Trans. Faraday Soc.* **32,** 1340 (1936).

30a. Evans, M. D., and Polany, M., *Trans. Faraday Soc.* **34,** 11 (1938).
31. Rehm, D., and Weller, A., *Ber. Bunsenges. Phys. Chem.* **73,** 834 (1969).
31a. Rehm, D., and Weller, A., *Isr. J. Chem.* **8,** 259 (1970).
32. Marcus, R. A., *J. Phys. Chem.* **72,** 891 (1968).
33. Levine, R. D., and Agmon, N., *Chem. Phys. Lett.* **52,** 197 (1977).
34. Agmon, N., and Levine, R. D., *J. Chem. Phys.* **71,** 3034 (1979).
35. Scandola, F., and Balzani, V., *J. Am. Chem. Soc.* **101,** 6140 (1979).
35a. Scandola, F., and Balzani, V., *J. Am. Chem. Soc.* **102,** 3663 (1980).
36. Scandola, F., Balzani, V., and Schuster, G. B., *J. Am. Chem. Soc.* **103,** 2519 (1981).
37. Kauzman, W., "Quantum Chemistry." Academic Press, New York, 1957.
38. Levich, V. G., *Adv. Electrochem. Electrochem. Eng.* **4,** 249 (1966).
39. Vorotyntsev, M. A., Dogonadze, R. R., and Kuznetsov, A. M., *Dokl. Akad. Nauk. SSSR Ser. Khim.* **195,** 1135 (1970).
39a. German, E. D., Dvali, V. G., Dogonadze, R. R., and Kuznetsov, A. M., *Elektrokhimiya* **12,** 639 (1976).
39b. Dogonadze, R. R., *in* "Reactions of Molecules at Electrodes" (N. S. Hush, ed.), p. 135. Wiley, London, 1971.
40. Kestner, N. R., Logan, J., and Jortner, J., *J. Phys. Chem.* **78,** 2148 (1974).
41. Ulstrup, J., and Jortner, J., *J. Chem. Phys.* **63,** 4358 (1975).
42. Efrima, S., and Bixon, M., *Chem. Phys.* **13,** 447 (1976).
43. Van Duyne, R. P., and Fisher, S. F., *Chem. Phys.* **5,** 183 (1974).
44. Siders, P., and Marcus, R. A., *J. Am. Chem. Soc.* **103,** 741 (1981).
45. Siders, P., and Marcus, R. A., *J. Am. Chem. Soc.* **103,** 748 (1981).
46. Buhks, E., Bixon, M., Jortner, J., and Navon, G., *J. Phys. Chem.* **85,** 3759 (1981).
47. Balzani, V., Scandola, F., Orlandi, G., Sabbatini, N., and Indelli, M. T., *J. Am. Chem. Soc.* **103,** 3370 (1981).
48. Balzani, V., and Scandola, F., Photochemical electron transfer reactions in homogeneous solution. *In* "Photochemical Conversion and Storage of Solar Energy" (J. S. Connolly, ed.), pp. 97–130. Academic Press, New York, 1981.
49. Redi, M., and Hopfield, J. J., *J. Chem. Phys.* **72,** 6651 (1980).
50. Jortner, J., *J. Am. Chem. Soc.* **102,** 6676 (1980).
51. Balzani, V., Bolletta, F., Gandolfi, M. T., and Maestri, M., *Top. Curr. Chem.* **75,** 1 (1978).
52. Bennet, L. E., *Prog. Inorg. Chem.* **18,** 1 (1973).
53. Pennington, D. E., *Adv. Chem. Ser.* No. 174, 466 (1978).
54. Bennet, L. E., and Taube, H., *Inorg. Chem.* **7,** 254 (1968).
55. Tendler, Y., and Faraggi, M., *J. Chem. Phys.* **57,** 1358 (1972).
56. Chou, M., Creutz, C., and Sutin, N., *J. Am. Chem. Soc.* **99,** 5615 (1977).
57. Marcus, R. A., and Sutin, N., *Inorg. Chem.* **14,** 213 (1975).
58. Newton, T. W., *J. Chem. Educ.* **45,** 571 (1968).
59. Walling, C., *J. Am. Chem. Soc.* **102,** 6854 (1980).
60. Bock, C. R., Meyer, T. J., and Whitten, D. G., *J. Am. Chem. Soc.* **96,** 4710 (1974).
61. Vogelman, E., Schreiner, S., Rauscher, W., and Kramer, H. E. A., *Z. Phys. Chem.* (*Neue Folge*) **101,** 321 (1976).
62. Kikuchi, K., Tamura, S. I., Iwenege, C., Kokubun, K., and Usui, Y. Z., *Z. Phys. Chem.* (*Neue Folge*) **106,** 17 (1977).
63. Breyman, V., Dreeskamp, H., Koch, E., and Zander, M., *Chem. Phys. Lett.* **59,** 68 (1978).
64. Ballardini, R., Varani, G., Indelli, M. T., Scandola, F., and Balzani, V., *J. Am. Chem. Soc.* **100,** 7219 (1978).

65. Martens, F. M., Verhoeven, J. W., Gase, R. A., Pandit, N. K., and de Boer, Th. J., *Tetrahedron* **34,** 44 (1978).
66. Brunschwig, B., and Sutin, N., *J. Am. Chem. Soc.* **100,** 7568 (1978).
67. Bock, C. R., Connor, J. A., Gutierrez, A. R., Meyer, T. J., Whitten, D. G., Sullivan, B. P., and Nagle, J. K., *J. Am. Chem. Soc.* **101,** 4815 (1979).
68. Nagle, J. K., Dressick, W. J., and Meyer, T. J., *J. Am. Chem. Soc.* **101,** 3993 (1979).
69. Amouyal, E., Zidler, B., Keller, P., and Moradpour, A., *Chem. Phys. Lett.* **74,** 314 (1980).
70. Rillema, D. P., Nagle, J. K., Barringer, L. F., Jr., and Meyer, T. Y., *J. Am. Chem. Soc.* **103,** 56 (1981).
71. Creutz, C., and Sutin, N., *J. Am. Chem. Soc.* **99,** 241 (1977).
72. Beitz, J. V., and Miller, J., *J. Chem. Phys.* **71,** 4579 (1979).
73. Kira, A., *J. Phys. Chem.* **85,** 3047 (1981).
74. Marcus, R. A., *Int. J. Chem. Kinet.* **13,** 865 (1981).
75. Indelli, M. T., Ballardini, R., and Scandola, F., *J. Phys. Chem.,* submitted.
76. Stynes, H. C., and Ibers, J. A., *Inorg. Chem.* **10,** 2304 (1971).
77. Buhks, E., Bixon, M., Jortner, J., and Navon, G., *J. Phys. Chem.* **85,** 3759 (1981).
78. Frese, K. W., Jr., *J. Phys. Chem.* **85,** 3911 (1981).
79. Matteson, D. S., and Bailey, R. A., *J. Am. Chem. Soc.* **91,** 1975 (1969).
80. Taube, H., *Adv. Chem. Ser.* No. 162, 127 (1977).
81. Chan, M. S., and Wahl, A. C., *J. Phys. Chem.* **82,** 2542 (1978).
82. Taube, H., Experimental approaches to electronic coupling in metal ion redox systems. *In* "Tunneling in Biological Systems" (B. Chance, D. C. De Vault, H. Frauenfelder, R. A. Marcus, J. R. Schrieffer, and N. Smith, eds.), pp. 173–197. Academic Press, New York, 1979.
83. Buhks, E., Bixon, M., Jortner, J., and Navon, G., *Inorg. Chem.* **18,** 2014 (1979).
84. Brown, G. M., Krentzien, H. J., Abe, M., and Taube, H., *Inorg. Chem.* **18,** 3374 (1979).
85. Weaver, M., and Yee, E. L., *Inorg. Chem.* **19,** 1936 (1980).
86. Lin, C. T., Böttcher, W., Chou, M., Creutz, C., and Sutin, N., *J. Am. Chem. Soc.* **98,** 6536 (1976).
87. Sutin, N., *J. Photochem.* **10,** 19 (1979).
88. Scandola, F., Ballardini, R., and Indelli, M. T., *in* "Photochemical, Photoelectrochemical, and Photobiological Processes" (D. O. Hall, and W. Palz, eds.), p. 66. Reidel, Dordrecht, 1981.
89. Harriman, A., Porter, G., Darwent, J. R., and Walters, P. C., *in* "Photochemical, Photoelectrochemical, and Photobiological Processes" (D. O. Hall, and W. Palz, eds.), p. 46. Reidel, Dordrecht, 1981.
90. Kiwi, J., and Grätzel, M., *J. Am. Chem. Soc.* **101,** 7214 (1979).
91. Maestri, M., and Sandrini, D., *Nouv. J. Chim.* **5,** 637 (1981).
92. Chan, S. F., Chou, M., Creutz, C., Matsubara, T., and Sutin, N., *J. Am. Chem. Soc.* **103,** 369 (1981).
93. Kiwi, J., and Grätzel, M., *Nature* (*London*) **281,** 657 (1979).
94. Grätzel, M., *Acc. Chem. Res.* **14,** 376 (1981).

2 Dynamics of Light-Induced Energy and Electron Transfer in Organized Assemblies

Pierre P. Infelta

Institut de Chimie Physique
École Polytechnique Fédérale
Lausanne, Switzerland

I. Introduction

All chemical reactions require that the participants be brought sufficiently close to each other to allow the interactions that may result in reaction. In liquid media, diffusion of the reactants in the bulk plays a vital role in the reaction process. Reactions induced by light require that one of the reactants interacts with photons received by the solution. Reactions of the photoexcited species with other reactants present is then possible. The relative distribution of the photoexcited species and the reactants as well

ENERGY RESOURCES THROUGH
PHOTOCHEMISTRY AND CATALYSIS

ISBN 0-12-295720-2

as the spatial distribution of all reactive species will be critical to the nature and the rate of the reactions that can take place (*45*). Those effects on reaction rates or reaction paths are of prime interest in relation to the conversion of light energy to chemical energy such as the decomposition of water into its elements.

A certain degree of organization of the reactants at the molecular level can provide advantageous spatial distribution of the reactants in assemblies such as micelles, microemulsions, vesicles, layers, or in certain functionalized molecules (*43*).

In micellar solution hydrophobic molecules may be solubilized in the hydrocarbon core. In the case of ionic micelles, the charge on the interface can be used to favor the approach of ionic hydrophilic reactants, via coulombic interaction, that should interact with other molecules solubilized in the core. Coulombic interaction with the interface as well as the hydrophilic or hydrophobic character of some of the products of the reaction can be used to achieve a certain degree of spatial separation following the reaction (*7, 8, 20*). This is an extremely important possibility because, in general, any energy-rich products will have a spontaneous tendency to react and waste the energy conversion previously achieved. Reactants may already be in close proximity before the photochemical interaction. This avoids a loss via deactivation of the photosensitive participant because diffusion to the other partner does not have to take place.

Microemulsions of the oil in water (O/W) type also provide a microheterogeneity that resembles that of micelles (*28*). A greater degree of freedom in the configuration of the system is achieved by the ability, for example, to select an oil that can have the needed solubilizing properties for one of the reactants. The coulombic charge at the interface is certainly less than that in ionic micelles, and thus interactions of a coulombic nature should be less noticeable.

One of the advantages of systems that are comparatively small in size with respect to molecular dimensions is that the reactants can easily be kept in the vicinity of each other. This is only of interest if stabilization of the products can be achieved, that is, if all undesirable reactions of the products can be prevented or at least slowed down considerably (*8, 33, 34*).

More complex and more rigid structures can be obtained by using vesicles. Some reactants can be included when the vesicles are prepared, and the various existing regions (double layer, core, etc.) offer a wide variety of locations and interactions, which are not present in micelles or microemulsions, for the reactants and the reaction products (*14*). Their larger geometric dimensions introduce a certain loss of some of the advantages offered by smaller systems because it implies a certain spatial randomization of the reactants, and intravesicular diffusion may play a role that is not negligible. The type of disadvantages just mentioned must be weighed against the possible advantages.

Monolayers can also provide some molecular arrangements of interest for the study of energy or electron transfer and can provide some very specific effects (*31, 32*). Their somewhat delicate structure as well as the small material content they can incorporate renders their industrial use somewhat difficult. Finally, information on reaction parameters, reaction distances, and the nature of a reaction may be obtained via the use of functionalized molecules that incorporate all of the reactants that are to be involved in the desired reaction (*47*).

Multiple variations of the preceding structures may be imagined in which one or several of the participants may be part of the structure of the microheterogeneous system or may simply be a "guest." Efficient design of such systems requires an in-depth understanding of the mechanisms that are taking place. We intend to review here a number of examples in which the organization of the reactants plays a major role in light-induced energy or electron transfer.

II. General Consideration of Organized Structure

Within the limited scope of this chapter, it is necessary, prior to the analysis of the reactions of interest, to develop a few considerations of the obtainment of synthetically organized assemblies as well as criteria that can lead to the selection of a particular organization to accomplish a specific objective. This last point will be reexamined in the conclusion after our survey of the existing data.

We shall examine three particular points: the ease with which the organized assemblies can be prepared, their stability (as organizate), and some of the specific properties that differentiate one type from the other.

A. Micelles

There is little doubt that micelles are the simplest assemblies to obtain (*15*). A solution of amphiphilic molecules (surfactant) at an appropriate concentration, larger than the critical micellar concentration (CMC), will provide spontaneous aggregation of the molecules into micelles; the aggregate may consist of a few tens to a few hundreds of molecules. The shape of the aggregates and the number of molecules that constitute them depends on the surfactant and other additives present in the solution. The CMC depends on the surfactant, and at concentrations slightly above the CMC, micelles are usually spherical. Micelle concentrations of the order of 10^{-4} M can easily be achieved, and the stability of such aggregates is excellent because they form spontaneously. The surfactant monomer is in dynamic equilibrium between the micelles and the bulk solution.

It is, however, difficult to vary the concentration in micelles because dilution eventually brings the concentration below the CMC, and an increase in surfactant concentration affects the structure of the micelle.

Selection of an ionic surfactant allows the micelle to have a positively or negatively charged interface. The electrical potential at or near the surface of the micelle can differ by a few hundred millivolts from that in the bulk of the solution (*30*). The site where solubilization of reactants takes place can vary widely. Hydrophilic ions or compounds may be solubilized preferentially near the interface, the interaction being then essentially electrostatic. The nature of the hydrophobic chain will play an important role in determining where the site of solubilization of the hydrophobic guest molecules can occur. The presence of ionic salts may affect coulombic effects between solute and micelle as well as the shape of the micelles and their aggregation number. Interactions between aggregates are also possible and may affect the reaction paths and/or the dynamics of some reactions. Such effects must be kept in mind. An increase in the concentration of micelles may bring about unwanted interaction between the aggregates, and exchange of solute molecules from micelle to micelle may take place directly. Such interactions can also take place as a consequence of higher ionic strength whereby coulombic effects between aggregates are lessened.

The solubilization process is a dynamic one, and it is clear that solubilized molecules are in dynamic equilibrium between the bulk aqueous phase and the micelle. The more hydrophobic a solute molecule is the longer its mean resident time in a micelle will be.

It is tempting to prepare micelles that can be polymerized, thus obtaining a more permanent structure. This would mean that spontaneous formation is no longer possible. However, once such an entity can be prepared, it would then be possible to work with more dilute or more concentrated solutions than those with normal micelles. Interference in the desired reactions by the monomer that is normally in the bulk could also be suppressed. In the case of ionic micelles, properties can also be altered by using specific counterions to create somewhat functionalized micelles that can be specifically adapted to certain energy or electron-transfer reactions (*33, 34*).

B. Microemulsions

Microemulsions like micelles form spontaneously if the constituents (oil, surfactant, cosurfactant, and water) are present in adequate proportions. Phase diagrams are usually necessary in order to be able to select a particular microemulsion composition such that it possesses stability as well as the desired properties. Microemulsions rich in water are believed to have a spherical structure resembling that of micelles (*30*). The size of the aggregates is slightly larger than that of micelles, and the ability to select the oil allows more flexibility in their solubilization properties.

The location of the solubilized molecules can vary as it does in micelles, but to a larger extent because the transition from water to oil comprises surfactant and cosurfactant. Again, as in micelles, the charged interface of ionic surfactants can be used to give rise to a surface potential that may be used to separate reactants or products.

The concentrations of the aggregates are again always in the same range. Increasing the proportion of oil (and other constituents to remain in the domain where the solution stays transparent) profoundly affects the structure up to the creation of water droplets in an oil bulk, delimited by surfactant and cosurfactant molecules. Such systems may prove to be of some interest because small pools of water can create unique reaction surroundings (*5*). The entire entities are again completely dynamic and are thus extremely stable in view of their spontaneous formation. Microemulsions in which the oil can be polymerized after they are formed seem to retain many of the properties of the unpolymerized microemulsion, but they present the advantage of being nondynamic entities (*6*). Hence their concentrations can be increased or decreased, leading to systems that cannot normally be achieved. These types of investigations are only at an early stage and may prove to be of great interest if enough developments follow.

C. Vesicles

Vesicles represent a simple model of membranes. They may be formed from natural material or, more simply, of double-chain amphiphiles. Appropriate concentrations of the constituents when heated in aqueous solution form large nonuniform vesicles (usually multicompartmental) (*26*). If ultrasonic power is applied to the solution, a certain uniformization of the vesicles' shape and size is possible. They have an ellipsoidal shape, and their actual size and shape are greatly influenced by the ionic environment and the inclusion of solutes. There are at least three regions: the outside, the limiting bilayer membrane, and the inside. The water pool inside may contain a number of reactants that are included in the bulk when preparing the vesicle solution. Once vesicles are formed, they seem stable enough to allow various chromatographic treatments whereby the inside water pool and the outside may be differentiated in ionic strength and chemical content. Such treatments allow for interesting studies, but eventually it seems that migrations from the outside to the inside occur reciprocally via osmosis until an equilibrium is reached. This type of aggregate is significantly larger than micelles or microemulsions. In view of their size, diffusion about the aggregates or in the bilayer will play a larger role in the reaction. The number of guest molecules per aggregate may also be quite large, and a certain randomization of reactants is thus achieved, but possibly in large local concentrations.

D. Monolayers

Systems of superposed monomolecular layers on a support can be assembled to obtain specific properties that cannot be obtained from solutions. For example, alternate layers of acceptor and donor can be obtained for extreme efficiency of electron or energy transfer.

E. Functionalized Molecules

Certain molecules can be multifunctional and thus include all the necessary characteristic properties to carry out light-induced energy or electron transfer. This type of molecules can be extremely useful in obtaining information on various reaction parameters that control the reaction and its rate (*24, 35*).

As the distance between two functional groups on a given molecule is varied, it is possible to know when the distance is too large to allow energy or electron transfer or when, on the contrary, the rate becomes limited by the actual transfer. Instead of preparing bifunctional molecules, it may be advisable to conceive more elaborate systems in which reactants are bound to some sort of a colloidal support and in which the support can play an active role in one of the phases of the reaction, for example, as a catalyst.

III. Kinetic Processes in Micellar Media

Selecting various sensitizers and quenchers allows the gathering of a large amount of information concerning the structure of the micelles (aggregation number), the location in or about the micelle of the different reaction partners, as well as the different interactions that are taking place. A thorough understanding of all these processes is necessary for the successful implementation of solar to chemical energy conversion in which some reactions are highly needed and others totally undesirable. In a micellar solution, the micellar volume represents but a small fraction of the total volume of the solution (~1/1000). Relatively low concentrations of solutes may still represent very high local concentrations if they are naturally confined to the core or near the micellar interface. Therefore, interactions of the type that might be expected in solution more concentrated by a factor of 1000 can occur. Because such solutions are normally impossible to prepare, we have means of obtaining effects that might not otherwise be observable. The role of dimensionality and spatial extent in micellar kinetic processes has been investigated (*16*). For highly simplified cases, the reaction of two molecules follows a first-order rate law, for reactions "inside" the micelles as well as reactions corresponding to surface diffusion of the reactants. Solutes are in dynamic equilibrium between bulk and micelles, and the average residence time of a solute molecule will be dependent on its interaction with the micelles and the bulk (hydrophobicity, charge of the solute, and charge of the micellar interface). Such a situation implies that any solute will be statistically distributed among the micelles. The nature and the rate of the observed processes will be dependent on the distribution. Therefore, it is of great importance to be able to decide on the applicable statistical distribution.

A number of models have been developed that correspond to a number of experimental occurrences. Some of these have been partially corroborated by experiments in which their applicability has been clearly demonstrated, and their application has led to a thorough understanding of the processes that are taking place.

We shall now describe some of the results that have been obtained, together with their interpretation.

A. Energy-Transfer Quenching

The investigation by Infelta *et al.* (*22*) of the decay kinetics of the pyrene excited state solubilized in sodium lauryl sulfate (NaLS) micelle via quenching by methyleneiodide represents one of the first attempts to take into account quantitatively the effect of the dispersion of the micellar phase (*11–13*). Pyrene is solubilized almost exclusively in the micelles, whereas CH_2I_2 is in dynamic equilibrium between the micelle and the bulk. The selection of this quencher was made such that the dynamics of the equilibrium quencher–micelle–bulk occurred on a time scale comparable to the lifetime of the excited state. The kinetics of the process were detected by observing the decay of the characteristic pyrene fluorescence at 400 nm (Fig. 1). In general, two distinct decay regions were observed. In a simplified way, we can say that we first observed a rapid decay that corresponds to a static quenching, that is, a quenching in which one or more quencher molecules are present in the micelle when the excited state is created, remaining there until the reaction occurs. A second, slower decay, which corresponds to the case in which an excited state is generated, is a quencher-free micelle, and the controlling kinetic phenomenon is the approach of the micelles by the quencher. The equilibrium of the quencher with the micelles is described by

$$Q_W + M \underset{k'}{\overset{n}{\rightleftarrows}} Q_M \tag{1}$$

The fluorescence intensity is proportional to the excited-state concentration $P^*(t)$, and taking into account that, at $t = 0$, $P^*(t) = P_\circ$, the initial concentration of excited state is

$$P^*(t) = P_\circ \exp\left(-\left(k_1 + \frac{k_q n[\mathrm{Q}]_W}{k' + k_q}\right)t - \frac{nk_q^2[\mathrm{Q}]_W}{k'(k' + k_q)^2}\{1 - \exp[-(k' + k_q)t]\}\right). \tag{2}$$

The rate k_q corresponds to what the decay rate of the fluorescent probe would be if it could not decay by any other means but by quenching and were imprisoned in one micelle with one quencher molecule, and k_1 is the

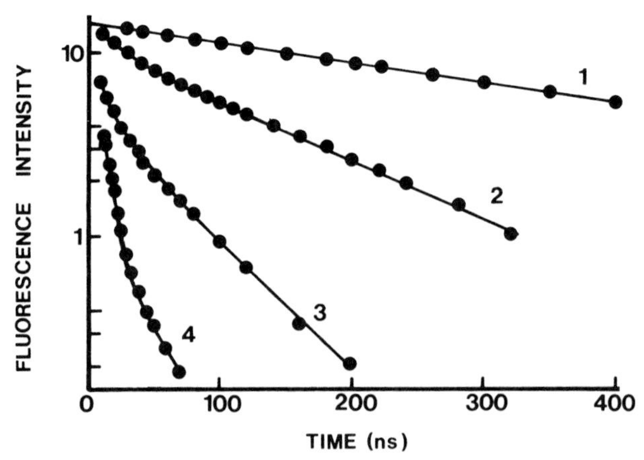

Fig. 1. Semilogarithmic plots of pyrene fluorescence decay curves. [Pyrene] = 10^{-4} *M*, [NaLS] = 0.1 *M*, $[CH_2I_2]_w$ = 0 (1); 2.9 × 10^{-4} *M* (2); 5.8 × 10^{-4} *M* (3); 1.16 × 10^{-3} *M* (4).

natural decay rate of the excited state. The entry rate n for CH_2I_2 was found to be 2.5×10^{10} M^{-1} s^{-1}, the exit rate k', 9.5×10^6 s^{-1}; and the quenching rate k_q, 7.5×10^7 s^{-1}.

Quenching of pyrene and pyrene derivatives by metal ions in sodium dodecyl sulfate (SDS) micelles has been investigated. The interpretation of the result required a refinement of the previous model whereby quenchers may be directly exchanged between micelles without going through the bulk (*9, 10*). A large range of rates has been observed mostly within a factor of 10 of those mentioned for CH_2I_2. Trivalent ions seem to be bound for longer times to the micelles, and it is in this case that direct exchange between micelles can play an important role regarding the kinetics of the quenching process. This type of model provides a very accurate description of the available quenching data. One must ensure that the concentration of the quenchers are kept low enough so as not to affect noticeably the CMC or the aggregation number. These types of experiment have allowed a certain number of kinetic parameters to be determined. It should be noted that equilibrium (1), where the association of a quencher to the micelle is independent of the quencher content of the micelle, implies a Poisson distribution (*3, 21, 41*) of the quencher among the micelles. We shall subsequently see a number of examples that all concur, showing that this distribution function is quite adequate in most cases (for a limited about of solute).

Infelta (*19*) and, simultaneously, Yekta *et al.* (*46*) have shown that steady-state measurements of fluorescence quenching can be quantitatively evaluated to yield a certain amount of information on the behavior of the fluorescent probes and the quencher. If the quenching rate k_q is much larger than the decay rate of the excited probe, then, in most cases, all of the quenching is static and the fluorescence intensity is given by

$$I = I_\circ \exp(-\tilde{n}_q), \tag{3}$$

where $I_\circ$ is the fluorescence intensity in the absence of quencher and $\tilde{n}_q$ the average number of quencher molecules per micelle. This provides a very simple determination of the aggregation number. As has been pointed out, however, one can only obtain reliable results if the dynamics of the probe–quencher system is known to fulfill all of the necessary requirements (*36, 37*).

B. Excimer Formation

The use of a probe that gives rise to the formation of excimers can prove very useful. Prior to the photochemical event only one type of solute molecule is present. Hence, at low enough concentrations, all of these molecules should be solubilized in the same type of environment. Infelta and Grätzel (*21*) have given a detailed kinetic investigation of the pyrene excimer formation in sodium hexadecyl trioxyethylene sulfate (CTOES) micelles. At low pyrene concentrations, the pyrene molecules are isolated in the compartments that constitute the micelles, and a fluorescence spectrum of pure monomer excited-state pyrene is observed. At large pyrene concentration, the excited-state pyrene is very likely to have a ground-state pyrene molecule as a neighbor, and this very readily leads to the formation of an excimer. The kinetic scheme is summarized in the

$$
\begin{aligned}
P &\xrightarrow{h\nu_0} P^* \\
P^* &\xrightarrow{k_r} P + h\nu_1 \\
P^* &\xrightarrow{k_{nr}^{m}} P \\
P_2^* + (i-1)P &\underset{}{\overset{(i-1)k_e}{\rightleftharpoons}} P_2^* + (i-2)P \\
P^* &\xrightarrow{k_r^{e}} 2P + h\nu_2 \\
P_2{}^* &\xrightarrow{k_{nr}^{e}} 2P
\end{aligned}
\qquad (4)
$$

accompanying reactions, where k_r^m and k_r^e are the radiative rate constants for the monomer and excimer, respectively, k_{nr}^m and k_{nr}^e describe all of the decay processes of the excited species other than fluorescence, and k_{-e} is the dissociation rate of the excimer. Finally, k_e is a first-order rate constant describing the formation of an excimer by reaction of one excited probe with a ground-state probe inside the same micelle. The rate of formation of the excimer P_2^* is supposed to be linearly dependent on the number of unexcited probes left in the micelle to react with the excited one. If one denotes by $[P_i^*]$ the excited-state monomer concentration and $[P_i^e]$ the excimer concentration, then the kinetic equations governing the *local concentration* of the various species at a point where the exciting light intensity is I are

$$d[P_1^*]/dt = 2.3\varepsilon[\mathrm{M}]I - (k_{nr}^m + k_r^m)[P_1^*],$$

and for $i = 2, 3, \ldots,$

$$
\begin{aligned}
d[P_i^*]/dt &= i2.3\varepsilon[\mathrm{M}_i]I + k_{-e}[P_i^e] - (k_{nr}^m + k_r^m + (i-1)k_e)[P_i^*], \\
d[P_i^e]/dt &= (i-1)k_e[P_i^*] - (k_{nr}^e + k_r^e + k_{-e})[P_i^e].
\end{aligned}
\qquad (5)
$$

For steady-state irradiations, solutions to the preceding system of equations are easily found that depend on the geometry of the experimental setup.

Assuming that the pyrene molecules are indeed distributed among the micelles according to a Poisson distribution, Infelta and Grätzel (*21*) have found that such a model describes very precisely the observed fluorescence intensities and allows the determination of the aggregation number of CTOES ($\nu = 96$). Selinger and Watkins (*40*) obtained similar expressions using probability theory and concluded that the monomer and excimer fluorescence in cetyltrimethylammonium bromide (CTAB) micelles is well accounted for by such a model. This is strongly in favor of stating that the distribution of solubilized micelles is a Poisson distribution and, as we shall see, is substantiated by a number of other results. The result is in contradiction with the interpretation of Dorrance and Hunter (*11*), and the reasons for such a difference are not clear. A more complete kinetic knowledge of the system is obtained from its time-dependent behavior. This has been done for pyrene in CTOES micelles (*21*) and for pyrene and sodium 5-(1-pyrenyl)pentanoate in cetyltrimethylammonium chloride (CTAC) and SDS micelles (*4*).

For the kinetic model to be applicable, it is necessary that only one excited probe molecule be present per micelle at any time. We have shown that if the probes are distributed according to Poisson's Law, then at the end of an infinitely short excitation pulse, the excited-state monomers (no excimer has had time to form) are also distributed according to a Poisson distribution, namely,

$$[P_i^*](0) = P^*(0) \frac{\bar{n}^{(i-1)}}{(i-1)!} \exp -\bar{n}, \tag{6}$$

where $[P_i^*](0)$ represents the excited-state monomer concentration in a micelle that contained i probe molecules prior to the excitation and $P^*(0)$ is the total excited-state monomer formed. The average number of probes per micelle is $\bar{n}$. Then explicit solution of the system of differential Eq. (5) can be found, and fitting the data allows the determination of the excimer formation and dissociation rate. The pyrene excimer rate of formation is found to be $\sim 10^7$ s^{-1} in CTOES, CTAC, and SDS. In our work we found a dissociation rate for the excimer of $\sim 4 \times 10^6$ s^{-1}, which is comparable to the natural decay rate of the monomer (2.8×10^6 s^{-1}). Most authors seem to have neglected a priori the excimer dissociation in their numerical evaluations (*4, 11, 12*). Pyrene excimer formation in micelles and the kinetics associated with it seem to be well understood.

C. Triplet Energy Transfer

The study of triplet energy transfer in micelles has provided a large amount of information on the aggregates, the solubilization process, and the energy transfer. Using *N*-methylphenothiazine as a donor, Rothenberger *et al.* (*38*) have studied in CTAB micelles the irreversible energy transfer to *trans*-stilbene and the reversible energy transfer to naphthalene. The reaction in a micelle containing i donor molecules (D) and j acceptor molecules (A) can be written as

$$D^T + (i-1)D + jA \underset{ik_{-q}}{\overset{jk_q}{\rightleftharpoons}} A^T + iD + (j-1)A \tag{7}$$

where the rate of transfer is proportional to the number of acceptors and the transfer for a single set of reactants (one D^T and one A or one D and one A^T) is described by a first-order rate. The reactants are solubilized nearly exclusively in the micelles. Irradiation conditions are such that, at most, one D^T is present initially in the micelles. In the two systems selected, the time scale on which the triplet energy transfer occurs is much shorter than the lifetime of either triplets, and there is no exchange of D or A between micelle and bulk or from micelle to micelle. This concept leads to the following differential time laws for D^T and A^T, where $i = 1, 2, 3, \ldots,$ and $j = 0, 1, 2, \ldots,$

$$\begin{aligned} d[D^T{}_{ij}]/dt &= -jk_q[D^T{}_{ij}] + ik_{-q}[A^T{}_{ij}], \\ d[A^T{}_{ij}]/dt &= ik_{-q}[A^T{}_{ij}] + jk_q[D^T{}_{ij}], \end{aligned} \tag{8}$$

where $[D^T{}_{ij}]$ and $[A^T{}_{ij}]$ represent the concentration of triplet of D and A, respectively, at time t in micelles that prior to excitation contained iD and jA molecules. The concentration $[D^T]$ present at time t in the solution is

$$[D^T] = \sum_{i=1}^{\infty} \sum_{j=0}^{\infty} [D^T{}_{ij}], \tag{9}$$

and the time-dependence law will depend on the distribution of the D and A molecules. Explicit expressions can be obtained. A simple and interesting case is that of the irreversible transfer ($k_{-q} = 0$). After all possible transfer has taken place, only triplet D^Ts, which are in micelles that do not contain any A molecules, are present. For a Poisson distribution of the A molecules among the micelles, the triplet concentration at "a long time" ($t \to \infty$) is given by

$$[D^T](\infty) = [D^T](0) \exp -\bar{n}_A, \tag{10}$$

where $\bar{n}_A$ is the average number of A molecules per micelle. This behavior (where D is *N*-methylphenothiazine and A is *trans*-stilbene) is displayed in Fig. 2. The long-time behavior of the system shows that a Poisson distribution among the micelles, for the acceptor, provides an extremely accurate description of the system. It also yields the aggregation number ($\nu = 92$ for CTAB) that is in agreement with literature values. The analysis

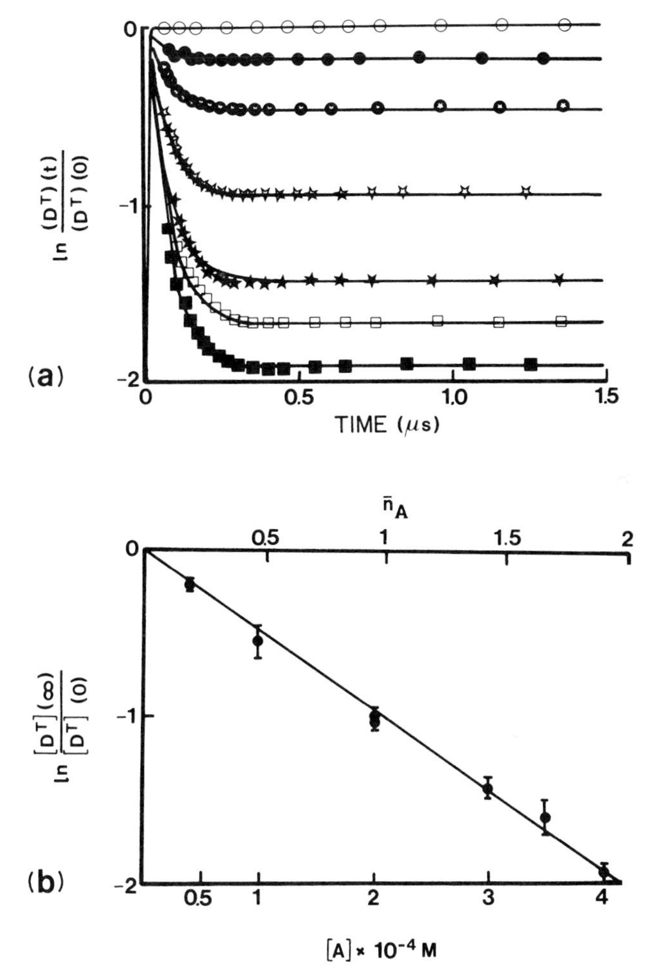

Fig. 2. (a) Kinetics of the irreversible energy transfer in micellar CTAB (2×10^{-2} *M*) solution from MPTH triplet to *trans*-stilbene. [MPTH] = 1.6×10^{-4} *M*; [*trans*-stilbene] = (○) 0.0 *M*; (●) 4×10^{-5} *M*; (✪) 1×10^{-4} *M*; (☆) 2×10^{-4} *M*; (★) 3×10^{-4} *M*; (□) 3.5×10^{-4} *M*; (■) 4×10^{-4} *M*. The curves are calculated. (b) Long-time behavior for the irreversible transfer. [A], *trans*-Stilbene. The solid line corresponds to Poisson's distribution [Eq. (10)]. $\bar{n}$, Average number of A molecules in a micelle.

of the case in which the transfer is reversible, although more complicated, corroborates these results. It has been suggested that other statistics may be applicable (*18, 42*). It is clear that there should be an upper limit to the number of molecules that can be solubilized in a micelle. As the average number per micelle of solubilized molecules (ions) is increased, changes in the micellar structure may be such that the application of these models become illusory. At comparatively low average occupancy (2 to 3), it seems that a Poisson distribution provides an excellent description of most systems. The time-dependent part of the evolution of such systems leads to a transfer rate of $k_q = 1.5 \times 10^7\ s^{-1}$ in the irreversible case (*trans*-stilbene) and $k_q = 2.8 \times 10^6\ s^{-1}$ and $k_{-q} = 3.3 \times 10^6\ s^{-1}$ for the reversible case (naphthalene). The difference in the transfer rate is expected, because in the case of irreversible transfer every encounter of the participants is efficient, whereas in the reversible transfer reaction will *certainly* not occur on each encounter.

D. Triplet–Triplet Annihilation

Rothenberger *et al.* (*39*) have examined the dynamics and statistics of triplet–triplet annihilation (TTA) in cationic (CTAB) micelles using 1-bromonaphthalene as a probe. Contrary to most other studies in which great care is taken to ensure that at most one excited species is present in each micelle, here it was attempted to get several of them per micelle. The exciting source was a 15-ns pulse from a frequency-quadrupled (265 nm) Nd laser. The initial distribution of the excited states is, of course, dependent on the distribution of the probes. Assuming that the probes are distributed according to Poisson, the fraction of the micelles containing i probes prior to excitation is

$$I = (\bar{n}^i/i!) \exp -\bar{n}, \tag{11}$$

where $\bar{n}$ is the average number of probes per micelle. Then if a is the probability of formation of an excited state (during the laser pulse), then the fraction of the micelles that contain x triplets (irrespective of the number of ground-state probes) also follows Poisson law and is

$$F = [(\bar{n}a)^x/x!] \exp -\bar{n}a, \tag{12}$$

where the average number of excited triplet states per micelle is $\bar{n}a$. Because the reactants once formed react pairwise, a distinction will arise between micelles containing an odd or even number of triplet states. Reaction can go to completion when an even number is present, whereas one triplet state remains in micelles that initially contained an odd number of triplets. We have shown that the annihilation corresponds to

$$^3M^* + {}^3M^* \longrightarrow \text{product} \tag{13}$$

where the product does not ultimately lead to the formation of a triplet. The mathematical treatment of these kinetics allows an accurate description of the time dependence of the observed disappearance. At high laser fluence, a fast decay is observed that corresponds to the intramicellar

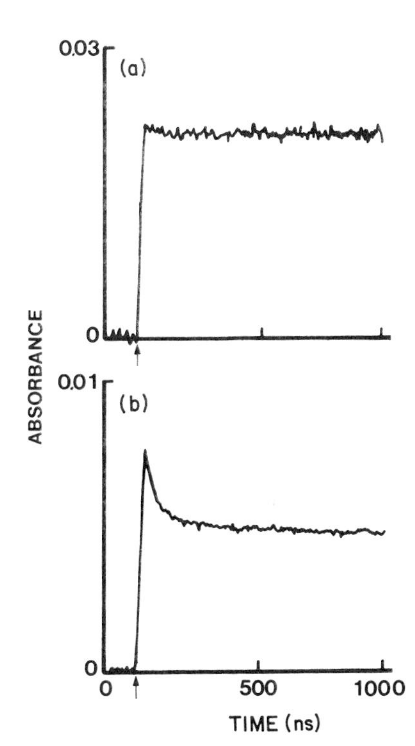

Fig. 3. Transient absorbance observed at λ = 425 nm of a solution containing 5.46×10^{-5} *M* 1-bromonaphthalene and 4.55×10^{-3} *M* CTAB ($\bar{n}$ = 1.45). The solution was saturated with N_2O at 1.3 atm. Fluence of the exciting laser pulse (arrow): (a) 10 and (b) 130 mJ cm^{-2}.

TTA. The longer lived remaining absorbance is due to triplets that are isolated in micelles. At low laser fluence no fast decay occurs, indicating that, at most, one triplet state molecule is present per micelle. These typical behaviors are displayed in Fig. 3. The rate constant for triplet disappearance in a micelle occupied by only two triplets is found to be $2.8 \times 10^7\ s^{-1}$. Half of this rate (two triplets disappear simultaneously) corresponds to the rate observed for irreversib' triplet energy transfer from MPTH to *trans*-stilbene. This suggests tha. .n average time of ~70 ns is necessary for an encounter of two reactant molecules in a CTAB micelle.

E. Electron Transfer

Photochemically induced electron-transfer reactions are important with respect to their potential applications to solar-energy conversion. The redox properties of the products of the electron-transfer reaction may be exploited to lead to energy-rich final products. Micelles can be used to facilitate an electron-transfer reaction or, alternatively, to slow down the energy-wasting back electron transfer, by providing a longer time span to carry out energy-storing reactions.

1. Photoionization Can Be Promoted Using Micelles

Phenothiazine and tetramethylbenzidine (*1, 2*) can be photoionized by a single 347-nm photon in anionic micelles. The resulting cation is then stabilized "inside" the micelle. The aqueous electrons produced are prevented from combining with the cations via coulombic interaction with the micellar interface.

A surfactant derivative of phenothiazine, when micellized, can be photoionized by 347-nm photons, and the process seems to be monophotonic (*17*). This is a typical case in which the assemblies play a major role in the reaction path. Thus, when a solution of the same compound at a concentration less than the CMC is irradiated under similar conditions, only the triplet state of the molecule can be observed. Photoionization no longer takes place.

The relatively high-energy photons needed for this type of reaction and the reactivity of the e_{aq}^- with a large number of species render the use of such reactions for solar-energy conversion somewhat improbable.

2. Photoredox Reactions

As in photoionization, photoredox reactions can be promoted by favorable arrangements of the reactants such as those offered by micellar systems. It has been shown (*29*), using a copper lauryl sulfate solution, that ruthenium tris-bipyridinium [$Ru(bipy)_3^{2+}$] ions are located near the micellar interface of anionic micelles. Using *N*-methylphenothiazine (MPTH) as an electron donor in tetradecyltrioxyethylene sulfate (TTOES) micelles, $Ru(bipy)_3^{2+*}$ produced by 530-nm photons is quenched reductively ($k_q = 3 \times 10^6\ s^{-1}$). The back electron transfer occurs very readily because in this case both of the products remain in the vicinity of each other ($k_b = 6 \times 10^6\ s^{-1}$). Stabilization of the charge separation could be obtained in the present case by the subsequent reduction by $Ru(bipy)_3^+$ of a hydrophobic uncharged species that would become negatively charged, thus becoming sufficiently soluble to allow it to be ejected from the micelle and stabilized in the bulk via coulombic interaction at the interface.

The micellar assemblies can be functionalized and participate in the redox reaction (*33, 34*). On photoexcitation MPTH acts as an electron donor to Eu^{3+} ions according to

$$\mathrm{MPTH^* + Eu^{3+} \longrightarrow MPTH^+ + Eu^{2+}} \tag{14}$$

with MPTH solubilized in mixed micelles of europium decyl sulfate and zinc lauryl sulfate. The schematic of the processes that subsequently take place is given in Fig. 4. The back electron transfer (first-order rate k_b)

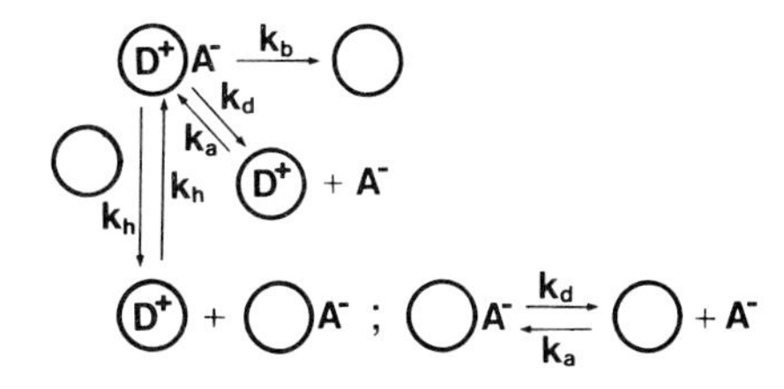

Fig. 4. Schematic illustration of the elementary kinetic processes contributing to the back transfer of an electron from a reduced acceptor to an oxidized donor in mixed micelles. ○, Micelle; A, electron acceptor; D, electron donor.

corresponds to the loss of the energy stored in the separated ions. The reduced acceptor A^- (here Eu^{2+}) can detach (first-order rate k_d) from the host micelle to go into the bulk, but it can also be directly transferred to a neighboring micelle (second-order rate constant k_h), the actual hopping rate being proportional to the micelle concentration. In laser flash photolysis experiments, two distinct time domains are observed. First, there is a rapid decay owing to an intramicellar back electron transfer in which Eu^{2+} ions reduce the very $MPTH^+$ ions from which they were formed. The intramicellar back electron transfer competes with the escape of Eu^{2+} from its native micelle by simple detachment from the micellar surface or by direct transfer to an adjacent aggregate. Then a slower decay takes place, owing to intermicellar reaction in which Eu^{2+} ions diffuse via the bulk or by micelle hopping to any $MPTH^+$-ion-containing micelle. This latter decay occurs according to a second-order rate law for which the rate constant is a function of the previously mentioned parameters as well as of the micelle concentration. Using the bivalent ion Zn^{2+} in the mixed counterion atmosphere of the micelles enhances the escape process and produces a retardation of the undesirable intermicellar reaction.

In the two previous examples, the close proximity of the reactants played a key role in promoting the electron-transfer reaction, and in the latter a certain slowing down of the back electron transfer was achieved. A somewhat more drastic effect in terms of the slowing down of the back electron transfer has been achieved (*7, 8, 20*). We investigated a system containing CTAC as a cationic micelle-forming surfactant, $Ru(bipy)_3^{2+}$ as a sensitizer, and *N*-tetradecyl-*N*-methylviologen ($C_{14}MV^{2+}$) an amphiphilic viologen as an electron acceptor. $Ru(bipy)_3^{2+}$ and most of the $C_{14}MV^{2+}$ ions are in the bulk of the solution where the primary redox reaction takes place. On reduction, the viologen $C_{14}MV^+$ becomes highly hydrophobic and becomes trapped inside the micelles. Subsequently, the coulombic interactions between the $Ru(bipy)_3^{3+}$ and the micellar interface are sufficient to prevent interactions between the oxidized sensitizer and the reduced acceptor. This unwanted reaction only takes place in the bulk of the solution and the accompanying reaction scheme is an accurate description of the reactions that are taking place (The subscript w indicates species solubilized in the bulk of the aqueous phase, whereas the subscript M indicates a species solubilized in a micelle) (*20*). The decay of the

$$
\begin{aligned}
Ru(bipy)_3^{2+} &\xrightarrow{h\nu} Ru(bipy)_3^{2+*} \\
Ru(bipy)_3^{2+*} &\xrightarrow{k_f} Ru(bipy)_3^{2+} \\
Ru(bipy)_3^{2+*} + C_{14}MV_w^{2+} &\xrightarrow{k_q} [Ru(bipy)_3^{3+} \cdots C_{14}MV_w^{+}] \\
&\xrightarrow{\phi} Ru(bipy)_3^{3+} + C_{14}MV_w^{+} \\
&\xrightarrow{1-\phi} Ru(bipy)_3^{2+} + C_{14}MV_w^{2+} \\
Ru(bipy)_3^{3+} + C_{14}MV_w^{+} &\xrightarrow{k_b} Ru(bipy)_3^{2+} + C_{14}MV_w^{2+} \\
C_{14}MV_w^{+} + M &\xrightarrow{k_a} C_{14}MV_M^{+} \\
C_{14}MV_M^{+} &\xrightarrow{k_d} C_{14}MV_w^{+} + M
\end{aligned}
\tag{15}
$$

$C_{14}MV^{+}$ and $Ru(bipy)_3^{3+}$ ions is slowed down considerably in the presence of micelles. These species disappear following a second-order rate law with rate constant k_{obsd} that, in the low micelle-concentration region, is given by

$$k_{obsd} = \frac{k_b}{1 + K_r[M]},$$

where k_b is the second-order rate for back electron transfer in homogeneous solution, K_r the equilibrium constant for the reduced species between bulk and micelles, and [M] the micelle concentration:

$$K_r = \frac{[C_{14}MV_M^{+}]}{[C_{14}MV_W^{+}][M]} = 2.7 \times 10^5 \quad M^{-1}$$

We found that in a 10^{-2} *M* CTAC solution *at equilibrium* 91% of the $C_{14}MV^{2+}$ would be in the bulk of the solution and 96% of $C_{14}MV^{+}$ would be inside the micelles. A very significant decrease in the observed rate constant can be obtained with a corresponding increase in the lifetime of the species.

IV. Kinetics in Other Organized Assemblies

A. *Microemulsions*

Most of the considerations developed for micelles and mentioned previously can probably be extended to spherical W/O microemulsions. It is obvious, however, that in larger aggregates, diffusion within the aggregate plays a more important role, and the first-order rate for encounter becomes smaller. The number of guest molecules that can be accommodated per aggregate can be larger than it is in micelles. Sufficiently large numbers of guest molecules produce a certain randomization in the interactions, and as aggregates become larger the kinetic behavior should approach that of an homogeneous phase. Rate enhancements can still be expected because the volume of the microemulsion oil phase is much smaller than the bulk volume. The dispersed nature of the aggregate would, of course, still play a very important role in cases in which some of the reactants or products do not reside in the same phase. Examples in which the kinetic processes have been analyzed in detail are few. Owing to the larger size of the aggregates, intradroplet reactions become slower. Thus intradroplet reactions, interdroplet reactions, and exchange of reactants with the bulk blend the kinetic steps that were more distinct in the case of micelles. Moreover, the limited lifetime of the probes (natural or owing to impurities) also limits the sensitivity of the experiments. In a study by Lianos *et al.* (*27*), the time evolution of the pyrene monomer fluorescence in a W/O microemulsion (SDS, 1-pentanol, dodecane, or toluene), was used to determine the concentration of the aggregates. In the domains in which the aggregates are probably spherical, it has been found that the number of surfactant molecules per aggregate varies over a wide range in an unpredictable manner. A quantitative interpretation of quenching energy or electron transfer in microemulsions seems to be within reach at this time only for the simplest microemulsion system.

B. Vesicles

Relatively sophisticated molecular arrangements can be obtained in specifically designed synthetic vesicle systems (*14, 26*). These large aggregates can contain a high proportion of reactants and thus create high local concentrations, leading to high reaction rate. The charges on the vesicle surface may favor separation of the products and/or slow down or impair their recombination. A certain dissymmetry will be introduced between the inside water pool and the bulk of the solution. This can be done at the time of preparation or by adding electrolyte to the bulk of the solution. This may affect electric potentials and, thereby, species migrations. Using dioctadecyldimethylammonium chloride (DODAC) vesicles containing $RuC_{18}(bipy)_3^{2+}$, a long-chain derivative of $Ru(bipy)_3^{2+}$ and MPTH, we have shown (*23*) that the amount of $MPTH^+$ formed via photooxidation by $RuC_{18}(bipy)_3^{2+}$ is substantially increased by addition in the bulk of 10^{-3} *M* NaCl. At this concentration of added electrolyte, the surface potential seems to still be sufficient to slow down substantially the recombination process. We should mention here that polymerized vesicles can be prepared (*44*) and have a much improved stability compared with the unpolymerized material.

C. Other Assemblies

Monolayer or multilayer systems can be arranged in a variety of fashions. Aggregates can be formed that show photophysical properties resulting from the molecular arrangement (J aggregates), allowing their detection (*25, 31, 32*). The large delocalization of the excitation energy in such aggregates is illustrated by the fact that at very low ratios of quencher–chromophore, a still very efficient quenching occurs, indicating that delocalization extends to 5000–10,000 dye molecules. These properties render such structures worthy of investigation.

All of the assemblies we have described include organic hydrophobic molecules. In general, it is clear that in most colloidal systems the arrangement and structure of the colloids will play a fundamental role in the determination of the reactions that can take place, and colloidal systems are the subject of intensive investigation both from a fundamental point of view and toward their use in achieving solar-energy conversion.

V. Conclusion

It is clear that great progress has been accomplished toward an in-depth understanding of the dynamics of light-induced processes in micellar systems. For reasonable concentrations of solute, the observed phenomena are very well described using Poisson's Law as the statistical distribution of solute or ions among the micelles. Hydrophylic or hydrophobic properties of some solute and/or coulombic effects can be used to prepare systems in which these properties strongly influence the reaction paths as well as the rates of the reactions. It is obvious that similar types of considerations can provide us with many fascinating systems involving other types of microheterogeneous media such as microemulsions, vesicles, functionalized multilayers, and even specific colloidal systems. A thorough knowledge of the types of organized systems that can be obtained (often very easily) will lead to interesting applications in the field of light-induced energy and electron transfer.

References

1. Alkaitis, S. A., and Grätzel, M., *J. Am. Chem. Soc.* **98,** 3549 (1979).
2. Alkaitis, S. A., Beck, G., and Grätzel, M., *J. Am. Chem. Soc.* **97,** 5723 (1975).
3. Almgren, M., Grieser, F., and Thomas, J. K., *J. Am. Chem. Soc.* **101,** 279 (1979).
4. Atik, S. S., Nam, M., and Singer, L. A., *Chem. Phys. Lett.* **76,** 75 (1979).
5. Atik, S. S., and Thomas, J. K., *J. Am. Chem. Soc.* **103,** 3543 (1981).
6. Atik, S. S., and Thomas, J. K., *J. Am. Chem. Soc.* **103,** 4279 (1981).
7. Brugger, P.-A., and Grätzel, M., *J. Am. Chem. Soc.* **102,** 2461 (1980).
8. Brugger, P.-A., Infelta, P. P., Braun, A. M., and Grätzel, M., *J. Am. Chem. Soc.* **103,** 320 (1981).
9. Dederen, J. C., Van der Auweraer, M., and De Schryver, F. C., *Chem. Phys. Lett.* **68,** 451 (1979).
10. Dederen, J. C., Van der Auweraer, M., and De Schryver, F. C., *J. Phys. Chem.* **85,** 1198 (1981).
11. Dorrance, R. C., and Hunter, T. F., *J. Chem. Soc. Faraday Trans. 1* **68,** 1312 (1972).
12. Dorrance, R. C., and Hunter, T. F., *J. Chem. Soc. Faraday Trans. 1* **70,** 1572 (1974).
13. Dorrance, R. C., and Hunter, T. F., *J. Chem. Soc. Faraday Trans. 1* **74,** 1891 (1978).
14. Fendler, J. H., *Acc. Chem. Res.* **13,** 7 (1980).
15. Fendler, J. H., and Fendler, E. J., "Catalysis in Micellar and Macromolecular Systems." Academic Press, New York, 1975.
16. Hatlee, M. D., Kozak, J. J., Rothenberger, G., Infelta, P. P., and Grätzel, M., *J. Phys. Chem.* **84,** 1508 (1980).
17. Humphry-Baker, R., Braun, A. M., and Grätzel, M., *Helv. Chim. Acta* **64,** 2036 (1981).
18. Hunter, T. F., *Chem. Phys. Lett.* **75,** 152 (1980).
19. Infelta, P. P., *Chem. Phys. Lett.* **61,** 88 (1979).
20. Infelta, P. P., and Brugger, P.-A., *Chem. Phys. Lett.* **82,** 462 (1981).
21. Infelta, P. P., and Grätzel, M., *J. Chem. Phys.* **70,** 179 (1979).
22. Infelta, P. P., Grätzel, M., and Thomas, J. K., *J. Phys. Chem.* **78,** 190 (1974).
23. Infelta, P. P., Grätzel, M., and Fendler, J. H., *J. Am. Chem. Soc.* **102,** 1479 (1980).
24. Kong, J. L. Y., Spears, K. G., and Loach, P. A., *Photochem. Photobiol.* **35,** 545 (1982).
25. Kuhn, H., *Pure Appl. Chem.* **51,** 341 (1979).
26. Kunitake, T., *J. Macromol. Sci. Chem.* **13,** 587 (1979).
27. Lianos, P., Lang, J., Strazielle, C., and Zana, R., *J. Phys. Chem.* **86,** 1019 (1982).
28. Mackay, R. A., *Adv. Colloid Sci.* **15,** 131 (1981).
29. Maestri, M., Infelta, P. P., and Grätzel, M., *J. Chem. Phys.* **69,** 1522 (1978).
30. Mittal, K. L. (ed.), "Micellization, Solubilization and Microemulsions," Vols. 1 and 2. Plenum, New York, 1977.
31. Möbius, D., *Ber. Bunsenges. Phys. Chem.* **82,** 848 (1978).
32. Möbius, D., *Acc. Chem. Res.* **14,** 63 (1981).
33. Moroi, Y., Braun, A. M., and Grätzel, M., *J. Am. Chem. Soc.* **101,** 567 (1979).
34. Moroi, Y., Infelta, P. P., and Grätzel, M., *J. Am. Chem. Soc.* **101,** 573 (1979).
35. Netzel, T. L., Bergkamp, M. A., and Chang, C. H., *J. Am. Chem. Soc.* **104,** 1952 (1982).
36. Rodgers, M. A. J., and Baxendale, J. H., *Chem. Phys. Lett.* **81,** 347 (1981).
37. Rodgers, M. A. J., Da Silva, E., and Wheeler, M. F., *Chem. Phys. Lett.* **53,** 165 (1978).
38. Rothenberger, G., Infelta, P. P., and Grätzel, M., *J. Phys. Chem.* **83,** 1871 (1979).
39. Rothenberger, G., Infelta, P. P., and Grätzel, M., *J. Phys. Chem.* **85,** 1850 (1981).
40. Selinger, B. K., and Watkins, A. R., *Chem. Phys. Lett.* **56,** 99 (1978).

41. Tachiya, M., *Chem. Phys. Lett.* **33,** 289 (1975).
42. Tachiya, M., *J. Chem. Phys.* **76,** 340 (1982).
43. Thomas, J. K., *Chem. Rev.* **80,** 283 (1980).
44. Tundo, P., Kippenberger, D. J., Klahn, P., Prieto, N. E., Jao, T.-C., and Fendler, J. H., *J. Am. Chem. Soc.* **104,** 456 (1982).
45. Turro, N. J., Grätzel, M., and Braun, A. M., *Angew. Chem. Int. Ed. Engl.* **19,** 675 (1980).
46. Yekta, A., Aikawa, M., and Turro, N. J., *Chem. Phys. Lett.* **63,** 543 (1979).
47. Zachariasse, K. A., Kuehnle, W., and Weller, A., *Chem. Phys. Lett.* **73,** 6 (1980).

3 Molecular Engineering in Photoconversion Systems

Michael Grätzel

Institut de Chimie Physique
École Polytechnique Fédérale
Lausanne, Switzerland

I. Introduction

Fossil energy resources are provided by photosynthesis, a fascinating process that has been widely investigated. Several key features as to how photosynthetic energy conversion operates are known. Light-induced charge separation is achieved through judicious spatial arrangement of the pigments and elements of the electron-transport chain in the thylakoid membrane. Cooperative interaction between these components allows the electron transfer to proceed in a vectorial fashion. Enzymes play the role of catalysts that couple the charge separation events to fuel-generating reactions.

Strategies to design artificial photoconversion devices should not attempt to imitate all the intricacies of natural photosynthesis. However, it

ISBN 0-12-295720-2

is inconceivable that the challenging task of driving endergonic chemical reactions, such as the cleavage of water into hydrogen and oxygen, by visible light can be accomplished without suitable engineering on the molecular level. Simple homogeneous solution systems have no prospect of being applied in such artificial devices because the rate of light-driven electron-transfer processes is limited by the diffusion of the reactants. In addition, there is no barrier to impair the thermal back electron transfer that degrades light energy into heat. Finally, the solution reaction is almost always a single electron-transfer event, whereas multielectron redox processes are frequently required in a fuel-generating reactions, such as the cleavage of water into H_2 and O_2.

These problems can be overcome only by using microheterogeneous solution systems. It has been demonstrated that molecular assemblies, such as micelles or colloidal semiconductors, can harvest light energy and convert it rapidly and efficiently to oxidation and reduction equivalents (electron–hole pairs). Catalytic sites introduced at the surface of the aggregates serve to transform the chemical potential available from these redox equivalents into energy-rich products. Research in this area requires an interdisciplinary effort and has been conducted for only a relatively short time. Nevertheless, astonishing progress has already been achieved. Many of the results have been summarized in earlier reviews (*7, 18, 20, 21, 33, 39, 53, 57*). This chapter therefore concentrates on several very recent and interesting observations made mainly in the author's laboratory.

II. Self-Organization and Light-Induced Charge Separation in Solutions of Amphiphilic Redox Chromophores

There are many examples in the literature of how light-induced charge separation can be brought about in aqueous solutions of molecular assemblies. Simple ionic micelles, for example, present a microenvironment suitable for this purpose (*2, 16, 17, 54, 60, 63*). The charged interface between the micellar and aqueous phases constitutes a microscopic electrostatic barrier that can be used to achieve local separation of oxidizing and reducing equivalents produced via light-driven electron transfer. In the photo redox reaction between a sensitizer (S) and a relay compound (R),

$$S + R \underset{\Delta}{\overset{h\nu}{\rightleftarrows}} S^+ + R^- \quad (1)$$

the goal is to maximize the rate of the forward electron transfer and at the same time impair the spontaneous thermal back reaction. This can best be achieved by utilizing functional micellar assemblies (*5, 6, 29, 30, 44, 45, 52*). The latter are distinguished from simple micelles by the chromophore's or the electron relay's being chemically linked to the surfactant molecule that constitutes the micellar aggregate. Consider, for example, the amphiphlic ruthenium complex, which forms micellar assemblies in

ClO_4^{2-} $[Ru(bipy)_2(bipy \cdot 2\,C_{13})^{2+}]$

$R^1 = R^2 = C_{13}H_{27}$

aqueous solution. Figure 1 shows a plot of the surface tension versus the logarithm of $Ru(bipy)_2(bipy \cdot 2C_{13})^{2+}$ concentration. Typical for the onset of micelle formation is the appearance of a cutoff point above which the surface tension remains constant and is unaffected by variations in the

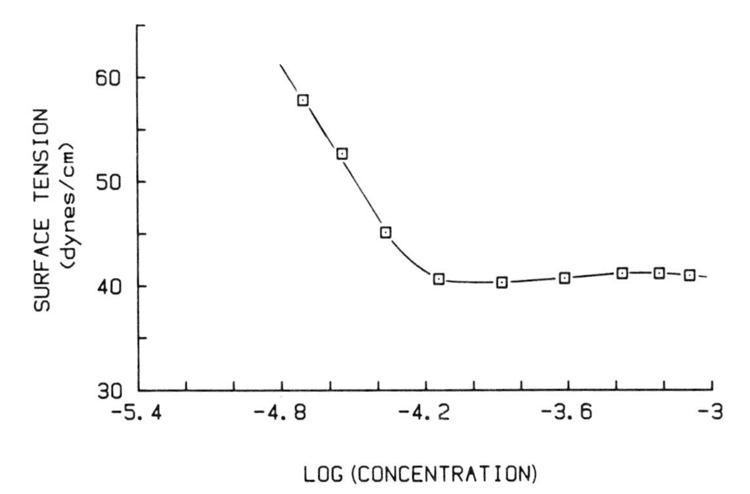

Fig. 1. Determination of the critical micelle concentration of $Ru(bipy)_3(bipy \cdot 2C_{13})^{2+}$. Gibbs plot of the surface tension versus concentration of the chromophore.

surfactant concentration. From these data the critical micelle concentration (CMC) of $Ru(bipy)_2(bipy \cdot 2C_{13})^{2+}$ is derived as 1.6×10^{-4} *M*, and the application of the Gibbs adsorption isotherm to the points below the CMC yields for the surface requirement for one head group a value of 150 A?

Self-organization of $Ru(bipy)_2(bipy \cdot 2C_{13})^{2+}$ has remarkable consequences for its photoredox behavior. An illustrative example is the reductive quenching of the complex by methyl phenothiazine (MPTH):

$$MPTH + Ru(bipy)_2(bipy \cdot 2C_{13})^{*2+} \longrightarrow MPTH^+ + Ru(bipy)_3(bipy \cdot 2C_{13})^+ \quad (2)$$

This reaction has attracted considerable attention (*32, 42, 43, 56*) because it is very efficient in converting radiant energy to chemical energy. From the redox potentials of the two couples involved in reaction (2), one can calculate that more than 90% of the energy required to excite the Ru complex is transformed into chemical potential of the product ions.

Although MPTH is insoluble in water, it is readily solubilized by $Ru(bipy)_2(bipy \cdot 2C_{13})^{2+}$ micelles. Laser photolysis analysis of the electron-transfer events occurring in such assemblies reveals the following characteristics. (i) The electron transfer from solubilized MPTH to excited chromophore occurs so rapidly that it is completed within the 15-ns laser pulse. (ii) Subsequent to electron transfer, $MPTH^+$ is ejected from the micellar into the aqueous phase. Back reaction with $Ru(bipy)_2(bipy \cdot 2C_{13})^+$ is impaired by the positive micellar surface potential. The advantage of using functionalized molecular assemblies becomes readily apparent if these results are compared to data obtained in homogeneous solution. To achieve a reaction time of less than 10 ns in homogeneous solution, one would have to use an MPTH concentration of at least 0.1 *M*, which exceeds the solubility of MPTH in most organic solvents. A further disadvantage of the homogeneous medium is that the back electron transfer occurs ~500 times more rapidly in it than in the organized assemblies, rendering the coupling of the photoredox event more difficult with subsequent fuel-generating reactions.

The MPTH/$Ru(bipy)_3^{2+}$ system was used to illustrate the way in which suitable molecular engineering can be accomplished to achieve light-induced charge separation. One key element is the effect of self-organization of the redox chromophore into micelles brought about by attaching alkyl chains of suitable length to one ligand moiety of the $Ru(bipy)_3^{2+}$ complex. These aggregates solubilize MPTH in their interior, thereby bringing the participants of the photoreaction into close proximity and enhancing the rate of the forward electron-transfer event. Another important feature of the microheterogeneous system is the buildup of an ionic double layer during self-assembly of the redox chromophore into micelles. Typically, the surface potential of a cationic micelle is around 100 mV (*13*). This positive potential of the micellar with respect to the aqueous phase is the reason for the rapid ejection of $MPTH^+$ from the micelle and the inhibition of its back reaction with the reduced $Ru(bipy)_3^{2+}$ complex.

III. Water-Cleavage Cycles and Development of Artificial Analogs of Photosystem II of Green Plants

There are a number of microheterogeneous systems that have been shown to split water into hydrogen and oxygen under visible and uv-light illumination. They all operate according to a common concept. Light is first used to generate reduction and oxidation equivalents. The former could be a reduced chemical species (R^-) or the conduction-band electron (e_{cb}^-) in a semiconductor particle, whereas the latter could be an oxidized chemical species (S^+) or the valence-band hole (h^+) in a semiconductor particle. The light reaction is coupled to dark (catalytic) processes producing hydrogen and oxygen from water and regenerating the starting chemicals (Fig. 2). Because the topic of light-induced H_2 generation via the sequence of reactions (3) and (4) will be treated extensively in other

$$\mathrm{S, R} \xrightarrow{h\nu} \mathrm{S^+ + R^-} \tag{3a}$$

$$\mathrm{Semiconductor} \xrightarrow{h\nu} \mathrm{h^+ + e_{cb}^-} \tag{3b}$$

$$\mathrm{R^- + H_2O} \longrightarrow \mathrm{\tfrac{1}{2}H_2 + OH^- + R} \tag{4a}$$

$$\mathrm{e_{cb}^- + H_2O} \longrightarrow \mathrm{\tfrac{1}{2}H_2 + OH^-} \tag{4b}$$

$$\mathrm{2S^+ + H_2O} \longrightarrow \mathrm{2H^+ + \tfrac{1}{2}O_2 + 2S} \tag{5a}$$

$$\mathrm{2h^+ + H_2O} \longrightarrow \mathrm{2H^+ + \tfrac{1}{2}O_2} \tag{5b}$$

chapters of this book, we shall concentrate here on the water oxidation part, that is, reaction (5). The formation of oxygen from water by a single electron oxidant (S^+) presents a formidable problem because it proceeds in four subsequent steps involving high-energy intermediates such as OH· radicals, hydrogen peroxide, and superoxide radicals. In order to avoid formation of these reactive intermediates, electron-storage catalysts are required. In 1978 it was discovered in our laboratory (*19, 36*) that noble-metal oxides such as PtO_2, IrO_2, and RuO_2 in macrodispersed or colloidal form are capable of mediating water oxidation by such agents as Ce^{4+}, $Ru(bipy)_3^{3+}$, and $Fe(bipy)_3^{3+}$. RuO_2 has subsequently been the most widely investigated (*34, 37, 38, 40, 41, 49*). An impression of the improvement of the activity of RuO_2-based catalysts may be gained from the

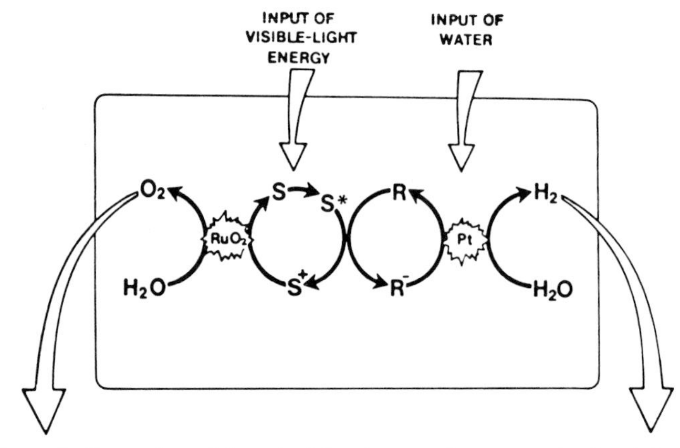

Fig. 2. Schematic illustration of the principle of water-cleavage cycles involving sensitizer, electron relay, and two redox catalysts.

following comparison. In 1978, we required 1 g/liter RuO_2 powder to effect water oxidation by $Ru(bipy)_3^{3+}$ within a period of several minutes.

$$4Ru(bipy)_3^{3+} + 2H_2O \longrightarrow 4Ru(bipy)_3^{2+} + O_2 + 4H^+ \quad (6)$$

Today, by using an ultrafine deposit of RuO_2 on colloidal TiO_2 particles, a half-life of 5–10 ms can be obtained (*30*) using only 3 mg/liter RuO_2. Thus, by decreasing the particle size of RuO_2 and stabilizing the catalyst on a suitable carrier, more than a 10^6-fold increase in the catalytic activity can be achieved.

Application of combined flash photolytic and fast conductometric techniques made it possible to probe the mechanistic details of the oxygen-evolution reaction. Thus $Ru(bipy)_3^{3+}$ was produced via the photoredox reaction (*30*)

$$2Ru(bipy)_3^{2+} + S_2O_8^{2-} \xrightarrow{h\nu} 2Ru(bipy)_3^{3+} + 2SO_4^{2-} \quad (7)$$

and the kinetics of oxygen production via reaction (6) were studied in the presence of a catalyst consisting of a transparent TiO_2 sol loaded with RuO_2. A comparison was made of the temporal behavior of the $Ru(bipy)_3^{3+}$ absorption decay and the increase in conductivity associated with water oxidation. Both events were found to occur simultaneously, indicating that hole transfer from $Ru(bipy)_3^{3+}$ to RuO_2 is immediately followed by release of protons and oxygen from water. Thus the particle couples reduction of $Ru(bipy)_3^{3+}$ to water oxidation. This experimental observation provides a direct proof of the concept of redox catalysis in which the RuO_2 particle is considered to be a microelectrode (*34*). The microelectrode acts as a local element (*55, 59*), coupling the reduction of the electron acceptor [$Ru(bipy)_3^{3+}$] to the oxidation of water. This process involves formation of surface-bound hydroxyl radicals, as pointed out by Trasatti and Buzzanca (*58*) and Galizzioli *et al.* (*15*). From the current–potential characteristics of these two redox processes on RuO_2, the potential of the particle and the overall rate of the reaction can be predicted. Such analysis is discussed in Chapter 4 by McLendon.

The action of RuO_2 as a microelectrode may be compared to the water-splitting enzyme in photosystem II (PS-II) present in chloroplasts. In the thylakoid membrane, oxidation equivalents are provided for by chlorophyll radical-cations associated with the reaction centers of PS-II. It was of interest to determine whether porphyrin II radical-cations, which are closely related to chlorophyll, would afford water oxidation to O_2 in the presence of a RuO_2/TiO_2 redox catalyst. A porphyrin having suitable redox potential is $ZnTMPyP^{4+}$ (zinc tetramethylpyridyl porphyrin, $E° =$ 1.18 V versus NHE) (*50*). At pH > 1 the reaction

$$4ZnTMPyP^{5+} + 2H_2O \longrightarrow 4H^+ + O_2 + 4ZnTMPyP^{4+} \tag{7a}$$

is thermodynamically downhill. The question is whether or not it can be brought about with very active RuO_2 catalysts. Indeed, Borgarello *et al.* (*4*) reported that acidic solutions of $ZnTMPyP^{4+}$ containing Fe^{3+} ions and a RuO_2–TiO_2 catalyst produced O_2 under visible-light illumination. This was ascribed to the reaction of porphyrin triplet states with Fe^{2+} followed

$$ZnTMPyP^{4+}(T_1) + Fe^{3+} \longrightarrow Fe^{2+} + ZnTMPyP^{5+} \tag{8}$$

by reaction (7a). However, these results were questioned in a subsequent investigation by Harriman *et al.* (*26*), who failed to observe O_2 formation in a similar experiment. In view of this conflicting evidence and the general importance of water oxidation by chlorophyll analogs we probed reaction (7a) directly using spectrophotometric and conductometric detection techniques. In a first series of experiments, the porphyrin radical-cation was generated by pulse radiolysis of aqueous $ZnTMPyP^{4+}$ solution containing 10^{-2} *M* Br^-, as described by Neta (*48*). The behavior of $ZnTMPyP^{5+}$ was monitored optically at 700 nm, at which the porphyrin radical-cation has a characteristic absorption maximum (*4*). Figure 3 illustrates the temporal behavior of the 700-nm absorption in solutions of pH 3. The formation of $ZnTMPyP^{5+}$ from the porphyrin and Br_2^- radicals produced by the radiation pulse is a step function on the time scale of observation. There is no noticeable decay of the absorption over 0.5 s in solutions containing porphyrin with and without TiO_2 colloid [120-Å-sized TiO_2 particles were prepared according to Moser and Grätzel (*46*)]. However, if the TiO_2 colloid is charged with 10% RuO_2, a rapid decay of ZnTM-

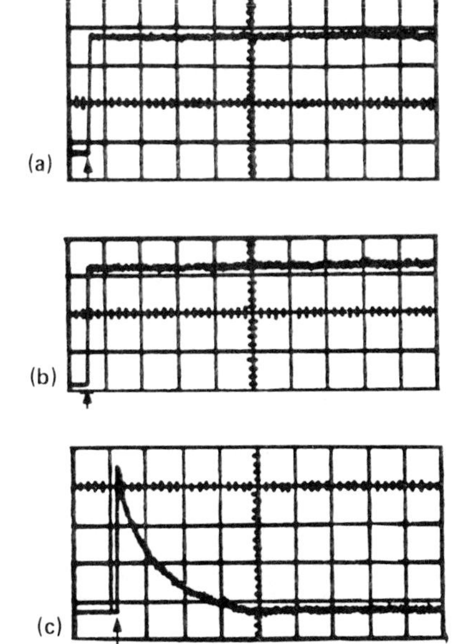

Fig. 3. Pulse radiolysis of 10^{-4} *M* $ZnTMPyP^{4+}$ in the presence of 10^{-3} *M* Br^-. Solutions saturated with N_2O at pH 3. The temporal evolution of the 700-nm absorption of the $ZnTMPyP^{5+}$ radical-cation is shown in the absence (a) and presence (b, 1 g/liter TiO_2; c, 1 g TiO_2 and 10% RuO_2) of catalyst. Pulse (arrows) λ = 700 nm. Time units (abscissa; ms): a, b, 50; c, 2.

PyP^{5+} is observed, which has a half-life of only 1.8 ms. Use of the conductometric detection technique showed that this reaction does not simply reflect injection of holes from $ZnTMPyP^{5+}$ into the catalyst; rather, it was shown to be integral to the process of water oxidation. Formation of protons was found to occur concomitantly with the 700-nm absorption decay, as should be the case if the stoichiometry of reaction (7a) is obeyed.

These results were confirmed in a second set of experiments in which laser photolysis was used to produce triplet-state $ZnTMPyP^{4+}$, which in the presence of Fe^{3+} reacts to give $ZnTMPyP^{5+}$ according to reaction (8). The kinetics of $ZnTMPyP^{5+}$ reaction was examined in the absence and presence of a RuO_2–TiO_2 colloidal catalyst. The water-oxidation reaction under the experimental conditions employed (pH 2) was found to occur with a specific rate of 2×10^2/s.

The case of water oxidation by porphyrin radical-cations in the presence of colloidal TiO_2–RuO_2 is of particular interest because it shows that the ultrafine particles can be considered to be artificial analogs of the water-splitting enzyme of PS-II. In both cases, oxidation equivalents produced through light-initiated electron transfer of the porphyrin moiety are transferred through the catalyst to water-releasing oxygen and protons. This has proven to be extremely valuable in energy-conversion systems in which water is the source of photodriven uphill electron flow (*21*). Thus, without exception, all microheterogeneous water-cleavage devices operating on visible light presently employ RuO_2-based catalysts in the oxidative part of the overall reaction. These will be discussed in more detail in subsequent chapters of this book.

IV. Colloidal Semiconductors

Colloidal semiconductors exhibit several advantageous features that make them attractive candidates as light-harvesting units in solar-energy-conversion devices. They possess high absorption coefficients, and by suitable choice of the material, optimal exploitation of the solar flux can be achieved. Quantum yields for electron–hole pair separation following band-gap excitation are usually unity. Furthermore, in semiconductor particles of colloidal dimension the charge-carrier diffusion to the interface is very rapid, competing efficiently with electron–hole recombination. Particularly intriguing is the possibility of surface modification of the semiconductor particle by chemisorption, chemical derivatization or catalyst deposition to couple light-induced charge separation with subsequent fuel-generating dark reactions. By use of aqueous dispersions of ultrafine semiconductor particles, striking effects of electron storage and hydrogen generation occur, as demonstrated in the following sections.

A. *Interfacial Electron- and Hole-Transfer Reactions in Colloidal Semiconductor Dispersions*

This section covers some basic features of electron-transfer reactions in colloidal semiconductor dispersions. It is highly desirable to obtain experimental information on the dynamics of photoinduced conduction or valence-band processes involving species in solution. In the case of solid semiconductor electrodes, a direct study of the kinetics of these reactions is made difficult by the response of the electrical circuit to phenomena unrelated to the electron-transfer event at the interface. Similarly, the study of macrodispersions of semiconductor powders is hampered by the high turbidity of these systems, rendering impossible the observation of transient species by fast kinetic spectroscopy.

In connection with our study of the light-induced cleavage of water into hydrogen and oxygen, we became intrigued with the idea of investigating semiconducting particles of size small enough to render scattering of light negligibly small. These particles have the advantage of yielding clear dispersions, thus allowing ready kinetic analysis of interfacial charge-transfer processes by laser photolysis technique. The method consists of exciting the colloidal semiconductor particle directly by a very short flash of band-gap radiation. The subsequent reaction of conduction-band electrons and valence-band holes with reactants in solution is followed using fast kinetic spectroscopy or conductometry. The competitive trapping of electrons by noble-metal catalysts, deposited onto the semiconductor particle to mediate hydrogen generation from water, can also be investigated. Finally, this study establishes a technique to determine the Fermi potential of ultrafine colloidal semiconductors.

1. Basic Kinetic Considerations

We shall first consider the question of light absorption by these ultrafine particles. The maximum absorption coefficients for the semiconductor materials investigated is of the order of 10^5/cm, corresponding to an absorption length of at least 1000 Å. In view of their small dimension (50–100 Å), light traverses many particles before complete extinction has occurred, thus producing electron–hole pairs spatially throughout the particles along the optical path. This distinguishes the colloidal particles from semiconductor powders or electrodes in which charge carriers are created mainly near the surface.

Consider a situation in which a colloidal semiconductor particle, such as TiO_2, is excited by a short (~10 ns) laser pulse, resulting in the generation of electron–hole pairs (Fig. 4). The reaction of the conduction-band

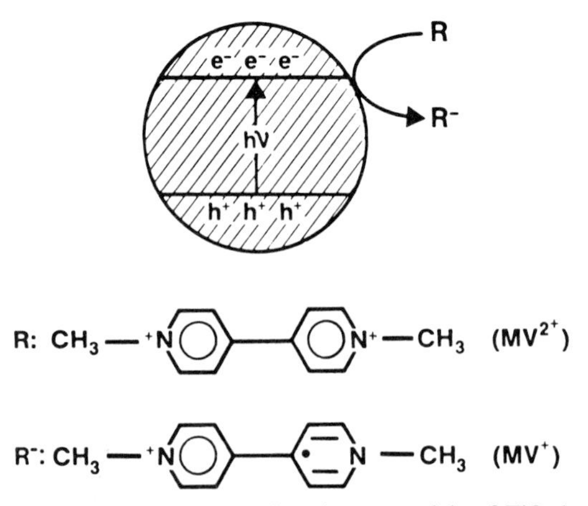

Fig. 4. Laser excitation of a colloidal semiconductor particle of TiO_2 by a short pulse of band-gap radiation (λ_{exc} = 347 nm, pulse width 10 ns) and subsequent electron transfer to a relay compound R. Concentration of electron–hole pairs produced by the laser pulse: ~10^{19}/cm^3 (E_s° − 0.45 V, λ_{max} = 602 nm).

electrons with a relay compound R present in the bulk solution is subsequently observed by kinetic spectroscopy. The hole is trapped readily by surface hydroxyl groups, leading to water oxidation; the back charge transfer into the particle is thus assumed to be negligible. In the case presented here, methylviologen was used as an electron relay. The overall process of reduction of MV^{2+} by conduction-band electrons is described by three elementary steps (Fig. 5).

a. Diffusion of Charge Carriers from the Particle Interior to the Interphase. It is of interest to calculate the average transit time for a charge carrier from the interior to the surface of the colloidal semiconductor. Neglecting electrical field gradients, one obtains for the average transit time (*23*)

$$\tau = r_0^2/\pi^2 D, \tag{9}$$

where D is the diffusion coefficient for electrons (or holes) in the semiconducting particle; D is related to the carrier mobility via

$$D = \mu(kT/e). \tag{10}$$

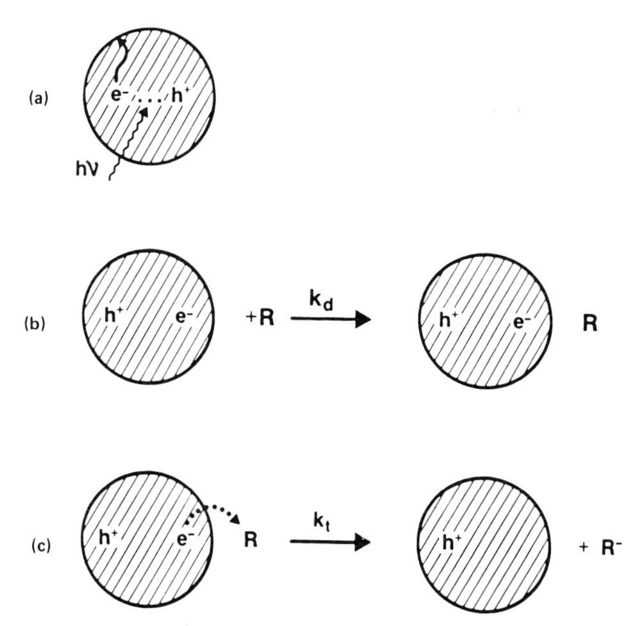

Fig. 5. Elementary processes involved in the charge transfer from the semiconductor particle to the relay compound: (a) excitation and diffusion of charge carriers to the particle surface (average diffusion time $\tau = R^2/\pi^2 D$), (b) encounter-complex formation with relay compound R, and (c) interfacial electron transfer.

For TiO_2 particles with $r_0 = 100$ Å and an electron mobility of 0.5 cm^2/(Vs), one calculates $\tau = 10$ ps. This transit time is much faster than the estimated value of 100 ns for recombination, calculated for bulk recombination in a direct-band-gap semiconductor with a majority carrier density of 10^{17} cm^3. A similar estimation (*62*) yields a recombination time of 100 ps at a majority carrier density of 2×10^{19} cm^3. Note that τ increases with the square of the particle radius. In the case of TiO_2 powders, a typical particle size would be 1 μ, which corresponds to $\tau = 100$ ns. It is noteworthy that for the 100-Å-radius particles, the transit time of charge carriers to the surface is much faster than the recombination time. This would indicate that for such small particles an internal electrostatic field is not necessary for separation of the photogenerated electron–hole pair.

b. Encounter-Complex Formation with the Electron (or Hole) Acceptor Present in Solution. In order that electron transfer occur, an encounter complex must first be formed between the semiconductor particle and the electron relay present in the bulk solution. The rate of this process is diffusion limited and, hence, determined by the viscosity of the medium and the radius of the reactants. Note that this diffusional displacement will play no role in systems in which the relay adheres to the particle surface.

c. Interfacial Electron Transfer. This is a heterogeneous process that involves the movement of an electron from the conduction band of the semiconductor particle across the Helmholtz layer to the acceptor molecule. The rate constant of this process (k_{et}) is expressed consequently in units of centimeters per second.

The sequence of encounter-complex formation between the semiconductor particle and relay and subsequent electron transfer can be treated kinetically by solving Ficks second law of diffusion (*1*). One obtains for the observed bimolecular rate constant for electron transfer the expression

$$\frac{1}{k_{obs}} = \frac{1}{4\pi r^2}\left(\frac{1}{k_{et}} + \frac{r}{D}\right), \tag{11}$$

where r is the reaction radius corresponding to the sum of the radii of the semiconductor particle and electron relay and D is the sum of their respective diffusion coefficients. The structure of Eq. (11) suggests two limiting cases:

$$k_{et} \ll D/r,$$

in which case the reaction is mass-transfer limited, and Eq. (11) reduces to the familiar Smolucholoski expression

$$k_{obs} = 4\pi r D. \tag{12}$$

This condition implies that for semiconductor particles with a radius of 100 Å and a relay with a diffusion coefficient of $\sim 10^{-5}$ cm^2/s, k_{et} will only be sufficiently large if greater than 10 cm/s. This is a large electrochemical rate constant, and many redox couples will require substantial overvoltage to reach values as high as this. The other limiting case is

$$k_{et} \ll D/r.$$

Here the heterogeneous electron transfer at the particle surface is rate limiting, and Eq. (11) simplifies to

$$k_{obs} = 4\pi r^2 k_{et}. \tag{13}$$

As k_{et} depends on the potential of the semiconductor particle, one would expect that under these conditions k_{obs} would also be a function of the overvoltage available to drive the interfacial electron-transfer event.

2. *Dynamics of Methylviologen Reduction by Conduction-Band Electrons of Colloidal TiO_2 Particles and Fermi Potential of Colloidal TiO_2*

These considerations will now be applied to analyze the dynamics of electron transfer from colloidal TiO_2 to methylviologen (MV^{2+}). The latter is known to undergo a reversible one-electron reduction with a well-defined and pH-independent redox potential (E° = −440 mV versus NHE). Furthermore, the reduced form (MV^+) can be readily identified by its characteristic blue color, corresponding to an absorption maximum of 602 nm (ε = 11,000/*M*/cm).

Upon exposure of a deaerated TiO_2 sol to near-uv light in the presence of MV^{2+}, the formation of this blue-colored species becomes readily apparent. Both the growth kinetics of MV^+ as well as its yield were found to be strongly pH dependent (*11*). Figure 6 illustrates the effect of pH on the yield of MV^+ after completion of the electron transfer. [In these experiments, the colloidal TiO_2 particles (500 mg/liter) were excited with a 347.1-nm ruby laser flash, and the MV^{2+} concentration was invariably 10^{-3} *M*.] This curve exhibits a sigmoidal shape; no reduction of MV^{2+} occurs at pH $\leq$ 2. The yield of MV^+ increases steeply between pH 2.5 and 5 and attains a plateau for neutral and basic solutions in which $[MV^+] = 2.5 \times 10^{-5}$ *M*.

The results depicted in Fig. 6 can be used to derive the Fermi potential (E_{fb}) of the colloidal TiO_2 particle for which we find

$$E_{fb} = -120 - 59 \text{ pH (mV versus NHE)}. \tag{14}$$

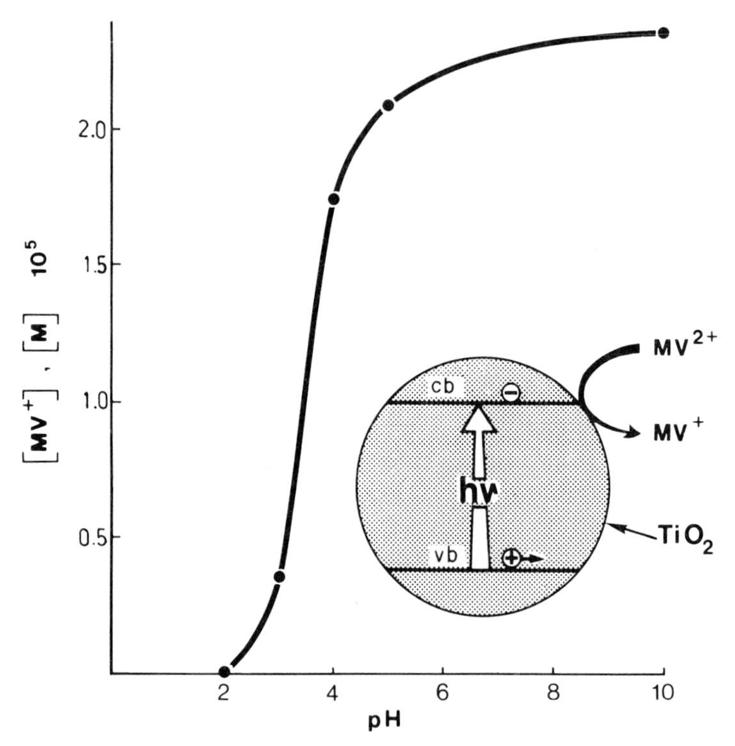

Fig. 6. Effect of pH on the yield of MV^+ (λ_{max} = 602 nm) after completion of interfacial electron transfer from colloidal TiO_2; $[TiO_2]$ = 500 mg/liter, $[MV^{2+}] = 10^{-3}$ *M*.

The position of the conduction-band edge of the colloidal TiO_2 particle greatly influences the rate of the interfacial electron transfer. Figure 7 plots the effect of pH on the observed rate constant for the reduction of MV^{2+} and the amphiphilic viologen:

$$CH_3-(CH_2)_{13}-\overset{+}{N}\langle\bigcirc\rangle-\langle\bigcirc\rangle\overset{+}{N}-CH_3 \quad (C_{14}MV^{2+})$$

Both relays give linear relationships in a pH domain of $4 \leq pH \leq 10$ when $\log k_1$ is plotted against pH. At pH > 10 the rate constant for MV^{2+} attains a plateau ($k_1 = 10^7$/s), but that of $C_{14}MV^{2+}$ continues to increase.

These phenomena may be rationalized in terms of the shift of the conduction-band position of the TiO_2 particle with pH, affecting the match of electronic donor levels in the semiconductor (D_{sc}^{occ}) with that of acceptor levels in the electrolyte (D_{el}^{unocc}). The overlap integral of these two density-of-state functions determines the rate constant for heterogeneous electron transfer:

$$k = v \int_0^\infty D_{sc}^{occ} D_{el}^{unocc} \, dE, \tag{15}$$

where v is the tunneling collision frequency. The term D_{sc}^{occ} is identical with the density of occupied electronic levels in the conduction band of the TiO_2 particle, which peaks sharply at the conduction-band edge E_{cb}. The term D_{el}^{unocc} is the distribution function for unoccupied levels in the $MV^{2+/+}$ redox system which has a maximum at $E°(MV^{2+/+}) - \lambda$, where λ is the reorganization energy. At low pH the position of the TiO_2 conduction band is relatively positive; hence there is poor overlap of occupied states in the particle with unoccupied ones in the electrolyte. The situa-

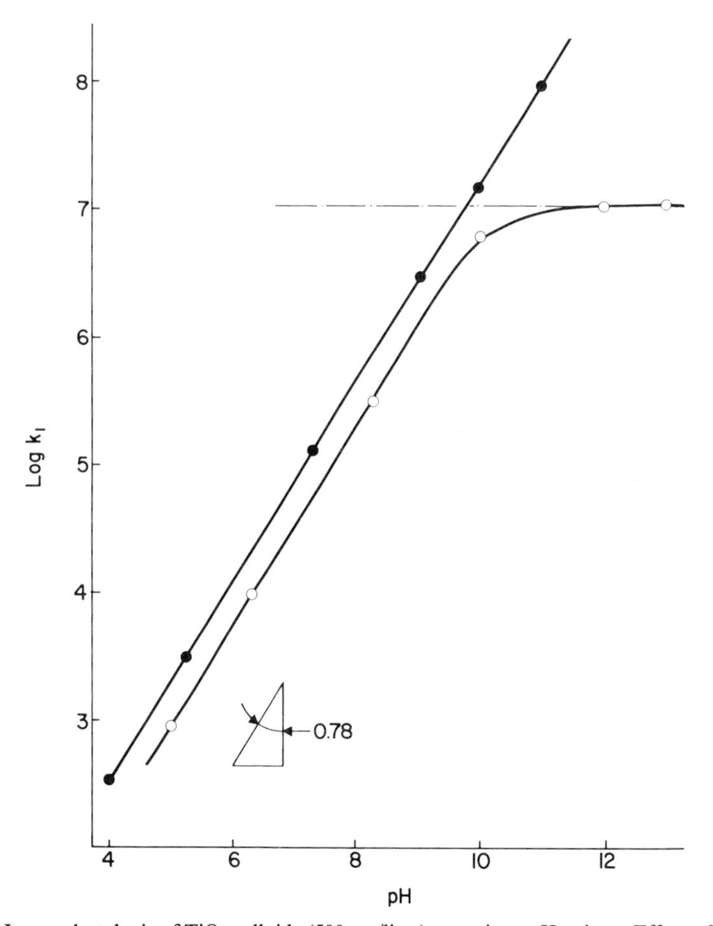

Fig. 7. Laser photolysis of TiO_2 colloids (500 mg/liter) at various pH values. Effect of pH on the observed rate constant (–·–·–; $k_l = 10^7\ s^{-1}$) of electron transfer from the conduction band to MV^{2+} and the amphiphilic viologen $C_{14}MV^{2+}$. $2 \times 10^{-4}\ M$: ●, $C_{14}MV^{2+}$; ○, MV^{2+}.

tion improves with increasing pH, with which the conduction-band position shifts to more negative values.

Using Eq. (11) one can quantitatively evaluate k_{et} as a function of pH from the measured bimolecular rate constants for MV^{2+} reduction. This evaluation (*23*) shows that a Tafel relationship is obeyed at pH < 10:

$$\log(k_{et}/k_{et}^{\circ}) = -(1 - \alpha)(nF/2.303RT)\eta. \tag{16}$$

The overvoltage can be expressed in terms of the standard redox potential of the $MV^{2+/+}$ couple and the potential of the TiO_2 conduction band:

$$\eta = E_{cb}(TiO_2) - E^{\circ}(MV^{2+/+}). \tag{17}$$

Inserting the values for E_{cb} and $E^{\circ}(MV^{2+/+})$ in Eq. (17) gives

$$\eta = 0.315 - 0.059 \text{ pH} \tag{18}$$

and

$$\log(k_{et}/k_{et}^{\circ}) = (1 - \alpha) \text{ pH} - (1 - \alpha)5.34, \tag{19}$$

where 5.34 is the pH at which $\eta = 0$.

From these considerations the interpretation of the data displayed in Fig. 7 is as follows. For reduction of simple methylviologen by conduction-band electrons of TiO_2, the interfacial electron transfer is rate controlling at lower pH, when the overvoltage available to drive the reaction is small. At higher pH, mass-transfer effects become increasingly important, and the reaction becomes diffusion controlled at pH > 10, at which the bimolecular rate constant for MV^{2+} reduction is $5 \times 10^{10}/M$/s. In the case of $C_{14}MV^{2+}$, the rate of reduction by e_{cb}^- continues to increase in alkaline solution, with k_1 far exceeding the diffusion-controlled limit. This indicates that mass-transfer effects play no role in the electron-transfer process and, hence, that $C_{14}MV^{2+}$ in contrast to MV^{2+} is strongly adsorbed to the surface of the TiO_2 particles. (Note, however, that in both cases the same transfer coefficient, $\alpha = 0.22$, is observed.)

The effect of $C_{14}MV^{2+}$ adsorption to the TiO_2 particle may be exploited to achieve multielectron storage, is shown later.

3. Multielectron Storage and Hydrogen Generation with Functional Viologen in Colloidal Semiconductor Dispersions

Illumination of an aqueous TiO_2 sol containing $C_{14}MV^{2+}$ with $\lambda > 300$ nm light results in immediate appearance of a blue color that can be attributed to formation of viologen radical-cations ($C_{14}MV^{+}$) (Fig. 8). However, in alkaline medium (pH > 4.5), this color does not persist under

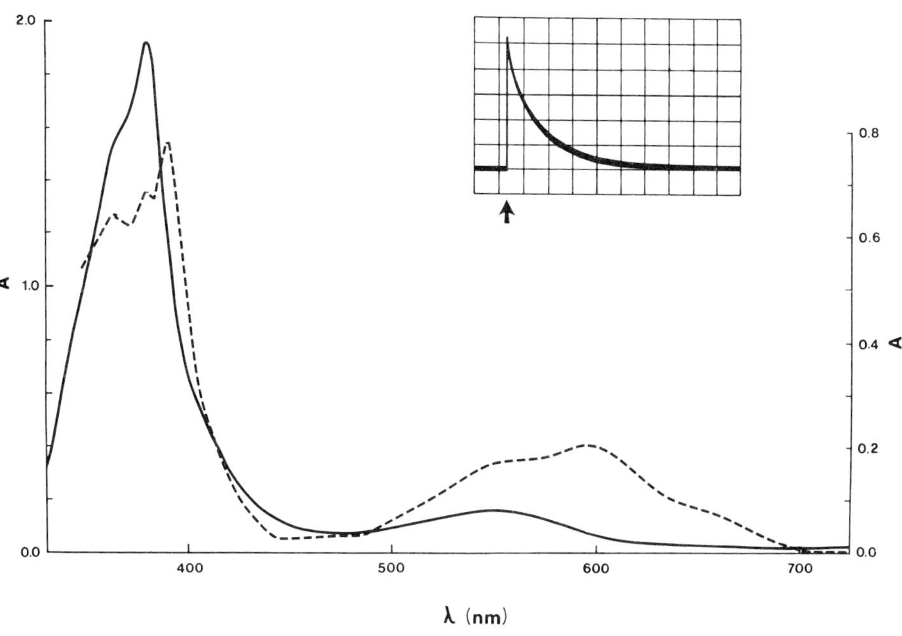

Fig. 8. Spectral changes observed upon illumination of a solution of TiO_2 (500 mg/l) containing 2×10^{-4} *M* $C_{14}MV^{2+}$ and 1 g/liter PVA Mowiol 10-98 with $\lambda > 300$ nm light (450 W): ——, 1.5 min (left ordinate); ---, 10 s (right ordinate). A solution of TiO_2 (500 mg/liter; pH 11.0) was used in the reference light beam. Insert: 347.1-nm laser flash photolysis (arrow, laser pulse) of the preirradiated (1.5 min) solution, temporal behavior of the 602-nm absorption of the $C_{14}MV^+$ radical (time units, abscissa, 20 μs). All solutions flushed with nitrogen prior to irradiation.

continuing exposure to light. A striking and rather rapid change from blue to intense yellow is noted (*24*). The resultant spectrum, also shown in Fig. 8, exhibits a weak absorption at 500 nm and a strong peak at 380 nm. The former absorption can be attributed to viologen dimer radicals, $(C_{14}MV^+)_2$ (*3*), present at low concentration, but the latter was identified as doubly reduced viologen, i.e., 1-methyl-1-tetradecyl-1,1′dihydro-4,4′-bipyridyl ($C_{14}MV^0$). The two-electron reduction of MV^{2+} by chemical and electrochemical means (*8, 12, 31, 61*) yields dihydrobipyridyl with similar spectral features. To substantiate the assignment of the 380-nm peak, $C_{14}MV^0$ was further synthesized using dithionite as a reductant (*24*).

The mechanism of light-induced $C_{14}MV^0$ formation involves band-gap excitation of TiO_2 particles, generating electron–hole pairs (Fig. 9). Conduction-band electrons reduce $C_{14}MV^{2+}$ first to the radical-cation, as discussed in Section IV,A,2,

$$e_{cb}^- + C_{14}MV^{2+} \xrightarrow{k_1} C_{14}MV^+ \tag{20}$$

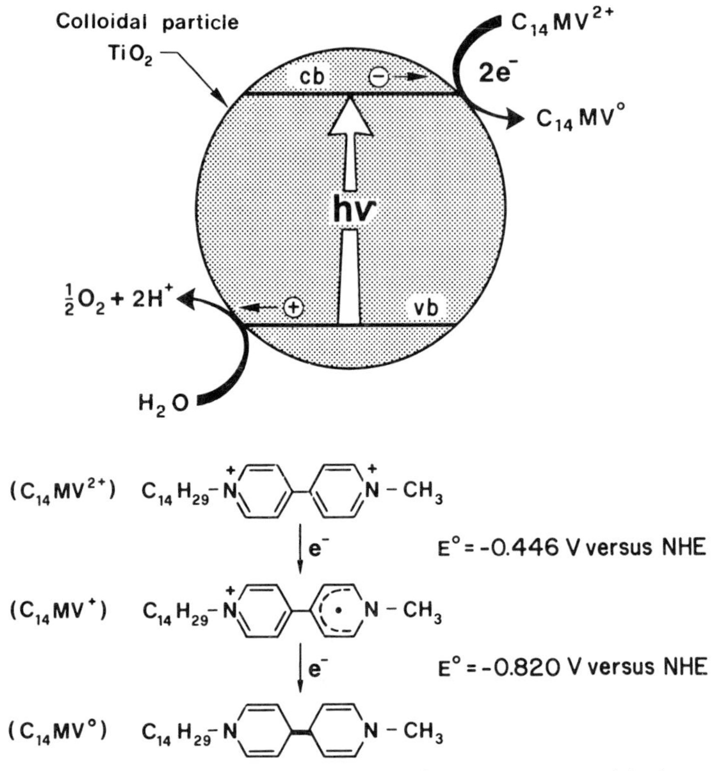

Fig. 9. Scheme for two-electron reduction of $C_{14}MV^{2+}$. The redox potentials given refer to simple methylviologen (MV^{2+}).

and subsequently to the dihydrobipyridyl:

$$C_{14}MV^{+} + e_{cb}^{-} \xrightarrow{k_2} C_{14}MV^{0} \tag{21}$$

Adhesion of the amphiphilic relay to the surface of TiO_2 particles allows both electron-transfer processes to occur at a very rapid rate. Application of laser photolysis technique yields for pH 11 the rate constants $k_1 = 10^8/s$ and $k_2 = 5 \times 10^4/s$. Two-electron reduction is thus completed in less than 100 μs. This contrasts sharply with the behavior of simple methylviologen, which does not undergo two-electron reduction at pH ≤ 12.5. Only at high alkalinity does slow formation of dihydrobipyridyl become apparent. In addition to kinetic reasons, there are thermodynamic reasons for this drastically different behavior of the two-electron relays. The second reduction of methylviologen requires a potential of -0.83 V versus NHE. Taking into account that the conduction-band position of the TiO_2 particle changes with pH according to $E_{cb} = -0.13 - 0.059$ pH, one can predict that a pH of at least 12 is required to render this process thermodynamically feasible. In the case of the amphiphilic viologen, adsorption to the surface of TiO_2 apparently shifts the potential of the $C_{14}MV^{+}/C_{14}MV^{0}$ redox couple by more than 100 mV, facilitating the second reduction step.

The valence-band process occurring in parallel with the reaction of conduction-band electrons comprises oxidation of water to oxygen (Fig. 9). This reaction is known to take place on TiO_2 (*14*). Oxygen is chemisorbed as O_2 to the TiO_2 surface (*47*) and in this state appears to be unreactive toward reduced viologen. Competing with water oxidation is the reaction of holes with the polyvinyl alcohol (PVA) polymer used to stabilize the sol, which has been shown, however, to be rather inefficient on TiO_2 colloids (*46*). A third valence-band process that begins after major fraction of $C_{14}MV^{2+}$ has undergone two-electron reduction is reoxidation of the dihydrobipyridyl to the viologen radical:

$$h^{+} + C_{14}MV^{0} \longrightarrow C_{14}MV^{+} \qquad (22)$$

The laser photolysis technique was applied to *preirradiated* $C_{14}MV^{2+}/TiO_2$ solutions in order to monitor this reaction. The insert in Fig. 8 shows the temporal behavior of the absorption at 602 nm. The sharp rise in the signal after the laser pulse is due to $C_{14}MV^{+}$ formation via reaction (22), whereas the subsequent absorption decay indicates the time course of $C_{14}MV^{+}$ reduction by conduction-band electrons [Eq. (21)]. In accordance with this interpretation, one finds that the 380-nm absorption behaves as an exact mirror image of the optical events at 602 nm; rapid bleaching after the laser pulse is followed by recovery of the original absorption, which obeys the same time law as the 602-nm decay. One can conclude that under illumination of TiO_2–$C_{14}MV^{2+}$ solutions, a state is approached in which cyclic electron transfer, that is, the sequence of reactions (2) and (3) occurring on individual TiO_2 particles maintains the electron relay in the doubly reduced form.

Attempts to exploit this two-electron storage system to achieve light-induced hydrogen generation in alkaline aqueous medium gives encouraging results. TiO_2 particles (500 mg/liter) loaded with 1.5% Pt were employed in these experiments. Solution volume was 5 ml, and the pH was adjusted to 12 with NaOH. Vigorous hydrogen generation (0.5 ml/h) was observed in samples containing $C_{14}MV^{2+}$, whereas blank experiments with TiO_2/Pt dispersions in the absence of $C_{14}MV^{2+}$ yielded only 35 μl/h H_2. Apparently, the cooperation of functional electron relay and catalyst on the surface of the semiconductor particle gives much better yields of hydrogen than catalyst deposits alone. This provides a good example for successful molecular engineering in light-energy conversion systems.

4. *Dynamics of Hole Transfer from Colloidal TiO_2 to Reductants in Solution*

Using a similar laser photolysis technique, the reaction of valence-band holes in colloidal TiO_2 particles (h^+) with electron donors such as halide ions or SCN^- has also been investigated (*11, 27, 46*). The oxidation of these species follows the reaction

$$X^- \xrightarrow{h^+} X\cdot \xrightarrow{+X^-} X_2^- \tag{23}$$

and thus results in the formation of halide or $(SCN)_2^-$ radical-anions. These species can be readily analyzed by their characteristic absorption spectrum. Using fast optical analysis it was found that reaction (23) occurs very rapidly and is completed within the ~10-ns laser pulse. The hole transfer was shown to involve only species adsorbed to the surface of the colloidal TiO_2 particle. The efficiency of the process follows the sequence $Cl^- < Br^- < SCN^- \sim I^-$ and hence is closely related to the redox potential of the X^-/X_2^- couple. Further studies showed that the yield of X_2^- decreases strongly with increasing pH, and this effect has been attributed to the competition of water oxidation with reaction (23). When RuO_2 is deposited onto the colloidal TiO_2 particles, yields of Cl_2^- and Br_2^- are increased significantly. This shows that the RuO_2 deposit acts as a hole-scavenger, promoting oxidation of halide ions adsorbed to the particle (*46*).

5. *Interfacial Charge Transfer in Colloidal CdS Solutions*

Cadmium sulfide has been extensively investigated as an electrode material in photoelectrochemical cells (see Chapter 13 by Hodes), but studies on suspensions of CdS powders have been comparatively scarce (see Chapter 7 by Kalyanasundaram). They have gained considerable momentum since light-induced H_2 (*9*) and cogeneration of H_2 and O_2 (*35*) was found to occur on noble-metal-loaded CdS particles. Water and H_2S cleavage can also be achieved with CdS sols; for this reason some laser studies with colloidal solutions of this material are briefly discussed.

The uv–visible absorption spectrum of the CdS sol, prepared by precipitation from a $Na_2S–Cd(No_3)_2$ solution in the presence of hexametaphosphate as stabilizing agent, is distinguished by a sharp band edge rising steeply toward the blue below 520 nm. The onset coincides exactly with the band gap of crystalline CdS electrodes, that is, 2.4 eV. Electron-diffraction studies show that the colloidal particles (hydrodynamic radius of ~150 Å) are mostly single crystals of Wurtzit structure. Excitation of the CdS sol leads to luminescence with spectral characteristics that are displayed in Fig. 10. This emission is distinguished by a broad maximum around 700 nm. Similar results have been obtained by Henglein (*28*). Using a single-photon counting technique, we determined the lifetime of the luminescence as $\tau = 0.3\ (\pm 0.02)$ ns.

The dashed line in Fig. 10 represents the excitation spectrum for the luminescence measured at 640 nm. It is identical with the absorption spectrum of the CdS sol. The onset is at 520 nm, the band rising sharply toward the uv. From the perfect coincidence of the excitation with the absorption spectrum, one can infer that the luminescence arose truly from internal transitions of the CdS particles following electron–hole pair for-

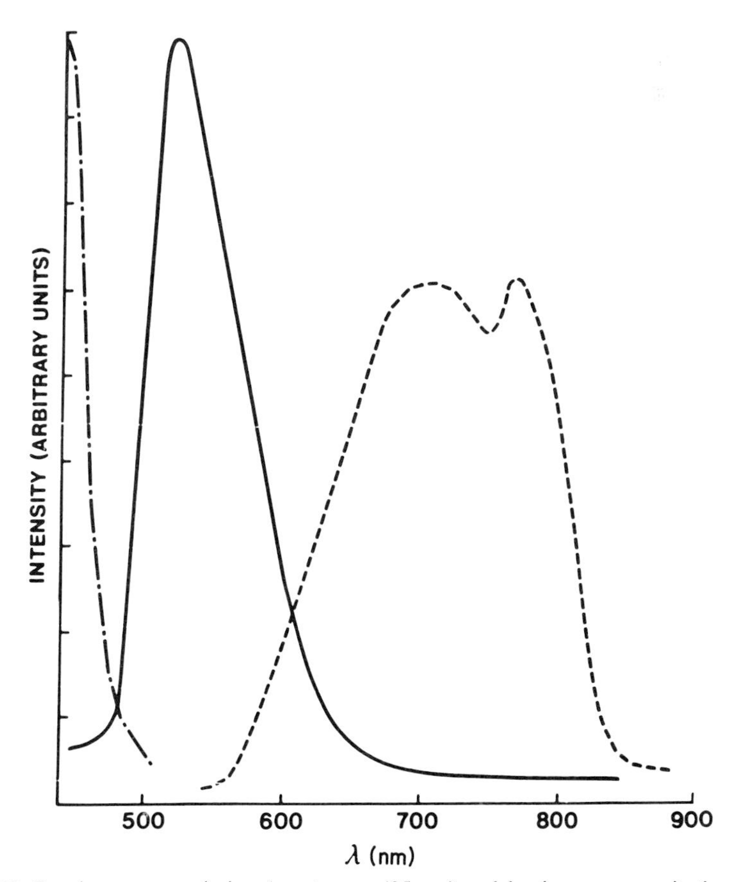

Fig. 10. Luminescence emission (---; λ_{exc} = 425 nm) and luminescence excitation (–·–; λ_{em} = 640 nm) spectrum of colloidal CdS (concentration = 3×10^{-3} *M*). ——, Luminescence spectrum observed in the presence of methylviologen.

mation by light excitation. The red shift in the luminescence with respect to the energy of the band gap is caused by the presence of surface states that can act as radiative recombination sites. This is shown by the effect of methylviologen addition, which produces a blue shift and sharpening of the emission peak. Thus at 10^{-2} *M* MV^{2+} the maximum of the luminescence is located at 524 nm, corresponding almost exactly to the energy difference between valence and conduction band. A tentative interpretation of this effect is as follows. At pH 7 the CdS sol is negatively charged, and because of coulombic attraction, MV^{2+} will be adsorbed onto the particle surface. The interaction with surface states, such as dangling bonds, causes them to split much in the same way as Ru^{3+} ions split the surface state levels of GaAs (*51*). These states are therefore no longer available for radiative recombination, which takes place at this time between conduction and valence band.

From the short lifetime of an electron–hole pair in colloidal CdS, one could infer that e_{cb}^- could only reduce species adsorbed to the surface of the particle. Diffusional displacement required for reaction with acceptors in the bulk would be too slow to be able to compete with $e_{cb}^- \cdots h^+$ annihilation. This idea is fully borne out by experimental results. Thus, in the presence of MV^{2+} as an acceptor, the blue color of MV^+ promptly appears within the laser pulse, and slow growth of the 602-nm absorption is not observed, as is the case for TiO_2. In the absence of O_2 the signal of MV^+ remains stable. An intense blue color is developed under continuous exposure to visible light, which is in agreement with the observations made by Harbour and Hair (*25*) on macrodispersed CdS powders.

The yield of MV^+ obtained by flash irradiation of the CdS sol with 487-nm laser light was found to increase with MV^{2+} concentration. Increasing the latter from 0.2 to 5×10^{-3} *M* results in an augmentation of $[MV^+]$ from 0.3 to 2×10^{-5} *M*. At the higher MV^{2+} concentration the coverage of the CdS particle surface with MV^{2+} is apparently increased. This enhances the rate of interfacial electron transfer, which can compete more efficiently with electron–hole recombination.

An interesting application of the laser technique is the study of electron and hole reaction with catalysts deposited on the surface of the semiconductor particle. Thus, in the case of the MV^{2+} reduction by CdS conduction-band electrons, the yield of MV^+ is reduced if the particles are loaded with Pt. This is due to the trapping of e_{cb}^- by Pt sites on the surface, which occurs in competition with MV^{2+} reduction.

$$e_{cb}^- + MV^{2+} \xrightarrow{k(MV^{2+})} MV^+ \tag{24}$$

$$e_{cb}^- + Pt \xrightarrow{k(Pt)} Pt^- \xrightarrow{-H_2O} \tfrac{1}{2}H_2 + OH^- \tag{25}$$

Electrons undergoing reaction (25) can no longer contribute to the yield of MV^+ as they are used to generate hydrogen from water under the prevailing pH conditions. This technique is potentially useful in that it allows for rapid screening and optimization of hydrogen-evolving colloidal semiconductor systems (*11*).

B. Light-Induced Water Cleavage through Direct Band-Gap Excitation of Colloidal Semiconductors

In this section we shall describe a water photolysis system based on band-gap excitation of semiconductor dispersion. This provides an illustrative example for the application of the concepts derived in the preceding chapters to develop photolytic, fuel-producing systems. A semiconductor particle is charged with both a catalyst for water oxidation (RuO_2) and reduction (Pt). Band-gap excitation produces an electron–hole pair in the particle. Both charge carriers diffuse to the interface, where hydrogen and oxygen generation occurs (Fig. 11). TiO_2 was the first material to be employed in such a system (*10*). The domain of photoactivity has been displaced in the visible by using *n*-CdS particles as carriers for Pt and RuO_2. An undesirable property of this material is that it undergoes photocorrosion under illumination. Holes produced in the valence band migrate to the surface, where photocorrosion occurs, that is,

$$CdS + 2h^+ \longrightarrow Cd^{2+} + S \tag{26}$$

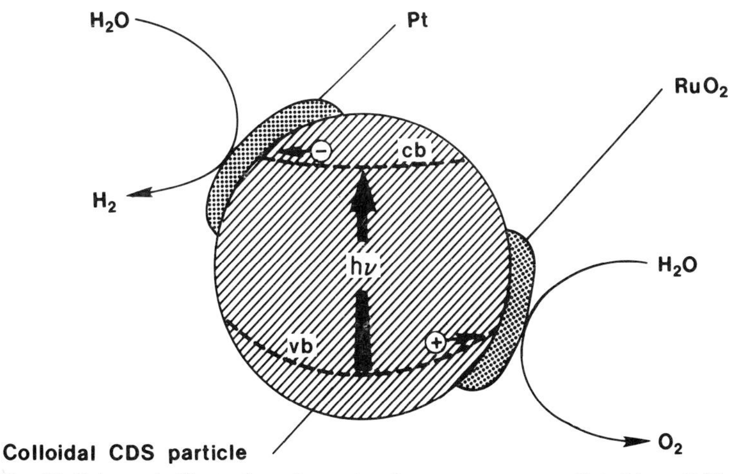

Fig. 11. Schematic illustration of a water-cleavage process mediated by a CdS particle.

We observed (*4*; see also Chapter 14 by Frank) that loading of CdS particles with an ultrafine deposit of RuO_2 prevents photodecomposition through promotion of water oxidation according to

$$4h^+(CdS) + 2H_2O \xrightarrow{(RuO_2)} O_2 + 4H^+ \tag{27}$$

Sustained water cleavage by visible light is observed when CdS particles loaded simultaneously with Pt and RuO_2 are used as a photocatalyst. Hydrogen and oxygen are generated by conduction-band electrons and valence-band holes, respectively, produced by band-gap excitation. This water-cleavage phenomenon can be illustrated unambiguously by gas chromatographic (gc) analysis. Figure 12 shows results obtained from visible-light ($\lambda > 420$ nm) irradiation of 25 ml of aqueous solution (pH 13) containing 10^{-3} *M* H_2PtCl_6 and 25 mg of CdS loaded with 3 mg of RuO_2 [deposition of RuO_2 was carried out via thermal decomposition of $Ru_3(CO)_{12}$ in Ar at 300°C for 30 min]. Initially there were only traces of H_2 generated because the reduction of H_2PtCl_6 occurs in preference over H_2O reduction. However, the peaks of H_2 and O_2 grew concomitantly after this induction period. Note that the ratio of the N_2 and O_2 peak heights decreases with irradiation time, indicating that the oxygen signal does not arise from air contamination. Under stationary conditions such a system delivers ~50 μl/h of H_2 and 25μl/h of O_2 (p = 1 atm at room temperature), corresponding to a quantum yield of H_2O cleavage of ~0.2%.

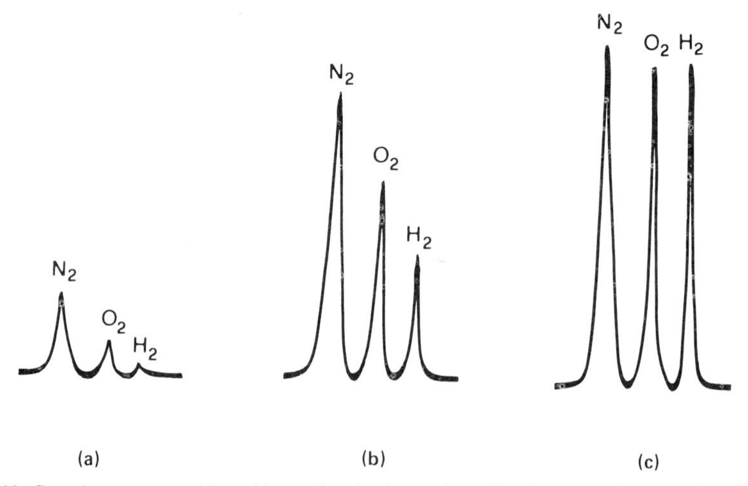

Fig. 12. Gas chromatographic evidence for the formation of hydrogen and oxygen by visible-light irradiation of CdS particles loaded with RuO_2. [CdS] = 1 g/liter, [RuO_2] = 150 mg/liter, [Li_2CO_3] = 0.1 *M*, [$PtCl_6^{2-}$] = 10^{-3} *M*, pH = 13. *t* (min): a, 5; b, 25; c, 110.

C. Cleavage of Hydrogen Sulfide by Visible Light

Apart from water, other agents such as hydrogen sulfide have the potential to play an important role as an alternative source for hydrogen production from sunlight. Sulfides occur widely in nature, and H_2S is produced in large quantities as an undesirable by-product in coal- and petroleum-related industries. Therefore, catalytic systems in which the process

$$H_2S \xrightarrow{h\nu} H_2 + S \tag{28}$$

is driven by visible light should be of significant interest. As the standard enthalpy of reaction (28) is 9.4 kcal/mol, this provides a method for energy conversion and storage as well as for recycling of H_2 employed in hydrodesulfurization processes.

When RuO_2-loaded CdS particles are dispersed in aqueous sulfide solutions and illuminated with visible light, hydrogen is generated at an astonishingly high rate. Results obtained from irradiating 25 ml of solution containing 25 mg of CdS loaded with 0.025 mg of RuO_2 in the presence of 0.1 *M* Na_2S (pH 3) showed that after a brief induction period the H_2-generation rate established itself a 3.2 ml/h until almost all the H_2S had been consumed.

A detailed investigation established that the efficiency of hydrogen generation depends on both solution pH and RuO_2 loading (*35*). At pH 13 and 0.5% RuO_2 loading of the CdS particles, the rate of H_2 generation is 10 ml/h, corresponding to a quantum yield of 35% [$= \phi(\frac{1}{2}H_2)$].

These observations may be rationalized in terms of band-gap excitation of the CdS particles, producing electrons in the conduction and holes in the valence band. The former migrate to the interface, where reduction of water to hydrogen occurs,

$$2\,e_{cb}^- + 2H^+ \longrightarrow H_2 \tag{29}$$

whereas the holes react with H_2S (or HS^- depending on pH) under sulfur formation:

$$H_2S + 2h^+ \longrightarrow 2H^+ + S \tag{30}$$

The overall reaction corresponds to splitting of H_2S into hydrogen and sulfur by four quanta of visible light.

Sulfur formed concomitantly with H_2 in the photoreaction does not seem to interfere with water reduction. It appears, therefore, that the reduction of sulfur by conduction-band electrons of CdS, though thermodynamically favored, is strongly inhibited for kinetic reasons. This explains why the H_2S photocleavage can proceed almost to completion without a decrease in the reaction rate.

In contrast, oxygen can compete with water reduction by conduction-band electrons. Thus, at pH 13 and 0.3% loading of CdS with RuO_2, the rate of H_2 evolution is decreased from 9 to 8 ml/h when air-saturated, as opposed to deaerated, suspensions are illuminated. Still, this competition of oxygen is surprisingly inefficient.

Apart from its importance in solar-energy research and in the treatment of H_2S-containing waste streams, the H_2S-cleavage reaction mimics in an intriguing way the function of photosynthetic bacteria, which frequently use sulfides as electron donors for the reduction of water to hydrogen.

V. Conclusions

Molecular assemblies such as micelles, colloidal semiconductors, and ultrafine redox catalysts provide suitable microscopic organization to accomplish the difficult task of light-energy harvesting and subsequent fuel generation. New kinetic models had to be conceived to provide a basis for the understanding of the dynamics of electron-transfer processes in these aggregates. In this chapter basic features of these processes are summarized and the general applicability of the concepts to different types of colloidal assemblies illustrated. As for the application of these systems in light-induced water-cleavage devices, work in the future will be directed toward improving the efficiency of such devices by identifying new photocatalysts and solving the problem of hydrogen–oxygen separation. Colloidal semiconductors will certainly play a primary role in this development. Along with other functional systems, they have the key advantage that light-induced charge separation and catalytic events leading to fuel production can be coupled without intervention of bulk diffusion. Thus a single colloidal semiconductor particle can be treated with appropriate catalysts so that different regions function as anodes and cathodes. It appears that this wireless photoelectrolysis could be the simplest means of large-scale solar-energy harnessing and conversion.

References

1. Albery, W. J., and Bartlett, P. N., *Discuss. Faraday Soc.* **70,** 421 (1980).

2. Alkaitis, S. A., and Grätzel, M., *J. Am. Chem. Soc.* **98,** 3549 (1976).

3. Bard, A. J., Ledwith, A., and Shine, H. J., *Adv. Phys. Org. Chem.* **13,** 155 (1977).

4. Borgarello, E., Kalyanasundaram, K., Okuno, Y., and Grätzel, M., *Helv. Chim. Acta* **64,** 1937 (1981).

5. Brugger, P.-A., and Grätzel, M., *J. Am. Chem. Soc.* **102,** 2401 (1980).

6. Brugger, P.-A., Braun, A. M., and Grätzel, M., *J. Am. Chem. Soc.* **103,** 320 (1981).
7. Calvin, M., *Photochem. Photobiol.* **23,** 425 (1976).
8. Carey, J. G., Cairns, J. F., and Colchester, J. E., *J. Chem. Soc. Chem. Commun.*, p. 1280 (1969).
9. Darwent, J. R., and Porter, G., *J. Chem. Soc. Chem. Commun.* p. 145 (1981).
10. Duonghong, D., Borgarello, E., and Grätzel, M., *J. Am. Chem. Soc.* **103,** 4685 (1981).
11. Duonghong D., Ramsden, J., and Grätzel, M., *J. Am. Chem. Soc.* **104,** 2977 (1982).
12. Elofson, R. M., and Edsberg, R. L., *Can. J. Chem.* **35,** 647 (1957).
13. Fernandez, M. S., and Fromherz, P., *J. Phys. Chem.* **81,** 1755 (1977).
14. Fujishima, A., and Honda, K., *Nature (London)* **238,** 97 (1972).
15. Galizzioli, D., Tantardini, F., and Trassatti, S., *J. Appl. Electrochem.* **5,** 203 (1975).
16. Grätzel, C. K., and Grätzel, M., *J. Phys. Chem.* **86,** 2710 (1982).
17. Grätzel, M., and Thomas, J. K., *J. Phys. Chem.* **78,** 2248 (1974).
18. Grätzel, M., *in* "Micellization and Microemulsions" (K. L. Mittal, ed.), Vol. 2, p. 531. Plenum, New York, 1978.
19. Grätzel, M., *in* "Dahlem Conferences on Light Induced Charge Separation" (H. Gerischer, and J. J. Katz, eds.), p. 299. Verlag Chemie, Weinheim, 1978.
20. Grätzel, M., *Ber. Bunsenges. Phys. Chem.* **84,** 981 (1980).
21. Grätzel, M., *Acc. Chem. Res.* **14,** 376 (1981).
22. Grätzel, M., *ACS Symp. Ser.* No. 177, 131 (1982).
23. Grätzel, M., and Frank, A. J., *J. Phys. Chem.* **86,** 2964 (1982).
24. Grätzel, M., and Moser, J., *Proc. Natl. Acad. Sci. USA* (1983).
25. Harbour, J. R., and Hair, M. L., *J. Phys. Chem.* **81,** 1791 (1977).
26. Harriman, A., Porter, G., and Walters, P., *J. Photochem.* **19,** 183 (1982).
27. Henglein, A., *Ber. Bunsenges. Phys. Chem.* **86,** 241 (1982).
28. Henglein, A., *Ber. Bunsenges. Phys. Chem.* **86,** 301 (1982).
29. Humphry-Baker, R., Grätzel, M., Tundo, P., and Pelizzetti, E., *Angew. Chem. Int. Ed. Engl.* **18,** 630 (1979).
30. Humphry-Baker, R., Lilie, J., and Grätzel, M., *J. Am. Chem. Soc.* **104,** 422 (1982).
31. Hünig, S., Gross, J., and Schenk, W., *Liebigs Ann. Chem.* 324 (1973).
32. Infelta, P. P., Grätzel, M., and Fendler, J. H., *J. Am. Chem. Soc.* **102,** 1479 (1980).
33. Kalyanasundaram, K., *Chem. Soc. Rev.* **7,** 453 (1978).
34. Kalyanasundaram, K., Micic, O., Pramauro, E., and Grätzel, M., *Helv. Chim. Acta* **62,** 2432 (1979).
35. Kalyanasundaram, K., Borgarello, E., and Grätzel, M., *Helv. Chim. Acta* **64,** 362 (1981).
36. Kiwi, J., and Grätzel, M., *Angew. Chem. Int. Ed. Engl.* **17,** 860 (1978).
37. Kiwi, J., and Grätzel, M., *Chimia* **33,** 284 (1979).
38. Kiwi, J., and Grätzel, M., *Angew. Chem. Int. Ed. Engl.* **18,** 624 (1979).
39. Kiwi, J., Kalyanasundaram, K., and Grätzel, M., *Struct. Bonding (Berlin)* **49,** 37 (1981).
40. Lehn, J. M., Sauvage, J. P., and Ziessel, R., *Nouv. J. Chim.* **3,** 423 (1979).
41. Lehn, J. M., Sauvage, J. P., and Ziessel, R., *Nouv. J. Chim.* **4,** 623 (1980).
42. Maestri, M., and Grätzel, M., *Ber. Bunsenges. Phys. Chem.* **81,** 504 (1977).
43. Maestri, M., Infelta, P. P., and Grätzel, M., *J. Chem. Phys.* **69,** 1522 (1979).
44. Moroi, Y., Braun, A. M., and Grätzel, M., *J. Am. Chem. Soc.* **101,** 567 (1979).
45. Moroi, Y. Infelta, P. P., and Grätzel, M., *J. Am. Chem. Soc.* **101,** 573 (1979).
46. Moser, J., and Grätzel, M., *Helv. Chim. Acta* **65,** 1436 (1982).
47. Munuera, G., Rives Arnau, V., and Saucedo, A., *J. Chem. Soc. Faraday Trans. 1,* 736 (1979).
48. Neta, P., *J. Phys. Chem.* **85,** 3678 (1981).

49. Neumann-Spallart, M., Kalyanasundaram, K., Grätzel, C. K., and Grätzel, M., *Helv. Chim. Acta* **63,** 1111 (1980).
50. Neumann-Spallart, M., and Kalyanasundaram, K., *Z. Naturforsch. B* **36,** 596 (1981).
51. Parkinson, B. A., Heller, A., and Miller, B., *J. Electrochem. Soc.* **126,** 454 (1979).
52. Pileni, M. P., Braun, A. M., and Grätzel, M., *Photochem. Photobiol.* **31,** 423 (1979).
53. Porter, G., and Archer, M. D., *ISR Interdiscip. Sci. Rev.* **1,** 119 (1976).
54. Razem, B., Wong, M., and Thomas, J. K., *J. Am. Chem. Soc.* **100,** 1629 (1978).
55. Spiro, M., and Ravno, A. B. *J. Chem. Soc. Faraday Trans. 1,* **75,** 1507 (1979).
56. Takayanagi, T., Nagamura, T., and Matsuo, T. *Ber. Bunsenges. Phys. Chem.* **84,** 1125 (1980).
57. Thomas, J. K., *Acc. Chem. Res.* **133,** 10 (1977).
58. Trasatti, S., and Buzzanca, G., *Electroanal. Chem.* **29,** (1971).
59. Wagner, C., and Traud, W., *Z. Elektrochem.* **44,** 391 (1938).
60. Waka, Y., Hamamoto, K., and Matuga, N., *Chem. Phys. Lett.* **53,** 242 (1978).
61. Weitz, E., and Ludwig, R., *Berichte* **55,** 795 (1922).
62. Williams, F., and Nozik, A. J., *Nature (London)* **271,** 137 (1979).
63. Wolff, C., and Grätzel, M., *Chem. Phys. Lett.* **52,** 542 (1977).

Photocatalytic Water Reduction To H_2: Principles of Redox Catalysis by Colloidal-Metal "Microelectrodes"

George McLendon

Department of Chemistry
The University of Rochester
Rochester, New York

I. Introduction

The basic principles of photoinduced redox catalysis based on homogeneous solutions or dispersions are discussed in detail by Grätzel (Chapter 3).

ENERGY RESOURCES THROUGH
PHOTOCHEMISTRY AND CATALYSIS

ISBN 0-12-295720-2

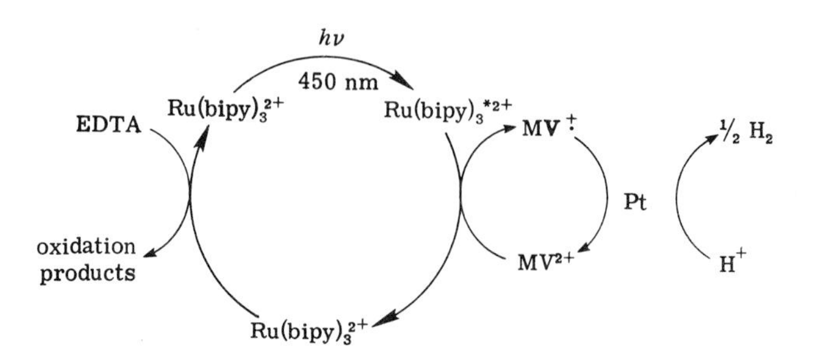

A recurring theme is the coupling of several system components, or reaction centers, as shown in Scheme 1. Previous chapters have focused on the mechanisms and design of the photoacceptor and electron acceptor, both in solution and at semiconductor interfaces. However, as the reducing equivalent Q^- is generated, it often cannot reduce water directly to hydrogen at a reasonable rate, even though the reaction may be thermodynamically feasible. For example, it was long known that mild reducing agents such as the methylviologen radical MV· ($E° = -0.42$ V) could be produced by electron transfer from excited-state donors (*1, 2*). Although MV· is thermodynamically capable of reducing water to H_2, in the absence of any catalyst, $MV^{\ddagger}$ is very stable. However, on addition of a noble-metal dispersion (e.g., colloidal Pt), rapid, quantitative H_2 formation ensues (*26*). This heterogeneous catalysis is the basis for most current schemes for photocatalytic water splitting (*3–30*).

In this chapter we shall focus on the mechanisms by which heterogeneous metal dispersions catalyze water reduction to H_2. We shall develop a qualitative model, based on simple electrochemical principles, that can quantitatively predict how catalytic rates will respond to solution conditions such as pH, catalyst composition, and the chemical nature of the electron acceptor (EA). The strengths and shortcomings of this approach are illustrated by experiments from our lab and others on catalytic water reduction.

II. The Electrochemical Model

When methylviologen radical ($MV^{\ddagger}$) is produced by photoinduced electron transfer, it is very stable in the absence of a secondary catalyst, producing dark-blue solutions. However, on addition of a noble-metal dispersion (PtO_2, colloidal platinum, colloidal gold, etc.), the blue color rapidly disappears, with concomittant vigorous production of H_2 gas. Clearly, the metal catalyzes the thermodynamically favored redox reaction: $MV^{\ddagger} + H^+ \rightarrow \frac{1}{2}H_2 + MV^{2+}$. The most obvious basis for this catalysis is that the metal provides a surface on which reaction intermediates like H· may be stabilized (as M—H). This greatly reduces the activation energy for water reduction. Such a mechanism is clearly analogous to the electrochemical reduction of water at a bulk cathode in a conventional electrolysis cell.

This analogy, that the metal colloids behave like bulk electrodes, was made independently and virtually simultaneously by a number of groups (*15, 20, 30*). Henglein (*30–35*) used the term *microelectrode* to underscore this analogy. The quantitative development and testing of this analogy has been somewhat slower.

A key step in the mechanistic work has been the development of stable, reproducible, and high-activity metal dispersions. Such work has antecedents in the earliest colloid chemistry, but it has been applied particularly to the production of H_2 by Grätzel and co-workers (*20–23*). Parallel work has proceeded elsewhere (*24–36*). Key developments included the preparation and kinetic characterization of ultrafine catalysts (radius <100 Å). The properties of these systems depend markedly on the preparative method and on the nature of the *protective agent* (a polymer added to inhibit flocculation), as discussed later.

Some of the first reports of photocatalytic H_2 production using these dispersions noted that H_2-production rates decrease with increasing pH (*2, 15, 27*). This parallels the Nernstian behavior of a bulk electrode. Such behavior was soon quantitatively confirmed by Miller (*15*). Using a homologous series of reductants based on 2,2′-dialkylbipyridines, it can be shown that H_2 production in a photocatalytic system depends smoothly on the overall redox potential for the reaction

$$\text{Donor}^{\ddagger} + H^+ \longrightarrow \tfrac{1}{2}H_2 + \text{donor}^{2+}$$

in quantitative accord with the Nernst equation (Fig. 1).

While catalyst-optimization proceeded, several groups continued mechanistic studies of this colloid catalysis. Henglein studied the reactivity of silver colloids with radiolytically produced organic radicals (*32–35*). These studies demonstrated a relatively show discharge step in water reduction by Ag colloids. Meisel and co-workers studied analogous reactions with colloidal gold (*27–29*). These studies were analyzed in terms of conventional homogeneous rate formalisms, for example,

$$M + e^- = M^- \overset{ne_-}{\rightleftharpoons} = M^{n-} \tag{1}$$

$$M^{n-} + mH^+ \rightleftharpoons M\text{—}H^{(n-m)-} \rightleftharpoons M + (m/2)H \tag{2}$$

For the highly active platinum systems, we wondered whether such a *homogeneous* reaction model was as appropriate as classical heteroge-

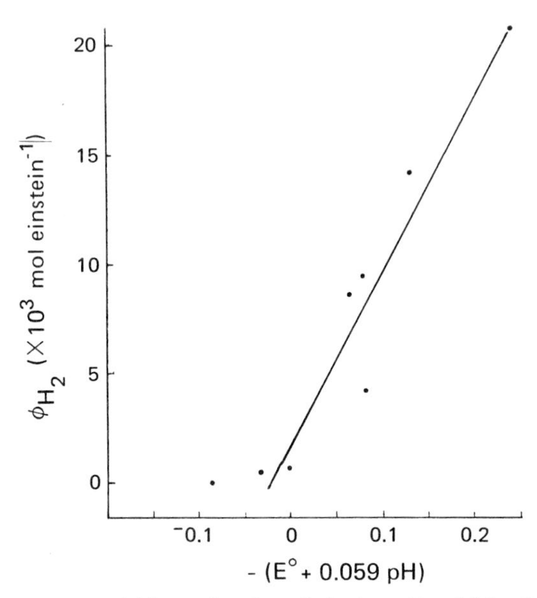

Fig. 1. Hydrogen quantum yield as a function of electromotive driving force. The system used contained $Ru(bipy)_3^{2+}$ ($\sim 10^{-5}$ *M*), an *N*,*N*′-dialkylbipyridyl mediator (10^{-3} *M*), EDTA (5×10^{-3} *M*), and an appropriate buffer (phosphate or borate). The points shown were obtained by varying the quencher, pH, or both. Data using different quenchers have been corrected for differential quenching rates.

neous electrochemical kinetics. The applicability of heterogeneous electrode kinetics to H_2 production at small particles had previously been invoked by Bard to explain some properties of semiconductor powder dispersions (*37–40*). Even earlier, Spiro (*40–43*) had pointed out that redox catalysis by colloidal dispersions could be qualitatively and quantitatively explained by a simple potential-matching model based on Wagner and Traud's classical theory of corrosion (*44, 46*).

III. A Simple Electrochemical Theory

In this model (for symbols, see Table I), the colloidal microelectrode, like a bulk electrode, serves several functions. It provides a medium by which the number of electrons (or equivalents) of the two half-reactions can be matched. It also provides a surface for lowering the activation energy of one or both of the half-reaction processes. In the process under consideration here, the one-electron oxidation of the donor $MV^{\ddagger}$ is matched to the overall two-electron process producing H_2. Moreover, the adsorption of

TABLE I

Symbols

Symbol	Meaning	Dimensions
α	Transfer coefficient (usually ~0.5)	None
A	Electrode area	cm^2
$C_j(\chi = 0)$	Concentration of species j at electrode surface	mol/cm^3
C_j^*	Bulk concentration of species j	mol/cm^3
E	Electrode potential	V
$E°$	Standard electrode potential	V
E_m	Mixed potential	V
f	$F/RT = 38.96$	1/V
F	Faraday, charge on 1 mol of electrons, 96,487	C
i	Current	A
i_l	Limiting current	A
i_m	Mixed current	A
j_o	Exchange current density	A/cm^2
$k°$	Standard (intrinsic) heterogeneous rate constant	cm/sec
m_j	Mass-transfer coefficient of species j	cm/sec
n	Electrons per molecule oxidized or reduced	None
v	Heterogeneous reaction rate	$mol/cm^2/sec$
v_m	Mixed heterogeneous reaction rate	$mol/cm^2/sec$
x_j	$z_j f(E - E_j°)$	None
z_j	Electrons transferred in the rate-determining step at the electrode (usually = 1)	None

hydrogen atoms on the surface of the metal lowers the energy required for the reduction step $H^+ \rightarrow \frac{1}{2}H_2$. A simple pictorial representation of the theory is provided by Fig. 2 for the specific case, $H^+ + MV^{\ddagger} \rightarrow \frac{1}{2}H_2 + MV^{2+}$. A general case is shown in Fig. 3.

The oxidation potential of the mediator ($MV^{\ddagger}$) is constant and pH independent and thus can be described by a single current–potential curve. The hydrogen half-reaction ($H^+ \rightarrow \frac{1}{2}H_2$) varies directly with pH according to the Nernst equation, and a family of parallel iE curves are obtained at all pH values. At *steady state*, the anodic current $i_a = \overline{bc}$ (the net rate of mediator oxidation) must equal the cathodic current $i_c = \overline{ab}$ (the net rate of proton reduction). The particle thus obtains a mixed potential (E_m) at which the mixed currents i_m define a rate v_m. This overall process is, therefore, a means of matching the donor and acceptor redox potentials and electron-transfer rates.

The overall rate v_m can be changed by adjusting the mixed potential E_m by changing the potential of either half-reaction. For example, lowering the pH would shift the mixed potential E_m toward the left in Fig. 2, until

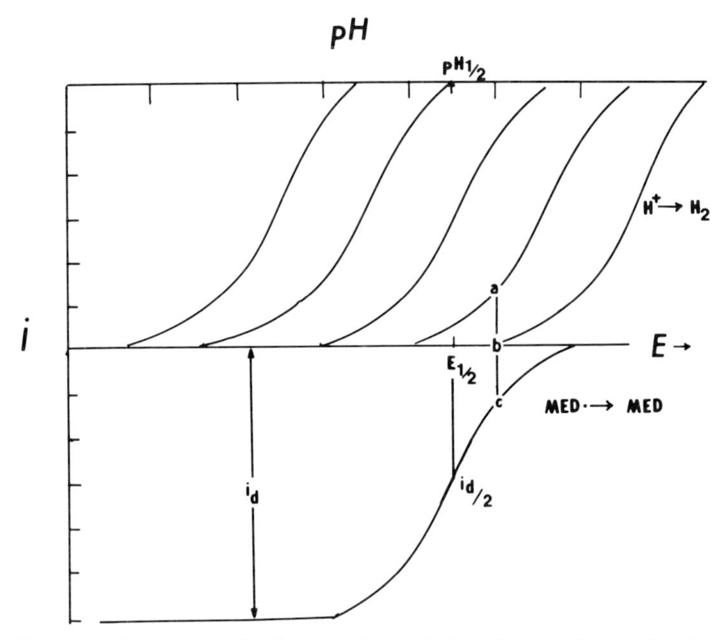

Fig. 2. Current–voltage curves for the potential matching of an anodic reaction (mediator$^{+\cdot}$ → mediator^{2+}) with a cathodic reaction ($H^+ \rightarrow \frac{1}{2}H_2$). The current, or rate of H_2 formation, at each pH is predicted by the vertical displacement between the zero-current line and the anodic mediator curve (line *bc*). At low pH, the rate i_{l,H_2} is assumed to be limited by mediator mass transport. The rate at half maximal velocity, $i_{1/2}$, defines the Nernst potential for the hydrogen reaction ($E_{1/2}$ or $pH_{1/2}$).

the cathodic and anodic currents were equal. The net rate would be increased by a predictable amount.

Several qualitative predictions for catalytic water reduction follow directly from this simple picture.

A. pH Effects

First, the current, or rate of H_2 production, will depend on the pH of the solution (and on the reduction potential of the mediator). For the mediated H_2-evolution reaction of interest here, the oxidation of the donor (e.g., $MV^{+\cdot}$) is rapid and occurs in a Nernstian reaction at the electrode. At a given [$med^{+\cdot}$]/[med^{2+}] ratio and mass-transfer rate, the anodic *iE* curve will remain fixed and be independent of pH (see Fig. 2). However, the cathodic current for proton reduction is pH dependent. The observed rate at a given pH can be found by considering both *iE* curves simultaneously. For example, at the pH labeled *a* in Fig. 2, the rate of H_2 production is predicted by line *bc*. This line extends to intersect the family

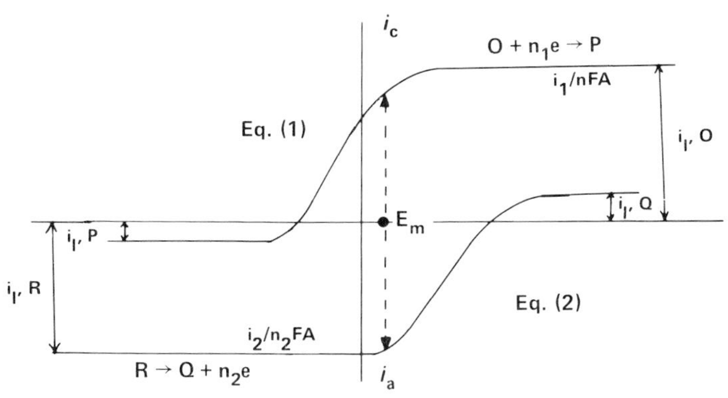

Fig. 3. Schematic illustration of the electrochemical parameters used in the calculations. See Eq. (1).

of cathodic water-reduction curves. At a given pH, the reduction rate must equal the oxidation rate, and one cathodic wave, determined by the pH-dependent overpotential, will be intersected such that *ab* equals *bc*. A convenient reference point in treating the effect of pH on the catalytic reaction is that pH at which the $i_m = \frac{1}{2} i_l$, called $pH_{1/2}$, which occurs at

$$E_m = E°(\text{med}) - (RT/nF) \ln(C_{ox}^*/C_{red}^*), \tag{3}$$

where i_l is the mass-transport limited current for mediator oxidation (see the Appendix). For the $MV^{2+}/MV^{\ddagger}$ system, $E_m = -0.44$ V versus NHE at $pH_{1/2}$. The value of i_m at this pH depends on the kinetics of the hydrogen-evolution reaction (HER). As shown in the Appendix, the determination of $pH_{1/2}$ may be useful in extracting the rate constant for this reaction; see Eq. (A14).

At higher pH values the i_m value approaches zero because the cathodic *iE* curve shifts to more negative values as the pH increases. If the HER follows the totally irreversible path implied by Eq. (A12), then the cathodic *iE* curve will shift ~120 mV toward more negative potentials for each unit increase in pH. At lower pH values the rate of H_2 production will increase sharply, and i_m ultimately is limited by diffusion of the mediator to the colloid surface.

B. Mediator Effects

In like manner, the effect of using homologous mediators with different standard potentials can be predicted easily. A change in mediator potential will shift the anodic (mediator) curve along the potential axis. More strongly reducing mediators will allow H_2 production at higher pH than mediators with lower potentials. The quantitative effect of this shift on rate (relative to the mass-transfer limit) under the conditions of Eq. (A12) is approximately a 0.12-V shift in $E_m{}^\circ$ being equivalent to a unit pH change. In like manner, a quantitative treatment of this problem correctly predicts the effect of changing the mediator concentration. As outlined in the following text, doubling the concentration of the $MV^{\dot{+}}$, MV^{2+} mediator couple should shift $pH_{1/2}$ by -0.3 pH units.

C. Metal Effects

Just as bulk-metal electrodes exhibit different rate constants for the HER, that is, different hydrogen overpotentials, so should potentiated colloids of different metals exhibit different catalytic efficiency. In the present model, the effect of an overpotential for H_2 production at a given metal will be to shift the pH curves along the potential axis. As the H_2-evolution rate constant decreases, the overpotential increases, so that a shift to lower pH will be required to maintain the same net driving force and thus to maintain the same rate. According to Eq. (A14), each tenfold decrease in $k_1{}^\circ$ (~120-mV increase in overpotential) must be compensated by a unit decrease in pH to maintain a constant i_m. By going to sufficiently low pH, it might be possible to produce H_2 at the mass-transfer-limited rate, even for metals with considerable hydrogen overpotentials (e.g., Ni or Ag). Thus, under these conditions, a relatively inexpensive metal could be substituted for Pt with no decrease in catalytic efficiency. The effect of colloid size on the rate of electrode reactions also can be simply predicted. Because the total current is proportional to area, the rate of H_2 production should increase linearly with colloid area. For spherical particles, the area will scale as $1/r$, the ratio of area/volume, assuming a constant total amount of metal. Finally, we note that theory predicts a complex dependence on colloid concentration. Increasing the concentration (number density) of particles can lead to a situation in which the distance between particles is small relative to the mediator diffusion distance. In this case, under non-steady-state conditions, the depletion layers around each platinum will strongly interact, and a greater than first-order dependence on [Pt] may be expected. A quantitative treatment of this problem might be modeled after multielectrode (thin-layer) electrolysis. We have not attempted such a quantitative treatment.

IV. Quantitative Aspects

A complete description of the application of potential-matching theory to colloid catalysis of water reduction is given in the Appendix and should be consulted for the details of the subsequently derived equations. Only the central features are repeated here. At steady state the reduction rate ν_1 for the reaction $H^+ + e^- \rightarrow \frac{1}{2}H_2$ equals the oxidation rate ν_2 for the reaction $MV^{\dagger} \rightarrow MV^{2+} + e^-$. The general equations are

$$\nu_1 = \frac{i_1}{n_1 FA} = \frac{C_O^* \exp(-\alpha_1 x_1) - C_P^* \exp[(1 - \alpha_1)x_1]}{(1/k_1^\circ) + [\exp(-\alpha_1 x_1)/m_O] + \{\exp[(1 - \alpha_1)x_1]/m_P\}}, \quad (4)$$

$$\nu_2 = \frac{i_2}{n_2 FA} = \frac{C_Q^* \exp(-\alpha_2 x_2) - C_R^* \exp[(1 - \alpha_2)x_2]}{(1/k_2^\circ) + [\exp(-\alpha_2 x_2)/m_Q] + \{\exp[(1 - \alpha_2)x_2]/m_R\}}, \quad (5)$$

where $x_1 f = z_1 f(E - E_1^\circ)$ and $x_2 = x_2(E - E_2^\circ)$.

The steady-state condition requires that $i_1/n_1FA = -i_2/n_2FA = V_n = i_m/n_1 n_2 FA$, where i_m is the mixed current at potential E_m. Although closed-form analytical solution for v_m is not possible, numerical solutions are straightforward. A numerical approach was used for comparing the theoretical prediction to the experimental results reviewed in the following text.

V. Preparation and Characterization of Active Metal Colloids

Noble-metal colloids that are active electrocatalysts for water reduction can be prepared in several ways. In general, a metal salt MX_n (e.g., $AuCl_3$ [*27, 47*] Na_2PtCl_4, or $AgNO_3$) (*30*) is reduced by an external reductant such as H_2 gas, citrate, or $NaBH_4$. The composition and size distribution of the colloidal suspension can depend markedly on the reductant used and on the concentrations of the metal salt and the reductant. For example, Frens (*47*) demonstrated that the specific size range of Au colloids could be controlled by varying the concentrations of $AuCl_3$ and the reducing agent, sodium citrate.

At high citrate–$AuCl_3$ ratios a greater number of nucleation sites form, producing many small particles. If the number of nucleation sites is kept low, by maintaining a low citrate–$AuCl_3$ ratio, fewer and larger particles are obtained. Similar approaches are available for the production of relatively monodispersed Pt or Ag.

Depending on the preparative method, uniform colloidal suspensions of Pt can be obtained with particle diameters ranging from 20 to 1000 Å. The (smooth) surface areas of such preparations can be astounding. The 300-Å particles generally used in our work have a calculated effective surface area of ~10^4 cm^2/g Pt. For the smaller 20-Å particles, this area increases to over 10^6 cm^2/g (*48*). The actual surface area, taking into account surface roughness, may be several times larger.

Once prepared, the colloid suspensions must be protected from aggregation and flocculation. These processes are generally thermodynamically favored. The rate of flocculation is increased by high ionic strength.

To minimize flocculation, a protective polymer is often adsorbed onto the colloid surface, after removing any extraneous electrolytes (such as citrate). Ideally, the polymer should contain both hydrophobic regions, which ensure strong polymer adsorption, and long "loops" of polymer chain that extend into the solution. The protective effect of the polymer is thought to involve these loops. As two particles approach, these loop regions of the two particles become intermingled, displacing solvent. This results in a local increase in osmotic pressure, which forces the particles back apart. A wide variety of polymers have been used, ranging from gelatin to synthetic polyvinyl alcohol. Based on these concepts, copolymers of styrene–maleic acid have been found to be particularly effective for stabilizing very small particles.

It is important to centrifuge the preparations to separate any residual flocculated material. The solutions obtained should be clear, with characteristic colors determined by particle light scattering. A well-known example is the color of ruby glass, which is owing to light scattering by colloidal gold particles of ~600 Å. As particle size decreases, the absorption maximum shifts to shorter wavelengths. Thus platinum sols of 100- to 300-Å radius give yellow solutions, whereas larger particles produce brown suspensions. Particles smaller than 80 Å in diameter preferentially scatter uv light and are virtually colorless. Thus, for example, the report of microscopically characterized 40-Å particles from brown suspensions should be taken cautiously (*8*).

Using these approaches, relatively monodispersed colloid suspensions can be prepared. The concentration, size distribution, and activity of the preparations must then be characterized. The size distribution is best measured by large scattering and transmission electron microscopy (tem). Light scattering should probe the hydrodynamic radius of both the metal particle "core" and the associated polymer, whereas tem detects only the electron-dense core. For example, for Pt particles prepared by H_2 reduction and stabilized with polyvinyl alcohol, we found that the radius estimated by light scattering was ~360 Å, whereas that found by direct tem measurements was 280 ± 30 Å. The total concentration of platinum is readily obtained by atomic absorption. The *molar particle concentration* is then $[\text{metal}]_{\text{total}}/[\text{number of atoms/particle}]$. The surface area can be estimated as [particle concentration] $\times N \times 4\pi r^2$.

VI. Assays of Activity for H_2O Reduction

Steady-state rates of hydrogen production have most commonly been monitored using photocatalytic hydrogen-production systems like those already shown in Fig. 1. The hydrogen produced can be assayed by microvolumetry *(50)*, electrochemical detection *(51)*, or, most commonly, by gas chromatography. We found optimal GC detection efficiency (10^{-10} mol H_2) using 40-Å molecular-sieve columns and N_2 carrier gas.

Electrochemical mediator reduction provides an attractive alternative to the photochemical assays. In this method, MV^{2+} (or an alternate mediator) is reduced to $MV\cdot$ at a bulk (Hg) electrode. This approach eliminates any extraneous reactions of the photoacceptor and the electron donor (e.g., EDTA). A cell design used in our electrochemical experiments is shown in Fig. 4. The radical diffuses into solution, where subsequent reactions with the platinum colloid and H_2O ensue. Interestingly, we found that negligible direct electron transfer occurs between the bulk Hg pool and the platinum. For example, the observed rate of H_2 production

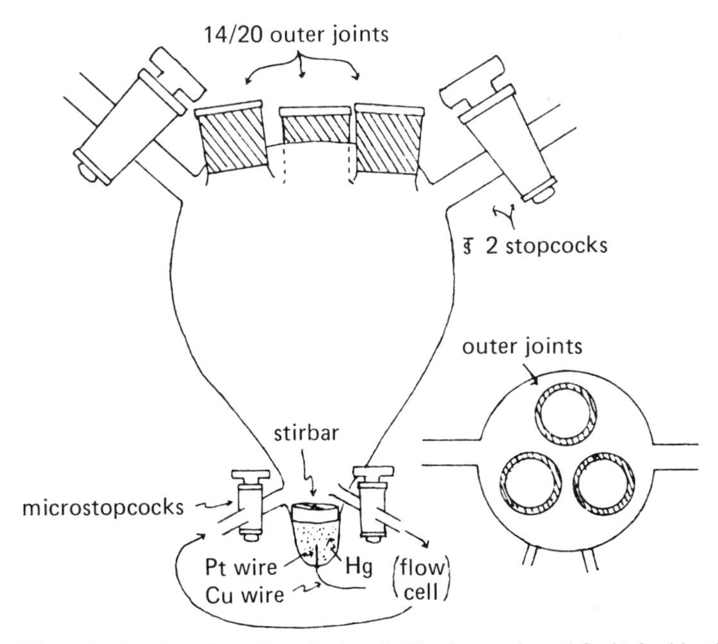

Fig. 4. Schematic drawing of an electrolysis cell (50-ml pear-shaped flask) for kinetic measurements using an electrogenerated reductant. Left, side view; lower right, top view.

depends directly on the steady-state concentration of reduced mediator and is independent of the potential of the Hg pool.

Complementary studies of the oxidation of a reducing agent at metal colloids have been carried out by electrochemical techniques, using cyclic voltammetry, and by direct visible-spectroscopy rate studies. The latter studies used flash photolysis to monitor the rate of MV^+ disappearance in the presence and absence of metal colloids. Finally, at sufficiently high pH or low [Pt] concentrations, conventional or stopped-flow mixing techniques have been used to measure radical disappearance. The interpretation of such data is not necessarily straightforward (see Figs. 5–7). Simple pseudo-first-order rate equations can describe most of the data: rate constants from 10^4 to 10^7 $M(\text{Pt})^{-1}$ s^{-1} have been obtained. Unfortunately, the dependence of these rates on [Pt] makes quantitative rate comparisons difficult. The steady-state kinetic treatment of HER developed in this chapter is inapplicable to experiments in which a radical species is instantaneously created and then irreversibly decays at the metal surface. Rather, such experiments are analogous to "potential jump" electrochemistry at bulk electrodes.

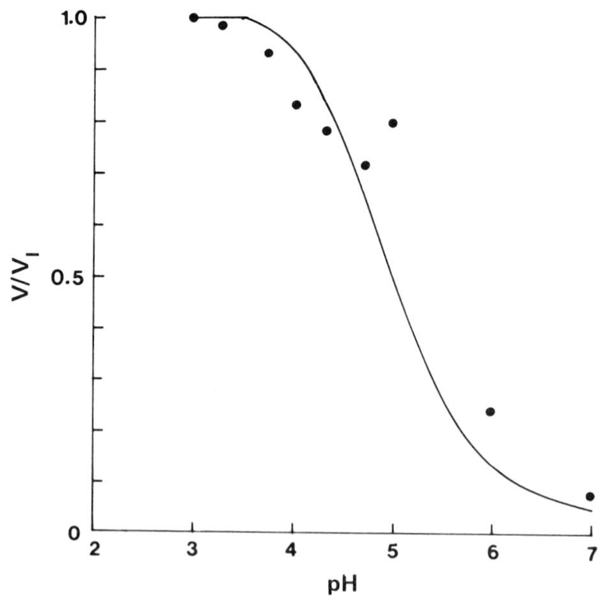

Fig. 5. Dependence of H_2-production rate (plotted as v/v_l, where v_l is the limiting rate) on pH, as determined using the electrogenerated reductant method. T = 298°C, $[MV^{2+}] \approx 2 \times 10^{-3}$ M, $[\text{Pt}]_{\text{tot}} \approx 10^{-6}$ M (~100-Å particles).

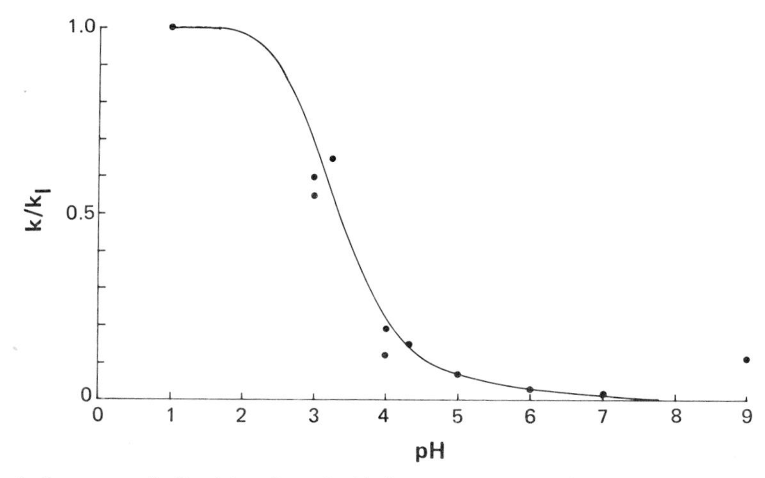

Fig. 6. Summary of all mixing data, [oxidation rate constants for $MV^{\ddagger}$ + Pt(λ = 600)], determined by stopped-flow spectrophotometry plotted as k/k_l, where k_l is the limiting rate. T = 298°C, pH adjusted with phosphate or borate buffer. [Pt] = 1–30 × 10^{-6} M (~100-Å particles), [$MV^{\ddagger}$] ≈ 1 × 10^{-4} M.

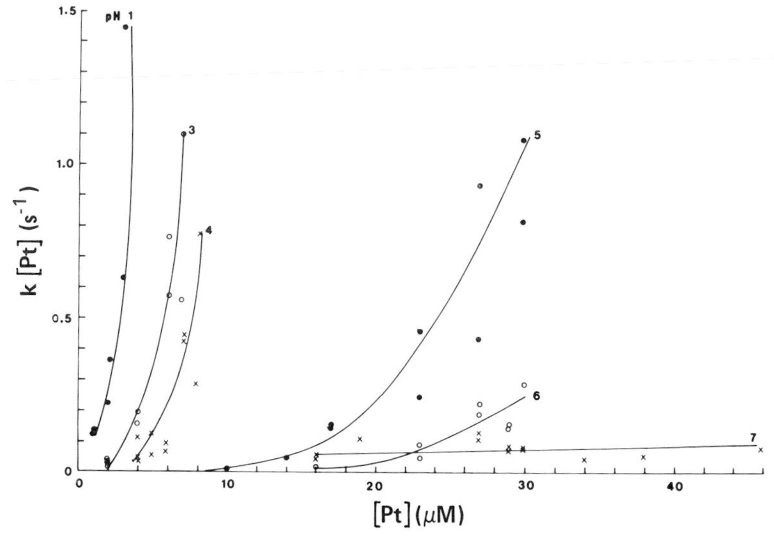

Fig. 7. Dependence of pseudo-first-order oxidation rate constants k[Pt] on [Pt] and pH as determined using mixing experiments, with conditions as in Fig. 6.

VII. Experimental Results

A. *The Dependence of Rate on pH: The "Acid" Test*

As discussed previously, the electrochemical model predicts a rapid change in rate around the pH($pH_{1/2}$) at which iE curves for H^+ reduction and mediator oxidation match ($E_m = E_{1/2}$ mediator). Below this pH, the rate of H_2 production will rise until it ultimately reaches a limit set by mediator mass transport. Beyond this point, increased acidity should not increase the rate (unless mass transport is simultaneously increased, by stirring, for example). These predictions are confirmed by the results shown in Figs. 5 and 6. Details of these experiments are presented elsewhere (*17*). Importantly, the results, when normalized as i/i_l, (where i_l is the mass-transfer-limited rate) are identical for both a photogenerated and an electrogenerated reductant. Thus, under the specific conditions used (*17, 17a*), the rate of H_2 formation is governed ultimately by the catalytic reaction chemistry of the metal colloid, and not by any photochemical steps.

In principle, the ratio $MV^{\ddagger}/MV^{2+}$ may change in the course of catalysis. If so, the anodic curve would shift, rather than remaining fixed, as shown in Fig. 2. In practice this shift is less than 0.1 pH units under the conditions shown in Figs. 5 and 6. However, deliberate manipulation of the concentrations of [$MV^{\ddagger}$] and [MV^{2+}] provides a quantitative, predictive test of the model. According to Eqs. (A13) and (A14), a twofold increase in mediator concentration should decrease $pH_{1/2}$ by 0.3 pH units. Indeed, experiments using electrochemically generated $MV^{\ddagger}$ have confirmed this prediction (Table II).

A different quantitative test is illustrated in Fig. 5, in which the predicted dependence of H_2-production rate, expressed as i/i_l versus pH for reasonable values of the parameters, follows the experimental data very closely. For the approximate conditions of the experiment in Fig. 2, $pH_{1/2} = 6.3$, yielding a rate constant for the H_2-evolution reaction $k_1°$ that can be related to the exchange current density for this reaction j_o by (*16, 17, 17a*)

$$j° = Fk_1°(C_{H_+})^{1/2}(C_{H_2})^{1/2},$$

where C_{H_2} is the solubility of H_2 at 1 atm, 7.7×10^{-7} mol/cm^3, α is taken as $\frac{1}{2}$, and the mechanism implied by Eq. (A3) is assumed; with $C_{H_+} = 10^{-3}$ mol/cm^3 and the preceding value of $k_1°$, log j_o -3.0 A/cm^2. The log j_o values quoted for Pt for the H_2-evolution reaction generally range from -2.6 to -3.6 (*45, 46*). Considering the assumptions and approximations

TABLE II

Dependence of Catalytic Rate ($pH_{1/2}$) on Mediator Concentration[a]

$[MV^+] \times 10^{-3}$ *M*	$pH_{1/2}$
0.08	5.0
0.14	4.7
0.30	4.4
0.60	4.2

[a] Conditions: Temperature = 25°C, ionic strength = 0.1 *M* KNO_3, MV^+ produced by reduction at a bulk Hg electrode held at −0.8 V versus SCE. The pH was restricted to the range 4–5. Below this pH artifacts owing to inefficient mass transport become significant.

made, this agreement is remarkable. A note of caution must be injected, however. The H_2-evolution reaction is more complicated than the preceding treatment implies, and the reaction mechanism may be different in acidic and alkaline solution and at different metals. Moreover, adsorption of organic compounds at electrode surfaces is known to affect the kinetics of the H_2-evolution reaction; both decreases (poisoning) and increases in the rate have been observed. Such adsorption will probably be a common occurrence in the types of systems employed in heterogeneous catalysis. Thus, although the qualitative and semiquantitative trends are worth noting, calculation of absolute rates from tabulated electrochemical parameters is risky.

B. Particle Size Effects

Some confusion has arisen in the literature about particle size effects in colloidal redox catalysis. Kiwi and Grätzel showed that H_2O-reduction rates at colloidal Pt increased markedly as particle size decreased (*20*). We subsequently pointed out that these rate data scaled inversely with the colloid radius (*15, 52, 53*). This scaling is consistent with an increase in surface area, which obviously scales as $1/r$ for spherical particles. As already noted, the rate of steady-state H_2 production should increase as the electrode area increases, as observed. Experiments with colloidal gold showed the same general trend (*52*) (Fig. 7). However, Keller *et al.* have questioned this observed size dependence (*8*). They reported no apparent differences in photocatalytic H_2-production rates (using Scheme 1) (*8*) between 1000- and 16-A particles. The reasons for these discrepancies are not obvious. However, we note that the Nord protocol (*54*) generally produces *much* larger particle diameters than the 16-Å colloids (*8*) claimed. Indeed, a diameter of ≥300 Å would be more consistent with the "light brown" color reported for the colloid suspension. In any event, Keller *et al.* have reiterated the importance of colloid concentration in making size comparisons. As already noted, colloidal metal catalysts (Pt, Ag, or Au) exhibit roughly a "second order" dependence on particle number. We have suggested that this anomaly reflects the complex diffusion equations expected when multiple electrodes are separated by short distances. Although this may prove to be true, our published arguments (*17a*) should be regarded solely as a suggested research direction rather than as a proven or even predictive model. Until this kinetic dependence on particle concentration is fully understood, any explication of particle size effects will be limited. As a final note, Frank and Stevenson (*48*) have reported that if *very* small (20 Å) Pt colloids are used at high concentration (6×10^{-5} *M* total Pt) in Scheme 1, the net reaction rate is limited by the rate of photogeneration of $MV^{\ddagger}$, rather than by H_2 production at Pt. This contrasts with the other investigations reviewed here (*52, 53*). In principle, of course, the Pt-catalysis step could again become rate determining merely by increasing the light flux.

These very small particles are not merely exceptional catalysts, they may prove to be of most interest in examining explicit surface reactions on Pt. Note that a 20-Å diameter corresponds to only ~100 surface Pt atoms (assuming a smooth spherical particle).

In this case, virtually *every* surface atom will be a "step" site. Furthermore, the comforting statistical reactivity distributions applicable to large surfaces will likely be inapplicable for these small aggregates. It will be interesting to see what anomolous chemistry (if any) is associated with these ultrafine particles, which are only marginally larger than the crystallographically characterized Pt clusters prepared by Chini *et al.* (*55*).

C. Metal Composition

Although our own work has focused on platinum, some interesting and important comparisons with other metals are available in the literature. In preliminary experiments (*52*), we found that the catalytic efficiencies of various metal colloids, as reflected by the $pH_{1/2}$ values, closely correlated with the known exchange current densities for the corresponding bulk-metal electrodes. This relationship is shown in Fig. 8. Given the likely differences in the rate-determining steps for H_2 evolution at these various metals, only the qualitative trend is important. Nonetheless, such a study does show that different metals can be substituted, if desired, for Pt, to

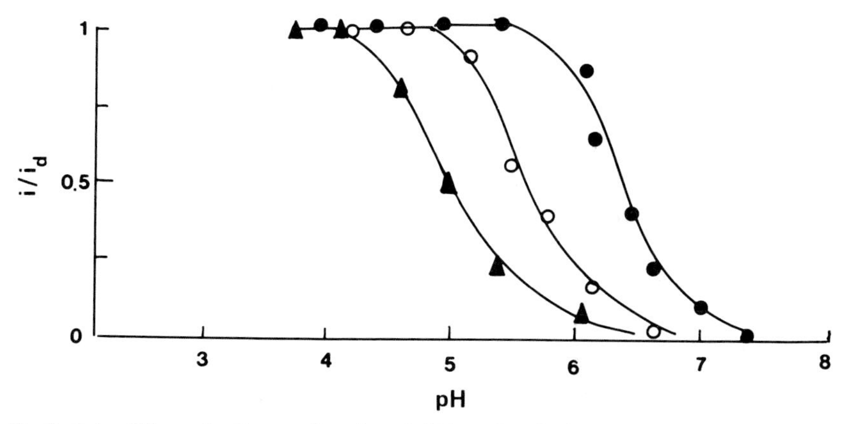

Fig. 8. Rate of H_2 production as a function of pH in a photolysis system. The system includes a methylviolgen mediator, *N*-phenylglycine electron donor, and $Ru(bipy)_3^{2+}$, as the photosensitizer. Results are normalized to maximum rate k_l to allow for ready comparison of data for Pt (●), Au (○), and Ag (▲) colloids.

mediate the reduction of water by reducing radicals. [As Henglein has noted (*30*), if the colloid intercepts these radicals, this may prevent normal free-radical reactions, for example, dimerization, from proceeding.] The interaction of the colloid and reducing agent may thus have interesting organic consequences apart from H_2 production.

The relative kinetic effectiveness of a metal-colloid catalyst can then be predicted qualitatively by its exchange current (or overpotential) for the process of interest (e.g., $H_2O \rightarrow H_2$). Thus dispersed nickel can serve as an efficient catalyst for H_2 evolution. However, because Ni^0 is thermodynamically unstable with respect to Ni(II), it does not afford a long-term stable catalyst.

More detailed studies of Ag and Au colloid catalysis have been carried out by Henglein (*30–36*) and Meisel (*27–29*), respectively. For both metals, the observations are consistent with simple electrochemical theory. However, some differences from Pt catalysis are noteworthy. For example, both electron discharge and H· combination are fast on Pt. Under favorable conditions the rate of H_2 evolution may ultimately be limited by H_2 mass transport away from the particle surface.

In contrast, electron discharge is fast on Ag, but H· combination is slow. (In Fig. 2, overpotential corresponds to a shift of the mediator *iE* toward lower potential.) This charge buildup has been directly measured by Henglein in an elegant experiment (*34*) in which Ag colloid particles equilibrate with radiolytically produced free radicals. The "charged" colloids can discharge the excess surface potential either by H_2 evolution or by charge transfer to a mercury pool. The measured potential of the Hg pool directly reflects the charge on the Ag particles. In this way, Henglein and Lilie calculated an apparent capacitance of 43 $\mu F/cm^2$ for 140-Å-diameter Ag particles (*34*). No such buildup is expected, or observed, for colloidal Pt. Finally, in a series of interesting papers, Meisel, Kopple, *et al.* have examined the redox chemistry of Au colloids in some detail (*27–29*). Two details of this work are particularly noteworthy. First, isotope-effect studies showed that the value $k_H/k_D = 6 \pm 0.4$ obtained with Au colloids closely matched the value found for the bulk Au electrode ($k_H/k_D = 6.4$ for high overpotentials). A second, surprising finding was that abstraction of H atoms for the protective polyvinyl sulfate polymer was observed in competition with H^+ donation from solvent.

In summary, the basic electrochemical concepts already discussed allow prediction of the relative catalytic efficiencies of different metals for H_2 evolution. However, a more rigorous application of the steady-state model cannot be made without significant changes to allow for differences in the reaction mechanism between different metals.

D. Dependence on Mediator Potential

We have shown elsewhere that specific mediators of defined potential may be designed to shift the pH range for photocatalytic water reduction (*15*). In the present model, changing the mediator corresponds to a shift of the anodic curve (labeled med in Fig. 2) along the potential axis. If the mediator potential $E_{1/2}$(med) is more negative, the anodic curve will shift until potential matching occurs at higher pH. If this potential is more positive, the potential match will occur at lower pH. In the region of potential matching ($pH_{1/2} \pm 1$ pH unit), a roughly linear dependence of rate on $E_{1/2}$(med) would be expected. Such behavior has indeed been observed; a linear plot of ($E_{mediator} - 0.059$ pH) versus rate of (photocatalytic) H_2 production has been reported (*15*) (Fig. 1). However, for pH values at which the rate is limited by mediator mass transport, that is, pH < pH(i_1), one predicts that mediator potential would not affect the rate. This leveling effect is indeed observed in an electrogenerated-mediator experiment. With the use of two mediators of different potential, $DMPB^{2+}$ ($E° = -0.6$ V) and MV^{2+} ($E° = -0.44$ V), large differences in H_2 generation rate are observed at pH 7, but not at pH 3, at which H_2 production is diffusion limited for both mediators. Analogous results were observed by Keller *et al.*, using photochemically reduced mediators.

In a catalytic system, the stability of the mediator is important. Keller (*8*), and others (*27*), have commented on the hydrogenation of methylviologen and related homologs at Pt or Au catalysts. The MV^{2+} is slowly reduced to 4,4′-demethylbipiperidine.

The severity of this reaction seems to depend on the mediator redox potential: strong reducing agents are more readily hydrogenated. This reaction is not surprising because Pt is commonly used as a hydrogenation catalyst. However, it seriously limits the practical utility of photocatalytic systems based on organic mediators. These problems can in principle be avoided by using a purely inorganic mediator (e.g., Eu^{2+} or Cr^{2+}), or by eliminating mediators, as in work done using platinized semiconductors (see Section VII).

VII. Related Systems

Finally, it seems appropriate to comment briefly on some related microelectrode reactions (platinized semiconductors and RuO_2–H_2O oxidation catalysis) in the context of the foregoing discussion. A complete discussion of these topics is contained elsewhere in this volume.

A. *Platinized Semiconductors*

Wide-band-gap semiconductors (e.g., TiO_2, $SrTiO_3$) have been used to promote photolysis of water to H_2 and O_2, using uv band-gap excitation, or using efficient band-gap sensitization. In 1980, A. Bard (*38*) showed that the efficiency of water splitting by semiconductor particles was greatly increased by photodeposition of a small amount of Pt on the TiO_2 surface. This catalytic effect is entirely consistent with the ideas already discussed. TiO_2 is a poor H_2 electrode, and the flat-band potential of TiO_2 is only slightly favorable for H_2O reduction (E_{fb} = −0.1 V for anatase TiO_2, E_{fb} = 0.2 V for rutile.) On this basis we would predict that unplatinized TiO_2 would produce minimal H_2. Rutile TiO_2 should be a poor catalyst, whereas anatase should be significantly better. Grätzel noted that the flat band potential can be further cathodically shifted by doping with NbO_2, thereby increasing the catalytic efficiency for H_2 formation. Similar approaches have been taken by Wrighton and Grätzel for small-band-gap semiconductors (Si and CdS). Details of such experiments appear elsewhere in this volume.

B. *Water Oxidation*

The same basic concepts that have been applied to water reduction to H_2 should be applicable to the mechanistically complex oxidation of water to O_2. Indeed, this analogy was noted by Grätzel when he showed that catalysis of water-oxidation rates by RuO_2 depends roughly on oxidant potential (*57, 57a*). Rapid progress has been made by Grätzel, Lehn, and others in increasing the catalytic efficiency of RuO_2 as a water-oxidation catalyst (*59–62*). A key element in this progress has been the development of small, high-surface-area catalysts in clear analogy with the work of H_2-productin catalysts. Kalyanasundaram later demonstrated (*62*) that using a more potent oxidant as a photocatalyst {ruthenium tris[4,4′-(carboxyiso-propylester)bipyridine]—$E^\circ_{Ru^{3+}\rightarrow Ru^{2+}} \cong 1.6$ V} increased O_2-production rates in a photocatalytic system. We had carried out parallel and independent investigations and reached similar conclusions (*17, 17a*). A caution should be noted, however. The oxidized ruthenium bipyridyls and phenanthrolines undergo a series of side reactions involving OH^- attack on the bipyridine rings. The rates of these undesired side reactions rapidly increase with increasing oxidation potential. Therefore, improving water-splitting yields by increasing catalyst oxidation potential may not be straightforward.

Appendix: Equations for Current–Potential Curves as Applied to Heterogeneous Catalysis

A. General Equations

Consider the reaction of two soluble species O and R at the catalyst surface to form the soluble products P and Q in the overall reaction

$$n_2\text{O} + n_1\text{R} \xrightleftharpoons{\text{(catalyst)}} n_2\text{P} + n_1\text{Q} \tag{A1}$$

This redox reaction occurs via the two half-reactions

$$\text{O} + n_1\text{e} \rightleftharpoons \text{P} \qquad E_1^\circ \tag{A2}$$

$$\text{Q} + n_2\text{e} \rightleftharpoons \text{R} \qquad E_2^\circ \tag{A3}$$

We assume that no species is adsorbed or precipitated at the electrode surface and that the concentrations of the species at a given temperature t are C_O^*, C_R^*, C_P^*, and C_Q^*. Because these concentrations change with time, the reaction velocity v (mol/s/cm) is also a function of time. We consider, therefore, an instantaneous velocity; a particularly convenient situation is at $t = 0$, where v is the initial velocity v°, which can be determined when the products are absent (i.e., $C_P^* = C_Q^* = 0$).

The general *iE* curves, shown in Fig. 5, are governed by the equations

$$v_1 = i_1/n_1FA = k_1^\circ\{C_O(x = 0)\exp[-\alpha_1 z_1 f(E - E_1^\circ)] - C_P(x = 0)\exp[(1 - \alpha_1)z_1 f(E - E_1^\circ)]\}, \tag{A4}$$

$$v_2 = i_2/n_2FA = k_2^\circ\{C_Q(x = 0)\exp[-\alpha_2 z_2 f(E - E_2^\circ)] - C_R(x = 0)\exp[(1 - \alpha_2)z_2 f(E - E_2^\circ)]\}, \quad \text{(A5)}$$

where $C_j(x = 0)$ represents the concentration of the jth species at the catalyst surface, k_1° and k_2° are the standard rate constants for the heterogeneous electron-transfer reactions (cm/s), and $f = F/RT = 38.96$ per volt. The mass-transfer-limiting currents for the different species are

$$i_{l,O} = n_1FAm_OC_O^*, \quad \text{(A6)}$$

$$i_{l,P} = -n_1FAm_PC_P^*, \quad \text{(A7)}$$

$$i_{l,Q} = n_2FAm_QC_Q^*, \quad \text{(A8)}$$

$$i_{l,R} = -n_2FAm_RC_R^*, \quad \text{(A9)}$$

where m_O, m_P, m_Q, and m_R are the mass-transfer coefficients (cm/s). The concentrations at the catalyst surfaces can be expressed in terms of these limiting currents as

$$C_j(x = 0) = [(i_{l,j} - i_1)/i_{l,j}]C_j^* \qquad j = \text{O,P}, \quad \text{(A10)}$$

$$C_j(x = 0) = [(i_{l,j} - i_2)/i_{l,j}]C_j^* \qquad j = \text{Q,R}. \quad \text{(A11)}$$

By combining Eq. (A4) with Eq. (A10) and Eq. (A5) with Eq. (A11) and solving for i_1 and i_2, the following general equations are obtained:

$$v_1 = \frac{i_1}{n_1FA} = \frac{C_O^*\exp(-\alpha_1x_1) - C_P\exp[(1 - \alpha_1)x_1]}{(1/k_1^\circ) + [\exp(-\alpha_1x_1)/m_O] + \{[\exp(1 - \alpha_1)x_1]/m_P\}}, \quad \text{(A12)}$$

$$v_2 = \frac{i_2}{n_2FA} = \frac{C_O^*\exp(-\alpha_2x_2) - C_R^*\exp[(1 - \alpha_2)x_2]}{(1/k_2^\circ) + [\exp(-\alpha_2x_2)/m_Q] + \{[\exp(1 - \alpha_2)x_2]/m_R\}}, \quad \text{(A13)}$$

where $x_1 = z_1(E - E_1^\circ)$ and $x_2 = z_2(E - E_2^\circ)$. The general solution to the steady-state rate of reaction for Eq. (A1) involves the condition

$$i_1/n_1FA = -i_2/n_2FA = v_m = i_m/n_1n_2FA, \quad \text{(A14)}$$

where v_m is the velocity of the mixed reaction and i_m the mixed current that occur at the mixed potential E_m. In principle, substitution of Eqs. (A12) and (A13) into Eq. (A14) at $E = E_m$ should allow calculation of E_m, and use of this value in either Eq. (A12) or (A13) would yield i_m.

A simple and important case obtains for Nernstian reactions:

$$i_1 = i_{l,O}/\{1 + \exp[-n_1(E - E_{1/2,1})]\}, \quad \text{(A15)}$$

$$i_2 = i_{l,R}/\{1 + \exp[-n_2(E - E_{1/2,2})]\}, \quad \text{(A16)}$$

$$E_{1/2,1} = E_1^\circ - (1/n_1)\ln(m_O/m_P), \quad \text{(A17)}$$

$$E_{1/2,2} = E_2^\circ - (1/n_2)\ln(m_Q/m_R). \quad \text{(A18)}$$

B. Results for the Hydrogen Ion–Methylviologen Reaction

The preceding treatment would apply to any electrode reaction by suitable modification of the general equations to account for adsorption, precipitation, etc. The H_2-evolution reaction at metal electrodes has been the subject of numerous investigations (*45, 46*). Although the form of the iE expression depends on the details of the electrode reaction mechanism, most mechanisms show a direct proportionality between i and C_{H^+}. For the calculations here we have employed the equation that applies under the conditions where the proton-discharge step is rate determining and the steady-state coverage by hydrogen atoms on the surface is small. Thus, with $a_1 z_1 = 0.5$ and $E_1° = 0.0$ V,

$$i_1/FA = k_1° C_{H^+} \exp(-0.5fE). \tag{A19}$$

The reduction of methylviologen is known to be rapid at Pt electrodes. Calculations for this couple were carried out using the general Eq. (A13) with $m_Q = m_R = 10^{-3}$ cm/s, $z_2 = m_2 = 0.5$; and $E_2° = -0.42$ V versus NHE. For $k_2°$ values of 10^{-2} cm/s or larger, the results are independent of $k_2°$ so that the reaction is essentially a Nernstian reaction [Eq. (A16)]. Because of the forms of Eqs. (A15) and (A16), the results for the conditions at which $E = E_m$ and Eq. (A14) applies can be given in terms of $im/i_{1,R}$ versus $\log k°_1 C_{H^+}$. Calculated values of these are shown in Fig. 5. Note that a convenient reference point is where $i_m/i_{1,R} = 0.5$ and $E_m = E_{1/2,2}$. This occurs at a particular pH, called $pH_{1/2}$; under the conditions assumed for the calculation in Fig. 5, the following equation is obtained:

$$pH_{1/2} = \log k_1° + 9.73. \tag{A20}$$

Acknowledgments

Professor Allen Bard provided stimulating discussion, valued friendship, and some of the key intellectual contributions to the work reviewed here. The work in the Appendix should be credited directly to Allen. Further debts are owed to Michael Grätzel and collaborators, and to Dan Meisel, Adi Mackor, and Mark Wrighton for continuing prepublication information exchanges. Finally, I wish to acknowledge the invaluable role of the A. P. Sloan Foundation Fellowship and a Dreyfuss Teacher Scholar award in providing the freedom to work in this area.

References

1. Bock, C., Meyer, T., and Whitten, D., *J. Am. Chem. Soc.* **96,** 4710 (1974).
2. Green, D., and Stickland, L., *Biochem. J.* **28,** 898 (1934).

3. Grätzel, M., *Acc. Chem. Res.* **14,** 376 (1982).
4. Koriakin, B., Dzahbiev, T., and Shilov, A., *Dokl. Akad. Nauk SSSR Ser. Khim.* **238,** 620 (1977).
5. Lehn, J. M., and Sauvage, J. P., *Nouv. J. Chim.* **1,** 441 (1977).
6. Brown, G., Brunschwig, B., Creutz, C., Endicott, J., and Sutin, N., *J. Am. Chem. Soc.* **101,** 1298 (1979).
7. Moradpour, A., Amouyal, E., Keller, P., and Kagan, H., *Nouv. J. Chim.* **2,** 547 (1978).
8. Keller, P., and Moradpour, A., *J. Am. Chem. Soc.* **102,** 7193 (1980).
9. Kalyanasundaram, K., and Grätzel, M., *Helv. Chim. Acta* **61,** 2720 (1978).
10. Krasna, A., *Photochem. Photobiol.* **29,** 267 (1979).
11. DeLaie, P., Whitten, D., and Gianotti, C., *Adv. Chem. Ser.* No. 173, 236 (1979).
12. Sutin, N., *J. Photochem.* **10,** 19 (1979).
13. Okura, I., and Kim Thuan, N., *J. Mol. Catal.* **6,** 227 (1979).
14. Kalyanasundaram, K., *Nouv. J. Chim.* **3,** 511 (1979).
15. Miller, D., and McLendon, G., *Inorg. Chem.* **20,** 950 (1981).
16. Miller, D., and McLendon, G., *J. Am. Chem. Soc.* **103,** (1981).
17. Miller, D., Bard, A., McLendon, G., and Ferguson, J., *J. Am. Chem. Soc.* **103,** 5336 (1981).
17a. Miller, D., Ph.D. Thesis, Univ. of Rochester, New York, 1981.
18. McLendon, G., and Miller, D., *J. Chem. Soc. Chem. Commun.,* p. 656 (1980).
19. Porter, G., Harriman, A., and Richoux, M., *J. Chem. Soc. Faraday Trans. 2* (1980).
20. Kiwi, J., and Grätzel, M., *J. Am. Chem. Soc.* **101,** 7214 (1979).
21. Brugger, P., Cuendet, P., and Grätzel, M., *J. Am. Chem. Soc.* **103,** 2423 (1981).
22. Kalyanasundaram, K., Micic, O., Pramauro, E., and Grätzel, M., *Helv. Chim. Acta* **62,** 2432 (1979).
23. Kalyanasundaram, K., and Grätzel, M., *Angew. Chem. Int. Ed. Engl.* **18,** 701 (1979).
24. Lehn, J. M., and Sauvage, J. P., *Nouv. J. Chim.* **5,** 291 (1981).
25. Turkevich, J., Aika, K., and Okura, I., *J. Res. Inst. Catal. Hokkaido Univ.* **24,** 54 (1976).
26. Turkevich, J., Stevenson, P., and Hillier, J., *Discuss. Faraday Soc.* **11,** 55 (1951).
27. Meisel, D., Mulac, M., and Matheson, D., *J. Phys. Chem.* **85,** 179 (1981).
28. Kopple, K., Meyerstein, D., and Meisel, D., *J. Phys. Chem.* **84,** 820 (1980).
29. Meisel, D., *J. Am. Chem. Soc.* **101,** 6133 (1979).
30. Henglein, A., *J. Phys. Chem.* **83,** 2209 (1979).
31. Henglein, A., and Tausch Treml, R., *J. Colloid Interface Sci.* **80,** 84 (1981).
32. Henglein, A., *Angew. Chem.* **91,** 449 (1979).
33. Tausch Treml, R., Henglein, A., and Lilie, J., *Ber. Bunsenges. Phys. Chem.* **82,** 1335 (1978).
34. Henglein, A., and Lilie, J., *J. Am. Chem. Soc.* **103,** 1059 (1981).
35. Henglein, A., and Westerhausen, J., *J. Phys. Chem.* **85,** 1627 (1981).
36. Lehn, J. M., Sauvage, J. P., and Ziessel, R., *Nouv. J. Chim.* **5,** 291 (1981).
37. Bard, A., *J. Photochem.* **10,** 59 (1979).
38. Bard, A., *Science (Washington, D.C.)* **207,** 139 (1980).
39. Krauetler, B., and Bard, A., *J. Am. Chem. Soc.* **100,** 2239 (1978).
40. Archer, M., and Spiro, M., *J. Chem. Soc.,* p. 73 (1970).
41. Spiro, M., and Ravno, A., *J. Chem. Soc. Faraday Trans. 1* **75,** 1507 (1979).
42. Spiro, M., and Ravno, A., *J. Chem. Soc.* 78 (1965).
43. Spiro, M., and Griffin, G., *J. Chem. Soc. Chem. Commun.,* p. 262 (1969).
44. Wagner, C., and Traud, W., *Z. Elektrochem.* **44,** 39 (1938).
45. Bard, A., and Faulkner, L., "Electrochemical Methods." Wiley, New York, 1980.

46. Bockris, J. O. M., and Reddy A. K. N., "Modern Electrochemistry." Plenum, New York, 1976.
47. Frens, G., *Nature (London)* **241,** 20 (1973).
48. Frank, J., and Stevenson, K., *J. Chem. Soc. Chem. Commun.,* p. 593 (1981).
49. Vincent, B., *Adv. Colloid Interface Sci.* **4,** 193 (1974).
49a. Feigen, R., and Napper, D., *J. Colloid Interface Sci.* **74,** 567 (1980).
50. Stevenson, K., *J. Chem. Educ.* (1973).
51. Porter, G., and Harriman, A., *Anal. Biochem.* (1981).
52. Miller, D., unpublished results.
53. Miller, D., and McLendon, G., *J. Am. Chem. Soc.* **103,** (1981).
54. Dunworth, W., and Nord, F., *Adv. Catal.* **6,** 125 (1954).
55. Chini, P., unpublished work.
56. Kawai, T., and Sakata, T., *J. Chem. Soc. Chem. Commun.,* p. 694 (1980).
57. Kiwi, J., and Grätzel, M., *Angew. Chem. Int. Ed. Engl.* **17,** 860 (1978).
57a. Kiwi, J., and Grätzel, M., *Angew. Chem. Int. Ed. Engl.* **18,** 624 (1979).
58. Borgarello, E., Kiwi, J., Pelicetti, E., Visca, M., Grätzel, M., *Nature (London)* **289,** 158 (1981).
59. Kalyanasundaram, K., Micic, O., Pramauro, E., and Grätzel, M., *Helv. Chim. Acta* **62,** 2432 (1979).
60. Kalyanasundaram, K., Borgarello, E., and Grätzel, M., *Helv. Chim. Acta* **64,** 362 (1981).
61. Alberts, A. H., and Mackor, M., personal communication.
62. Kalynasundaram, K., *J. Chem. Soc. Chem. Commun.,* p. 437 (1981).

5 Development of Molecular Photocatalytic Systems for Solar-Energy Conversion: Catalysts for Oxygen and Hydrogen Evolution from Water

K. I. Zamaraev
V. N. Parmon

Institute of Catalysis
Novosibirsk, U.S.S.R.

I. Introduction

World annual consumption of energy is presently the enormous value of about 10^{17} kcal, and it tends to increase, doubling every 15–20 yr (*97, 98, 101*). At the same time, the reserves of fossil fuels that now make the main contribution to the energy balance (~80% of primary energy resources, including 60% from petroleum and natural gas) are recognized as limited, the estimated total energy equivalent of their prospected reserves comprising ~10^{19}–10^{20} kcal (*1, 97, 100*). Thus the search for alternative sources of energy indeed becomes urgent.

It is anticipated that in the near future the energy demands of the world will be satisfied mainly through the growth of coal extraction and its processing, as well as nuclear sources of electric energy and heat. Both of these branches of energetics are generally agreed to have the raw material reserves for at least the next century. The existing level of science and

ENERGY RESOURCES THROUGH PHOTOCHEMISTRY AND CATALYSIS

ISBN 0-12-295720-2

technology also seems to be sufficient to provide rapid development of both coal and nuclear energy. However, a rapid increase of coal and nuclear fuel use can create some environmental problems. For example, the rapid growth of coal combustion and processing is expected to require special measures to prevent air and water pollution from sulfur oxides and from carcinogenic and other toxic substances. The problem of the so-called nuclear risk is also widely recognized (*90*). The danger of heat pollution resulting from anthropogenic evolution of heat from fossil and nuclear fuels, resulting in the unpredictable change of climate on our planet by the middle of the next century, has also been discussed (*26*).

From a long-term perspective, nuclear fusion and solar light seem to be the most promising sources of energy (*92*). A wide use of these practically inexhaustible sources is not possible at present because of the insufficient level of scientific and technological development in these fields. However, serious progress can be expected here. Although a controlled reaction of nuclear fusion has not yet been accomplished experimentally, the potential of nuclear fusion has been widely discussed in the literature and is well known to the educated public. In contrast to this, the potential of solar energy is much less known. Meanwhile, this energy is continuously received by the Earth in amounts largely exceeding the needs of our civilization. The only problem is how to utilize it rationally.

The annual flux of solar energy coming to the Earth is $\sim 1.34 \times 10^{21}$ kcal (*21, 97, 98, 101*), in other words, it exceeds the present world annual energy consumption by a factor of 10^4. About 35% of this energy is reflected by the atmosphere and the Earth's surface and thus is lost. The rest of the energy is absorbed by the atmosphere (2.5×10^{20} kcal), oceans and seas (4.4×10^{20} kcal), and land ($\sim 1.9 \times 10^{20}$ kcal).[1]

Of the energy absorbed, ~ 3–6×10^{17} kcal/yr is converted to chemical energy via photosynthesis in plants and microorganisms. However, only a small portion of this energy is successfully utilized at present: $\sim 8 \times 10^{15}$ kcal/yr as fuels and materials and $\sim 4 \times 10^{15}$ kcal/yr as food. Even smaller is the portion of solar energy used indirectly in the form of electricity produced by hydraulic and wind–electric power plants: $\sim 1.4 \times 10^{15}$ kcal/yr (*97, 101*).

The most important advantage of solar energy is its ecological purity, that is, it offers the possibility of accomplishing energy cycles without pollution of the environment and additional heating of the Earth. In contrast to this, the energy production from nuclear fusion would inevitably lead to additional heating of the Earth.

[1] Note that exact values for the fate of solar energy coming to the Earth as given by various authors are sometimes different. However, they seem to agree with each other in their order of magnitude.

However, wide use of solar energy is hindered mainly by three factors:

1. The density of the solar energy flux is low, even in the space at the outer edges of the Earth's atmosphere: 1373 ($\pm$20) W/m^2 = 12,030 kWh/m^2/yr = 3.28 $\times$ 10^{-5} kcal/cm^2/s; at the surface of the Earth the flux is lower still (see following discussion; also *21*).
2. The energy flux is dependent on the weather, season, and day-to-night variations.
3. It is necessary to convert the light energy into a form convenient for practical use.

Consider first the restrictions imposed by factors 1 and 2. In space a power of 1 GW, which is characteristic of modern electric power plants, is equal to the solar-energy flux falling on an 850 $\times$ 850 m square, which is comparable to the area occupied by a modern electric power plant of the same power. On the Earth's surface insolation causes this value to be far less because the sun shines during only part of a daily cycle and because solar light is reflected by and absorbed in the atmosphere. However, in the U.S.S.R., even at the latitude of Moscow and Novosibirsk, the total amount of the solar energy reaching the surface of the Earth exceeds 1000 kWh/m^2/yr. This is only 2.5 times lower than in the region of greatest solar irradiation, near the Red Sea (*49*). In Soviet mid-Asia, the total solar irradiation is 1600 kWh/m^2/yr. A simple estimate shows that an annual production of energy by a 1-GW electric power plant approximately equals the solar energy falling on squares of 3 $\times$ 3 and 2.3 $\times$ 2.3 km in central U.S.S.R. and mid-Asian U.S.S.R., respectively. It can easily be calculated that the energy of combustion of 600 $\times$ 10^6 tons of oil, which is the annual production scale in the U.S.S.R., approximately equals the amount of solar energy falling on a 67.4 $\times$ 67.4 km square in mid-Asia. Note that the area of the Kara Kum Desert, which is located in this region, exceeds the area of this square by more than 60 times. If 20% efficiency of solar-energy conversion were achieved, the energy of 1300 $\times$ 10^{9}kWh, which was the annual production of electricity in the U.S.S.R. in 1980, could be produced from the 64.7 $\times$ 64.7 km area of mid-Asia. For comparison, note that the area of hard covered roads in the U.S., the total length of which in 1972 was reported to be about 6 $\times$ 10^6 km, exceeds this area by several times.

Thus, despite the fairly low density of solar-energy flux, at the modern level of technology even a total switch to energy production from solar light is perhaps not as unrealistic as it might first appear.

Of course, in the near future the traditional energy sources will still play the major role in the total energy balance of the world. However, according to the prognosis of the International Energy Conference, by the year

2020 as much as 10–15% of world energy consumption will be provided by solar-energy conversion (*97, 101*).

Among the possible methods of solar-energy storage, its conversion into the energy of chemical fuels seems to be one of the most promising (e.g., see *7, 8, 10, 11, 15, 92*). In fact, storage of solar energy in the form of chemical fuels allows one to cope simultaneously with the difficulty associated with the dependence of the solar-energy flux on season, weather, and day-to-night variations and to produce energy in a concentrated and versatile form.

In principle, any chemical process that occurs with free energy or enthalpy storage can be used for solar-energy conversion. The most attractive at present are reactions of high-potential fuels produced from water and atmospheric gases, which are readily available, cheap, and almost inexhaustible resources. Some reactions that are discussed as promising ones for this purpose are listed in Table I. Fuels produced in these reactions (hydrogen, methane, methanol, ethanol, ammonia, or hydrazine) have high-energy capacity. These reactions are known also to satisfy the thermodynamic criteria for high efficiency under the action of solar light (*105*). According to theoretical estimates, in the most favorable conditions this efficiency can be as high as 30% (*85, 106*). Note that all the processes listed in Table I are known to occur *in vivo* via photosynthesis (with subsequent fermentation of the resulting products to H_2, CH_4, or alcohols) or via atmospheric-nitrogen fixation.

TABLE I

Thermodynamic Characteristics for Some Reactions that Can Be Used for Storage of Solar Energy in the Form of Chemical Energy

Reactions[a]	n[b]	ΔG°_{298} (kcal/mol)	ΔH°_{298} (kcal/mol)	ΔG°_{298} (per one electron, eV)	λ_{thres} (nm[c])
$H_2O \longrightarrow H_2 + \frac{1}{2}O_2$	2	56.71	68.25	1.23	1008
$2H_2O + CO_2 \longrightarrow CH_4 + 2O_2$	8	195.54	212.8	1.06	1176
$2H_2O + CO_2 \longrightarrow CH_3OH + \frac{3}{2}O_2$	6	167.92	173.64	1.21	1025
$3H_2O + 2CO_2 \longrightarrow C_2H_5OH + 3O_2$	12	318.34	336.82	1.15	1077
$\frac{3}{2}H_2O + \frac{1}{2}N_2 \longrightarrow NH_3 + \frac{3}{4}O_2$	3	81.09	91.44	1.17	1059
$2H_2O + N_2 \longrightarrow N_2H_4 + O_2$	4	181.33	148.74	1.97	629

[a] Thermodynamic data are given for liquid H_2O, CH_3OH, C_2H_5OH, and N_2H_4.
[b] n, Number of transferred electrons.
[c] λ_{thres}, Threshold wavelength for the n-photon mechanism.

However, outside living organisms these processes do not occur upon illumination by solar light. Therefore, to accomplish them it is necessary to develop *photocatalytic* systems involving *photocatalysts* that would provide occurrence of photochemical steps, and to develop *catalysts* that would provide occurrence of subsequent dark-reaction steps of an overall energy-storage chemical reaction.

A photocatalyst is a substance that upon absorption of light photons induces chemical transformations of the reactants via repeated intermediate interaction with the reaction participants, being regenerated after each reaction cycle. Note that the term *photosensitizer* is also sometimes used in literature in the same sense as the term *photocatalyst*.

Of the reactions listed in Table I, hydrogen production via photocatalytic water cleavage seems to be the most attractive at present. Hydrogen is known to be not only a good and ecologically pure chemical fuel, but also a valuable material for chemical industry. In addition, from the information accumulated to date about the mechanisms of chemical reactions, it may be expected that of all the energy-storage reactions, the photocatalytic cleavage of water will be the simplest one to accomplish.

Developments in use of photocatalytic systems for water cleavage into molecular hydrogen and oxygen are already achieving successes. These systems are conventionally divided into two types: *semiconductor* and *molecular* photocatalytic systems.

Semiconductor systems use massive or highly dispersed semiconductors as photocatalysts. Systems of this type have been developed with the efficiency of several percent of solar light conversion into the chemical energy of a hydrogen-plus-oxygen pair [see the pioneering work of Fujishima and Honda (*31*); also reviews in references *3, 33, 33a, 34, 79*]. It is expected that in the near future the efficiency of semiconductor photocatalytic systems for water cleavage can be substantially increased. A detailed discussion of the developments in this field can be found in the previously mentioned review articles as well as in other chapters of this book.

Molecular photocatalytic systems are based on modeling the scheme of natural photosynthesis. The work with these systems was started later than that with semiconductor systems and has not yet resulted in development of the systems that would provide water cleavage with the same efficiency as that already achieved with semiconductor systems. Nevertheless, because the efficiency of solar-energy conversion in molecular systems of natural photosynthesis is known to be quite high (as high as ~5–10%) for a very complicated reaction between H_2O and CO_2 (*9, 78*), we may hope that highly effective artificial converters of this type can be created in the future for a much more simple reaction of water cleavage.

The current state of study aimed at the development of molecular photocatalytic systems for water cleavage has been reviewed in detail by Dzhabiev and Shilov (*20*), Grätzel (*32, 33, 33a*), Maverick and Gray (*71*), Moradpour (*75*), Varfolomeev (*99*), and Bagdasaryan (*2*), as well as by Zamaraev and Parmon (*105*). Additional more recent papers by Grätzel *et al.* (see *32, 33, 33a, 39, 48*) should also be noted. These studies describe the first artificial molecular system that is capable of accomplishing—although as yet with low efficiency—the complete cycle of water cleavage into oxygen and hydrogen by visible light.

Most frequently discussed schemes of water cleavage include three steps (Fig. 1): charge separation, that is, formation of sufficiently strong oxidant (D^+) and reductant (A^-) under the action of solar light in the presence of a photocatalyst (PhC)

$$D + A \xrightarrow[\text{PhC}]{h\nu} D^+ + A^- \tag{1}$$

and subsequent catalytic reactions of oxygen evolution from water by oxidant D^+

$$4D^+ + 2H_2O \xrightarrow{\text{cat}_1} 4D + 4H^+ + O_2\uparrow \tag{2}$$

and of hydrogen evolution by reductant A^-

$$2A^- + 2H^+ \xrightarrow{\text{cat}_2} 2A + H_2\uparrow \tag{3}$$

The last two steps do not necessarily require light to be accomplished. The overall process is described by the equation

$$2H_2O \xrightarrow[\text{cat}_1,\ \text{cat}_2]{h\nu,\ \text{PhC, D, A}} 2H_2 + O_2$$

Thus, to accomplish water cleavage into hydrogen and oxygen by solar light in a molecular photocatalytic system, according to the above scheme, it is necessary to develop efficient catalysts for three types of processes: charge separation under the action of light, hydrogen evolution from water under the action of one-electron reductants, and oxygen evolution from water under the action of one-electron oxidants. For example, in the previously mentioned molecular photocatalytic system suggested by Kalyanasundaram and Grätzel (*39*) and Kiwi *et al.* (*48*), the roles of PhC, A, cat_1, and cat_2 were provided, respectively, by a trisbipyridyl complex of Ru(II) [$Ru(bipy)_3^{2+}$], organic bication dimethyl-4,4′-bipyridine (methylviologen or MV^{2+}), colloidal ruthenium dioxide, and colloidal platinum. The photocatalyst PhC served simultaneously as an intermediate donor D.

In this chapter we present results of work on developing and studying the catalysts for all three steps of water cleavage in molecular photocata-

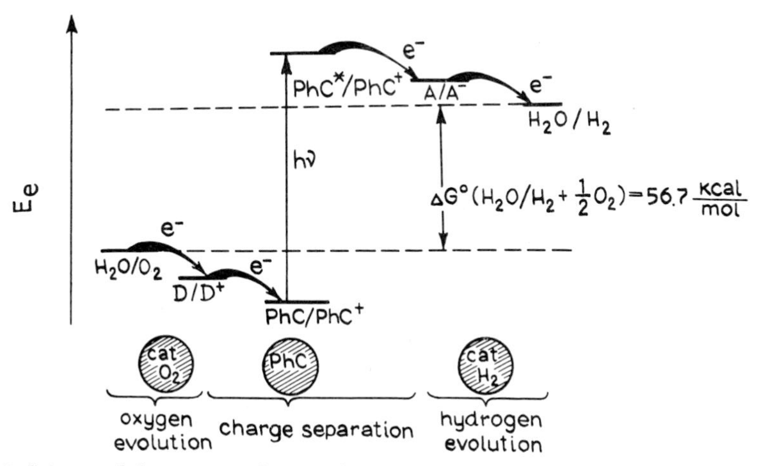

Fig. 1. Scheme of the process of water cleavage in a molecular photocatalytic system. E_e represents the energy of electrons; the curved arrows show paths for electron transfer.

lytic systems obtained in our laboratory subsequent to publication of our previous review article (*105*).

II. Hydrogen Evolution from Water

One-electron reductants A^- formed at the step of charge separation [reaction (1)] are able to evolve hydrogen from water via reaction (3) only in the presence of appropriate catalysts. Beginning with the pioneering work of Yu and Wolin (*104*), Benemann *et al.* (*5*), Krasnovsky *et al.* (*57*), Berezin *et al.* (*6*), Markiewicz *et al.* (*69*), Koryakin *et al.* (*51, 52*) and Lehn and Sauvage (*60*), a number of catalysts have been proposed for this process. Most often these are supported (*53, 54, 73, 77*) or colloidal noble metals and enzymes such as nitrogenases and hydrogenases (*50, 53, 55–57, 63, 81, 84, 89, 99*). To evaluate the prospective practical use of these catalysts in future photocatalytic systems, it is necessary to carry out systematic quantitative studies of their activity and stability as well as of the dependence of these properties on conditions of catalyst preparation. Such studies have been made for colloidal metal catalysts (*33, 33a, 35, 37, 37a, 38, 42, 43, 43a, 72, 74, 74a, 80*).

In this chapter we shall consider two types of catalysts for hydrogen evolution from water developed in our laboratory: (i) highly dispersed noble metals stabilized in polymeric globules and granules that are permeable for reductant particles and water molecules and (ii) heteropolyanions—compounds containing several ions of transition metals and capable of reversible multielectron reduction. Thus heteropolyanions can be considered as structural and functional models of hydrogenases and nitrogenases, enzymes which are also known to contain several metal ions and to be capable of reversible multielectron reduction.

A. Rhodium–Ethylenediamine and Rhodium–Polyethyleneimine Systems

We have found (*12*) that coordinatively unsaturated Rh(III) complexes with ethylenediamine (en) and its polymeric analogs (polyethyleneimines) are efficient catalysts for hydrogen evolution from water by such one-electron reductants as ${V_{aq}}^{2+}$, ${Cr_{aq}}^{2+}$, and the radical-cation of $MV^{\dagger}$ in reactions

$$2A^- + 2H^+ \longrightarrow 2A + H_2\uparrow \tag{3a}$$

or

$$2A^- + 2H_2O \longrightarrow 2A + H_2\uparrow + 2OH^- \tag{3b}$$

Of these the most interesting for possible use in photocatalytic converters are heterogeneous catalysts based on two polymers, namely, a branched polyethyleneimine (PEI) with a molecular mass of ~20,000 and a commercial ion-exchanging resin AN-221, which is granulated, partially cross-linked polyethyleneimine.

Spectral data, including epr, xps, and uv–vis, indicate (*13*) that in the initial forms of catalysts with monomeric ligands (ethylenediamine) practically all rhodium is in the form of Rh(III) complexes, whereas in the initial forms of polymeric catalysts as much as 5% of rhodium (depending on preparation conditions) is in the form of Rh(II) complexes. The remainder of the rhodium in these catalysts is also in the form of Rh(III) complexes. For monomeric ligands the initial complexes can have the composition Rh : N = 1 : 6, 1 : 4, and 1 : 2 (complexes of the latter composition can be obtained only in the mixture with the complexes of composition Rh : N = 1 : 4); however, for deprotonated polymeric ligands the initial complexes were found to have the composition Rh : N = 1 : 4 and 1 : 2. By varying the metal-to-ligand ratio during synthesis, it is possible to obtain heterogenized metal–polymeric complexes with predominant compositions of both Rh : N = 1 : 2 and Rh : N = 1 : 4.

Addition of rhodium complexes with ethylenediamine and polyethyleneimine to aqueous solutions of such reductants as ${V_{aq}}^{2+}$ (redox potential $E° = -0.25$ V versus NHE) and ${Cr_{aq}}^{2+}$ ($E° = -0.4$ V versus NHE) resulted in hydrogen evolution. Typical kinetic curves for H_2 evolution are shown in Fig. 2.

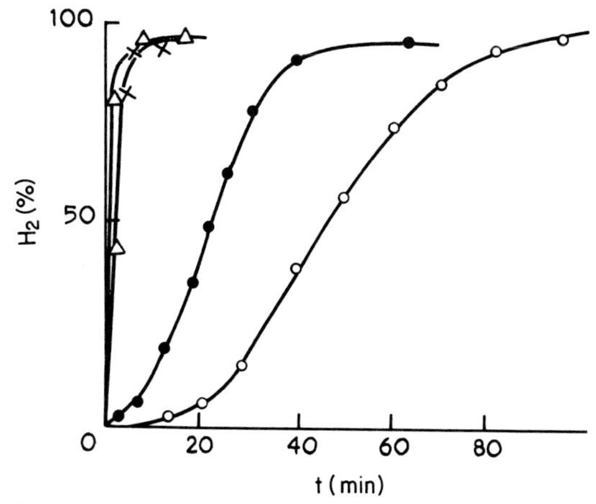

Fig. 2. Kinetics of hydrogen evolution (in percent of stoichiometric yield) from 1.5×10^{-2} *M* VSO_4 solution in water in the presence of various catalysts: X, $[Rh\text{–}en]Cl_3$; △, Rh–PEI with the composition Rh : N ≈ 1 : 2; ●, $[Rh(en)_2Cl_2]ClO_4$; ○, Rh–PEI with the composition Rh : N ≈ 1 : 4. The volume of the solution is 20 cm^3, and the amount of Rh in the solution is $\sim 10^{-5}$ mol; pHO is maintained by H_2SO_4 at 25°C.

These curves are often S-shaped: in the short-time region there is sometimes an induction period during which the reaction rate tends to gradually increase. During this period, active states of catalysts, which, most probably, are highly dispersed particles of metallic rhodium seem to be formed (see following discussion). The decrease of the reaction rate in the long-time region is apparently due to the complete consumption of the reductant. Indeed, in the plateau region in Fig. 2 the amount of evolved hydrogen is nearly 100% of the stoichiometric expectation via reactions (3a) or (3b). An addition of the reductant in the initial concentration after the reaction has been terminated again leads to hydrogen evolution, the slope of the kinetic curve after introduction of the second portion of the reductant being the same or even bigger than that of the initial curve (Fig. 3). As can be seen from Fig. 2, for both monomeric and polymeric ligands the catalysts prepared from complexes with composition Rh : N = 1 : 2 show far higher activity than the catalysts prepared from complexes with composition Rh : N = 1 : 4. In the presence of monomeric coordinatively saturated complexes with composition Rh : N = 1 : 6, no hydrogen is evolved under the action of V_{aq}^{2+} and Cr_{aq}^{2+}.

As ascertained by x-ray and xps analysis, activation of monomeric complexes of Rh : N = 1 : 2 composition leads to their complete reduction to finely dispersed metallic rhodium. Formation of small particles of metallic rhodium was also observed in all activated polymeric catalysts. According to electron microscopic data, species of metallic rhodium in poly-

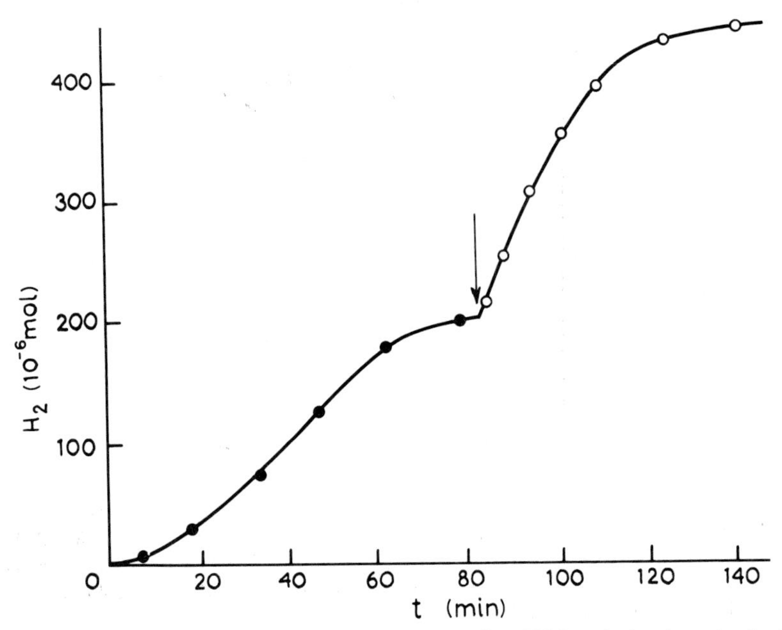

Fig. 3. Kinetics of hydrogen evolution from 1.5×10^{-2} *M* VSO_4 solution in water in the presence of polymeric catalyst Rh–PEI with the composition Rh : N ≈ 1:4. The arrow shows the moment of the addition of the second portion of VSO_4. The volume of the solution is 20 cm^3 and the amount of Rh in the solution ~10^{-5} mol; pHO is maintained by H_2SO_4 at 25°C.

meric catalysts have complex fractional compositions that depend on the catalyst preparation conditions. However, in all cases, notable amounts of small species of 10 to 30 A in size were observed, which partly grow together producing loose aggregates of larger size. Evidently, metallic rhodium is responsible for the catalytic activity of all catalysts based on ethylenediamines and polyethyleneimines. However, in activated catalysts with polyethyleneimines of composition Rh : N ≈ 1 : 4, only a portion of the rhodium ions are reduced to metallic state. In fact, the reflectance spectra of these catalysts in the uv–vis region indicate the presence of large amounts of initial complexes with composition Rh : N = 1 : 4 (*14*). x-Ray photoelectron spectra (xps) of these samples also indicate that on their surface rhodium is present as Rh(III). The structure of activated catalysts, which coincides well with the results of investigations with various physical methods, is schematically represented in Fig. 4. The size of highly dispersed metal particles in the polymer globule was found to depend on the conditions of catalyst preparation and activation. For example, the particle size appreciably decreases when the pH of the solution in which activation takes place is increased.

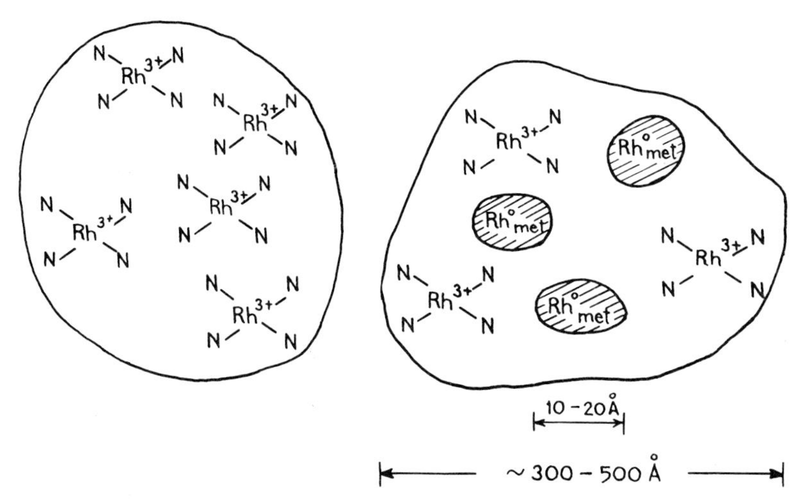

Fig. 4. Structure of the catalyst Rh–PEI with the composition Rh : N ≈ 1 : 4 before (left) and after (right) activation.

The maximum rate α of hydrogen evolution during the reaction (i.e., the rate measured in the region of the maximum slope of S-shaped curves) per 1 g-atom of introduced rhodium was taken as a quantitative characteristic of the catalytic activity. The values of α for various experimental conditions and for catalysts prepared by various methods are compared in Table II. For the most active systems we give only the lower estimates of α (see Table II). For these systems reaction (3) was almost completed during the time needed to take the first probe for the chromatographic analysis of the evolved H_2, so that only the lower limit of the reaction rate could be estimated.

It can be seen from Table II that the catalysts prepared from complexes with monomeric ligands having composition Rh : N = 1 : 2 show activity higher (by several orders of magnitude) than complexes with the same ligands having composition Rh : N = 1 : 4. It is likely that the lesser activity of these latter catalysts is the result of the lesser conversion of initial rhodium(III) to the highly dispersed metal. In fact, upon activation of the monomeric complex having composition Rh : N = 1 : 2, the formation of metallic rhodium in solutions was visually observed, whereas for complexes having composition Rh : N = 1 : 4 no metal particles were visually observed.

The dependence of the catalytic activity α on the concentration of H^+ ions has been studied for monomeric complexes of composition Rh : N = 1 : 4. It can be seen from Table II that α gradually increases with the increase of H^+ concentration in the solution. However, α increases much

TABLE II

Catalytic Activity α of Rhodium–Polyimine Complexes in the Reaction of Hydrogen Evolution[a]

Initial form of catalysts	pH	Initial ratio $[V^{2+}]/[Rh^{3+}]$	α (mol H_2/mol Rh/hr)
$[Rh(en)_3](ClO_4)_3$	0	30	<0.1
(Rh : N = 1 : 6)			
$[Rh(en)_2Cl_2](ClO_4)$	0	6	5.0
(Rh : N = 1 : 4)	1	6	2.9
	2	6	1.6
	3	6	0.75
	4	6	0.15
$[Rh\text{-}en]Cl_3$[b]	0–2	30	>300
(Rh : N ≈ 1 : 2)	2	240	>1700
Rh–PEI	0	30	16
(Rh : N ≈ 1 : 4)	2	30	0.82
Rh–PEI[b]	0	30	>400
(Rh : N ≈ 1 : 2)	2	30	>300
Rh–AN-221	2	120	8.7
(7% Rh; Rh : N ≈ 1 : 3.4)	2	120	640[c]

[a] From 1.5×10^{-2} *M* water solution of VSO_4 at 25°C; pH is maintained by H_2SO_4.
[b] For these most active systems, only the lower estimates for α are given.
[c] Catalyst granules are dispersed into powder with a particle size of <0.1 mm; the initial size of granules is ~2 mm.

more slowly than one might expect from the mass–action law for the reaction in which H^+ participates in the rate-determining step. Further investigations are needed to elucidate the nature of the observed dependency of α on $[H^+]$.

When passing from the polymer having the composition Rh : N = 1 : 4 to the Rh : N = 1 : 2 polymer, a notable increase in the catalytic activity just as it was in the case of the monomeric complex is observed. As for the monomeric catalyst with Rh : N = 1 : 2 composition, only the lower estimate for α was obtained for polymers with Rh : N ≈ 1 : 2.

For the Rh–AN-221 catalyst represented by granules with about 7% Rh content (Rh : N = 1 : 3.4), it was found that the rate of hydrogen evolution was dependent on the granule size, on the amount of the catalyst added, and on the concentration of both reagents, namely H^+ and V_{aq}^{2+}. When stirring of the solution was at a sufficiently high rate, the rate of hydrogen evolution was found to be independent of the stirring rate and to increase directly proportionally to the amount of the catalyst added. This indicates that the rate of hydrogen evolution is limited under our conditions by the processes assisted by the catalyst but not by the mass transfer in the solution or at the solution–gas phase boundary. We have observed, however, an appreciable increase in the rate with a decrease in the catalyst particle size (cf. $\alpha = 8.7$ for granules ~2 mm in diameter and $\alpha = 640$ for the same catalyst but finely dispersed with particle diameter <0.1 mm; see the last two entries in Table II). This fact proves that the rate of H_2 evolution is determined by the mass-transfer processes inside the polymer particles rather than by the catalytic reaction itself.

The study of α dependence on the concentration of V_{aq}^{2+} has shown that α increases with the increase of $[V_{aq}^{2+}]$ at low concentrations ($<5 \times 10^{-3}$ M) and, practically, does not depend on $[V_{aq}^{2+}]$ at higher concentrations (5×10^{-3}–2×10^{-2} M). This suggests that at low concentrations of V_{aq}^{2+} the rate-determining step of the process involves these ions. However, at high concentrations of V_{aq}^{2+}, the rate of this step increases so that it ceases to be rate determining. Thus the diffusion inside a polymer particle of some other reaction participant (product or reagent) becomes the rate-determining step. The dependence of α on $[H^+]$ was studied only at low $[V_{aq}^{2+}]$. It was found to be described by the relationship $\alpha \sim [H^+]^{0.2}$, provided the ionic strength of the solution was maintained constant at $I = 2$ M. These data have been obtained by measuring at various values of pH the values of α for the same catalyst sample, which was preactivated at pH 2 and 25°C in the solution with 10^{-2} M VSO_4 at $[V_{aq}^{2+}]/[Rh^{3+}] = 20$. This dependence of α on $[H^+]$ again turned out to be much weaker than that expected from the mass-action law for the step limited by the reaction involving H^+. For a deeper insight into the nature of the observed dependence of α on $[H^+]$ for this catalyst, further investigations are needed (likewise in the case of monomeric catalysts).

From these data it follows that at high $[V_{aq}^{2+}]$ the rate of H_2 evolution assisted by anionite rhodium catalysts is determined by the diffusion of the products of reaction (3) (i.e., V^{3+} cations or hydrogen molecules) inside polymer particles.

All rhodium catalysts under consideration were found to retain their activity in hydrogen evolution from water when used in the classical photocatalytic system based on trisbipyridyl complex of Ru(II) (the photocatalyst), MV^{2+} (the intermediate acceptor–electron carrier), and anions of ethylendiaminetetraacetic acid ($EDTA^{2-}$; the irreversibly consumed electron donor):

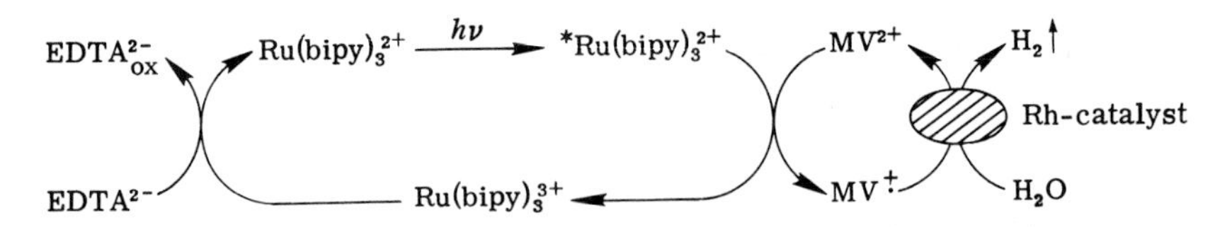

Here and later $EDTA_{ox}{}^{2-}$ means the products of irreversible oxidation of $EDTA^{2-}$ (*69*). When operating with rhodium or other catalysts that contain dispersed metals after activation, this photocatalytic system undergoes rather rapid deactivation, most probably as a result of hydrogenation of the electron carrier MV^{2+}.

We shall now compare the activity of our catalysts for hydrogen evolution from aqueous solutions with the activity of hydrogenase in the same reaction. For hydrogenases α ranges from 10^4 to 10^7 mol H_2/mol enzyme/h (*50*), which significantly exceeds α for our catalysts. This difference becomes less pronounced if we compare the activities defined by α' referred not to g-mol but to 1 g of rhodium catalyst and of enzyme. In fact, for the most efficient rhodium–anionite catalysts studied, α' exceeds about 1 mol H_2/g catalyst/h, whereas for hydrogenases α' ranges from 0.2 to 200 mol H_2/g enzyme/h (*50*). The activity of these rhodium catalysts is controlled by diffusion and thus can be further improved by decreasing the size of the catalyst granules and perhaps by varying the number of links in the polymeric carrier.

Being less active than hydrogenases, the rhodium polymeric catalysts exceed them in stability, especially in acid media, in which hydrogenases are almost immediately deactivated. In contrast to hydrogenases, activated rhodium–anionites and Rh–PEI catalysts are stable under storage in air.

The complexes of PEI with ruthenium (*12*), palladium, and platinum (*76*) were also found to be catalytically active in hydrogen evolution from water under the action of one-electron reductants. The practical advantages of the metal–polymer catalysts are their stability in aqueous solutions over a wide range of pH and the possibility of preparing them in the form of granules and films with controlled geometric characteristics.

B. Heteropolyanion Systems

A class of compounds of potential use in photocatalytic systems for hydrogen evolution from water seems to be that of heteropolycompounds. As previously mentioned, in their redox properties these compounds resemble active centers of hydrogenases and nitrogenases. It is known that reduced states of 12-heteropolyacids (HPAs) often have negative or close to zero (versus NHE) redox potentials, that is, they satisfy the thermodynamic requirements for hydrogen evolution from acid aqueous solutions. For example, redox electrochemical potential $E°$ of $H_3[PMo_{12}O_{40}]$ (I) reduced by six electrons and $H_4[SiMo_{12}O_{40}]$ (II) reduced by four electrons seems to be about 0.1 V, redox potentials of $H_3[PW_{12}O_{40}]$ (III) and $H_4[SiW_{12}O_{40}]$ reduced by two electrons are, respectively, about 0 and −0.2 V (*18, 30, 82*). We have found that all of these reduced forms of HPA can spontaneously evolve hydrogen from acid aqueous solutions at a notable rate, even when unassisted by additional catalysts (*86, 87*).

We have studied this process in more detail for solutions of HPA IV (12-silicontungsten acid) (*86*). The structure of this acid is schematically represented in Fig. 5. The HPA IV reduced by two electrons (KH_2) on zinc amalgam rapidly evolved hydrogen from aqueous solutions with a yield close to 100% of the stoichiometrically expected one for

$$KH_2 \xrightarrow{k_{eff}} KH + \tfrac{1}{2}H_2\uparrow \tag{4}$$

The kinetics of this process are described by the scheme

$$KH_2 \xrightarrow{k_1} K + H_2\uparrow \tag{4a}$$
slow evolution of H_2

$$KH_2 + K \longrightarrow 2KH \tag{4b}$$
rapid disproportionation

that leads to the first-order kinetics of the overall reaction. In this scheme K and KH are initial oxidized and one-electron-reduced forms, respectively, of HPA IV. The rate constant of the overall reaction (k_{eff}) measured in three independent sets of experiments both from the disappearance of KH_2 (using optical methods) and accumulation of H_2 (using chromatographic methods) or KH (using esr) is $k_{eff} = 2k_1 = (7 \pm 4) \times$

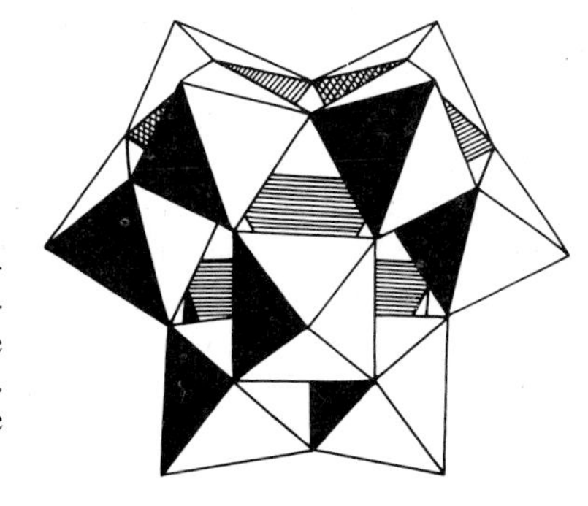

Fig. 5. So-called Keggin spatial structure of heteropolyanions that is typical for 12-heteropolyacids. Metal cations are located inside the neighboring octahedra formed by oxygen anions. The heteroatom is located in the center of the heteropolyanion.

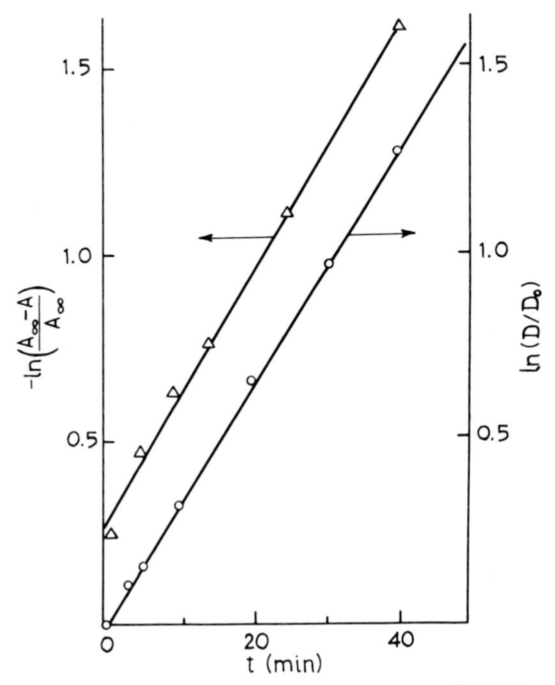

Fig. 6. Kinetics of hydrogen evolution and of the disappearance of KH_2 from water solutions of reduced 12-silicotungsten heteropolyacid at pH 2 and room temperature in anaerobic conditions. *A*, Height of the hydrogen chromatographic peak; A_∞, height for the total amount of evolved hydrogen. The total initial concentration of heteropolyacid is 2.2×10^{-3} *M*. *D*, Optical density of the solution at $\lambda = 650$ nm and for an optical path length of 1 cm; D_0, optical density for the initial moment. The total initial concentration of heteropolyacid is 5×10^{-4} *M*.

10^{-4}/s at ambient temperature.[2] As an example, the results of measuring k_{eff} from the disappearance of KH_2 and from the accumulation of H_2 are compared in Fig. 6. The presence of the rapid step of disproportionation [reaction (4b)] is confirmed by the decrease of the H_2 yield upon addition of an oxidized form, K, with the rate constant for H_2 evolution remaining unchanged.

The rate constant of hydrogen evolution k_{eff} was found to be independent of pH over the pH range 1–4. This implies that the protonation of the anions of HPA IV reduced by one and two electrons cannot be the rate-determining step. The so-called internal protons, that is, the protons bound to the HPA frame, have been revealed by nmr (*40*). The rate-deter-

[2] Because this value has now been averaged over more kinetic runs, it is slightly different from the value $k_{eff} = (1.2 \pm 0.7) \times 10^{-3}$/s reported by us earlier (*86*).

mining step of the overall reaction is perhaps the reduction of these protons. The mechanism of this reaction is still under study; however, some interesting features of it can be outlined. For example, we have detected an appreciable isotopic effect for reaction (4). The rate of this reaction decreases approximately twofold in passing from H_2O to D_2O (*88*). This observation proves that the rate-determining step of reaction (4) is the process that includes the participation of protons. We have also found that this reaction has a low activation energy, $E^\ddagger = 6 \pm 2$ kcal/mol, but a high negative activation entropy, $\Delta S^\ddagger \approx -50$ eu. For the reduction of internal protons as the rate-determining step, the high value of $-\Delta S^\ddagger$ is not surprising because this process is a basic but complicated step, during which two electrons are transferred from the KH_2 frame to two internal protons that must come near each other in the process of migration within the frame, thus leading to the formation of the hydrogen molecule. This reaction can be notably accelerated under the action of visible (including red) light. This light is known to excite the bands of intervalence charge transfer in reduced forms of HPA (*30, 96*), which seems to make the electron transfer from the KH_2 frame to its internal protons more facile.

Hydrogen evolution was also observed from water–ethanol (0.1–10.0 *M* of alcohol) solutions of phosphoromolybdenum acid (HPA I) reduced by six electrons, silicomolybdenum acid (HPA II) reduced by four electrons, and phosphorotungsten acid (HPA III) reduced by two electrons (*87*). The values of $k_{eff} \gtrsim 10^{-4}$/s for these HPAs are close to that for HPA IV in aqueous solution, $k_{eff} = (7 \pm 4) \times 10^{-4}$/s. This suggests that the reaction of hydrogen evolution by all the heteropolyacids studied has perhaps the same afore-mentioned rate-determining step.

The well-known ability of heteropolyacids to be easily reduced by various reductants and the previously described ability of the reduced forms of HPA to evolve hydrogen from aqueous solutions both open up the possibility of using HPAs as homogeneous catalysts for hydrogen evolution from water in the presence of various reductants.

In fact, HPAs I–II were found to catalyze hydrogen evolution from acid aqueous solutions of various two-electron reductants, including such moderate reductants as tin(II) ($E° = 0.15$ V versus NHE). Under the action of these reductants on HPAs I–II, no appreciable amounts of the forms reduced to the maximum of aforementioned extents were observed, and the rate of hydrogen evolution was significantly smaller than that of hydrogen evolution by highly reduced HPA forms under identical conditions. We may, therefore, suggest that the observed slow (for tens of hours) hydrogen evolution occurs either via reactions involving less reduced forms of HPAs I–II themselves (e.g., via the reaction $KH + KH \rightarrow 2K + H_2$), which can proceed from the left to the right side at small H_2

pressures, or via disproportionation of these forms, producing small amounts of reduced HPA forms, which (as shown previously) are indeed capable of evolving hydrogen.

Heteropolyacids can also provide the role of photocatalysts for endoergonic reactions of hydrogen evolution from water by irreversibly consumed donors—"sacrifice." For example, we observed hydrogen evolution under uv irradiation of oxidized HPA forms in water–ethanol (0.1–10.0 *M* of alcohol) solutions with pH 1–4 (*87*). The typical kinetic curve for this process is shown in Fig. 7. Hydrogen evolution starts after the induction period and occurs at a constant rate. The rate of H_2 evolution remains constant at a prolonged irradiation when the amount of evolved H_2 considerably exceeds that of HPA in irradiated solution. Thus hydrogen evolution is, in fact, a catalytic process. The results of the analysis of optical absorption spectra in the vis–uv region and fast redox titration with $KMnO_4$ solution suggest that during the induction period under the action of light with wavelength $\lambda < 400$ nm, HPA I is reduced consecutively by two, four, and six electrons, HPA II by two and four electrons, and HPAs III and IV by one and two electrons. For $\lambda = 333$ nm the quantum yield of the reduced forms of HPA in the initial parts of the kinetic curves of Fig. 7 comprises several percent for HPAs I and II and about 20% for HPAs III and IV. Thus the induction period for hydrogen evolution appears to result from the accumulation of rather highly reduced forms of HPA capable of spontaneous evolution of hydrogen from acid water–ethanol solutions. These reduced forms are produced by ethanol oxidation. In agreement with this conclusion, acetaldehyde in a ratio

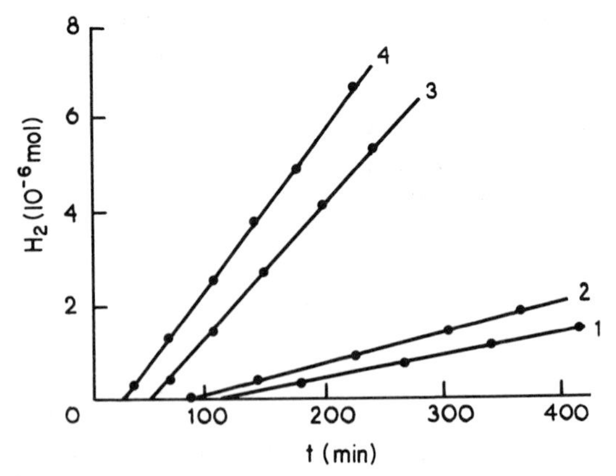

Fig. 7. Kinetics of hydrogen evolution from 8 cm^3 of an 8.7 *M* solution of ethanol in water containing also 0.25 *M* H_2SO_4 and (1) 1.3×10^{-4} *M* HPA I, (2) 10^{-4} *M* HPA II, (3) 5×10^{-4} *M* HPA III, and (4) 4×10^{-4} *M* HPA IV, on irradiation by filtered uv light with $\lambda < 400$ nm from 1-kW mercury lamp of superhigh pressure.

of about 1 : 1 to the evolved hydrogen was detected in the solution after termination of the photocatalytic reaction.

A severalfold decrease of the uv-light intensity makes the induction period longer. However, the rate of hydrogen evolution in the linear part of its accumulation curve falls only slightly. Accumulation of hydrogen continues after the light is removed but the process gradually slows down because of the consumption of the reduced HPA forms.

For the water–ethanol solutions of HPA IV, the rate constant of hydrogen evolution was estimated to be $k_1 \approx 5 \times 10^{-4}$/s in the linear part of the accumulation curve in Fig. 7. This constant was calculated as the ratio of H_2 evolution rate to the steady-state concentration of KH_2 (the form of HPA IV reduced by two electrons). As is seen, the value of k_1 is close to $k_1 = k_{eff}/2$ for dark reactions of H_2 evolution from acid water solutions of KH_2 HPA IV (see previous discussion).

These facts allow us to suggest the following overall scheme of the photocatalytic production of H_2 from water–ethanol solutions of oxidized forms of HPA:

$$C_2H_5OH + K \xrightarrow{h\nu} CH_3CHO + KH_2 \tag{5a}$$

$$KH_2 \longrightarrow K + H_2\uparrow \tag{5b}$$

reaction (5b) being the rate-determining step. Here KH_2 represents the highly reduced forms of HPAs that are capable of evolving hydrogen. This reaction occurs spontaneously even in the absence of light, but as we mentioned, it can be accelerated upon irradiation by visible light into the bands of intervalence charge transfer. In contrast, reaction (5a) can occur only under irradiation. Reaction (5a) is assumed to be complex, that is, to consist of several steps (*88*). Further investigations are needed to clarify the relative role of two-electron (e.g., $C_2H_5OH + K \xrightarrow{h\nu} CH_3CHO + KH_2$) and one-electron processes (e.g., $C_2H_5OH + K \xrightarrow{h\nu} CH_3\dot{C}HOH + KH$, $C_2H_5OH + KH \xrightarrow{h\nu} CH_3\dot{C}HOH + KH_2$, $CH_3\dot{C}HOH + K \xrightarrow{h\nu} CH_3CHO + KH$, $CH_3\dot{C}HOH + KH \longrightarrow CH_3CHO + KH_2$, and $K + KH_2 \rightleftharpoons 2KH$) in providing reaction (5a).

Thus, in the photochemical process, HPA takes part in cyclic transformations that include the steps of its reduction by two and more electrons and the subsequent production of hydrogen by these reduced forms of HPA, leading also to regeneration of the initial oxidized forms of HPA. In this case HPA simultaneously provides both functions: that of the photocatalyst and that of the catalyst of the dark reaction of hydrogen production. The photocatalytic evolution of hydrogen in the system under study

$$C_2H_5OH \xrightarrow[PhC]{h\nu} CH_3CHO + H_2\uparrow$$

is endoergonic and leads to a storage of light energy (~0.2 eV at each step of H_2 molecule evolution).

III. Oxygen Evolution from Water

Oxygen evolution is significantly more difficult to accomplish than hydrogen evolution because four electrons must be transferred in reaction (2). The systematic search for catalysts for this reaction has begun only recently. We have found *(22–24)* a series of homogeneous catalysts containing metal complexes for reaction (2) in which $Ru(bipy)_3^{3+}$, $Fe(bipy)_3^{3+}$, and $Fe(phen)_3^{3+}$ serve as oxidants D^+. Some of these catalysts allow oxygen evolution not only in alkaline and neutral solutions, but also in slightly acidic ones (up to pH 3). A series of homogeneous catalysts for the step of oxygen evolution have been suggested by Shilov and co-workers *(44, 44a, 94)*. Kiwi and Grätzel *(46, 47, 47a)*, Lehn *et al.* *(61, 62)*, Pramauro and Pelizzetti *(83)*, Calvin *(16)*, Harriman and Mills *(35)*, Mackor *et al.* *(68)*, Shafirovich *et al.* *(93, 95)*, and Kiwi *(45)* have also found a group of heterogeneous catalysts for this step, for example, oxides of ruthenium, platinum, iridium, cobalt, and manganese.

We have studied the two types of compounds that are homogeneous catalysts for reaction (2): (i) partially hydrolyzed or aqueous complexes of various transition metals with nitrogen- or oxygen-containing ligands *(22, 23, 25;* see also *44, 44a)* and (ii) water-soluble phthalocyanins and porphyrins of various metals *(24)*. The catalysts of the first type were found to be more efficient; iron, cobalt, and copper complexes demonstrated the highest activity (exceeding one molecule of O_2 per one metal ion per min at 20°C), but they were less stable than catalysts of the second type. As shown by Khannanov and Shafirovich *(44, 44a)*, Shafirovich *et al.* *(93, 95)*, and Harriman *et al.* *(36)*, catalysts of the first type are catalytically active not only in dark evolution of oxygen from water, but also in model photocatalytic systems in which oxygen evolution is associated with an irreversible reduction of some electron acceptors (sacrifice).

The study of the mechanism of reaction (2) in the presence of homogeneous catalysts is still in progress; however, some important details have been clarified. There are some difficulties that one faces when studying reaction (2). When the previously mentioned complexes with bipyridine and phenantroline are used as oxidants D^+, this reaction is sensitive to the state of the walls of the reaction vessel. For example, in glass vessels oxygen can sometimes evolve via reaction (2) without special catalysts, with the amount evolved being dependent on both the glass composition and the roughness of its surface. However, vessels made from polyethylene were found to be inert and thus suitable for our experiments.

Another difficulty that one faces when studying the mechanism of catalytic reaction (2) is the absence of sufficiently rapid as well as convenient methods to record the kinetics of oxygen evolution. In fact, under typical experimental conditions the characteristic length of time for the rapid stages of reaction (2) in the presence of catalysts (see following discussion) usually last a few seconds. This is shorter than the characteristic time of 20–30 s for the most rapid available methods of detecting oxygen in solution, that is, via the use of the Clark electrode. As a result, it is difficult to obtain the most informative data on the kinetics of oxygen evolution and thus on the catalytic activity per se. In this situation we characterize the quality of a catalyst by the ratio of the amount of evolved oxygen to its amount expected from the stoichiometry of reaction (2). The value of this ratio, β (expressed as a percentage), will be called the *catalyst efficiency*.

It is possible to achieve a catalyst efficiency of about 100% only in exceptional cases; usually, the observed efficiency is notably smaller. This must be accounted for by the partial consumption of the oxidant D^+ in fast oxidation side reactions of the organic ligands in both the catalyst and the oxidant. In the absence of catalysts, the reduction of oxidants occurs only because of these side reactions. Note that the side reaction in the absence of catalysts leads to substantial reduction of D^+ (~40% in acidic and ~90–95% in alkali media; see following discussion) within a period shorter than several seconds. No oxygen is evolved in this situation. Thus the role of the catalysts in providing reaction (2) is to direct the process of D^+ reduction away from ligand oxidation and toward oxygen evolution.

In the presence of catalysts after addition of alkali to acidic (pH 1–2) solutions containing initial reagents (oxidant and catalyst) up to the desired pH value (which was the procedure used to start the reaction of oxygen evolution in our experiments) (*22–24*), only a part of the oxidant is rapidly consumed because of the fast (several seconds) oxygen evolution and ligand oxidation. The residual amount of oxidant D^+ is reduced more slowly (tens of minutes). For some, but not all, catalysts this slow process of oxidant reduction can also be followed by oxygen evolution from water via reaction (2).

The fraction of an oxidant consumed in the rapid step of the process largely depends on the pH of the solution. For example, after addition of alkali in amounts providing an initial pH of 3, typically about half of the amount of oxidant is rapidly reduced, whereas at an initial pH of 9 about 80–90% of the oxidant is usually reduced.

Note that at pH 3 and at the optimum concentration of a catalyst (see following discussion), the efficiencies β when calculated per the amount of the oxidant reduced in rapid reactions have values close to most of the homogeneous catalysts studied (Table III). Some of the homogeneous catalysts have a complicated behavior in solutions having various pH values. For example, at pH 3 all cobalt complexes studied show approximately equal efficiency in catalysis (see Table III), but at pH 9–10 their efficiency changes in the series:

$$\begin{array}{llllll} \text{Catalyst:} & Co(NH_3)_6^{3+} < & Co(NH_3)_5Cl^{2+} < & Co(NH_3)_5(H_2O)^{3+} < & Co_{aq}^{2+} < & \\ \beta\ (\%): & 10 & 39 & 42 & 60 & \\ & & & & Co(NH_3)_4(H_2O)_2^{3+} < & Co(bipy)_{3-x}(H_2O)_x^{3+} \\ & & & & 66 & 70 \end{array}$$

This series seems to reflect the role of the coordination of water molecules or hydroxyl ions to the catalysts during oxygen evolution under these conditions.

TABLE III

Oxidant Consumption Q in a Rapid Stage and the Efficiencies β of Various Homogeneous Catalysts[a]

Catalyst	Catalyst concentration (10^{-4} M)	$Q \pm 5$ (%)	$\beta \pm 5$ (%)
No catalyst	0	40	No O_2
$CoCl_2$	0.5	60	85
	2.2	55	75
	4.3	30	65
	6.4	20	30
cis- and *trans*-$[Co(en)_2Cl_2]Cl$, *trans*-$[Co(en)_2Cl(H_2O)]Cl_2$	0.5	60	80
	2.2	60	75
	8.3	60	75
	20.0	45	70
$[Co(NH_3)_5Cl]Cl_2$	5.0	50	80
	10.0	50	80
$[Co(NH_3)_5(H_2O)]Cl_3$	0.9	30	45
	8.3	55	85
	30.0	55	85
$[Co(NH_3)_4(H_2O)_2]Cl_3$	2.0	50	70
	20.0	40	70
Tetrasulfophthalocyanin of Co(II)	0.07	55	70
	0.22	60	60
	0.64	65	40
$[(bipy)_2FeOFe(bipy)_2]Cl_4$	0.25	60	85
	1.0	60	85
	10.0	50	90

[a] In a 8×10^{-4} M solution of $Ru(bipy)_3^{3+}$ with initial pH 2 after addition of alkali to provide final pH 3.0–3.5; 25°C.

The complicated character of reaction (2) in the presence of our catalysts and trisbipyridyl and trisphenantroline complexes as oxidants is evidenced by the notable effect on the value of β by the pH, the duration of the period between the moment of preparation of the catalyst solution and the start of the reaction, as well as by the procedure of mixing the reagents. For example, the values of β differ if a catalyst is introduced into the acid solution of an oxidant and an alkali solution is subsequently added to this reaction mixture or if a catalyst is introduced into an alkali solution, which is subsequently mixed with an acid solution of an oxidant. These data can be understood if one assumes an essential role for the hydrolysis processes in activation of the catalysts. The rate and direction of hydrolysis processes can depend significantly on the pH of the media.

The role of hydrolysis in conversion of initial complexes to the species active in catalysis has been demonstrated for the systems with the complexes of the type $Fe_m(bipy)_n(OH)_p{}^{(3m-p)+}$ or $Fe_m(phen)_n(OH)_p{}^{(3m-p)+}$ used as catalysts at pH > 4. The bipyridyl and phenatroline complexes of Fe(III) containing OH^- ions as ligands were found to be more efficient in providing catalysis than the same complexes without OH^- ligands. Comparison of the data on the efficiency of such complexes in catalysis with those on the mechanism of their hydrolysis suggests that binuclear di-μ-hydroxo (**1**) and/or di-μ-oxo (**2**) structures can perhaps play the role of

$$\mathrm{M}\langle{}^{\mathrm{OH}}_{\mathrm{OH}}\rangle\mathrm{M} \quad (\mathbf{1}) \qquad \mathrm{M} = \text{metal ion} \qquad \mathrm{M}\langle{}^{\mathrm{O}}_{\mathrm{O}}\rangle\mathrm{M} \quad (\mathbf{2})$$

catalytically active intermediates in the oxygen evolution reaction. Oxygen evolution seems to begin with a set of four consecutive one-electron reactions of catalyst oxidation by particles D^+, each M(OH) fragment of the binuclear catalyst being able to donate two electrons or less. At least for some of the catalysts studied, the subsequent reaction of oxygen evolution and regeneration of the catalyst active forms do not proceed via formation of free (uncoordinated) forms of intermediate products of water oxidation, such as $\dot{O}H$ radicals, H_2O_2 molecules, and $H\dot{O}_2$ radicals (*22–24*). For example, for the catalytic action of partially hydrolyzed bipyridyl complexes of Fe(III), the following two alternative schemes can be suggested:

$$\mathrm{M}\langle{}^{\mathrm{OH}}_{\mathrm{OH}}\rangle\mathrm{M} + 2D^+ \longrightarrow \mathrm{M}\langle{}^{\mathrm{O}}_{\mathrm{O}}\rangle\mathrm{M} + 2H^+ + 2D$$

$$\mathrm{M}\langle{}^{\mathrm{O}}_{\mathrm{O}}\rangle\mathrm{M} + 2D^+ \xrightarrow{2H_2O} \mathrm{M}\langle{}^{\mathrm{OH}}_{\mathrm{OH}}\rangle\mathrm{M} + 2H^+ + 2D + O_2\uparrow \qquad (6)$$

or

$$\mathrm{M{<}^{OH}_{OH}{>}M} + 2\mathrm{D}^+ \xrightarrow{2\mathrm{H_2O}} \mathrm{M_2(H_2O)_2}\ [\text{or } 2\mathrm{M(H_2O)}] + 2\mathrm{H}^+ + 2\mathrm{D} + \mathrm{O_2}\uparrow$$

$$\mathrm{M_2(H_2O)_2}\ [\text{or } 2\mathrm{M(H_2O)}] + 2\mathrm{D}^+ \longrightarrow \mathrm{M{<}^{OH}_{OH}{>}M} + 2\mathrm{H}^+ + 2\mathrm{D} \tag{7}$$

For catalysts with M = Co(III) or Fe(III), oxygen evolution from the oxidized center of type **1** can proceed via formation and dissociation of the binuclear μ-peroxo complex

$$\mathrm{M{-}O_2^{2-}{-}M} \rightleftharpoons 2\mathrm{M}^- + \mathrm{O_2}$$

or the mononuclear peroxo complex

$$\mathrm{M{-}O_2^-} \rightleftharpoons \mathrm{M^-{-}O_2} \rightleftharpoons \mathrm{M}^- + \mathrm{O_2}$$

as intermediates. In fact, such processes of reversible oxygenation of Co(II) and Fe(II) complexes in alkaline solutions are well known (*4, 17, 102, 103*).

Among the catalysts of the second group, the tetrasulfophthalocyanins and tetrasulfophenylporphine of Co(II) demonstrated the highest efficiency ($\beta \approx 70\%$) (*24*). When $\mathrm{Ru(bipy)_3^{3+}}$ is used as the oxidant D^+, the efficiency of tetrasulfophthalocyanins of various metals at pH 9 decreases in the series

Metal:	Co(II) >	Fe(II) >	Zn(II) >	Cu(II) >	Ni(II) >	Mn(II) >	Cr(II)
β (%)	65	47	25	21	18	14	5

Phthalocyanins without a metal ion have almost zero activity. The mechanism of action of the second-group catalysts is not clear at present. However, the catalytic activity of metal phthalocyanins does not seem to result from their dissociation with the formation of aqueous complexes of metal ions. Indeed, at about 10^{-5} *M* concentration the aqueous ions of Cr(III), Mn(II), Ni(II), and Al(III) are inactive in water oxidation by $\mathrm{Ru(bipy)_3^{3+}}$, in contrast to phthalocyanins with the same metal ions, which do show an appreciable catalytic activity under the same conditions (*24*). In sufficiently acidic solutions when reaction (2) does not occur, we have observed very rapid and quantitative two-electron oxidation of some metallophthalocyanins accompanied by one-electron reduction of $\mathrm{Ru(bipy)_3^{3+}}$. In strongly alkaline solutions these twice-oxidized forms of metallophthalocyanins will most likely be intermediate products of oxygen evolution.

An interesting feature of the studied catalysts, which also reflects the complexity of the catalytic process (2), is the existence in the region pH > 4 of sharp maxima for the catalyst efficiency as a function of its concentration. For example, at pH 9 the optimum concentration (C) is $\sim 10^{-4}$ M for the catalysts of the first group and $\sim 10^{-5}$ M for the catalysts of the second group. The notable influence of the catalyst concentration on its efficiency is also observed at pH 3 (see Table III); however, the dependence of β on C in this case usually demonstrates no sharp maxima.

The role in reaction (2) of the oxidant D^+ used in our studies may be more complicated than that envisaged by the hypothetical schemes (6)–(7), in which they simply serve as acceptors of electrons from catalyst particles. Indeed, we have observed that addition to the reaction solution of various redox-inactive anions (chlorides, chlorates, sulfates, carbonates, acetates, borates, etc.) in the form of sodium salts suppresses both the process of oxygen evolution in reaction (2) and the side process of D^+ reduction due to oxidation of organic ligands. However, the efficiency β calculated per the amount of the reduced oxidant D^+ does not change significantly in the presence of these anions. This suggests that reaction (2) and the side reaction have common steps that are sensitive to the interaction of the anions with the oxidant D^+. In the search for such interactions we have studied the pmr spectra of D^+. It was found that the pmr spectra of $Ru(bipy)_3^{3+}$ show a rather strong chemical shift of 6- and 6′-protons of bipyridyl upon addition of various anions (Fig. 8), which seems to be induced by the outer-sphere coordination of anions with the bipyridyl ligands. It can be anticipated that such coordination of anions creates obstacles for an interaction of the ligands of the oxidant with water molecules or H^+ and OH^- and that such coordination may play some yet unknown role in both reaction (2) and the side process of the ligand oxidation. Study of this phenomenon is underway.

In addition, the inner-sphere coordination of water to Ru(III) complex cations via the substitution of coordinating nitrogen atoms of the ligands seems not to be an intermediate step of reaction (2). For example, we have found that mixed-ligand complexes $[Ru(bipy)_2(py)_2]^{3+}$ (py-pyridine), for which the ligand-substitution processes are expected to have rates different from those for $Ru(bipy)_3^{3+}$ when used as D^+, provide oxygen evolution from water in the presence of our homogeneous catalysts with the same efficiency as does $Ru(bipy)_3^{3+}$ (*25*). Moreover, the $[Ru(bipy)_2$-$(py)(H_2O)]^{3+}$ complex that contains water in the first coordination sphere is unable to evolve O_2 from water because of its insufficient oxidation potential.

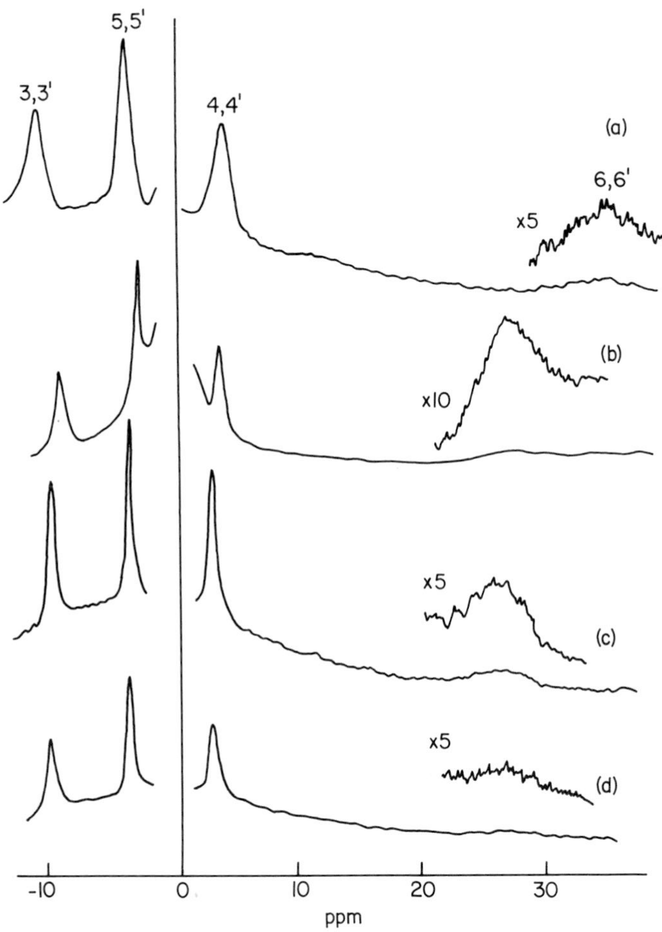

Fig. 8. The ft–pmr (300 MHz) spectra of 10^{-3} *M* solution of $Ru(bipy)_3^{3+}$ in D_2O containing also: (a) 0.01 *M* DCl, (b) 1 *M* DCl, (c) 1 *M* NaCl, and (d) 1 *M* Na_2SO_4. The temperature is 20°C; 200 scans. The chemical shifts are measured versus the signals of water protons. The peaks are numbered according to the positions of protons in 2,2′-bipyridine:

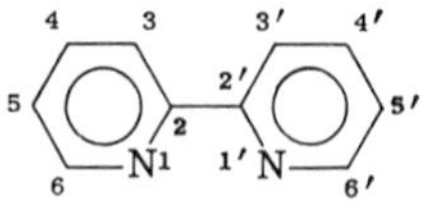

One may hope that further investigations of the homogeneous catalysis of reaction (2) with the use of $Ru(bipy)_3^{3+}$ as an oxidant will result in development of catalysts efficient and stable enough for oxygen evolution that will be able to work in a closed cycle of water cleavage involving a popular $Ru(bipy)_3^{2+}$ complex as a photocatalyst.

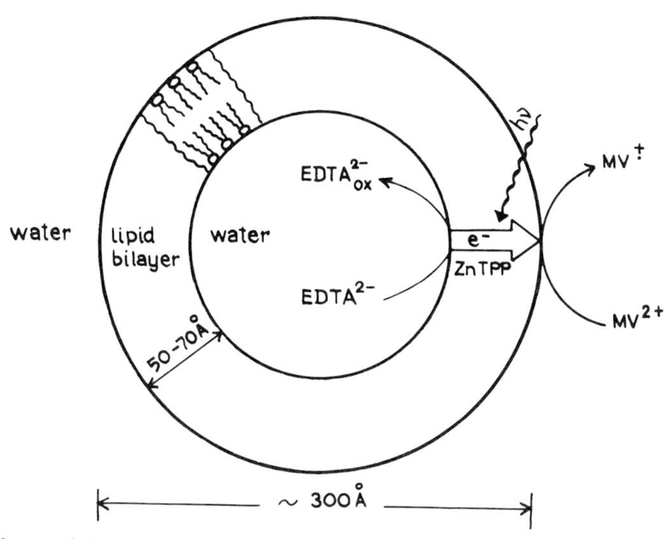

Fig. 9. Scheme of electron phototransfer through the membrane of lipid vesicles containing ZnTPP as a photocatalyst.

IV. Photochemical Charge Separation

The absence of efficient ways to carry out reaction (1) is at present the main obstacle that hinders the development of photocatalytic systems that could accomplish the entire closed cycle of water cleavage (i.e., without irreversible consumption of something other than water reagents). The main problem is how to suppress the back recombination reaction $D^+ + A^- \rightarrow D + A$, which is a simple exothermic bimolecular process and therefore typically proceeds much more rapidly than complex catalytic reactions (2) and (3).

We are now studying several ways to suppress the back process in our laboratory, two of which are discussed here. The first method is based on the use of the lipid vesicles, and the second on the use of microemulsions.

A vesicle, the structure of which is shown schematically in Fig. 9, consists of a shell formed by a lipid bilayer membrane. To prepare the vesicles we used natural egg lecithin or synthetic dipalmitoylphosphatidylcholine. The inner volume of the vesicles contains an aqueous solution of electron donors, namely, $EDTA^{2-}$ or NADH, and no electron acceptor, whereas the outer volume contains only an electron acceptor (MV^{2+}) and no electron donor. As was shown in pioneering work by Ford *et al.* (*28, 29*), Kurihara *et al.* (*58, 58a*), Matsuo *et al.* (*70, 70a*), Calvin (*16*), and

Laane *et al.* (*59*), under illumination of a solution of vesicles containing photocatalysts in their membranes, the electrons can be transferred through the membrane, thus leading to the formation of D^+ and A^- particles that are membrane separated and therefore do not recombine with each other. Compounds with long hydrophobic groups such as chlorophyll (*58, 58a*) and surface-active derivatives of zinc porphyrin (*70, 70a*) or of ruthenium trisbipyridyl complex (*16, 28, 29*) have been used as photocatalysts in these works. In our experiments we used a simpler compound having no such groups, namely, zinc tetraphenylporphine (ZnTPP).

Under illumination of the vesicle solution both in the Soret band of ZnTPP ($\lambda \approx 425$ nm) and in longwave absorption bands of ZnTPP ($\lambda > 500$ nm), we observed the phototransfer of electrons from donor particles inside the vesicles to acceptor particles outside the vesicles (see data in Fig. 10). In this case metalloporphyrin participates repeatedly in the cycle of MV^{2+} reduction by the electron donors, that is, it plays the role of a photocatalyst (*65*). The quantum yield of MV^{2+} reduction in the system with the natural lecithin ranges from 10^{-1} to $10^{-2}\%$.

That electron transfer takes place even upon irradiation in longwave absorption bands indicates the important role of lower ZnTPP first-triplet excited states in electron transfer.

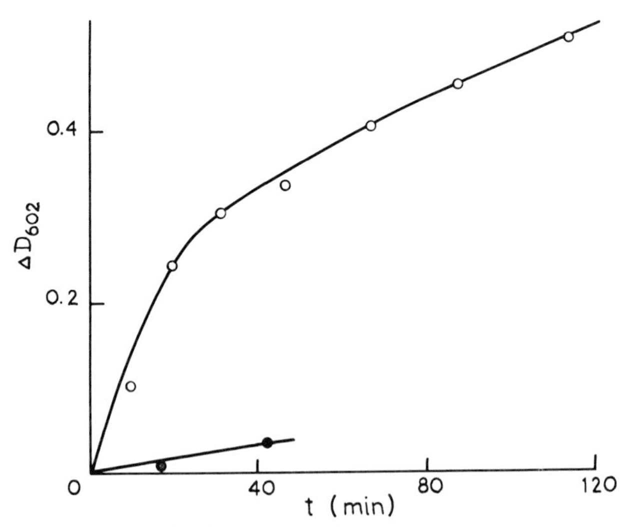

Fig. 10. Change in the optical density at 602 nm due to accumulation of $MV^{\ddagger}$ during illumination by light with $\lambda > 500$ nm from a 1-kW mercury lamp of the solution of vesicles containing ZnTPP: ●, in the absence of an electron donor; ○, in the presence of 0.2 *M* of Na_2EDTA inside vesicles. The concentrations of lecithin, zinc tetraphenylporphine, and MV^{2+} are 2.4×10^{-3} gm/cm³, 6×10^{-5} *M*, and 2×10^{-2} *M*, respectively; 25°C.

Pulse photolysis of the vesicle solution containing only ZnTPP in the intermediate optical spectrum gave absorption bands attributed to the triplet electronically excited state of the porphyrin ($ZnTPP^T$) and to the radical-cation of the porphyrin ($ZnTPP^{\dot{+}}$). A very fast deactivation of the triplets occurred, probably because of their annihilation:

$$2ZnTPP^T \longrightarrow 2ZnTPP + \Delta E$$

A significantly higher rate of this process in the solution of vesicles (as compared to the homogeneous solution containing the same total amount of porphyrin) is caused by a strong local concentration of porphyrin inside the lipid bilayer (as much as $\sim 2.5 \times 10^{-2}$ *M*). The observed radical-cation $ZnTPP^{\dot{+}}$ seems to result also from the triplet annihilation:

$$2ZnTPP^T \longrightarrow ZnTPP^{\dot{+}} + ZnTPP^{\dot{-}} \quad (8)$$

Unfortunately, we could not observe $ZnTPP^{\dot{-}}$ because of the strong absorption of triplet states in the region of characteristic absorption bands of $ZnTPP^{\dot{-}}$.

In the presence of MV^{2+} outside the vesicles, the intense long-lived absorption ascribed to the radical-cations $MV^{\dot{+}}$ and $ZnTPP^{\dot{+}}$ is observed immediately after the flash. The decay of the products of flash photolysis, which is caused by their recombination, is described by the second-order kinetic equation with the rate constant $k = (1.9 \pm 0.2) \times 10^8/M/s$. The concentration of $MV^{\dot{+}}$ and $ZnTPP^{\dot{+}}$ observed immediately after a flash of $\sim 10^{-5}$-s duration comprises about a one-fourth of the ZnTPP concentration in the sample, which is notably higher than the initial concentration of $ZnTPP^{\dot{+}}$ after the same flash in the absence of MV^{2+}. Appreciable amounts of produced MV^{2+} and its decay via a second-order reaction, that is, via recombination with other than geminate radical-cations $ZnTPP^{\dot{+}}$, points to a high efficiency for both the formation of $MV^{\dot{+}}$ and its escape from the reaction cage after the act of the electron phototransfer.

The presence of $EDTA^{2-}$ in the form of sodium salt inside the vesicles does not affect the experimental results of pulse photolysis of the vesicles containing ZnTPP both in the presence and absence of MV^{2+} in the outside solution.

Two alternative mechanisms for electron phototransfer through membranes of lipid vesicles, namely, one- (*16, 28, 29, 58, 58a*) and two-quantum (*70, 70a*) mechanisms, have been discussed in the literature. For both the natural and synthetic lipids at temperatures 14–60°C we observed the quadratic dependence of the initial accumulation rate of $MV^{\dot{+}}$ on the intensity of the light (Fig. 11). This evidence favors the two-quantum mechanism for the phototransfer assisted by ZnTPP over the whole temperature range studied. The quadratic dependence is observed for both the photolysis in the Soret band and in the longwave absorption band of ZnTPP.

Two schemes for the electron phototransfer are proposed that simultaneously satisfy the data on the stationary and pulse photolysis:

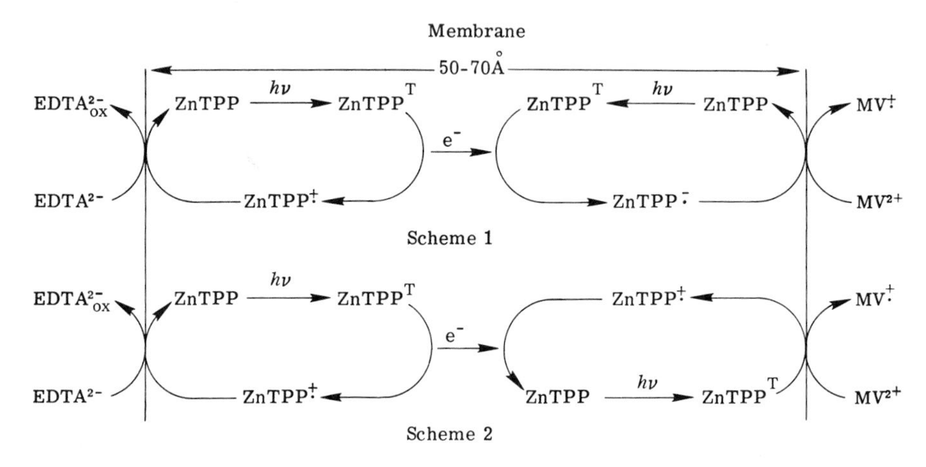

($EDTA_{ox}^{2-}$ denotes products of irreversible oxidation of $EDTA^{2-}$.) Scheme 1 involves annihilation of two $ZnTPP^T$s, producing $ZnTPP^{\dagger}$ and $ZnTPP^{\overline{\cdot}}$ and subsequent interaction of these radical-ions with $EDTA^{2-}$

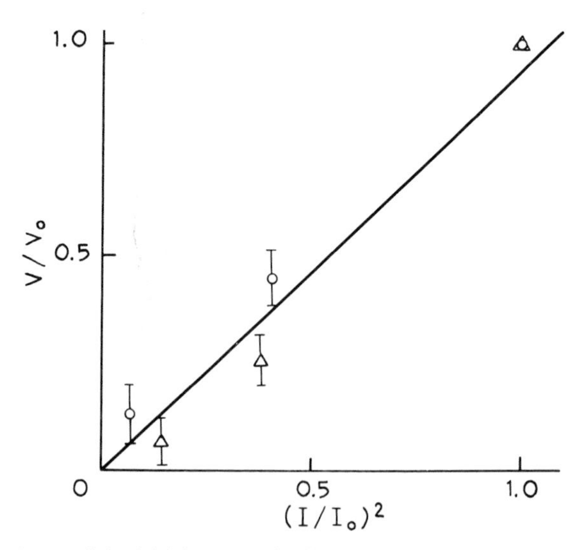

Fig. 11. Dependence of the initial rate V of $MV^{\dagger}$ accumulation in the solution of the lipid vesicles on the light intensity I at irradiation in Soret (○) and longwave (△) absorption bands of ZnTPP. The system is the same as in Fig. 10.

and MV^{2+} as the processes responsible for the electron transfer through the membrane. In Scheme 2 such processes are the reaction of $ZnTPP^T$ with MV^{2+} to give $MV^{\dagger}$ and $ZnTPP^{\bar{\cdot}}$ located near the outer surface of the vesicle. This is followed by further transfer to this $ZnTPP^{\dagger}$ of an electron from the second $ZnTPP^T$ particle located on the internal surface or in the volume of the vesicle membrane, and finally by the reaction of the newly formed particle $ZnTPP^{\dagger}$ with $EDTA^{2-}$. In both cases the quadratic dependence of the MV^{2+}-accumulation rate on the light intensity can be observed. As follows from the data on redox potentials (*91*) of particles involved in the electron transfer, both schemes are thermodynamically allowed.

As shown previously, the presence of MV^{2+} outside the vesicles leads to an appreciable increase of the initial concentration of $ZnTPP^{\dagger}$ observed at pulse photolysis. This means that the electron transfer from $ZnTPP^T$ to MV^{2+} at the outer vesicle surface is, under these conditions, a more efficient route of $ZnTPP^T$ decay than reaction (8). Therefore, Scheme 2 seems to be more probable. The diffusion of an electrically charged intermediate particle of photolysis, $ZnTPP^{\dagger}$, across a nonpolar lipid membrane does not appear to be a very probable process. In this situation electron transfer through the membrane can be provided, for example, by electron exchange between $ZnTPP^T$ and $ZnTPP^{\dagger}$ via the tunnel mechanism. In this context it is of interest that below the melting point of the lipid the rate of the electron transfer through the membrane remains above zero (Fig. 12) (*67*). A similar phenomenon, that is, the occurrence

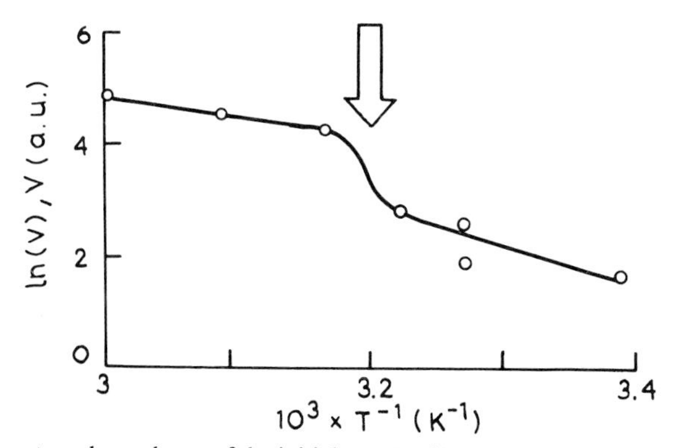

Fig. 12. Temperature dependence of the initial rate V of $MV^{\dagger}$ accumulation in the solution of dipalmitoylphosphatidylcholine vesicles. The concentrations of lipid, ZnTPP, and MV^{2+} are 1.5×10^{-3} gm/cm³, 3.7×10^{-5} M, and 2×10^{-2} M, respectively; the concentration of Na_2-EDTA inside the inner volume of vesicles is 0.2 M. Illumination is carried out at the longwave absorption band of ZnTPP. The arrow shows the temperature of the phase transition of the lipid (~41°C).

of electron transfer at a notable rate at temperatures below the freezing point of the system, has been observed for reactions of recombination of light-separated charges in the photosystems of plants and bacteria and has been accounted for by the tunnel mechanism of electron transfer [see, e.g., the pioneering work of De Vault and Chance (*19*) and also the paper by Ke *et al.* (*41*)].

Note a clear analogy between the mechanisms of charge photoseparation via scheme 2 and via the generally adopted Z scheme of plant photosynthesis. This permits one to consider vesicles containing metalloporphyrin as analogs of biological systems for charge separation. Thus, in analogy with biological systems, there is the possibility that the quantum yield of this process can actually be increased by the appropriate choice of a photocatalyst, an electron donor, an electron acceptor, and a lipid.

The process of MV^{2+} photoreduction with electron donors of $EDTA^{2-}$ or NADH leads to energy storage, for example, 0.12 eV of energy being stored per each act of phototransfer in the case of NADH. For illumination in the longwave absorption bands of ZnTPP this corresponds to storage of about 3% of the total energy of two light photons used for electron phototransfer.

Thus, by introducing a rather simple photocatalyst (zinc tetraphenylporphine) into the lipid membrane, it is possible to accomplish electron transfer through the membrane with the help of visible light, although as yet with a small quantum yield. The reduced form of the electron acceptor used, namely, $MV^{\dagger}$, is known to be capable of evolving hydrogen from water even at neutral pH when assisted by a catalyst. We have succeeded in observing this process upon introduction of the hydrogenase from *Tiocapsa roseopersicina* ($\sim 10^{-6}$ *M*) into the outer solution of the vesicles. Despite the fact that the oxidation potential of oxidized forms of the final electron donor is insufficient to evolve oxygen from water, the oxidation potential of the intermediate product of the process, $ZnTPP^{\dagger}$, [+1.0 V versus NHE (*91*)] is enough for this purpose, provided the inner solution has pH $\gtrsim 5$. Thus, in principle, it seems possible to accomplish the complete cycle of water cleavage by way of selecting appropriate electron donors in the system based on ZnTPP and similar metalloporphyrins.

The recombination of reaction (1) products can also be slowed down in water-in-oil-type microemulsions, the structure of which is schematically represented in Fig. 13. We used the microemulsion having the composition *p*-xylene (60.8 wt%), *n*-pentanol (22.0 wt%), sodium dodecylsulfate (3.6 wt%), and water (13.6 wt%), which is a transparent solution consisting of water droplets of about 100 Å in diameter in *p*-xylene (*27*). These droplets are surrounded by a layer of sodium dodecylsulfate (SDC) and *n*-pentanol forming the approximately 20-Å-thick interphase layer.

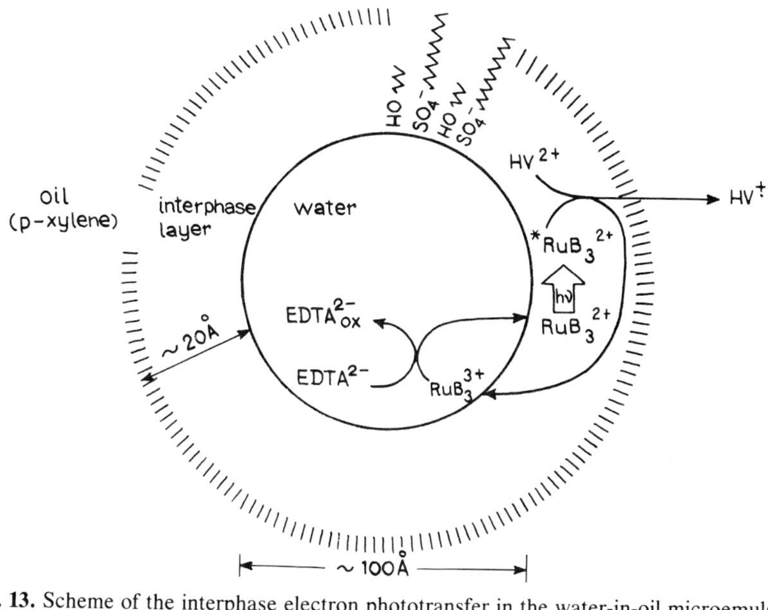

Fig. 13. Scheme of the interphase electron phototransfer in the water-in-oil microemulsion containing HV^{2+}, $Ru(bipy)_3^{2+}$, and $EDTA^{2-}$.

The photocatalyst $Ru(bipy)_3^{2+}$ or its surface-active analog $Ru(C_{17}\text{-bipy})(bipy)_2^{2+}$, which differs from the former by the presence of one cetyl substituent in the fourth position in one bipyridyl ligand (*66*), was introduced into the microemulsion. Comparison of the rates of Ru(II)-luminescence quenching in the homogeneous water solution and in microemulsions by compounds easily soluble in water and poorly soluble in *p*-xylene (Table IV) suggests that ruthenium complexes are mainly located in the interphase layer. The considerable decrease in the quenching rate constant that occurs when passing from homogeneous solution to microemulsion indicates the localization of photocatalysts outside the water phase. However, they cannot be located mainly in the organic phase because they still have such high rate constants of quenching ($k_q \approx 10^7/M/s$) by compounds that are located in water phase.

Illumination with visible ($\lambda > 450$ nm) light of microemulsions, which also contain the electron acceptor heptylviologen HV^{2+} (3×10^{-4} *M*) and the electron donor Na_2EDTA (6×10^{-3} *M*), located primarily in the water phase, along with the photocatalyst $Ru(bipy)_3^{2+}$ (6×10^{-5} *M*), leads to the accumulation of radical-cations $HV^{\dot{+}}$ with a quantum yield of about 1%. During the photolysis of the homogeneous water solution containing $Ru(bipy)_3^{2+}$, HV^{2+}, and Na_2EDTA, a violet precipitate of water-insoluble $HV^{\dot{+}}$ is observed, which can be readily and quantitatively extracted by a

TABLE IV

Rate Constant k_q for the Quenching of the Ru(II) Complex Luminescence in a Homogeneous Water Solution and in a Microemulsion at 25°C

	Quencher (A)	[A][a] (10^{-3} *M*)	k_q[b] ($10^8/M/s$)
$Ru(bipy)_3^{2+}$ in water	MV^{2+}	5	4.7
	HV^{2+}	5	8.0
$Ru(bipy)_3^{2+}$ in the microemulsion	MV^{2+}	4	0.8
		25	0.8
	HV^{2+}	4	0.6
		25	0.5
$Ru(C_{17}\text{-bipy})(bipy)_2^{2+}$ in the microemulsion	MV^{2+}	4	0.6
		25	0.5
	HV^{2+}	4	0.6
		25	0.9

[a] For the local concentrations of the water-soluble MV^{2+} and HV^{2+} given for the microemulsion, it is assumed that they are located only in the water phase.
[b] The effective bimolecular rate constants of quenching at quencher concentrations are given.

mixture of *p*-xylene and *n*-pentanol in the same ratio as in microemulsion. Thus one can propose that $HV^{\ddagger}$ produced by photolysis of microemulsion is also located mainly inside the organic phase. The electron donors ($EDTA^{2-}$ anions) do not seem to penetrate either into the anion-saturated interphase layer or into the nonpolar organic phase, that is, they seem to be located only in the water phase of the microemulsion. Because HV^{2+} can be almost completely reduced to $HV^{\ddagger}$ during photolysis of the microemulsion, we conclude that $Ru(bipy)_3^{2+}$, which is present in an amount five times less than that of HV^{2+} and which is located in the interphase layer, can indeed serve as a photocatalyst for electron transfer from the donor $EDTA^{2-}$ located in the water phase to the acceptor HV^{2+}, the reduced form of which is extracted into the organic phase.

At pulse photolysis of the homogeneous water solution containing only $Ru(bipy)_3^{2+}$ and HV^{2+}, the radical-cation $HV^{\ddagger}$ is produced, its concentration gradually falling because of the recombination with the oxidant $Ru(bipy)_3^{2+}$ formed simultaneously with $HV^{\ddagger}$. The decay of $HV^{\ddagger}$ is described by second-order kinetics, with the rate constant $k = (1.5 \pm 0.1) \times 10^9/M/s$ up to the 99% conversion. In contrast to this, the decay of the optical absorption of $HV^{\ddagger}$ at pulse photolysis of the microemulsion in the

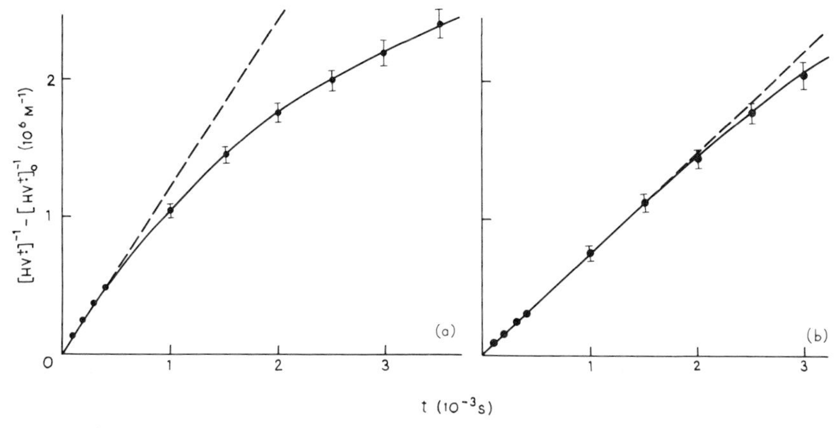

Fig. 14. Kinetics of $HV^{\ddagger}$ recombination with (a) $Ru(bipy)_3^{3+}$ and (b) $Ru(C_{17}\text{-bipy})(bipy)_2^{3+}$ after flash photolysis of the microemulsion containing 3.1×10^{-3} *M* HV^{2+} and 2.5×10^{-4} *M* of the ruthenium complexes; 25°C. The dashed lines correspond to the second-order kinetics with the rate constants obtained from the initial part of the kinetic curves.

same conditions has a more complicated character (Fig. 14a). During short times the decay of $HV^{\ddagger}$ follows second-order kinetics with $k = (1.2 \pm 0.1) \times 10^9/M/\text{s}$, whereas starting from $\sim 10^{-3}$ s the process is slowed down.

This behavior of the reaction kinetics can be explained as follows (for more details see *64*). Assume that one of the products formed in the interphase layer in the primary electron-transfer process, namely $HV^{\ddagger}$, is very rapidly distributed (within the time of duration of the light flash), according to its solubility, between the interphase layer and the organic phase, whereas the second product, $Ru(bipy)_3^{3+}$, moves more slowly into the water phase to oxidize $EDTA^{2-}$ (if $EDTA^{2-}$ is introduced into the system). Then the initial part of the recombination kinetics described by the second-order kinetics corresponds to the recombination of $HV^{\ddagger}$ with $Ru(bipy)_3^{3+}$ that has not escaped into the water phase. The observed slowdown of recombination during long times reflects the fact that some of the $Ru(bipy)_3^{3+}$ complexes escapes into the water phase, so that their recombination is slowed down.

Interpreting the data in Fig. 14a in this way, one might expect that when substituting $Ru(bipy)_3^{2+}$ with the more lipophilic complex $Ru(C_{17}\text{-bipy})(bipy)_2^{2+}$, the deviation of the recombination kinetics from the simple second-order kinetics would be less pronounced because of additional difficulties created for the oxidized photocatalyst movement from the interphase layer into the water phase. This has been found to actually be the case (see Fig. 14b).

Thus, due to the rapid extraction of $HV^{\dagger}$ into the organic phase and the far slower transition of $Ru(bipy)_3^{3+}$ into the water phase, portions of the primary products of the electron transfer move to different phases, which slows down their recombination. Hence, one may hope that by varying the hydrophilic and oleophilic properties of the primary electron donor and acceptor via introduction of appropriate substituents into their molecules, it might be possible in the future to extract efficiently the products of the primary electron phototransfer into different phases of microemulsions.

For creation of efficient photocatalytic systems, it is necessary to slow down the recombination of D^+ and A^- particles formed in reaction (1) to such an extent that processes (2) and (3) leading to hydrogen and oxygen evolution will become more rapid than reaction of D^+ with A^-. It should be emphasized that the reduced form of HV^{2+}, like MV^{2+}, can provide hydrogen evolution from water in the presence of catalysts described in Section II. The ability of $Ru(bipy)_3^{3+}$ and its analogs to evolve oxygen from water has already been mentioned in Section III.

Evaluating the situation as a whole, we may conclude that the development of the systems providing efficient charge separation seems to be at present the "rate-determining" step in creation of molecular photocatalytic converters of solar energy into chemical energy on the basis of water cleavage. It may be hoped that once such systems become available, efficient decomposition of water by solar light can be achieved under laboratory conditions.

Acknowledgments

We wish to thank our co-workers Drs. E. R. Buyanova, G. L. Elizarova, M. I. Khramov, A. I. Kokorin, N. V. Lozhkina, S. V. Lymar, L. G. Matvienko, V. M. Nekipelov, S. S. Saidkhanov, E. N. Savinov, and I. M. Zvetkov for their contributions to the work the results of which are summarized in this review.

References

1. Antropov, P. Ya., "Toplivno-Energeticheskij Potenzial Zemli." VINITI, Moscow, 1976.
2. Bagdasar'yan, Kh. S., *Khim. Fiz.* **1,** 391 (1982).
3. Bard, A. J., *Science* (*Washington, D.C.*) **207,** 139 (1980).
4. Basolo, F., and Pearson, R. G., "Mechanisms of Inorganic Reactions." Wiley, New York, 1967.

5. Benemann, J. R., Berenson, J. A., Kaplan, N. O., and Kamen, M. D., *Proc. Natl. Acad. Sci. USA* **70,** 2317 (1973).
6. Berezin, I. V., Varfolomeev, S. D., and Zajtsev, S. V., *Dokl. Akad. Nauk SSSR* **229,** 94 (1976).
7. Bockris, J., "Energy: The Solar Hydrogen Alternative." Architectural Press, London. 1975.
8. Bockris, J., *Int. J. Hydrogen Energy* **6,** 223 (1981).
9. Bolton, J. R., *Science (Washington, D.C.)* **202,** 705 (1978).
10. Broda, E., *Int. J. Hydrogen Energy* **3,** 119 (1978).
11. Broda, E., *Int. J. Hydrogen Energy* **5,** 453 (1980).
12. Buyanova, E. R., Matvienko, L. G., Kokorin, A. I., Elizarova, G. L., Parmon, V. N., and Zamaraev, K. I, *React. Kinet. Catal. Lett.* **16,** 309 (1981).
13. Buyanova, E. R., Kokorin, A. I., Shepelin, A. P., Zhdan, P. A., and Parmon, V. N., *J. Mol. Catal.*, in press.
14. Buyanova, E. R., Shepelin, A. P., Zhdan, P. A., Plyasova, L. M., and Parmon, V. N., *in* "Fotocatalitícheskoe Preobrazovanie Solnechnoj Energii" (V. N. Parmon, ed.), p. 52. Institute of Catalysis, Novosibirsk, 1983.
15. Calvin, M., *Int. J. Energy Res.* **1,** 299 (1977).
16. Calvin, M., *Discuss. Faraday Soc.*, No. 70, p. 383 (1981).
17. Caraco, R., Braune-Steinle, D., and Fallab, S., *Coord. Chem. Rev.* **16,** 147 (1975).
18. Van Cheu, N., and Polotebnova, N. A., *Zh. Neorg. Khim.* **18,** 2189 (1973).
19. DeVault, D., and Chance, B., *Biophys. J.* **6,** 825 (1966).
20. Dzhabiev, T. C., and Shilov, A. E., *Zh. Vses. Khim. Ova.*, No. 5, p. 503 (1980).
21. Eddi, J., *in* "The Solar Output and its Variation" (O. R. White, ed.), p. 33. Colorádo Univ. Press, Boulder, 1977.
22. Elizarova, G. L., Matvienko, L. G., Parmon, V. N., and Zamaraev, K. I., *Dokl. Akad. Nauk SSSR* **249,** 863 (1979).
23. Elizarova, G. L., Matvienko, L. G., Lozhkina, N. B., Parmon, V. N., and Zamaraev, K. I., *React. Kinet. Catal. Lett.* **16,** 191 (1981).
24. Elizarova, G. L., Matvienko, L. G., Maizlish, V. E., Lozhkina, N. B., and Parmon, V. N., *React. Kinet. Catal. Lett.* **16,** 285 (1981).
25. Elizarova, G. L., Matvienko, L. G., Lozhkina, N. B., and Parmon, V. N., *React. Kinet. Catal. Lett.* **21**(3–4) (1982).
26. "Energy and Climate." (Soviet translation; G. Gruz, ed.). Natl. Acad. Sci., Washington D.C., 1977.
27. Fendler, J. H., *J. Phys. Chem.* **87,** 1000 (1978).
28. Ford, W. E., Otvos, J. M., and Calvin, M., *Nature (London)* **274,** 507 (1978).
29. Ford, W. E., Otvos, J. M., and Calvin, M., *Proc. Natl. Acad. Sci. USA* **76,** 3590 (1979).
30. Fruchart, J. M., Herve, G., Lanney, J. P., and Massart, R., *J. Inorg. Nucl. Chem.* **38,** 1627 (1976).
31. Fujishima, A., and Honda, K., *Nature (London)* **238,** 37 (1972).
32. Grätzel, M., *Ber. Bunsenges. Phys. Chem.* **84,** 981 (1980).
33. Grätzel, M., *Acc. Chem. Res.* **14,** 376 (1981).
33a. Grätzel, M., *Discuss. Faraday Soc.*, No. 70, p. 359 (1981).
34. Gurevich, Y. Y., and Pleskov, Yu. V., *in* "Itogy Nauki i Techniki, Elektrokhimia," Vol. 18, p. 3. Moscow, VINITI, 1982.
35. Harriman, A., and Mills, A., *J. Chem. Soc. Faraday Trans. 2* **77,** 2111 (1981).
36. Harriman, A., Porter, G., and Walters, P., *J. Chem. Soc. Faraday Trans. 2* **77,** 2373 (1981).
37. Henglein, A., and Lilie, J., *J. Phys. Chem.* **85,** 1246 (1981).

37a. Henglein, A., and Lilie, A., *J. Am. Chem. Soc.* **103,** 1059 (1981).
38. Henglein, A., Lindig, B., and Westerhausen, J., *J. Phys. Chem.* **85,** 1627 (1981).
39. Kalyanasundaram, K., and Grätzel, M., *Angew. Chem. Int. Ed. Engl.* **18,** 701 (1979).
40. Kazansky, L. P., *in* "Issledovanie Svoistv i Primenenie Heteropolykislot v Katalize" (K. I. Mateev, ed.), p. 129. Institute of Catalysis, Novosibirsk, 1978.
41. Ke, B., Demeter, S., Zamaraev, K. I., and Khairutdinov, R. F., *Biochim. Biophys. Acta* **545,** 265 (1979).
42. Keller, P., and Moradpour, A., *J. Am. Chem. Soc.* **102,** 7193 (1980).
43. Keller, P., Moradpour, A., Amoyal, E., and Kagan, H. B., *J. Mol. Catal.* **7,** 539 (1980).
43a. Keller, P., Moradpour, A., Amoyal, E., and Kagan, H. B., *Nouv. J. Chim.* **4,** 377 (1980).
44. Khannanov, N. K., and Shafirovich, V. Ya., *Kinet. Katal.* **22,** 248 (1981).
44a. Khannanov, N. K., and Shafirovich, V. Ya., *Dokl. Akad. Nauk SSSR* **260,** 1418 (1981).
45. Kiwi, J., *J. Chem. Soc. Faraday Trans. 2* **78,** 339 (1982).
46. Kiwi, J., and Grätzel, M., *Angew. Chem. Int. Ed. Engl.* **17,** 11 (1978).
47. Kiwi, J., and Grätzel, M., *Chimia* **33,** 289 (1979).
47a. Kiwi, J., and Grätzel, M., *Angew. Chem. Int. Ed. Engl.* **18,** 624 (1979).
48. Kiwi, J., Borgarello, E., Pelizetti, E., Visca, M., and Grätzel, M., *Angew. Chem. Int. Ed. Engl.* **19,** 646 (1980).
49. Kondrat'ev, K. Ya., "Aktinometriya." Gidrometizdat, Leningrad, 1965.
50. Kondrat'eva, E. N., and Gogotov, I. N., "Molekulyarnyj Vodorod v Metabolizme Mikroorganizmov." Nauka, Moscow, 1981.
51. Koryakin, B. V., Dzhabiev, T. S., and Shilov, A. E., *Dokl. Akad. Nauk SSSR* **233,** 620 (1977).
52. Koryakin, B. V., Rodin, V. V., and Dzhabiev, T. S., *Izv. Akad. Nauk,* No. 10, *SSSR Ser. Khim.* No. 12, p. 2788 (1977).
53. Krasna, A. I., *Photochem. Photobiol.* **29,** 267 (1979).
54. Krasna, A. I., *Enzyme Microb. Technol.* **1,** 165 (1979).
55. Krasnovsky, A. A., *in* "Research in Photobiology" (A. Gaslland, ed.), p. 361. Plenum, New York, 1977.
56. Krasnovsky, A. A., Ni, C. V., Nikandrov, V. V., and Brin, G. P., *Plant Physiol.* **66,** 925 (1980).
57. Krasnovsky, A. A., Nikandrov, V. V., Brin, G. P., Gogotov, I. N., and Oschepkov, V. P., *Dokl. Akad. Nauk SSSR* **225,** 711 (1975).
58. Kurihara, K., Sukigara, M., and Toyoshima, Y., *Biochim. Biophys. Acta* **547,** 117 (1979).
58a. Kurihara, K., Sukigara, M., and Toyoshima, Y., *Biochem. Biophys. Res. Commun.* **88,** 320(1979).
59. Laane, C., Ford, W. E., Otvos, J. W., and Calvin, M., *Proc. Natl. Acad. Sci. USA* **78,** 2017 (1981).
60. Lehn, J. M., and Sauvage, J. P., *Nouv. J. Chim.* **1,** 449 (1977).
61. Lehn, J. M., Sauvage, J. P., and Ziessel, R., *Nouv. J. Chim.* **3,** 423 (1979).
62. Lehn, J. M., Sauvage, J. P., and Ziessel, R., *Nouv. J. Chim.* **4,** 355, 623 (1980).
63. Likhtenstein, G. I., "Mnogoyadernye Okislitelno-Vosstanovitelnye Metallofermenty." Nauka, Moscow, 1979.
64. Lymar, S. V., Khramov, M. I., Parmon, V. N., and Zamaraev, K. I., *Khim. Fiz.* **2,** 550 (1983).
65. Lymar, S. V., Zvetkov, I. M., Parmon, V. N., and Zamaraev, K. I., *Khim. Fiz.* **1,** 405 (1982).

66. Lymar, S. V., Khramov, M. I., and Parmon, V. N., *Khim. Fiz.*, in press.
67. Lymar, S. V., Zvetkov, I. M., and Parmon, V. N., *in* "Fotokatalicheskoe Preobrazovanie Solnechnoj Energii" (V. N. Parmon, ed.), p. 100. Institute of Catalysis, Novosibirsk, 1983.
68. Mackor, A., Alberts, A. H., Timmer, K., and Noltes, J. G., *Proc. Int. Conf. Photochem. Conversion Storage Sol. Energy 3rd,* p. 213 (1980).
69. Markiewicz, S., Chan, M. S., Sparks, R. H., Evans, C. A., and Bolton, J. R., *Proc. Int. Conf. Photochem. Conversion Storage Sol. Energy 2nd,* p. E6 (1976).
70. Matsuo, T., Takuma, K., Thutsui, Y., and Nishijima, T., *Chem. Lett.* No. 9, p. 1009 (1980).
70a. Matsuo, T., Takuma, K., Thutsui, Y., and Nishijima, T., *J. Coord. Chem.* **10,** 187 (1980).
71. Maverick, A. W., and Gray, H. B., *Pure Appl. Chem.* **52,** 2339 (1980).
72. Meisel, D., Mulac, W. A., and Matheson, M. S., *J. Phys. Chem.* **85,** 179 (1981).
73. Mićić, O. I., and Nenadović, M. T., *J. Chem. Soc. Faraday Trans. 1* **77,** 919 (1981).
74. Miller, D. S., and McLendon, G., *J. Am. Chem. Soc.* **103,** 6791 (1981).
74a. Miller, D. S., and McLendon, G., *Inorg. Chem.* **20,** 950 (1981).
75. Moradpour, A., *Actual. Chim.*, No. 2, p. 7 (1980).
76. Muradov, N. Z., Bazhutin, Yu. V., Bezuglaya, A. G., Izakovich, E. N., and Rustamov, M. I., *React. Kinet. Catal. Lett.* **18,** 355 (1981).
77. Nenadović, M. I., Mićić, O. I., and Kosanić, M. M., *Radiat. Phys. Chem.* **17,** 159 (1981).
78. Nichiporovich, A. A., *in* "Preobrazovanie Solnechnoi Energii" (N. N. Semenov, ed.), p. 49. Chernogolovka, AN SSSR, 1981.
79. Nozik, A., *Discuss. Faraday Soc.*, No. 70, p. 7 (1981).
80. Okura, I., Kobayashi, M., Kim-Thuan, N., Nakamura, S., and Nakamura, K. I. *J. Mol. Catal.* **8,** 385 (1980).
81. Okura, I., Nakamura, K.-I., and Nakamura, S., *J. Mol. Catal.* **5,** 315 (1979); **6,** 227, 261, 299 (1979).
82. Pope, M. T., and Varga, G. M., *Inorg. Chem.* **5,** 1249 (1966).
83. Pramauro, E., and Pelizzetti, E., *Inorg. Chim. Acta Lett.* **45,** 131 (1980).
84. Rosen, M. M., and Krasna, A. I., *Photochem. Photobiol.* **31,** 259 (1980).
85. Ross, R. T., and Hsiao, T. L., *J. Appl. Phys.* **48,** 4783 (1977).
86. Savinov, E. N., Saidkhanov, S. S., Parmon, V. N., and Zamaraev, K. I., *React. Kinet. Catal. Lett.* **17,** 407 (1981).
87. Savinov, E. N., Saidkhanov, S. S., and Parmon, V. N., *Kinet. Katal.* **24,** 68 (1983).
88. Savinov, E. N., Saidkhanov, S. S., Parmon, V. N., and Zamaraev, K. I., *Dokl. Akad. Nauk SSSR* **269**(1–3) (1983).
89. Schlegel, H. G., and Schneider, K. (eds.), "Hydrogenases: Their Catalytic Activity, Structure and Function." Goltze, Göttingen, 1978.
90. Schneider, A., *Chem. Eng. Prog.* **76,** 13 (1980).
91. Seely, G. R., *Photochem. Photobiol.* **27,** 639 (1978).
92. Semenov, N. N., "Nauka i Obschestvo," 2nd ed. Nauka, Moscow, 1973.
93. Shafirovich, V. Ya., Khannanov, N. K., and Shilov, A. E., *J. Inorg. Biochem* **15,** 113 (1981).
94. Shafirovich, V. Ya., Khannanov, N. K., and Strelets, V. V., *Nouv. J. Chim* **4,** 81 (1980).
95. Shafirovich, V. Ya., Khannanov, N. K., and Strelets, V. V., *Dokl. Akad. Nauk SSSR* **260,** 1197 (1981).

96. So, H., and Pope, M. T., *Inorg. Chem.* **11,** 1441 (1972).
97. Starshinov, Yu. N. (ed.), "Mirovaya Energetika. Prognoz Razvitiya do 2020 Goda." Energiya, Moscow, 1980.
98. Styrikovich, M. A., and Shpilrain, E. E., "Energetika. Problemy i Perspektivy." Energiya, Moscow, 1981.
99. Varfolomeev, S. D., "Konversiya Energii Biokatalyticheskimi Systemami." Moscow Univ. Press, Moscow, 1981.
100. Warman, H. R., *in* "Future Sources of Organic Raw Materials" (L. E. St.-Pierre and G. R. Brown, eds.), p. 1. Pergamon, New York, 1980.
101. World Energy Conference, "World Energy: Looking Ahead to 2020." A.P.C. Sci. & Technology Press, 1978.
102. Yatsimirsky, K. B. (ed.), "Uspekhi Khimii Koordinatsionnykh Soedinenij." Naukova Dumka, Kiev, 1975.
103. Yatsimirsky, K. B. (ed.), "Biologicheskie Aspekty Koordinatsionnoj Khimii." Naukova Dumka, Kiev, 1979.
104. Yu, L., and Wolin, M. J., *J. Bacteriol.* **98,** 51 (1969).
105. Zamaraev, K. I., and Parmon, V. N., *Catal. Rev.* **22,** 283 (1980).
106. Zhdanov, V. P., Parmon, V. N., and Zamaraev, K. I., *Dokl. Akad. Nauk SSSR* **259,** 1385 (1981).

6 The Role of Porphyrins in Natural and Artificial Photosynthesis

Anthony Harriman

Davy Faraday Research Laboratory
The Royal Institution
London, England

I. Introduction

There is considerable interest in industrial, academic, and general audiences in the development of systems capable of collecting and storing solar energy. In principle, this is an extremely important area of research because the total amount of solar energy that falls on the surface of the earth each year far outweighs our present or near-future energy demands, and a practical system that works, even at a modest operating efficiency, will be of genuine significance. However, there are many barriers that

ENERGY RESOURCES THROUGH
PHOTOCHEMISTRY AND CATALYSIS

ISBN 0-12-295720-2

must be overcome before such devices become reality, and we are still at a preliminary stage of development. In fact, to date we have not yet identified all the problems involved in the construction of a practical solar-energy storage device, and many of the problems that have been identified remain unsolved. However, significant advances have been made and the rate of progress is rapid, so we may look forward to the future with some degree of optimism.

One system that has been optimized and perfected to work on a large scale, often under adverse conditions, is the natural photosynthetic process that successfully converts sunlight into a storable form of chemical fuel. The overall process, as displayed by green plants, required some 2.3×10^9 yr of evolution, but the final version is the only really practical solar-energy storage device available to mankind. Thus mankind can learn a great deal from studying the natural system and, it is hoped, produce a limited version of photosynthesis by 1990 or so. Obviously, the natural process is an extremely complex one, and the carbohydrate fuel produced by plants is not an ideal one for our present needs, but even so it should be possible to construct a simple *in vitro* photosynthetic apparatus capable of the production of a useful chemical fuel. This chapter is intended to portray such an idea and evaluate the status of our present models that *mimic* photosynthesis. In particular, it is hoped that we can highlight the areas where further research is urgently required. Similar proposals will be outlined in other chapters of this volume, but here we concentrate on the use of homogeneous solutions as the reaction media. This is the simplest possible system to use, and it has the most applications for large-scale development, but perhaps it is the most difficult to realize in practice, and in the final version it may prove necessary to introduce some degree of order into the structure.

II. Photosynthesis

In real terms, research into photosynthesis began in about 1650 when van Helmont grew a willow tree, which began as a 5-lb seedling planted in 200 lb of soil. After 5 yr, the tree weighed 570 lb and the soil weighed 199 lb. Well before the law of conservation of matter (by Lomonosov in 1748 and Lavoisier in 1760), van Helmont guessed that most of the weight of the tree must have come from the water that had been added to the soil. A century later, Bonnet recorded that leaves submerged in water developed gas bubbles when placed in the sun, and in 1771 Joseph Priestley discovered that the gas evolved by green plants during the daytime was oxygen.

Some time later, Ingenhousz showed that plants need their green parts and light to evolve oxygen and that at night they "spoil" the air. Then in 1782 Senebier reported that plants need carbon dioxide to produce oxygen, and this finding prompted Ingenhousz to suggest that CO_2 was the source of all organic matter in the plant. Finally, in 1804 de Saussure confirmed van Helmont's guess that most of the weight of a plant comes from water (and from CO_2). By that time, all the essential ingredients of photosynthesis had been identified, and it only remained for Mayer to point out in 1845 that the energy taken up as sunlight was stored, in part, as chemical energy in the organic matter.

The complexity of the whole photosynthetic unit of green plants is beyond any possibility of total synthesis at present. By necessity, *in vitro* studies have to be restricted to relatively small parts of the whole cell, but they provide a means of checking separately each part of the proposed mechanism. It is not impossible that, by putting together these separate parts, a much simplified *in vitro* system may be synthesized that is capable of carrying out the essential features of photosynthetic units in a manner more suited to our present needs.

There are two major varieties of photosynthetic organisms found in nature. First, there are the photosynthetic bacteria that reduce CO_2 to carbohydrate but that are incapable of oxidizing water to O_2. Instead, they use organic and sulfur compounds as electron donors and store relatively little energy. Second, there are the green plants and algae (red, green, and blue-green) that, in addition to reducing CO_2, oxidize water to O_2. For practical solar-energy storage in combustible form, the oxidation of water to O_2 is the key; the nature of the fuel is of much less importance, provided that it can be made to regenerate exergonically the original compound on reaction with O_2 (*1–3*).

The formation of one molecule of O_2 from two molecules of water requires the transfer of four electrons, as does the reduction of one molecule of CO_2 to the level of glucose:

$$2H_2O \longrightarrow O_2 + 4H^+ + 4e^- \quad (1)$$

$$4e^- + 4H^+ + CO_2 \longrightarrow (CH_2O) + H_2O \quad (2)$$

$$H_2O + CO_2 \longrightarrow (CH_2O) + O_2 \quad (3)$$

$$\Delta G° = 502 \quad kJ\ mol^{-1}$$

In plant and algal photosynthesis, the electron transfers occur in two stages, each requiring absorption of a photon, so that the overall quantum requirement is eight photons for each molecule of O_2 produced. The manner in which these two photochemical reactions operate in series was first suggested by Hill and Bendall in 1960 and the Z scheme that they proposed (*4*), although continually modified, is generally accepted today.

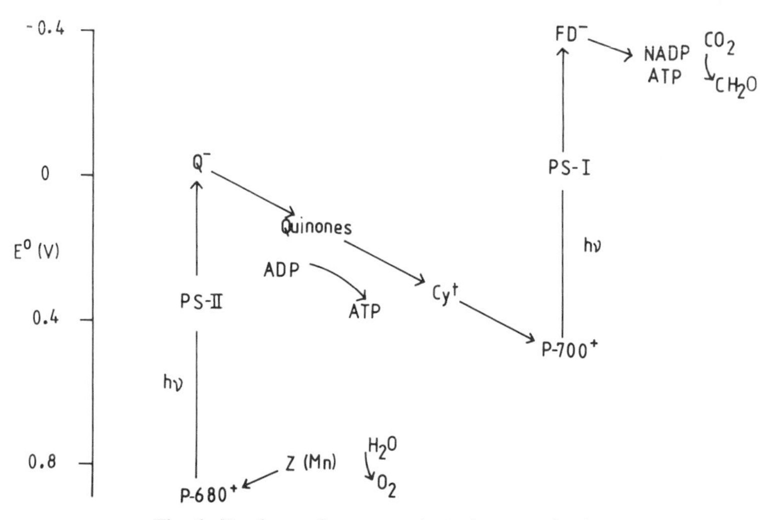

Fig. 1. Z scheme for green-plant photosynthesis.

The Z scheme (Fig. 1) is composed of two principal parts, photosystem I (PS-I) and photosystem II (PS-II), and it is followed by the Calvin cycle in which two molecules of NADPH and three molecules of ATP, made in the light reactions of PS-I and PS-II, reduce one molecule of CO_2 to carbohydrate in the dark. The two systems, PS-I and PS-II, are believed to be three-dimensionally separated in the chloroplast, and they can be isolated in partly separated form by mechanical and chemical treatment of the chloroplast. They are linked together by a pool of quinone, and each system contains an array of antenna pigments, of which chlorophyll is the principle component, the function of which is to harvest the absorbed light and transfer the energy to a site, known as the reaction center, where the primary electron transfer occurs. There are typically 300 pigment molecules to each reaction center.

The simplest sequence of reactions that accounts for the operation of photosynthesis is as follows:

1. Light harvesting. (a) Several pigments absorb light and transfer the energy to chlorophyll *a* (heterogeneous transfer). (b) Energy is transferred between chlorophyll *a* molecules until it reaches a reaction-center complex (homogeneous transfer).
2. Charge separation. In the reaction-center complex, an electronically excited molecule, which is most probably a chlorophyll dimer, transfers an electron to a molecule of chlorophyll or pheophytin (Q) to give the one-electron transfer products in very high quantum efficiency.

$$\mathrm{Chl_2^* + Q \longrightarrow Chl_2^{+\cdot} + Q^{-\cdot}} \qquad (4)$$

3. Charge transport. The redox ions produced in the primary photochemical step undergo reaction with secondary donors and acceptors so that the oxidizing and reducing equivalents are removed to remote sites.

4. Water oxidation. The chlorophyll dimer radical-cation oxidizes water by a process very probably involving manganese and necessarily involving the transfer of four electrons per O_2 liberated. If the donor is designated as Z, the reaction may be written, purely formally, as

$$4Chl_2^{+\cdot} + Z \longrightarrow 4Chl_2 + Z^{4+} \quad (5)$$

$$Z^{4+} + 2H_2O \longrightarrow Z + 4H^+ + O_2 \quad (6)$$

This part of the overall photosynthetic process is termed PS-II (Fig. 2), and it is among the least understood of all biological processes. To date, the role of Mn in PS-II remains obscure, and there is no real evidence to implicate Mn as a catalyst for O_2 formation.

5. Fuel formation. Through PS-I, CO_2 is fixed and reduced to the level of carbohydrate, which is the plant's fuel. For our present purposes, it would be beneficial if the reducing equivalents produced in the charge-separation step could be used to reduce water to hydrogen, and, in fact, the thermodynamic requirements for water and CO_2 reduction are not too dissimilar. Thus we can envisage our modified version of PS-I as occurring via the sequence of reactions shown in Fig. 3.

This short summary provides a brief account of the essential features of green-plant photosynthesis, and in the following sections we follow the progress made since 1975 or so in elaborating the mechanism by which photosynthesis functions. Emphasis has been given to *in vitro* model systems, especially those originating from the Davy Faraday Research Laboratory, and in many places the models deviate considerably from plant photosynthesis. This is especially true in two respects. First, we have concentrated heavily on studies carried out in fluid solution, whereas nature makes extensive use of membranes, proteins, etc. to hold the reactants in place. Second, for O_2 and fuel (H_2) production we have concentrated on catalysts that are not present within the natural environment.

III. Light Harvesting

A. *Kinetic Studies* in Vivo

Picosecond fluorescence and flash photolysis techniques, made possible by the mode-locked laser, have opened up the possibility of time resolution of the primary processes that follow immediately on light ab-

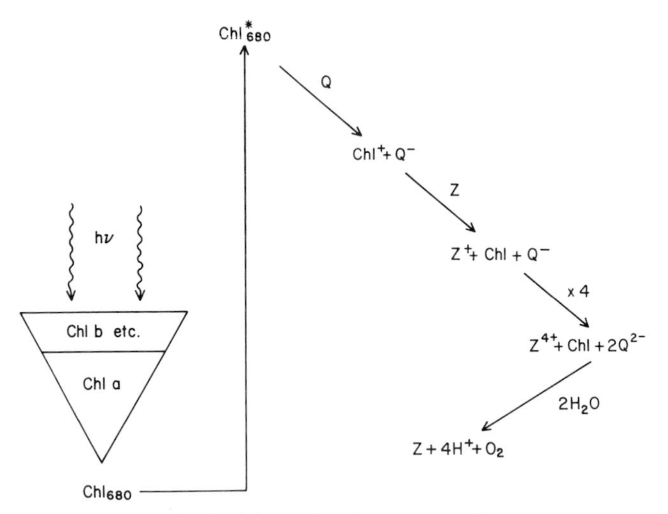

Fig. 2. Scheme for photosystem II.

sorption. Most of the work carried out has been in two areas: (a) transient absorption studies on the isolated reaction-center complexes of photosynthetic bacteria and (b) fluorescence studies of the light-harvesting process in chloroplasts and algae.

Light harvesting by ancillary pigments and transmission of the excitation to chlorophyll *a* has been studied (*5*) using *Porphyridium cruentum*,

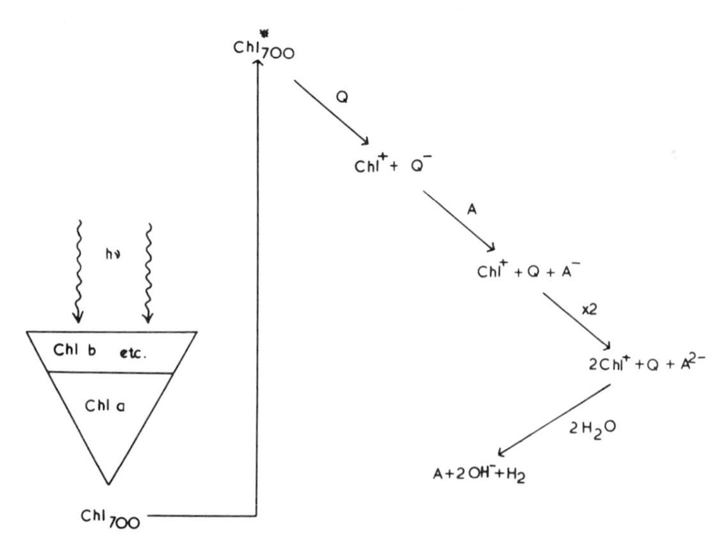

Fig. 3. Scheme for photosystem I.

which is a unicellar red alga that utilizes chlorophyll *a* as the last and lowest-energy component of the antenna system. In addition, it possesses water-soluble accessory light-harvesting pigments, having a noncyclic tetrapyrrole structure, that are contained within structures known as phycobilisomes attached to the thylakoid membrane. Phycobilisomes contain three main pigments that, in decreasing order of first excited singlet energy, are B-phycoerythrin, R-phycocyanin and allophycocyanin-B. The first absorption bands of these pigments and of chlorophyll *a* are, respectively, at the wavelengths 561, 622, 651, and 680 nm, and each pigment fluoresces in a wavelength region sufficiently different from those of the others that the four fluorescences can be resolved (*5*).

Steady-state studies indicate that the phycobilisomes preferentially serve PS-II, and the energy-transfer pathway is

$$\text{B-phycoerythin} \longrightarrow \text{R-phycocyanin} \longrightarrow \text{allophycocyanin} \longrightarrow \text{chlorophyll } a$$

With a single-pulse Nd laser, the frequency-doubled pulse at 530 nm conveniently falls near the absorption maximum of the highest-energy pigment, B-phycoerythin, and there is minimal direct excitation of the other pigments. Consequently, it has been possible to study this system with picosecond time resolution (*5*), and the time-resolved fluorescence intensities of intact algae at these four emission wavelengths are shown in Fig. 4. The rise time of B-phycoerythin corresponds to that of the excitation pulse and the detector system profile, but the other three pigments show rise times that are significantly longer. The 1/e times for rise and decay of each of the four fluorescences are given in Table I, and it is seen

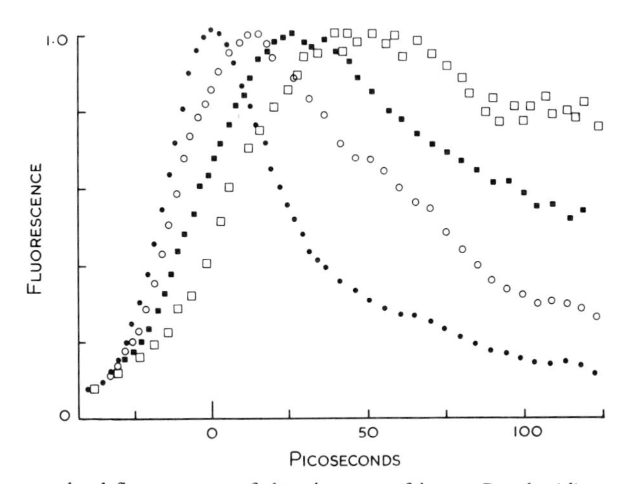

Fig. 4. Time-resolved fluorescence of the pigments of intact *Porphyridium cruentum:* ●, B-phycoerythrin; ○, R-phycocyanin; ■, allophycocyanin; □, chlorophyll *a*. From Porter (*2*).

TABLE I

Fluorescence of *Porphyridium cruentum*

Species	λ (nm)	τ(1/e) (rise) alga (ps)	τ(1/e) (decay) alga (ps)	τ(1/e) (decay) phycobilisome (ps)
Phycoerythrin	578	0	70	70
Phycocyanin	640	12	90	—
Allophycocyanin	660	22	118	4000
Chlorophyll *a*	685	52	175	—

that excitation energy is transferred sequentially between the pigments and, ultimately, from chlorophyll *a* to the reaction-center complex. When the phycobilisomes are isolated from the algae, to remove chlorophyll *a* and the whole reaction-center complex, the lifetime of allophycocyanin is very much longer (close to that of the pigment in dilute solution), whereas the lifetime of B-phycoerythin is unchanged (*6*).

This system illustrates the efficiency of interpigment energy transfer in the photosynthetic unit. Quantum yields of fluorescence, calculated from the measured lifetimes, are all below 1%, implying an efficiency of energy transfer greater than 99%. Thus a large fraction of the solar spectrum can be collected and the excitation energy transferred rapidly and efficiently to the reaction-center complex. The long lifetime of allophycocyanin in the phycobilisome compared with that in the intact alga containing chlorophyll *a* is also noteworthy because it shows that, wihin the allophycocyanin antenna, concentration quenching is virtually absent.

Most of the fluorescence of whole chloroplasts and *Chlorella* at room temperature is emitted by the light-harvesting chlorophyll *a* associated with PS-II, and this emission has been well studied by picosecond-flash photolysis techniques. The basic concepts that have been evaluated to date include

1. Heterogeneous energy transfer between donor and acceptor pigments and homogeneous transfer between molecules of the same kind often occurs with nearly unit efficiency.
2. In the absence of traps or acceptors, lifetimes approach those of the pigment in dilute solutions, even though the average concentration of the pigment within the harvesting unit might be as high as 0.1 *M*.

3. The lifetime of the excitation before transfer to another pigment or acceptor is much longer than that of a coherent exciton, and the use of a hopping model, involving weak Förster-type resonance interaction, or the stronger orbital overlap interaction of Dexter, is justified.

B. In Vitro *Models*

The simplest system in which to study the individual steps of photosynthesis is a homogeneous fluid solution, and because little is known of the organization within the chloroplast and because the primary steps are too fast for diffusion to intervene, this may not be so irrelevant a model as might at first be supposed. Even so, models in which chlorophyll is dissolved in lipid mono- and multilayers on slides and in vesicles or liposomes in aqueous suspension are better because chlorophyll is miscible with lipids in almost any proportion, and, therefore, it is possible to use concentrations as high as those in the chloroplast. The absence of diffusion over longer times is also important in some studies, such as charge separation, and, finally, although at least half of the chlorophyll is attached to protein much of it resides in a lipid environment.

Heterogeneous energy transfer between different pigments (e.g., chlorophyll *b* to chlorophyll *a*) is readily observed *in vitro,* in fluid solutions, in rigid matrices, and in lipid multilayers. The relative yield of fluorescence of donor is given by the Förster expression

$$\phi/\phi_o = 1 - \sqrt{\pi}\gamma \exp \gamma^2(1 - \operatorname{erf} \gamma), \tag{7}$$

where $\gamma = C/C_o$ and C_o is the critical concentration of acceptor, corresponding to one molecule of acceptor in a sphere of radius R_o, where R_o is the critical transfer distance at which transfer and fluorescence probabilities are equal (5.8 nm for chlorophyll *b* to chlorophyll *a*). Studies of energy transfer in multilayer lipid matrices between chlorophyll *b* and chlorophyll *a* agree with this expression both in the form and the absolute magnitude of the quenching curves (*7*).

However, when pure chlorophyll *a* (or pure *b*) is investigated, the fluorescence yield falls as the concentration increases. This homogeneous "concentration quenching" occurs in a concentration range similar to that of heterogeneous energy transfer, but the form of the quenching curve is different, and, in many systems, including multilayers (*8*), liposomes (*8*), vesicles (*8*), monolayers (*9*), and fluid solutions (*10*), it has been found to conform to the empirical expression first found by Watson and Livingstone (*10*):

$$\phi/\phi_o = 1/(1 + A[\mathrm{Chl}]^2) \tag{8}$$

Typical self-quenching data obeying this relationship for chlorophyll *a* in lecithin vesicles are shown in Fig. 5 and half-quenching concentrations in a variety of environments are collected in Table II.

This concentration quenching, which seems to occur in all model systems, is in complete contrast to the situation found in the chloroplast, where concentrations of chlorophyll *a* as high as 0.1 *M* exist. There are several important mechanisms whereby concentration quenching can occur in model systems, including ground-state dimerization, excimer formation, impurity quenching, ionization reactions, or statistical-pair quenching. The term *statistical pair* refers to two chlorophyll molecules in the random distribution that happen, on purely statistical grounds, to be closer than the average near-neighbor distance and are close enough that, when one of them is electronically excited, interaction occurs, leading to quenching (*11*). This interaction may, in favorable circumstances, lead to collapse to a true equilibrium excimer, although this is not necessary for quenching to occur, but no evidence of excimer fluorescence in chlorophyll (or metalloporphyrins) has yet been found.

From our many model systems, we are led to the conclusion that the only arrangement that can account for the high-energy transfer probabil-

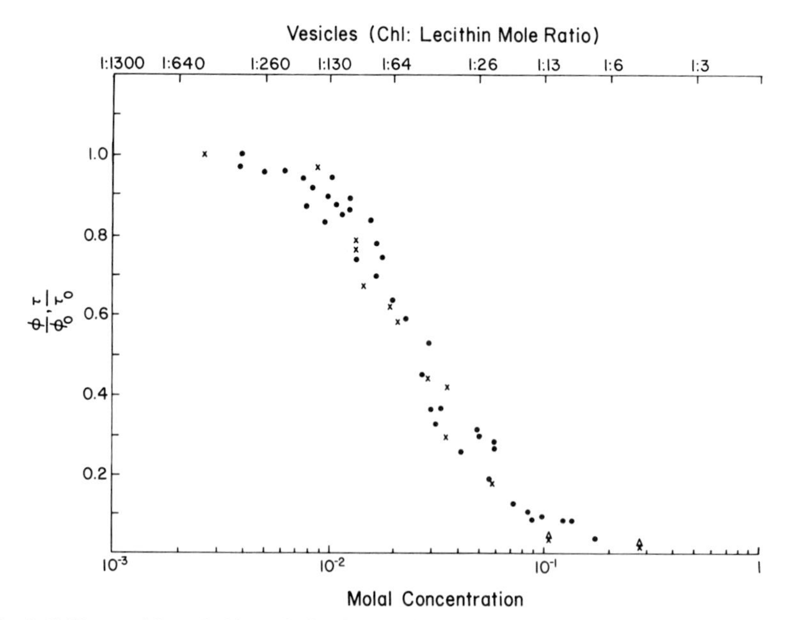

Fig. 5. Self-quenching of chlorophyll *a* in lecithin vesicles. ϕ/ϕ_o (●) is the yield and τ/τ_o (x) the lifetime of fluorescence referred to infinite dilution.

TABLE II

Concentration Quenching in Chlorophyll *a*

Medium	$C_{1/2}$ (molal)	Chl : lipid mole ratio
Lecithin multilayers (dry)	1.4×10^{-3}	—
Ether solution	1.6×10^{-2}	—
Lecithin vesicles	2.8×10^{-2}	1 : 43
Mono-digalactosyl diglyceride 3 : 1 vesicles	3.9×10^{-2}	1 : 30
Lecithin liposomes	4.6×10^{-2}	1 : 25
Mono-digalactosyl diglyceride 3 : 1 liposomes	7.0×10^{-2}	1 : 17

ity of the chloroplast is one in which the molecules are closely spaced (and the average concentration in the chloroplast must be 0.1 *M* or greater) and yet no significant proportion of molecules are close enough to their nearest neighbors to permit quenching by excimer or similar interactions. The organization of chlorophyll within the chloroplast is not known with much certainty, but it probably involves chlorophyll–protein complexes.

Several different types of chlorophyll–protein complexes have been identified, mostly by absorption spectroscopy (*12*). Whenever a chlorophyll *a* species with an absorption maximum below 670 nm is observed, it is usually attributed to monomeric chlorophyll *a*; chlorophyll *a* species absorbing at ~670–680 nm are considered to be aggregates formed by chlorophyll–chlorophyll interactions, and any chlorophyll *a* species that absorbs at wavelengths longer than 680 nm is most likely to be an aggregate of chlorophyll but aggregated by a bifunctional ligand (such as water) that cross-links the chlorophyll. Water is a particularly good cross-linking agent (*13, 14*), and large aggregates can be formed; the chlorophyll *a* hydrate that absorbs at ~740 nm is the largest red shift found for any chlorophyll species. On the basis of absorption spectroscopy, the chlorophyll *a* found in the antenna of most chloroplasts is in the form of an oligomer (Chl $a)_n$ in which the chlorophyll *a* molecules are orthogonal so as to optimize the directionality of the keto CO···Mg coordination bonds and to minimize overlap repulsions (*15*). In this structure (Fig. 6), the transition density–transition density interaction energy between adjacent molecules will be almost zero and the red shift is a result primarily of environmental effects and is independent of aggregate size. That is, the absorption maximum for a small oligomer ($n > 3$) is about the same as for

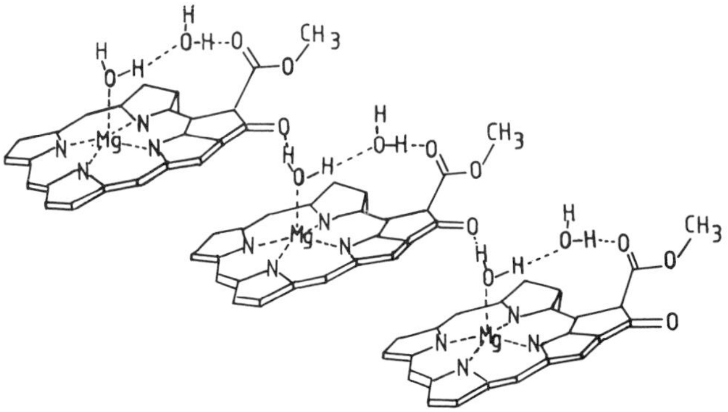

Fig. 6. Model for the assembly of chlorophylls within the plant.

a large one. Even though the absorption spectral profiles of the antenna chlorophyll *a* are entirely consistent with chlorophyll–chlorophyll interactions there can be no doubt that the chlorophyll is intimately linked with protein. Consequently, the detailed structure of the antenna chlorophyll *a* remains to be elaborated. It is hoped that the work of Boxer *et al.* (*16*), reporting the synthesis and properties of synthetic chlorophyll–myoglobin complexes, will go some way toward solving some of these problems.

Electronic energy transfer from molecule to molecule along a chain of chlorophyll *a* oligomer should proceed by a Förster-type mechanism (*17*). It is believed (*18*) that chlorophyll *a* dimers are nonfluorescent, but there is some experimental evidence that chlorophyll *a* oligomers are more strongly fluorescent because of their greater rigidity (*19*). Even so, fluorescence is strongly quenched in the oligomer, which suggests that Förster energy transfer may not be efficient. However, the individual steps in energy transfer along the chain are very rapid, and many events can occur within the chlorophyll lifetime, even when this lifetime is relatively short.

C. Synthetic Models

There have been many studies of energy transfer using a metalloporphyrin, such as chlorophyll *a*, as donor or acceptor. Many of these studies have been concerned with porphyrin dimers and trimers, some of which are shown in Table III. These dimer structures range from the simple μ-oxo dimers (*20*) to linear porphyrin dimers (*21*) to the complex *strati*-bisporphyrins (*22*). In fact, attempts to prepare synthetic chlorophyll-based dimers began with the work of Boxer and Closs (*23*), who obtained a covalently linked pyrochlorophyllide *a* dimer (Fig. 7). Addition of hydroxylic ligands, such as ethanol, causes the dimer to fold in such a

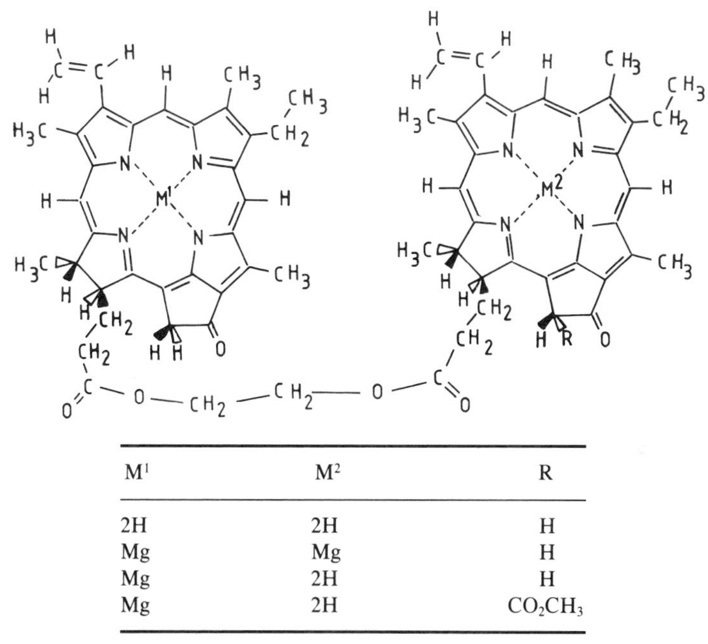

M^1	M^2	R
2H	2H	H
Mg	Mg	H
Mg	2H	H
Mg	2H	CO_2CH_3

Fig. 7. Structure of the chlorophyllide and pheophorbide dimers prepared by Boxer *et al.* (*23*).

way that the two macrocycles are held parallel by two bridging ligands, each of which hydrogen bonds to the carbonyl at position 9 of one macrocycle and coordinates to the central metal of the other. Subsequently, covalently linked chlorophyllide and bacteriochlorophyllide dimers were prepared that exhibited identical aggregation properties in solution. Porphyrin derivatives having two or four covalent bridges to produce rigid cofacial or hinged dimers have been reported (Fig. 8) (*22*).

TABLE III

Synthetic Porphyrin Dimers

Type of dimer	References
Linear porphyrin dimers	*21, 75–77*
μ-Oxo dimers	*20, 66*
Nucleophile bridged	*23, 78*
Cofacial diporphyrins	*23, 79*
"Capped" porphyrins	*80–84*
strati-Bisporphyrins	*22, 85*

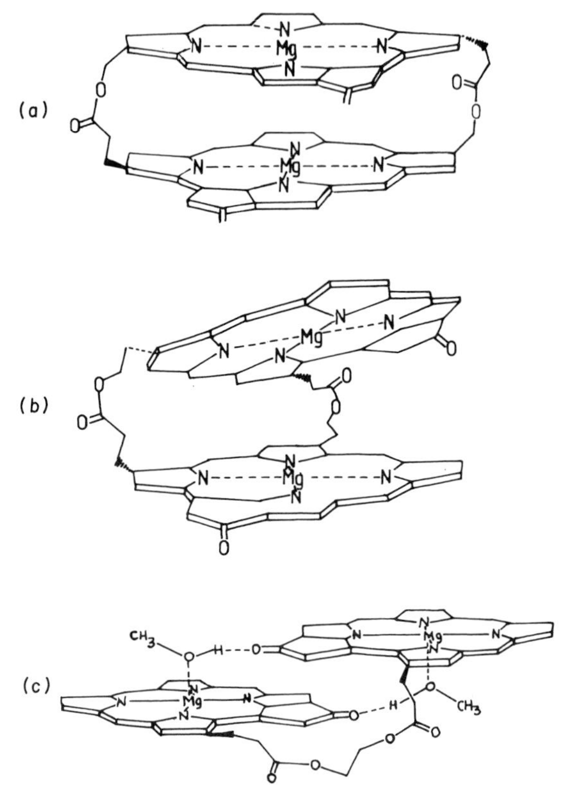

Fig. 8. Chemical structures and schematic conformations for the dimers discussed in the text. (a) Cofacial. (b) Hinged. (c) Single link.

Rather surprisingly, the fluorescence quantum yields and singlet excited-state lifetimes of the dimers were found to be only slightly lower than those of the corresponding monomers. This effect is demonstrated in Fig. 9, which shows the relative fluorescence yields for the *strati*-bisporphyrin and the monomer. Often with these systems, the dimers show small red shifts, but there are no dramatic spectral changes.

The synthesis and characterization of many of the preceding compounds present severe experimental difficulties, and, it often requires rigorous treatment to obtain pure materials. Although of considerable academic interest, these compounds are no less fragile than chlorophyll *a*, and they are just as expensive. Thus they have little or no application in large-scale solar-energy storage devices, and there have been several attempts to construct porphyrins that have built-in antenna units. The idea behind such materials is to attach, via covalent linkages, substituents that

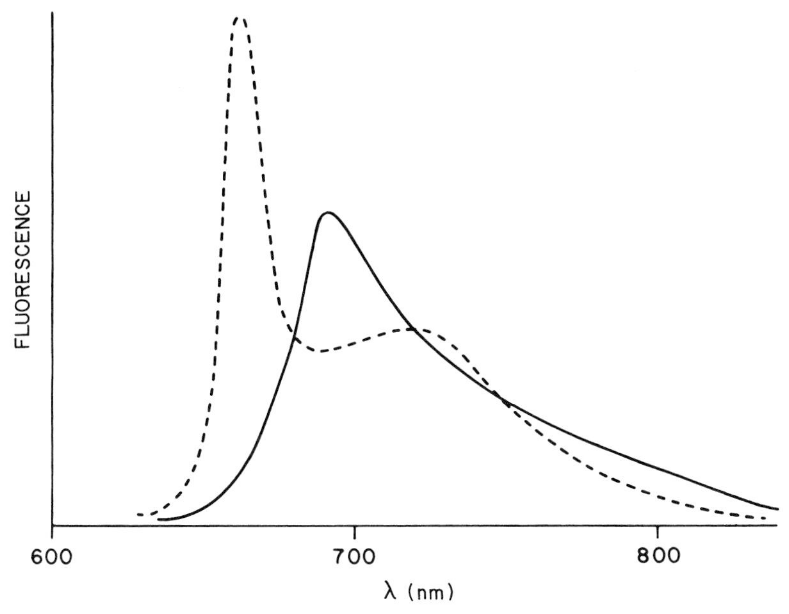

Fig. 9. Fluorescence spectra of the *strati*-bisporphyrin (——) and the corresponding monomer (---).

absorb intensely in the visible region and channel the incident photons to the lowest excited singlet state of the porphyrin.

The simplest materials are those prepared by Harriman (*24*); they consist of a porphyrin ring substituted at each of the four meso carbon atoms with an aryl hydrocarbon [Fig. 10, (I)–(IV)]. These materials have increased absorption in the near-uv region, but absorption changes in the visible region are minimal. More interesting models have been reported by Moore *et al.* (*25, 26*) [Fig. 10, (V), (VI)], who prepared porphyrin–polyene compounds that had increased absorption in the spectral region around 500 nm. Relatively high-energy transfer efficiencies were found for electronic energy transfer from polyene to porphyrin, although the fluorescence yield of the porphyrin moiety was quenched substantially. Even more drastic spectral changes were observed with ferrocenyl-substituted porphyrins (*24*) [Fig. 10, (VII)], but no fluorescence could be seen for this compound. As such compounds become more popular, it should be possible to produce "spectrally tuned" materials that possess intense absorption throughout the entire visible region.

Triplet–triplet energy transfer involving metalloporphyrins has been investigated in several systems, especially in outgassed fluid solution. It has been established (*27*) that triplet–triplet energy transfer occurs from

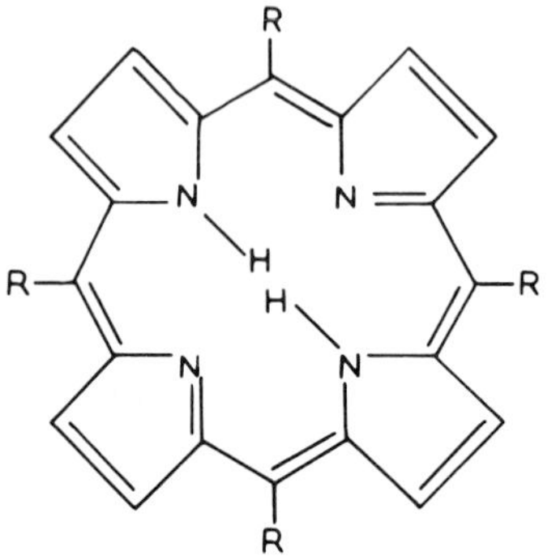

Fig. 10. Structures of meso substituted porphyrins used for intramolecular energy transfer. R: 1-phenyl (I), 2-naphthyl (II), 9-anthryl (III), 2-pyrenyl (IV), β-carotenyl carboxylate (V), 8-retinyl carboxylate (VI), 1-ferrocenyl (VII).

metalloporphyrin molecules to aromatic hydrocarbon acceptors with low-lying triplet levels (such as tetracene) and to metalloporphyrin molecules with triplet levels located below the levels of the donor. In most cases, the transfer rate constants were close to the diffusion-controlled rate limit.

IV. Charge Separation

A. In Vivo *Studies*

The first direct evidence of photoinduced charge separation in photosynthetic reaction-center complexes concerned Duysen's observation of the reversible light-induced bleaching of P-870 in purple bacteria. Later, Kok and Witt independently discovered the reversible bleaching of P-700 in chloroplasts, and Witt and co-workers reported the reversible bleaching of P-680, the primary electron donor in PS-II. The detection of light-induced absorbance changes for P-680 was more difficult because the oxidized form P^+ had a very short lifetime because of rapid reduction by secondary electron donors (see the following text).

The best evidence that $P \rightarrow P^+$ involves the transfer of a single electron comes from redox titrations of the P/P^+ couple. The midpoint potential of this couple has been measured by titration with ferri- ferrocyanide, and for purple bacteria it has a value of 0.4–0.5 V, independent of pH, with $n = 1$. The value $n = 1$ shows that P and P^+ differ by just one electron, as might be expected for a photoredox process. With this information, it was possible to determine extinction coefficients for P-870 and P^+ that, together with those already known for the other reactants in the charge separation process, have allowed a time profile to be compiled that describes the sequence of events.

Figure 11 gives such a time profile for photoinduced electron transfer

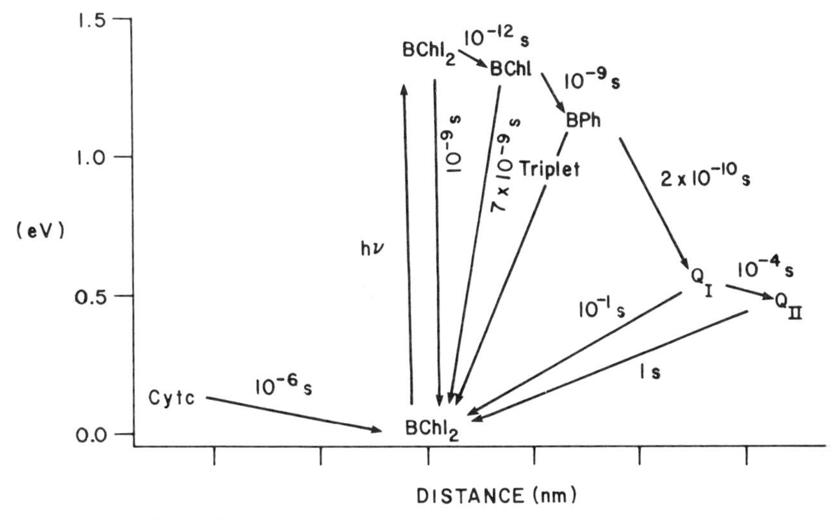

Fig. 11. Energy levels, intermolecular distances, and average lifetimes of substances associated with electron-transfer processes in bacterial reaction centers.

within the reaction center of a bacterial photosynthetic unit. Although there are still considerable doubts about the identity of the primary electron acceptor, it is relatively certain that this species is either a bacteriochlorophyll or a bacteriopheophytin molecule. All available spectrosopy data suggest that Q is a quinone. The overall quantum efficiency for the photoredox process is approximately unity, and the reaction appears to be a singlet excited-state reaction (the triplet excited state does not possess sufficient thermodynamic reducing power to drive the reaction).

The primary reactants in green-plant photosynthesis are not so well characterized as for the bacterial system, but there is relatively strong evidence to suggest that the two natural processes have much in common. Thus the primary electron donors in both PS-I and PS-II of green plants appear to be specialized pairs of chlorophyll *a* molecules, whereas the primary electron acceptors appear to be chlorophyll *a* or pheophytin molecules, and a quinone functions as a secondary electron acceptor.

B. In Vitro *Studies*

There have been many reports of photoredox processes in which chlorophyll (or another metalloporphyrin) is used to photooxidize an electron donor or photoreduce an electron acceptor, and these systems have been reviewed in great detail by Seely (*28*) and others (*29*). Very few of these studies have used chlorophyll or pheophytin molecules as the electron acceptor; usually an easily reduced compound such as a quinone (Q) has been used.

Electron transfer from electronically excited chlorophyll to quinones has been shown to occur readily, and the problem is to inhibit the rapid reverse electron transfer so that the redox ions can separate. Both singlet and triplet excited-chlorophyll molecules are quenched by quinones, and the bimolecular quenching rate constants have been determined for the singlet (by fluorescence yield or lifetime measurements) and the triplet (by kinetic absorption spectroscopy using flash photolysis techniques). However, it does not follow from quenching studies alone that separated ions Chl^{+} and Q^{-} are formed.

In fluid solution (*30*), the bimolecular rate constants for quenching excited chlorophyll *a* by duroquinone in ethanol are

$$S_{Chl^*} + Q \rightleftharpoons S_{(Chl^* \cdots Q)} \quad (9)$$

$$k_S = 9.2 \times 10^9 \quad M^{-1}\ s^{-1}$$

$$T_{Chl^*} + Q \rightleftharpoons T_{(Chl^* \cdots Q)} \quad (10)$$

$$k_T = 1.4 \times 10^9 \quad M^{-1}\ s^{-1}$$

The initially formed encounter complex (Chl*···Q), which has the same spin multiplicity as the excited state of the chlorophyll *a* reactant, may react as follows:

$$(Chl^* \cdots Q) \longrightarrow Chl + Q \quad (11)$$

$$\longrightarrow Chl^+ + Q^- \quad (12)$$

$$Chl^+ + Q^- \longrightarrow Chl + Q \quad (13)$$

There is clear evidence that the complex formed from triplet chlorophyll reacts via reaction (12) to form separated ion products in very high yield; both the chlorophyll *a* cation and the quinone anion have been detected by optical (*30*) and epr techniques (*31*), and their subsequent reverse reaction [via reaction (13)] is a diffusion-controlled process. In contrast, there is no real evidence (*32*) to show that the singlet complex dissociates into separated ion products. It has been suggested (*33*) that in the case of bacteriopheophytin–benzoquinone, the rate of conversion to ground-state reactants via reaction (11) in the singlet complex is very much faster than in the triplet complex, because of the spin-forbidden nature of the triplet process, so that it competes favorably with dissociation of the singlet complex into ions. This seems to be a good hypothesis, but it is not a general one because in many systems (e.g., dimethylaniline–aryl hydrocarbon) it is well established that separated ions are formed from the singlet-state reaction (*34*).

Whether or not separated ions can be formed via a singlet excited-state reaction of chlorophyll *a* in fluid solution is a subject of considerable academic interest, but there can be no doubt that rigid solvents and lipid solutions are better models of the photosynthetic unit. Such studies have been reported (*35*) for the chlorophyll–quinone system in lecithin, in which diffusion is totally inhibited during the lifetimes of the excited states, and there is no spectroscopic evidence to suggest ground-state complexation between the reactants.

Singlet states are quenched by quinone, and the half-quenching concentrations are given (*35*) in Table IV. Triplet states were observed by flash photolysis, with lifetimes of ~0.9 ms that remained unchanged when the quinone concentration was varied from 10^{-4} to 10^{-1} *M*. The singlet and triplet yields showed (*35*) the same dependence on quinone concentration. This drastic difference between singlet- and triplet-state quenching rates (by a factor of more that 10^5 when the relative lifetimes of the two states are considered) compares with a difference of only about 10 in fluid solvents. The explanation of this difference is to be found in the random distribution of molecules, which is frozen in rigid solutions, whereas in fluid solvents each molecule has time to sample a variety of near-neighbor distances. Consider, in the rigid solution, an excited chlorophyll molecule with nearest-neighbor quinone at a distance r. The dependence of quenching rate on r for an electron transfer between molecules is a very sharp one, approximating a Perrin active-sphere model. If r lies within this sphere, the singlet will be quenched before fluorescence or intersystem crossing occurs, whereas if it is outside the sphere it will fluoresce or form triplets, but these triplets (now lying outside the quenching sphere and having a somewhat lower quenching probability and hence a smaller quenching sphere) will be unaffected by the quinone, despite the longer triplet lifetime. For singlet-state quenching, typical transfer distances are

TABLE IV

Concentration of Quinone Needed to Quench the Fluorescence of Chlorophyll *a* to Half of Its Original Yield

Quinone	$E°$ (V)	$C_{1/2}/10^2$ (M)
2,5-Dichlorobenzoquinone	0.74	2.0
Benzoquinone	0.71	2.8
2,5-Dimethylbenzoquinone	0.60	3.8
Duroquinone	0.47	4.9
Plastoquinone-9	0.53	5.8
α-Tocopherylquinone	0.47	9.4

2.4 nm for benzoquinone and 2.0 nm for duroquinone and plastoquinone, and the probability of transfer falls off rapidly with increased distance (*35*).

The preceding conclusion (*2*) raises a problem concerning the mechanism of charge separation within the photosynthetic unit. Singlet excited states are quenched before triplets are formed in any significant amount by intersystem crossing, and other mechanisms for triplet formation, such as singlet–singlet annihilation, can be neglected under the normal conditions of photosynthesis. During the lifetime of fluorescence in PS-II (~500 ps) the triplet yield resulting from intersystem crossing must be less than 6%. Thus it appears that charge separation must occur via the singlet excited state, but the model-compound studies have shown that this is a difficult step to realize. For the model systems, the quenching of singlet chlorophyll *a* by quinones must involve charge separation to some degree (as shown in Table IV there is a good correlation between the redox potential of the quinone and the efficiency of the quenching process) because energy transfer is forbidden. Thus the most likely reason why separated ions are not found for the singlet reactions of chlorophyll *a* in model systems is that geminate recombination is very much faster than ionic separation of the complex (*33*). To account for the high efficiency of photosynthetic units, we must invoke the hypothesis that secondary electron transfer (e.g., from I^- to Q in Fig. 4) is even faster than geminate recombination.

One of the great advantages that the natural photosynthetic process possesses relative to the model systems is that nature makes extensive use of membranes and proteins to hold the reactants at optimum geometries so that electron transfer can proceed efficiently and rapidly. This realization has led to the synthesis of very complex organic structural units, aimed at providing models more closely related to the natural reaction-center complexes. Several of these model systems have been investigated by picosecond-flash spectroscopy.

Most of these model systems have centered on the use of porphyrin derivatives that possess covalently linked quinone units; several such materials have been reported. The first porphyrin–quinone complex was described by Kong and Loach (*36*) (the structure is given in Fig. 12), who reported, rather briefly, that the fluorescence of this material was somewhat quenched relative to the unsubstituted porphyrin and that the quenching act resulted in the formation of ionic products. Later, Bolton *et al.* (*37*), using epr spectroscopy, observed that in methanol solution at low temperature, steady-state illumination of the porphyrin–quinone complex resulted in transient formation of redox ions consistent with P^+ and Q^- species. These findings have been confirmed by Kong *et al.* (*38*),

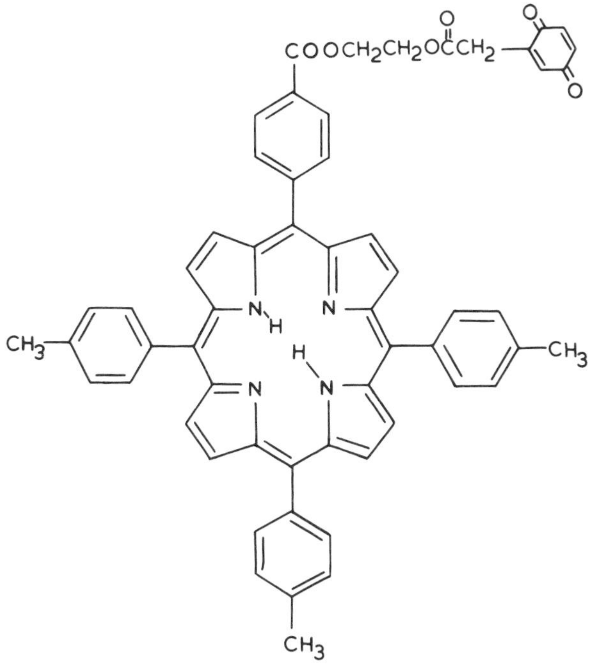

Fig. 12. Porphyrin–quinone compound used for intramolecular electron-transfer studies.

who also described a complex solvent dependence for the quenching process. Thus, in pentane or *N,N*-dimethylformamide, virtually no fluorescence quenching was observed, whereas in acetonitrile or dichloromethane both the fluorescence yield and the lifetime were reduced substantially (~75%) relative to the unsubstituted porphyrin, and in the latter solvents radical-ions were observed at 77 K by epr spectroscopy. These radical ions were very stable at 77 K but decayed rapidly when the solution was warmed (*38*). It was concluded that singlet excited-state light-induced charge separation occurred with this compound in CH_3CN or CH_2Cl_2 solutions, but in the absence of picosecond-time-resolved flash spectroscopy, this conclusion should be regarded with some caution. However, other work suggests that such evidence will be available (*39*).

Porphyrins with more than one quinone unit covalently attached were synthesized by Dalton and Milgrom (*40*), who reported that the tetrasubstituted derivative (Fig. 13) is nonfluorescent in CH_2Cl_2 solution. Later workers (*41, 42*) reported that this compound does exhibit some fluorescence, albeit with low yield (the fluorescence lifetime and yield were slightly dependent on the solvent), but the triplet excited state of the porphyrin could not be observed by optical methods using a 15-ns-flash

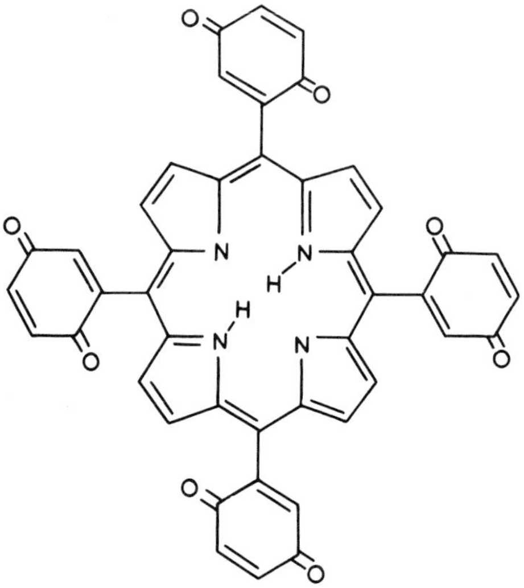

Fig. 13. Structure of *meso*-tetra(quinone)porphine.

excitation pulse. In CH_2Cl_2 solution at room temperature, picosecond-flash spectroscopy showed the transient formation of the porphyrin cation, but the high laser intensities used in this work may cause some experimental difficulties.

Complex organic units that closely resemble the primary reactants within the reaction center of bacterial photosynthetic units have been synthesized by Boxer *et al.* (*43*). These units consist of two pyrochlorophyll molecules linked by a covalent chain, and there is a single metal-free pyropheophytin molecule attached to one of the pyrochlorophyll macrocycles (Fig. 14). There is some evidence to suggest that intramolecular electron transfer is an important route for nonradiative deactivation of the "trimer," but the rate of electron transfer is slow ($<10^{10}$ s^{-1}) and the yield of redox products is low. These studies have shown that to duplicate the efficient processes of the naturally occurring charge-separation units, it is necessary to give considerable attention to the geometric arrangement of the reactants within the unit. Obviously, this is a very difficult challenge to the organic synthetic chemist.

Overall, these studies have shown, very clearly, that it is possible to induce some degree of charge separation from the singlet excited state of a metalloporphyrin, even in the absence of driving forces such as electrostatic potentials. It now remains for the ion products of this charge separation act to be used in secondary electron transfers so that the oxidizing and/or reducing equivalents can be transported to remote sites and used for useful chemical purposes. So far, this has not been achieved.

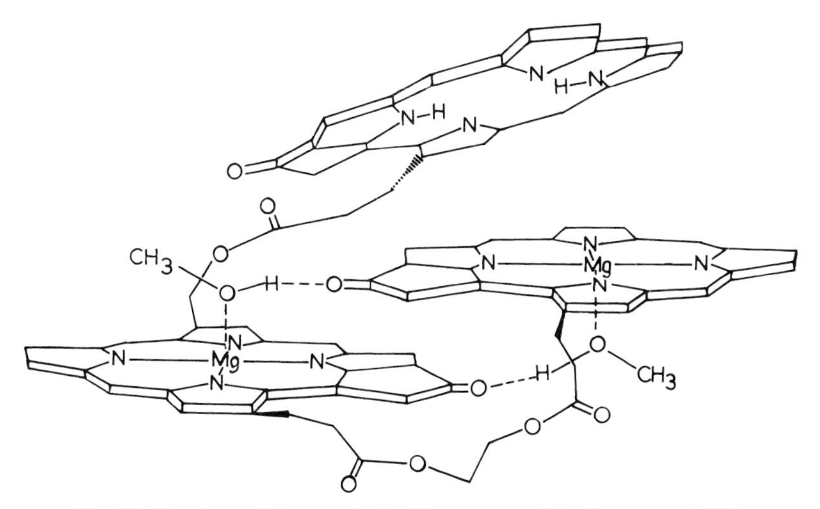

Fig. 14. Structure of the porphyrin trimer prepared by Boxer *et al.* (*43*).

Unfortunately, to obtain even a modest yield of redox ions from a singlet excited-state process involving a metalloporphyrin, it has proved necessary to use complex, fragile, and expensive porphyrin derivatives. Such materials cannot be used for large-scale solar-energy storage devices. However, it is well known that high yields of redox ions can be generated from the triplet excited state of many metalloporphyrins, and in the Section IV,C we consider (i) why the triplet state is a much better sensitizer than the singlet state and (ii) methods to optimize the yield of redox ions from the triplet state. Such information is useful, not only in regard to the ultimate construction of a solar-energy storage device, but also because results obtained from the triplet-state studies may have some bearing on the singlet-state reactions.

C. Triplet-State Reactions

The efficiency of light-induced charge separation depends on a large number of parameters, but perhaps the most important factor concerns the degree of ionic separation from the ion pair. Under normal circumstances, an ion pair is a key intermediate in a photoredox process; it can decay to either ground-state reactants or separated ion products, and it is this partition that reflects the overall yield of redox products. The rate constant for ionic separation from an ion pair can be calculated from a modified Fuoss equation (*44, 45*), and it depends on solvent viscosity, dielectric constant, and (in some cases) ionic strength, temperature, and electrostatic forces between the ions. Geminate reverse electron transfer within the ion pair is a type of internal conversion and depends on orbital overlap between the ions and the amount of electronic energy that must be dissipated in the form of heat. Furthermore, it seems probable that spin-selection rules are important here, and it has been proposed (*33*) that reverse electron transfer for a triplet ion pair is considerably slower than for the analogous singlet ion pair. Thus, in fluid solution, a triplet ion pair can dissociate to give separate ion products, whereas the singlet ion pair gives no redox ions (except under special circumstances).

To test the preceding hypothesis, we can consider the reaction between metalloporphyrins with different central metal ions and 1,4-benzoquinone (BQ) in outgassed fluid solution at room temperature. Although no ground-state interactions have been observed by absorption spectroscopy, quinones quench the triplet excited state of many different metalloporphyrins in ethanol solution, and in most cases net electron-transfer products have been observed. Consequently, the overall triplet-state photosensitized reduction of BQ, in polar solvents, can be described by the scheme (*46, 47*).

$$\mathrm{T_{P^*}} + \mathrm{BQ} \underset{k_{-\mathrm{D}}}{\overset{k_\mathrm{D}}{\rightleftharpoons}} \mathrm{T_{(P^*,BQ)}} \underset{k_{-\mathrm{A}}}{\overset{k_\mathrm{A}}{\rightleftharpoons}} \mathrm{T_{(P^+,BQ^-)}}$$
$$\mathrm{T_{(P^+,BQ^-)}} \xrightarrow{k_\mathrm{B}} \mathrm{P} + \mathrm{BQ} \qquad \mathrm{T_{(P^+,BQ^-)}} \underset{k_{-\mathrm{S}}}{\overset{k_\mathrm{S}}{\rightleftharpoons}} \mathrm{P^+} + \mathrm{BQ^-}$$

Here, $\mathrm{T_{P^*}}$ refers to the triplet excited state of the metalloporphyrin, and the first step in the quenching process involves formation of a reversible encounter complex in which there is practically no binding energy between the reactants. In competition with dissociation of the encounter complex, electron transfer can take place to form an ion pair in which the radical-ions are caged within a cavity in the solvent. This ion pair may dissociate to give separated ion products (P^+ and BQ^-) or it may recombine to form ground-state reactants (P and BQ). To treat the data in a simple (perhaps naive) manner, it will be assumed that the reactants are incompressible spheres for which any electronic charge is fully delocalized around the surface and that electron transfer within the encounter complex is irreversible (i.e., $k_{-\mathrm{A}}$ can be neglected). This latter assumption is questionable, but in most cases the thermodynamic driving force is high, as is the overall triplet-quenching efficiency, so that $k_\mathrm{A} \gg k_{-\mathrm{A}}$.

Where the binding energy within the encounter complex is very weak, it is often assumed (*33*) that the stability constant K for the complex can be related to the molar volumes of the reactants. For reaction between BQ and a metal tetraphenylporphyrin, K should have a value of about 14.3 M, and because

$$K = k_{-\mathrm{D}}/k_\mathrm{D}, \tag{14}$$

we can calculate k_{-D} if it is accepted that k_D is the diffusion-controlled bimolecular rate constant. For neutral reactants in ethanol solution, $k_D = 6.65 \times 10^9\ M^{-1}\ s^{-1}$, and, consequently, k_{-D} must have a value of $9.5 \times 10^{10}\ s^{-1}$. Furthermore, we can now estimate k_A because the overall triplet-state-quenching rate constants k_T, measured by flash photolysis techniques, can be expressed in the form

$$k_T = k_D k_A/(k_A + k_{-D}). \tag{15}$$

The derived values (*47*) are collected in Table V, together with the standard free-energy change for electron transfer within the encounter complex, calculated according to

$$\Delta G = -nF(E^\circ_{p^+/p} - E^\circ_{BQ/BQ^-}) - (e^2N/r\varepsilon) - E_T, \tag{16}$$

where E° refers to the standard redox potential for a particular couple, ε is the dielectric constant of the solvent, N Avogadro's number, and E_T the triplet energy of the metalloporphyrin (*48*).

As shown in Fig. 15, there is a correlation between k_A and ΔG in that the rate constant for electron transfer increases with more negative standard free-energy changes. In fact, assuming an outer-sphere reaction mechanism, the rate of electron transfer can be expressed in the form (*47*)

$$k_A = k_o e^{-\Delta G^\ddagger/RT}, \tag{17}$$

and there have been several attempts to relate the free energy of activation $\Delta G^\ddagger$ to ΔG. From Fig. 15, it is seen that as ΔG becomes more negative k_A tends toward a value of approximately $10^{11}\ s^{-1}$. This value is close to

TABLE V

Thermodynamic and Kinetic Parameters for Formation of the Ion Pair

Compound	$10^{-9} \times k_T$ ($M^{-1}\ s^{-1}$)	$10^{-10} \times k_A$ (s^{-1})	$-\Delta G$ (kJ mol^{-1})	$\Delta G^\ddagger$ (kJ mol^{-1})
MgTPP	3.1	8.4	75.3	0.43
CdTPP	3.0	7.8	73.3	0.61
RuTPP	3.0	7.8	69.5	0.61
ZnTPP	1.8	3.5	67.5	2.60
PdTPP	2.6	6.1	58.9	1.22
Chl *a*	2.2	4.7	58.9	1.87
CuTPP	1.9	3.8	56.9	2.40
CrTPP	2.0	4.1	50.2	2.21
AlTPP	1.0	1.7	40.5	4.39
H_2TPP	0.9	1.5	30.9	4.70

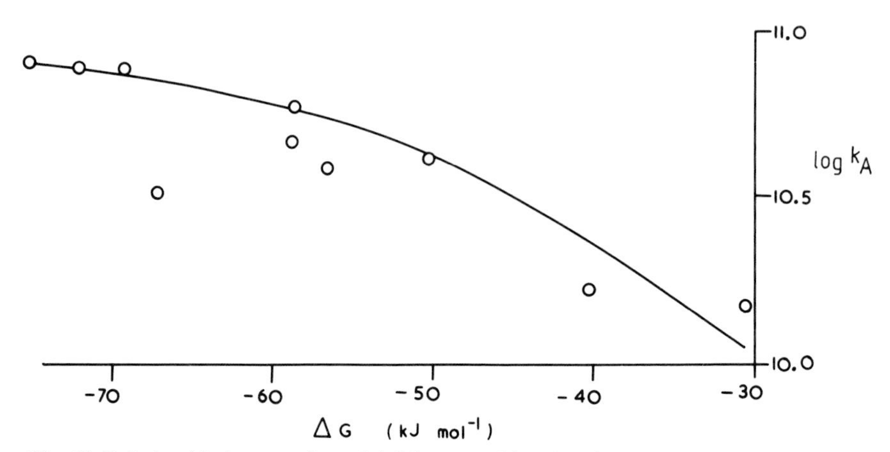

Fig. 15. Relationship between k_A and ΔG for quenching the triplet excited state of metalloporphyrins with benzoquinone.

the dielectric relaxation time of the solvent, suggesting that $k_\circ \approx 10^{11}\ s^{-1}$. Consequently, we can raise the assumption that

$$\Delta G^{\ddagger} = -RT \ln(k_A/10^{11}), \tag{18}$$

and, using this approximation, it is possible to estimate $\Delta G^{\ddagger}$ (Table V). As shown in Fig. 16, the derived $\Delta G^{\ddagger}$ values are a monotonous function of ΔG, and extrapolation of the curve gives a value for the intrinsic barrier to electron transfer ($\Delta G_\circ^{\ddagger}$) of ~9 kJ mol^{-1}. From these observations (*47*), it appears that the different triplet-quenching rate constants found when BQ is used to quench the triplet excited states of a variety of metalloporphyrins can be explained simply in terms of thermodynamic effects (*47*).

Once formed, the ion pair can dissociate into separate ion products or recombine to form ground-state reactants. In turn, the separate ions undergo reverse electron transfer, and the rate for this latter process is probably the diffusion-controlled rate limit. For this system, the rate constant for ionic separation of the ion pair k_S can be calculated from a modified Fuoss equation (*44, 45*)

$$k_S = f(9.5 \times 10^{10}), \tag{19}$$

$$f = W/RT(1 - e^{-W/RT}), \tag{20}$$

$$W = -Ne^2/\varepsilon r, \tag{21}$$

where the radius of the ion pair r is 8.8 Å, and for the system under investigation has a value of $2.0 \times 10^{10}\ s^{-1}$. Now, the efficiency of ionic separation from the ion pair (ϕ_S) can be expressed in the form

$$\phi_S = k_S/(k_S + k_B), \tag{22}$$

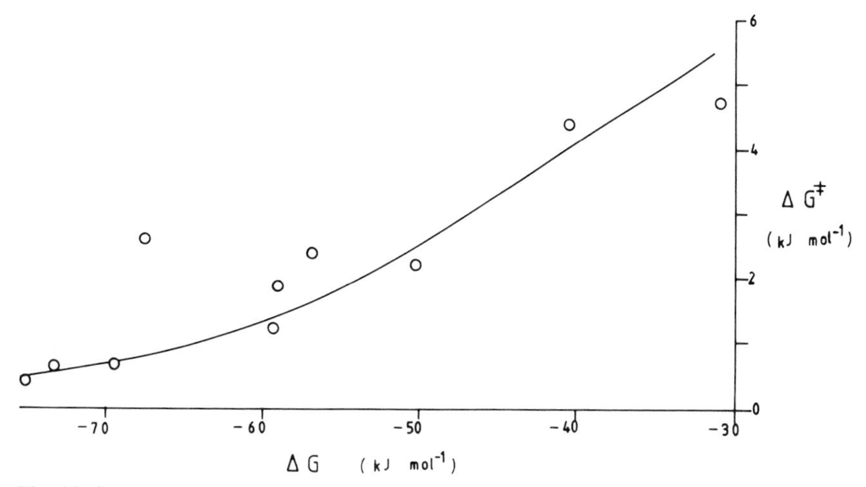

Fig. 16. Correlation between ΔG and the activation energy for quenching the triplet excited state of metalloporphyrins with benzoquinone.

and ϕ_S can be measured by flash photolysis techniques. Thus

$$\phi_S = \phi_{ions}/(\phi_T \phi_Q), \tag{23}$$

where ϕ_T is the quantum yield for formation of the triplet excited state of the metalloporphyrin (Table VI) and ϕ_Q the probability of quenching the triplet excited state at a particular concentration of quencher:

$$\phi_Q = k_T[BQ]/(k_T[BQ] + \tau_T^{-1}). \tag{24}$$

TABLE VI

Thermodynamic and Kinetic Parameters for Deactivation of the Ion Pair

Compound	ϕ_T	ϕ_{ions}	ϕ_S	$10^{-10} \times k_B$ (s^{-1})	$-\Delta G$ (kJ mol^{-1})
MgTPP	0.85	0.24	0.28	5.1	67.5
CdTPP	1.00	0.16	0.16	10.5	76.2
ZnTPP	0.90	0.22	0.24	6.3	83.9
CrTPP	—[a]	$<10^{-3}$	$<10^{-3}$	$>10^3$	91.7
RuTPP	1.00	0.07	0.07	26.6	94.6
CuTPP	1.00	$<10^{-3}$	$<10^{-3}$	$>10^3$	102.3
H_2TPP	0.88	0.084	0.095	19.0	107.1
AlTPP	0.85	0.21	0.25	6.0	113.9
PdTPP	1.00	0.055	0.055	34.4	113.9

[a] Assumed to be 1.0.

For metalloporphyrins with long excited triplet-state lifetimes (τ_T), it is easy to use sufficient BQ to ensure that ϕ_Q approach unity. This is not so easy when the triplet lifetime is very short, as it is with CuTPP and CrTPP (*49*), and here it is necessary to use concentrations of BQ as high as 1 *M* to obtain $\phi_Q \approx 1$ (with such high concentrations of BQ there is the possibility of ground-state complexation). The quantum yield for formation of redox ions (ϕ_{ions}) for each system was measured by nanosecond-flash photolysis by determining the concentration of metalloporphyrin π radical-cation formed after flash excitation of the metalloporphyrin in the presence of BQ ($\sim 10^{-3}$ *M*). The derived values are collected in Table VI.

The derived ϕ_S values are collected in Table VI; they vary between 0.28 for MgTPP and <0.002 for CuTPP and CrTPP. Thus the nature of the central metal ion has a strong influence on the yield of redox ions, and from Eq. (22), this influence seems to be concerned with k_B (Table VI). Reverse electron transfer within the ion pair is a type of nonradiative transition, because a large amount of electronic energy must be dissipated in the form of vibrational degrees of freedom. From classical kinetic theory, it is possible to express k_B in the form of Eq. (17), and according to this equation the factors that contribute toward the rate of geminate recombination include thermodynamic ($\Delta G^{\ddagger}$ and $\Delta G^{\ddagger}_{\circ}$), spin, and entropy effects connected with $k_{\circ}$.

First, let us consider the thermodynamic effects. The overall free-energy change associated with geminate recombination can be calculated according to

$$\Delta G = -nF(E^{\circ}_{BQ/BQ^-} - E^{\circ}_{p^+/p}) + e^2N/r\varepsilon, \tag{25}$$

and the calculated values (*47*) are given in Table VI. It is seen that recombination is extremely exergonic in all cases, but, as shown in Fig. 17, there does not appear to be a simple relationship between log k_B and ΔG. If we assume that the intrinsic barrier is of the same order as that involved in k_A and if we accept that $\Delta G^{\ddagger}$ should be a monotonous function of ΔG (see Fig. 16), then the variations in k_B observed for the different metalloporphyrins cannot be ascribed solely to thermodynamic effects. Certainly, we would expect k_B to increase with more negative ΔG values, but, from Fig. 15, with such large thermodynamic driving forces as we have here this increase should be relatively small (*47*).

If the variations in k_B cannot be explained by thermodynamic arguments, then we must invoke the hypothesis that $k_{\circ}$ varies for the different metalloporphyrins, and from absolute reaction-rate theory, we can formulate $k_{\circ}$ as

$$k_{\circ} = (\kappa kT/h)\exp(\Delta S^{\ddagger}/R)\exp(-\Delta H^{\ddagger}/RT), \tag{26}$$

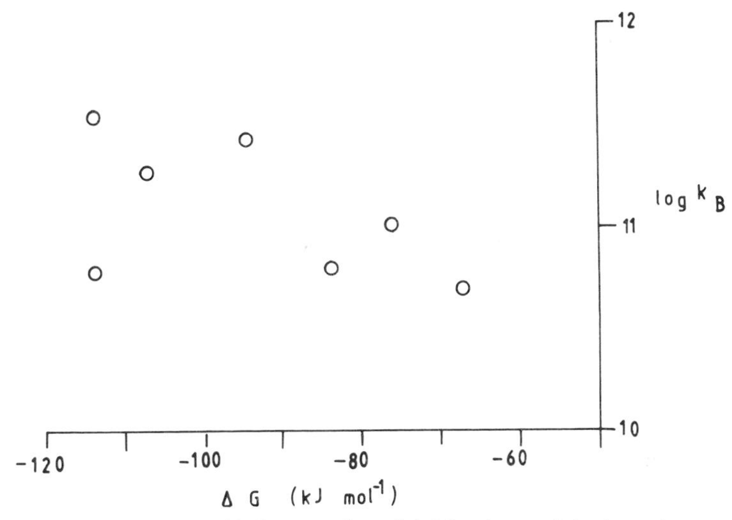

Fig. 17. Relationship between k_B and ΔG for decay of the ion pairs.

where k is the Boltzmann constant and κ a transmission coefficient. This implies that either the entropy term or κ is dependent on the nature of the central metal ion. The entropy term is rather vague, but geminate recombination should lead to an increase in entropy so that k_o for k_B should be higher than for k_A. Thus we are left with having to explain the observed k_B values in terms of κ.

In our simple model, geminate recombination is a spin-forbidden nonradiative process, and, as such, κ will be small (perhaps in the region of 0.01) (*47*). This suggestion is in agreement with the observed results for MgTPP, ZnTPP, and AlTPP, where k_B is relatively independent of ΔG and seems to be approaching a maximum value of 10^{11} s^{-1} (Fig. 17). With an estimated κ of 0.01, k_o would be $\sim 10^{13}$ s^{-1}, which is in keeping with that predicted by absolute reaction-rate theory when entropy factors do not inhibit reaction. For the other diamagnetic metalloporphyrins (Cd, Ru, Pd, and H_2TPP), the observed k_B is somewhat higher than 10^{11} s^{-1}, suggesting that here κ exceeds 0.01 because of spin–orbital coupling effects. The value of κ for these compounds should be a function of the spin–obital coupling constant of the metal ion, and the degree of overlap between the metal ion and the π-orbital system of the porphyrin π radical-cation (*47*).

In fact, some estimate of the relative efficiency of spin–orbital coupling for the different metalloporphyrin π radical-cations should be possible from epr spectroscopy because the g value is sensitive to intramolecular interactions (*21*). Although such values are not readily available and there may be complications resulting from ion pairing or solvent effects, it is known that the presence of atoms or ions with large spin–orbital coupling constants leads to distortions in the *g* values. This distortion is manifest in enhanced relaxation. Thus for PdTPP, RuTPP, and CdTPP the observed increase in k_B is probably a reflection of an enhanced κ resulting from spin–orbital coupling perturbations. However, this explanation does not hold for H_2TPP. With the paramagnetic metalloporphyrins (CrTPP and CuTPP), interaction between the unpaired electrons on the central metal ion and the porphyrin π system is strong [no epr signals have been observed for the π radical-cations (*21*)], so it is difficult to assign a formal spin number to the π radical-cations. This interaction has the effect of relaxing the spin selection rules governing geminate recombination and, hence, κ should approach unity. Consequently, for the paramagnetic metalloporphyrins, k_o should be about 10^{13} s^{-1}, and from Eq. (17), with a small activation energy we would expect that k_B be of the order of 10^{13} s^{-1}. If so, then ionic separation of the ion pair (k_S) would not compete with geminate recombination, and the yield of redox ion products would be minimal (*47*).

Thus according to this model k_B should depend on thermodynamic effects, spin–statistical factors, entropy changes, and the degree of overlap between orbitals on the two reactants. To optimize the yield of redox ion products, it is necessary to pay attention to all of the preceding parameters because, from Eq. (22), the efficiency of ionic separation from the ion pair depends on the magnitude of k_B. It depends also on the magnitude of the rate constant for ionic dissociation of the ion pair k_S, and, from Eq. (19), this rate constant depends on electrostatic forces between the ions (Z_AZ_B), ionic strength, dielectric constant, solvent viscosity, and temperature. Each of these latter parameters has been investigated, using the zinc(II) porphyrin-photosensitized reduction of methylviologen (MV^{2+}) in aqueous solution as a test system.

The importance of coulombic forces between the reactants was studied (*50*), rather neatly, by using a series of zinc porphyrins with different types of water-solubilizing group, as shown in Fig. 18, which allowed the Z_AZ_B term to be varied from 8+ to 8−. This Z_AZ_B term is important in two respects: first, it plays a significant role in determining both the rate of diffusional encounter between the reactants and the stability constant of the encounter complex, and second, it makes a considerable contribution toward the rate of ionic separation from the ion pair. In quantitative terms, strong coulombic repulsion between the reactants (e.g., when $Z_AZ_B = 8+$) should decrease k_D and the stability constant for the encounter complex and increase the rate of ionic separation from the ion pair. Strong electrostatic attraction between the reactants (e.g., $Z_AZ_B = 8-$)

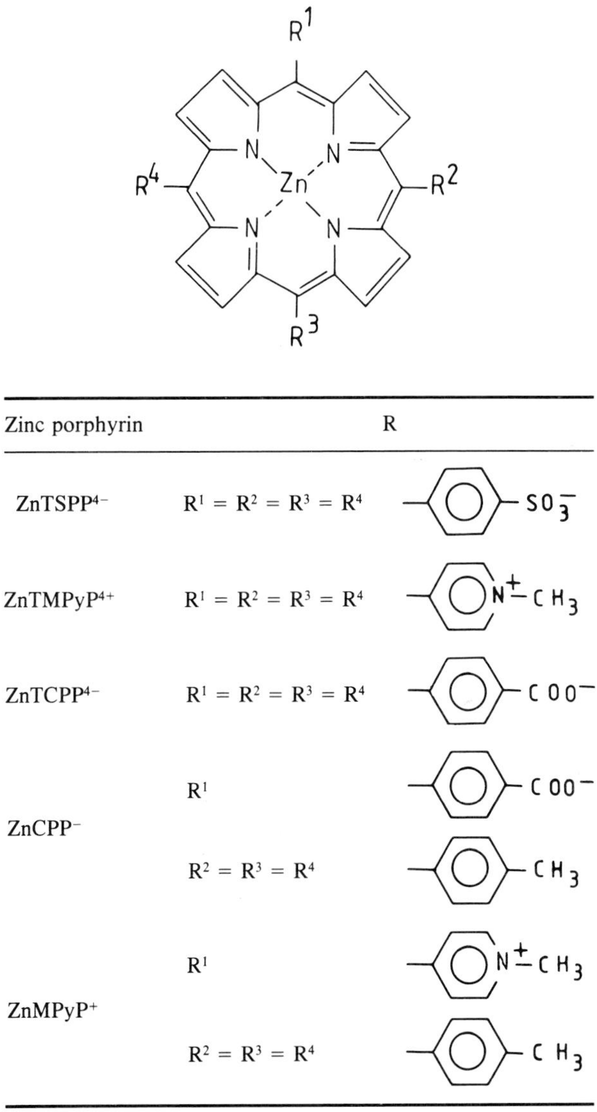

Zinc porphyrin		R
$ZnTSPP^{4-}$	$R^1 = R^2 = R^3 = R^4$	$-C_6H_4-SO_3^-$
$ZnTMPyP^{4+}$	$R^1 = R^2 = R^3 = R^4$	$-C_5H_4N^+-CH_3$
$ZnTCPP^{4-}$	$R^1 = R^2 = R^3 = R^4$	$-C_6H_4-COO^-$
$ZnCPP^-$	R^1	$-C_6H_4-COO^-$
	$R^2 = R^3 = R^4$	$-C_6H_4-CH_3$
$ZnMPyP^+$	R^1	$-C_5H_4N^+-CH_3$
	$R^2 = R^3 = R^4$	$-C_6H_4-CH_3$

Fig. 18. Structures of water-soluble zinc porphyrins.

should have much the opposite effect. These expectations are realized by the experimental data (*50*), as shown in Table VII, and, in particular, the quantum yield for formation of redox ions (ϕ_{ions}) depends markedly on the Z_AZ_B term. For $Z_AZ_B = 8+$, the ϕ_{ions} term is extremely high (which is essential for a practical solar-energy storage device), and in this case the

TABLE VII

Influence of Electrostatic Forces on the Efficiency of Photoreduction of MV^{2+}

Sensitizer	Z_AZ_B	$10^{-7} \times k_T$ (M^{-1} s^{-1})	ϕ_{ions}
$ZnTMPyP^{4+}$	8+	1.8	0.75
$ZnMPyP^{+}$	2+	2.1	0.10
$ZnCPP^{-}$	2−	100	0.037
$ZnTCPP^{4-}$	8−	1300	<0.01
$ZnTSPP^{4-}$	8−	1400	<0.01

low triplet-quenching rate constant (k_T) can be offset simply by increasing the concentration of the MV^{2+} quencher until all of the triplet state undergoes deactivation via electron transfer to MV^{2+} (this requires $[MV^{2+}] > 5 \times 10^{-3}$ *M*). This high ϕ_{ions} term can be attributed (*50*) to the strong coulombic repulsion within the ion pair, but it should be noted that changing the nature of the water-solubilizing group also changes the thermodynamic driving force for electron transfer within the encounter complex ($\Delta G°$). This term is important for both k_A and k_B (*50*).

The preceding study has shown (*50*) that $ZnTMPyP^{4+}$ is by far the most effective sensitizer for the reduction of MV^{2+} in aqueous solution. Because this reaction involves highly charged ions, the efficiency of charge separation from the ion pair should depend on the ionic strength. In fact, according to Eq. (19), k_S contains an ionic strength-dependent term, and Fig. 19 shows the influence of ionic strength on ϕ_{ions}. The effect is not too marked (*45*), but at ionic strengths greater than 0.1 *M*, ϕ_{ions} decreases steadily. This is an important finding in that the total ionic strength of a solution used for H_2 or O_2 production is determined by the need to use quite high concentrations of several ingredients, and in most cases, it will be high (probably >0.1 *M*). The influence of ionic strength on ϕ_{ions} will decrease for systems in which Z_AZ_B is not so highly positive (in fact, for a negative Z_AZ_B term, increasing the ionic strength should increase ϕ_{ions}), and in a final system some consideration should be given to this point (*45*).

The dielectric constant ε of the solvent is also involved in determining the magnitude of k_S [Eq. (19)] and the stability constant for the encounter complex. According to Eq. (19), for $Z_AZ_B = 8+$, decreasing the dielectric constant should lead to an increase in k_S and, hence, an increase in ϕ_{ions}, but, as shown in Fig. 20, this is not realized experimentally. In these experiments (*51*), a series of water–dioxane mixtures was used to vary ε,

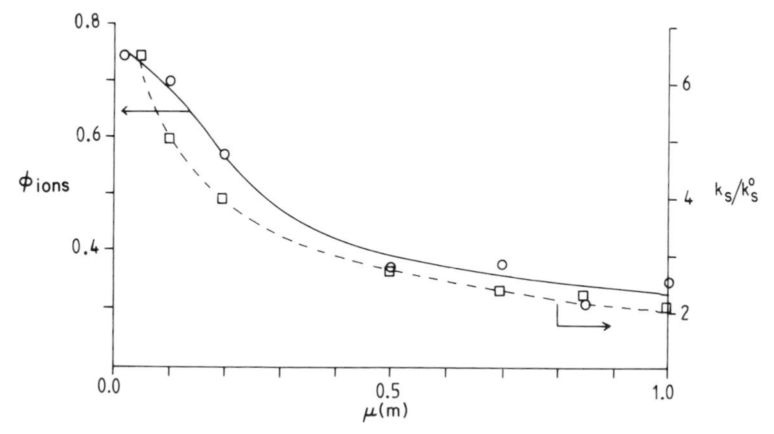

Fig. 19. Influence of ionic strength on the yield of ion products (○) and k_s (□), calculated according to Eq. (19).

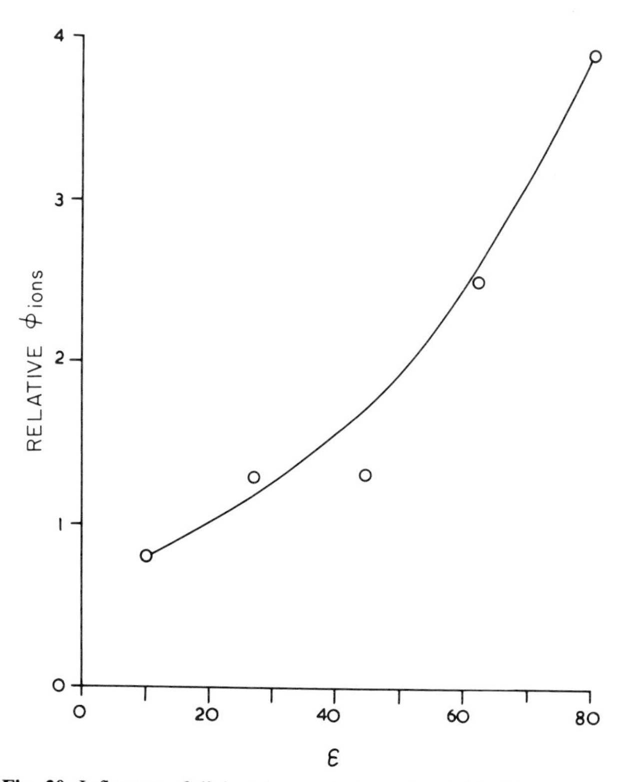

Fig. 20. Influence of dielectric constant on the yield of ion products.

and it was found that ϕ_{ions} decreased substantially with increasing concentration of dioxane. This effect was attributed to the decreasing stability of the encounter complex toward reverse dissociation, so that in nonpolar solvents electron transfer (step k_A) could not compete with dissociation (step k_{-D}). The importance of polar solvents in enhancing the yield of redox ions in systems where there is no electrostatic potentials between the initial reactants has been demonstrated previously.

Finally, for the $ZnTMPyP^{4+}$–MV^{2+} system in water at $\mu = 0.1\ M$, the effect of temperature on ϕ_{ions} was studied (*51*). This could be an important effect, because any practical solar-energy storage device by necessity will have to operate in sunlight so that the device may achieve modest operating temperatures. Fortunately, as shown in Fig. 21, the yield of redox ions increases (*51*) with increased temperature, whereas the rate constant for reverse electron transfer between the ions decreases with increasing temperature. This may not be so for all photoredox processes, because there are many individual steps in the overall reaction that possess temperature-dependent terms.

Overall, we have seen that the triplet excited state of a metalloporphyrin can generate high yields of photoredox products in fluid solution. The

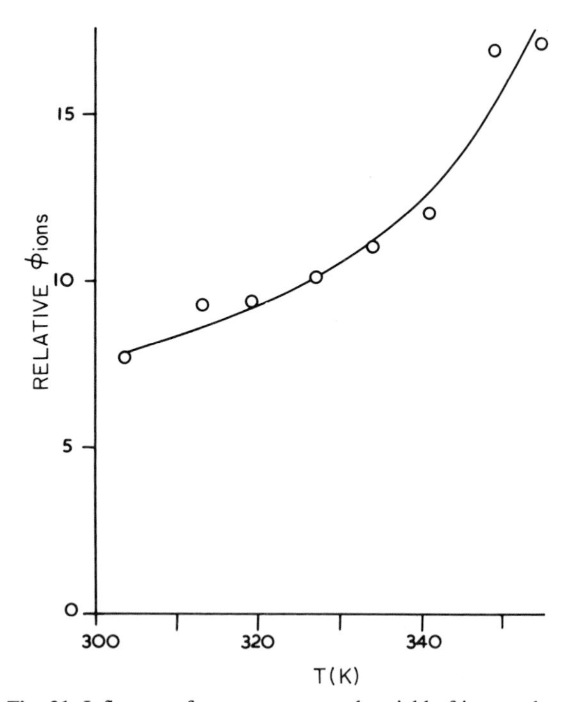

Fig. 21. Influence of temperature on the yield of ion products.

long excited-state lifetimes normally associated with these states allow many diffusional encounters with a quencher to occur before nonradiative deactivation takes place, so that low quenching rate constants or low quencher concentrations can be tolerated (this is not so with singlet excited-state reactions, which are usually limited by the short singlet lifetimes). However, the metalloporphyrin and the reaction conditions must be selected with great care if the maximum yield of redox ions is to be obtained. Assuming that electron transfer is allowed on thermodynamic grounds, then the central metal ion should be selected so that it possesses minimal spin–orbital coupling properties, and in this respect Mg, Al, and Zn porphyrins seem to be the most attractive (note that nature makes extensive use of Mg porphyrins). Then, depending on the type of quencher molecule used in the reaction, the yield of redox ions can be tuned, at least to some degree, by paying attention to electrostatic considerations and varying the nature of the solvent or the ionic strength. It is obvious from the preceding discussion that significant improvements in ϕ_{ions} can be obtained from these simple considerations, and by careful attention to detail, it should be possible to obtain ϕ_{ions} in the order of 50% or higher for most thermodynamically allowed systems.

The application of such considerations may allow modest yields of redox ions from singlet excited-state reactions. One of the major problems with singlet-state processes is that k_B is spin allowed and competes favorably with k_S, but according to the preceding discussion, there are several ways available to increase k_S. In fact, both Gouterman (*52*) and Harriman (*53*) and their co-workers have shown that it is possible to obtain redox ions from the singlet excited state of a metalloporphyrin when there is relatively strong electrostatic repulsion between the redox ions within the ion pair. So far, such systems have remained poorly studied, but there are many advantages to be gained by using the singlet rather than the triplet state as the active photosensitizer (notably thermodynamic ones), so further work in this area should be encouraged.

V. Charge Transport

A. In Vivo *Studies*

The consequence of light harvesting is that the captured photons are tunneled efficiently into the reaction-center complex where charge separation occurs. The redox ions so produced are thermodynamically unstable with respect to reverse electron transfer, but secondary electron transfers take place so that the oxidizing and reducing equivalents are removed from the reaction-center complex. This secondary electron transfer, which is a thermal process, is a most important feature of the natural photosynthetic process, and in many respects it is the key to efficient charge separation. Although secondary electron transfers should present no real difficulties in model systems, it is important that the amount of energy dissipated in these steps is minimal, or else the final oxidizing and reducing equivalents will have little application for useful chemical reactions.

Figure 22 shows the pathway for electron transport in green-plant photosynthesis, with the ordinate giving the approximate midpoint potential of each redox component. As can be seen from this figure, several of the important electron carriers remain unidentified, but I is almost certainly a pheophytin molecule, M quite possibly an Fe–S complex, and Q_a and Q_b are special quinones, both of which are magnetically coupled to iron(II). These quinones, which are commonly believed to be plastoquinones, are in some specialized environment that helps to stabilize their radical-anions (these are seen easily by epr spectroscopy), and they couple the one-electron redox chemistry of the reaction-center complex to the two-electron–two-proton chemistry of the plastoquinone pool. Various cytochromes function as important electron donors to PS-I, but the nature of the intermediate donor Z in PS-II (and any intermediate donors be-

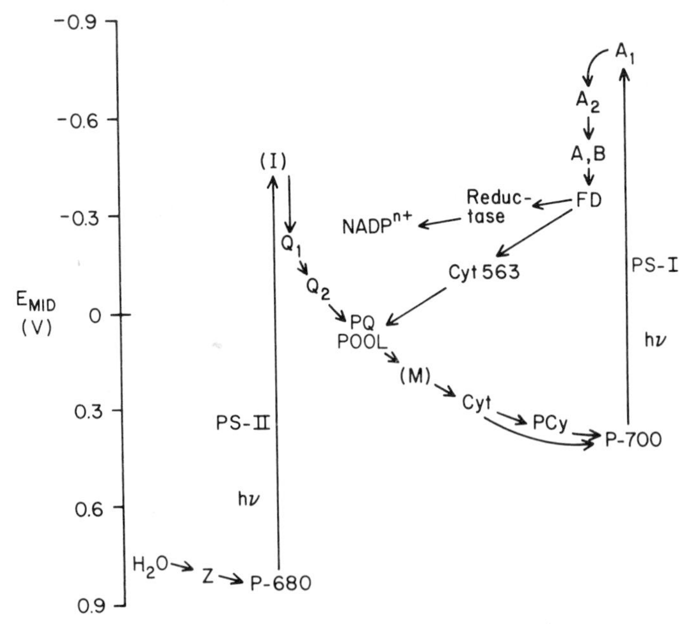

Fig. 22. Pathway of electron transport in green-plant photosynthesis.

tween water and Z or between Z and P-680) remains obscure. This area is about the least understood of all the processes that together make up photosynthesis, but there is a general acceptance that a manganese complex of some type may be involved in this part of the system (see Section VI,B). For a more detailed survey of the electron-transport pathway in bacterial and green-plant photosynthesis see references *2* and *3*.

The approximate time scale for charge transport throughout a bacterial reaction-center complex is given in Fig. 23. The vertical scale refers to the free-energy change in transferring an electron from the highest filled molecular orbital of a donor to the lowest empty molecular orbital of an acceptor. The distances were estimated from spectroscopic data assuming a linear chain of electron-transfer reactants with an average molecular distance of 0.5 nm. The average lifetimes and energies were obtained from published reports. The most significant feature of this figure is that the secondary electron-transfer steps are all much faster than the competing, energy-wastage steps. This explains why the quantum yield for charge separation is close to unity.

This high quantum yield for overall photoinduced charge separation, which is also an inherent feature of green-plant photosynthesis, is a consequence of the rapid charge transport, which also allows a rapid turnover with respect to the primary reactants within the reaction-center complex. If model systems are to adequately mimic the natural processes, then they must possess some means of quickly transporting the oxidizing and reduc-

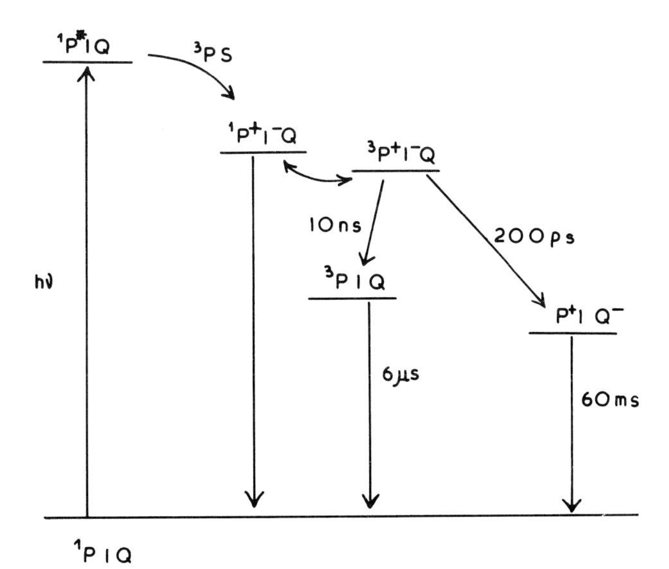

Fig. 23. Time scale for bacterial photosynthesis.

ing equivalents away from the initial reaction site so that geminate recombination is inhibited. However, all biological energy transduction apparatuses contain membranes; the chlorophyll reaction center uses photon energy to separate charges across a membrane, cytochromes reduce $NADP^+$ at a membrane interface, enzymes fix chemical species at membrane sites, and bacteriorhodopsin is a proton pump spanning a membrane. It seems relatively clear that these biological membranes play an indispensable role in the efficient functioning of energy-conversion processes in living systems, and it seems probably that totally artificial systems must also incorporate interfaces and/or membranes across which gradients can be developed and maintained.

B. In Vitro *Studies*

Despite its obvious importance in the natural photosynthetic process, there have been relatively few attempts to develop model systems capable of charge transport. This is because, at least in part, of the difficulties associated with the organization of reagents so that preferential electron transfer occurs in a unidirectional manner. In fluid solution, random diffusional encounters between the reactants can only be controlled by kinetic and concentration factors, whereas in more organized environments, such as monomolecular films, spatial separation and nearest-neighbor interactions are more closely controllable, but such systems are fragile and require highly sensitive spectroscopic analysis. Consequently, most model systems have been concerned with very limited versions of charge transport, and, as shown subsequently, many studies have involved the use of irreversible redox couples so that charge separation occurs on a permanent time scale.

The simplest, and in several respects the most elegant, type of charge transport for a metalloporphyrin involves self-exchange in which the oxidizing or reducing equivalents are transferred between similar species, for example,

$$\mathrm{Chl}^{\ddagger} + \mathrm{Chl} \rightleftharpoons \mathrm{Chl} + \mathrm{Chl}^{\ddagger} \quad (27)$$

Several investigations have been concerned with measuring the self-exchange rate constants for metalloporphyrin, and, in particular, cobalt and iron porphyrins have been studied in some detail. The magnitude of the rate constant for self-exchange is strongly dependent on the nature of any axially coordinated ligand, the spin state of the central metal ion, and whether the redox change involves the porphyrin ring or the metal ion (*54*, *55*). For $ZnTMPyP^{4+}$ in aqueous solution, the rate constant lies in the range of 10^8 M^{-1} s^{-1} (*56*).

More pertinent to the present discussion, there have been several attempts to study electron-transfer reactions involving the one-electron oxidation or reduction products of chlorophyll *a* in fluid solution. As reviewed by Seely (*28*), there are numerous examples of simple diffusion-controlled reverse electron-transfer processes involving chlorophyll *a*, and many other metalloporphyrins, but these need not be considered here.

$$\text{Chl } a^* + \text{Q} \longrightarrow \text{Chl } a^{\dot{+}} + \text{Q}^{\dot{-}} \qquad (28)$$

$$\text{Chl } a^{\dot{+}} + \text{Q}^{\dot{-}} \longrightarrow \text{Chl } a + \text{Q} \qquad (29)$$

Instead, Neta *et al.* (*57*) have shown from pulse radiolysis studies that reduced chlorophyll *a* rapidly transfers its extra electron to quinones in fluid solution,

$$\text{Chl } a^{\dot{-}} + \text{Q} \rightleftharpoons \text{Chl } a + \text{Q}^{\dot{-}} \qquad (30)$$

and in basic isopropanol the rate constant for electron transfer to 9,10-anthraquinone was found to be near $10^9\ M^{-1}\ s^{-1}$. Other metalloporphyrins show similar kinetic behavior (*57*), and Table VIII contains a collection of some rate constants for electron transfer from a metalloporphyrin π radical-anion (donor) to an electron acceptor in fluid solution. In neutral aqueous solution, the rate constants follow the order $MV^{2+} \approx$ anthraquinone–disulphonate < duroquinone < BQ, corresponding to the changes in their redox potentials, whereas in some experiments there was spectral evidence of a long-lived intermediate. One difficulty with such studies is

TABLE VIII

Rate Constants for Electron Transfer from Metalloporphyrin Radical-Anions

Donor[a]	Acceptor	pH	k ($M^{-1}\ s^{-1}$)
$H_3TCPP\cdot$	BQ	7.0	1.3×10^9
$H_3TCPP\cdot$	DQ	7.2	1.0×10^9
$H_3TCPP\cdot$	AQS	6.9	1.1×10^7
$H_2TCPP^{\dot{-}}$	AQS	11.4	3.3×10^8
$H_3TCPP\cdot$	MV^{2+}	7.0	$<10^8$
$H_3TCPP\cdot$	MV^{2+}	9.6	6×10^8
$Na_2TPP^{\dot{-}}$	AQ	—	1.8×10^9
$ZnTPP^{\dot{-}}$	AQ	—	2.6×10^9
Chl $a^{\dot{-}}$	AQ	—	8×10^8

[a] TCPP, *meso*-Tetra(4-carboxyphenyl)porphyrin.

that the porphyrin π radical-anions undergo a pK transition in alkaline solutions (around pH 8–10),

$$P^{\overline{\cdot}} + H^{+} \rightleftharpoons PH\cdot \tag{31}$$

so that, sometimes, it is a problem to ascertain the real nature of the donor moiety (*57*).

This is not a problem with the chlorophyll *a* π radical-cation, which is reduced by many electron donors in fluid solution. In fact, Darwent (*58*) has investigated this reaction in some detail, and Table IX gives a compilation of some rate constants, measured by flash photolysis techniques, for reduction of Chl $a^{\dot{+}}$ in solution:

$$\text{Chl } a^{\dot{+}} + D \rightleftharpoons \text{Chl } a + D^{\dot{+}} \tag{32}$$

As shown in Fig. 24, there is a good correlation between the rate constant for electron transfer k and the free-energy driving force $\Delta E°$, as expected for a redox process, and the observed data (*58*) can be explained in terms of the Marcus theory for outer-sphere electron-transfer reactions. Many other porphyrins exhibit identical behavior, and in aqueous solution ethylenediaminetetraacetate (EDTA) has been a commonly used electron donor because the $\text{EDTA}^{\dot{+}/0}$ redox couple shows irreversible character:

$$\text{EDTA}^{\dot{+}} \longrightarrow \text{products} \tag{33}$$

With such irreversible couples, charge separation is permanent.

A few important points should be noted from the preceding two studies. First, for Chl $a^{\dot{+}}$ reverse electron transfer is extremely rapid, and for any particular quinone–hydroquinone couple, the quinone radical-anion is a far better reductant than the hydroquinone. Thus, in fluid solution, hydroquinones are poor candidates for reversible reductants because the

TABLE IX

Rate Constants for Electron Transfer to the Chlorophyll *a* Radical-Cation

Donor	$E°$ (V)	k (M^{-1} s^{-1})
$BQH\cdot–BQH_2$	1.084	4.1×10^{3}
$DQH\cdot–DQH_2$	0.896	5.1×10^{6}
$BQ–BQ^{\overline{\cdot}}$	0.081	3.3×10^{9}
$DQ–DQ^{\overline{\cdot}}$	−0.264	7.5×10^{9}
Cysteine	0.13	4.3×10^{8}
Ascorbate	0.46	9.5×10^{8}
MV^{2+}	−0.44	8.1×10^{8}

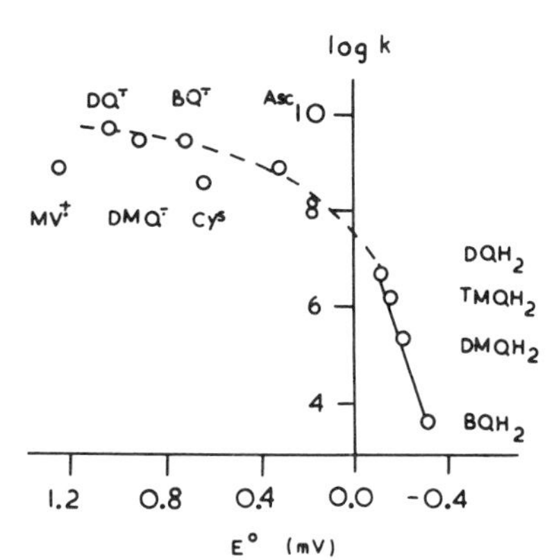

Fig. 24. Relationship between k and $\Delta E°$ for electron transfer from chlorophyll *a* radical-cation (*58*).

quinone and semiquinone products short-circuit the reaction. Consider a reaction scheme in which triplet chlorophyll *a* is used to photoreduce MV^{2+} in fluid solution and benzhydroquinone (BQH_2) is used to reduce the chlorophyll *a* π radical-cation so formed:

$$\text{Chl } a^* + MV^{2+} \longrightarrow \text{Chl } a^{\dagger} + MV^{\dagger} \quad (34)$$

$$\text{Chl } a^{\dagger} + MV^{\dagger} \longrightarrow \text{Chl } a + MV^{2+} \quad (35)$$

$$\text{Chl } a^{\dagger} + BQH_2 \longrightarrow \text{Chl } a + BQH\cdot + H^+ \quad (36)$$

$$2BQH\cdot \rightleftharpoons BQ + BQH_2 \quad (37)$$

$$BQH\cdot \rightleftharpoons BQ^{\overline{\cdot}} + H^+ \quad (38)$$

As shown in Tables VIII and IX, both BQ and $BQ^{\overline{\cdot}}$ can participate in the overall scheme—BQ by reducing triplet chlorophyll *a* and $BQ^{\overline{\cdot}}$ by reacting with Chl $a^{\dagger}$. Also, electron transfer from $MV^{\dagger}$ to BQ occurs so that there is no net buildup in the concentration of reduced viologen.

Second, irreversible reductants, such as cysteine, EDTA, or ascorbate, react efficiently with the chlorophyll *a* π radical-cation, and steady-state irradiation of chlorophyll *a*–MV^{2+} in fluid solution containing one of the preceding reagents as sacrificial electron donor results in a buildup of $MV^{\dagger}$. Depending on the experimental conditions, the reduced viologen may be used to reduce water to hydrogen, as described in a following section.

Third, electron transfer from reduced metalloporphyrins to quinones occurs with a high rate (*57*), and the process is irreversible. In most examples reported so far, the energy wastage in this step is very high; for example, electron transfer from Chl $a^{\overline{\cdot}}$ to anthraquinone involves an approximate $\Delta G°$ of -1 eV. Presumably, the rate of electron transfer for less exergonic processes will be much slower and the reaction more reversible.

It is interesting to note that pulse radiolysis studies with the porphyrin molecule containing four covalently attached quinone units, as described previously (Fig. 13), have shown (*59*) that the extra electron is delocalized over the entire molecule and does not reside on the quinone units. Thus on γ irradiation in methanol solution, both the solvated electron and the hydroxymethylene radical reduce the porphyrin–quinone molecule (*59*):

$$e^- + PQ \longrightarrow PQ^{\overset{\cdot}{-}} \quad (39)$$

$$k = 1.7 \times 10^{10} \quad M^{-1}\,s^{-1}$$

$$\cdot CH_2OH + PQ \longrightarrow PQH\cdot + CH_2O \quad (40)$$

$$k = 1.9 \times 10^{8} \quad M^{-1}\,s^{-1}$$

$$PQ^{\overset{\cdot}{-}} + H^+ \rightleftharpoons PQH\cdot \quad (41)$$

$$pK \approx 10$$

The observed absorption spectra (Fig. 25) are typical of porphyrin π radical-anions. Time-resolved measurements showed that the protonated

$$2PQH\cdot \rightleftharpoons PQ + PQH_2 \quad (42)$$

radical disproportionated, but there was no evidence to show that the added electron could be localized on the quinone unit (*59*).

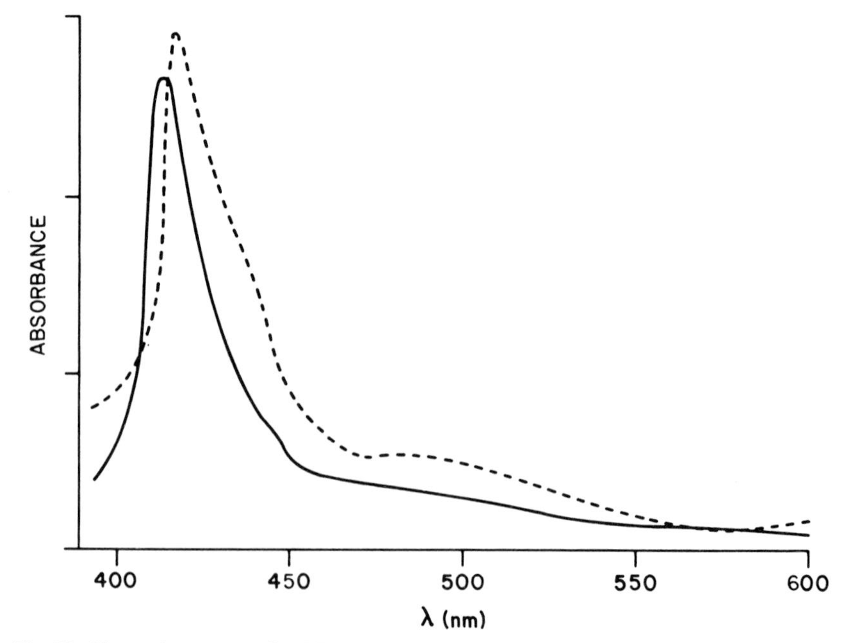

Fig. 25. Absorption spectra for TQP$^{\overset{\cdot}{-}}$ (——) and TQPH· (---) as observed by pulse radiolysis. TQP, *meso*-Tetra(quinone)porphine.

As just mentioned, it seems probable that any useful model system must possess an interface that assists three-dimensional separation of the charges. Several different types of interface can be considered. Lipid-bilayer membranes of living cells lack sufficient strength and stability outside the cell to be successfully incorporated into structures of reasonable size or to be operated for long periods of time. Even for applications requiring substantially smaller surface areas than would be needed for practical solar-energy storage devices, membrane fragility has been an insurmountable problem. However, totally artificial membranes such as polymer films generally have been too thick to display the charge-transport properties that membranes in living systems possess, and micelles and vesicles have not been convincing models. Attempts to stabilize lipid-bilayer membranes by interlacing with a polyamide or polycarbonate support seem to be highly promising because the final materials, although multilamellar, possess great stability and manipulability. Another type of potentially fruitful development is the manufacture of ultrathin polymer films by photopolymerization of monomer films impregnated with chromophores.

A type of assembly that may be able to replace the bilayer lipid membrane is the hollow-fiber membrane. Although cellulose acetate membranes have been in use for many years and hollow-fiber geometries can be readily produced, the major applications have been in dialysis and reverse osmosis. Most of the research in this area is proprietary. It has been suggested, however, that hollow-fiber membranes may have potential use in photosensitized redox reactions (*60*).

VI. Oxygen Formation

A. In Vivo *Studies*

Photosynthetic O_2 evolution is associated with PS-II, where the energy of a red quantum (~1.8 eV) is sufficient to induce charge separation in the reaction-center complex, yielding a strong oxidant $Z^{\ddagger}$ and a mild reductant $Q^{\overline{\cdot}}$:

$$\text{Z Chl Q} \xrightarrow{h\nu} \text{Z Chl}^{\ddagger}\ \text{Q}^{\overline{\cdot}} \longrightarrow \text{Z}^{\ddagger}\ \text{Chl Q}^{\overline{\cdot}} \qquad (43)$$

The reaction-center complex contains a special form of chlorphyll *a* (most probably a dimer), which is known as P-680. The primary electron acceptor Q is almost certainly either a monomeric chlorophyll *a* or a pheophytin molecule, but the nature of the donor Z is unknown. The PS-II charge separation takes place across the thylakoid membrane, with the oxidation process leading to O_2 formation occurring on the inner side.

To form one molecule of O_2 it is necessary to remove four electrons from two molecules of water and the average potential of the four oxidiz-

$$2H_2O \longrightarrow O_2 + 4H^+ + 4e \tag{44}$$

ing equivalents must be higher than +0.8 V. Moreover, at least four quanta are necessary to drive the four-electron process. Therefore, in green plants, either four reaction centers cooperate together to bring about the evolution of one molecule of O_2, or charge accumulation occurs within individual reaction centers. These two possibilities can be distinguished by monitoring O_2 evolution induced by short, bright flashes capable of bringing about single excitations of the PS-II reaction centers. Thus, for a dark-adapted system, if cooperation between reaction centers occurs, then O_2 should be evolved on the first flash, whereas further flashes would be required if charge accumulation was necessary. In fact, years ago several workers demonstrated (*3*) that there was no O_2 evolved on the first flash. The availability of short-duration flash lamps, together with very sensitive O_2-electrode detector systems, has enabled a thorough study of this type to be made. Joliot *et al.* (*61*) were the first to show that with successive flashes of light a characteristic pattern of O_2 evolution occurred. In essence, there is no O_2 evolved on the first or second flash, whereas the third flash gives the highest yield of O_2. Subsequent yields show a damped oscillation of four. Several hypotheses have been proposed to explain these results, but the most widely accepted is that given by Kok *et al.* (*62*). In this model, the O_2-evolving enzyme can exist in four photoactive charged states S.

In the S-state model (*62*) it is envisaged that oxidizing equivalents are accumulated by successive reduction of $Z^{\ddagger}$:

$$\text{Z P-680 Q} \xrightarrow{h\nu} \text{Z P-680}^+ \text{ Q}^- \longrightarrow \text{Z}^+ \text{ P-680 Q}^- \;\; (S_n \rightarrow S_{n+1}) \longrightarrow \text{Z P-680 Q}^- \tag{45}$$

The sequence leading to O_2 evolution is thought to be

$$\begin{array}{ccccccc} O_2 + 4H^+ & & S_0 \xrightarrow{h\nu} S_0^* \longrightarrow S_1 \xrightarrow{h\nu} S_1^* \\ \uparrow & & \uparrow \qquad\qquad\qquad\qquad \downarrow \\ & & S_4 \\ & & \uparrow \\ 2H_2O & & S_3^* \xleftarrow{h\nu} S_3 \longleftarrow S_2^* \xleftarrow{h\nu} S_2 \end{array} \tag{46}$$

For more information about the main properties of the S-state model, see the reviews by Joliot and Kok (*63*) and by Radmer and Cheniae (*64*).

According to the S-state model, the conversion of S_4 to S_0 should yield one O_2 molecule and four protons. Attempts have been made to prove that proton release follows O_2 evolution in the expected way, but it has been found that protons are released in several stages. There is considerable argument over the exact sequence of flash-induced proton release, but, certainly, the results cannot be explained in terms of the S_4-to-S_0 step releasing four protons.

Although there have been many attempts to isolate a biochemically active complex from photosynthetic tissue containing the basic functional unit required for the photooxidation of water, none has been successful. There is, however, some evidence to suggest (*3*) that the charge accumulator may involve manganese—a suitable candidate because it can exist in a wide range of different oxidation states. It is well known that when photosynthetic organisms are grown in the absence of manganese they lose the ability to evolve O_2 and also that this metal is essential for PS-II activity. However, it is possible to incorporate manganese into depleted cells so as to restore normal O_2 evolution. The reincorporation process only occurs with illuminated, intact living systems. For maximum activity, it seems that one manganese atom is required per 50–100 chlorophyll *a* molecules, in line with other evidence that there are about four manganese atoms per PS-II reaction center (*3*).

The involvement of manganese in O_2-evolving systems is also supported (*3*) by experiments in which isolated chloroplasts have been subjected to various treatments that remove this metal from its binding site. For example, treatment with hydroxylamine, washing with alkaline Tris, and gentle heating will release about two-thirds of the bound manganese, with a concomitant drop in the yield of evolved O_2. From a wide range of experiments, it appears that the extractable manganese that controls the O_2-evolving process is not required for the photooxidation of primary electron donors (e.g., Z and P-680) (*3*). Overall, the evidence suggesting that manganese is essential for photosynthetic O_2 evolution seems very convincing, but there is no real evidence to show that the active catalyst (i.e., the charge accumulator) is a manganese complex.

B. Manganese in Model Systems

The hypothesis that the natural PS-II utilizes some kind of manganese complex to oxidize water to O_2 has led to prolonged and detailed investigation into the possible role of manganese as a charge accumulator. In particular, the research groups of M. Calvin and G. Porter have long been active in this field. However, from a thorough review of the chemical and photochemical properties of manganese compounds, it was proposed (*65*) that the most suitable materials for *in vitro* model systems were the manganese porphyrins (MnP), which can undergo a series of well-defined redox changes. Thus it has been shown (*66*) that a Mn(II)P can be oxidized readily to the corresponding Mn(III)P that, in turn,, undergoes oxidation in alkaline solution to form the Mn(IV)P. This latter compound is stable only in relatively strong alkaline solution (pH $\geq$ 11), where it exists in the form on a μ-oxo dimer. The dimer can be oxidized further to give a monomeric Mn(V)P, but if the pH of the solution is decreased, the

$$\begin{array}{ccccccc} \mathrm{Mn(II)P} & \rightleftharpoons & \mathrm{Mn(III)P} & \rightleftharpoons & \mathrm{Mn(IV)P} & \rightleftharpoons & \mathrm{Mn(V)P} \\ & & \nwarrow & & \upharpoonleft\downharpoonright & & \\ & & & & \mathrm{Mn(IV)POMn(IV)P} & & \end{array} \qquad (47)$$

dimer undergoes rapid reduction to form the stable Mn(III)P. These interconversions (*66*) are shown in reaction (*47*) for details about the characterization and preparation of the individual compounds, see reference *66*.

The stability of the Mn(IV)P μ-oxo dimer is limited, even in alkaline solution, and in neutral or acidic pH its redox potential approaches (*66*) that necessary for liberation of O_2 from water. In fact, the rate of reduction for the dimer increases (*2*) with decreasing pH, although there still exists considerable doubt as to whether or not molecular O_2 is evolved during this reduction process. Definitive experiments are now in progress.

C. In Vitro *Studies*

Although most work has centered on the use of manganese porphyrins as oxidants for water, a great deal of attention has been focused on other metalloporphyrin photosensitizers. Some time ago, Wang (*67*) reported that zinc porphyrin coated on aluminum electrodes was able to photooxidize water to O_2. Fong and co-workers (*68*) have described the use of chlorophyll *a* hydrates, in the form of multilayers deposited upon platinum, as photooxidants for water, but the work is not at all convincing.

The photodissociation of water into H_2 and O_2 using visible-light excitation presents severe experimental difficulties, but if these problems can be solved, a successful system may have genuine application as a solar-energy storage device. At the present time, two major approaches toward the photodissociation of water in nonelectrochemical systems are being advocated. One approach involves irradiation of a semiconductor powder, either with band-gap excitation or with dye sensitization, but such systems generate a mixture of H_2 and O_2 that must be separated. The other approach, which has been less successful, involves coupling two separate photoreactions in much the same way that photosynthesis has evolved. One of the two photoreactions must be capable of the reduction of water to H_2, as shown in Section VII, and the other process must involve the oxidation of water to O_2, which is a particularly difficult process, and this has not been realized with much success. The only well-developed system capable of the photooxidation of water to O_2 in homogeneous solution uses tris(2,2′-bipyridyl)ruthenium(II) as the photosensitizer, persulfate or $Co(NH_3)_5Cl^{2+}$ as the sacrificial electron acceptor, and RuO_2 as the catalyst, but it suffers from several problems (*69–71*). Notably, the photosensitizer is expensive and collects only a small fraction of the solar spectrum, and the RuO_2 catalyst employed (*71*) is not sufficiently active to allow the sacrificial electron acceptor to be replaced with a reversible one. Thus there is an urgent need for us to identify some alternative O_2-producing systems, and Borgarello *et al.* (*72*) have reported that zinc(II) *meso*-tetra(*N*-methyl-4-pyridyl)porphine ($ZnTMPyP^{4+}$) in acidic solution is an efficient sensitizer of O_2 generation by visible light in the presence of electron acceptors, such as iron(III), and a colloidal RuO_2–TiO_2 catalyst, although we have not been able to confirm this finding.

The triplet excited state of $ZnTMPyP^{4+}$ is a powerful reductant ($E° \sim -0.4$ V versus NHE) and is oxidized by a wide variety of electron acceptors. Net electron transfer occurs in many cases, but even with a sacrificial acceptor where reverse electron transfer is inhibited, the lifetime of $ZnTMPyP^{5+}$ is short and decays over a time scale of a few seconds or less. In fact, the rate of decay increases with decreased pH throughout the range $2 < pH < 8$. At acidic pH irradiation of $ZnTMPyP^{4+}$ in the presence of a sacrificial acceptor results in formation of the dication $ZnTMPyP^{6+}$, and at higher pH there is some isoporphyrin formation.

The redox potential for the one-electron oxidation of $ZnTMPyP^{4+}$ in aqueous solution has been measured (*73*) as 1.18 V versus NHE, so that

$$ZnTMPyP^{5+} + e^- \longrightarrow ZnTMPyP^{4+} \tag{48}$$

the π radical-cation is a strong oxidant. On thermodynamic grounds it should be capable of O_2 evolution from water at $pH > 1$ because the redox potential for oxidation of water can be expressed as

$$E°_{H_2} = 1.23 - 0.059\ pH \quad V. \tag{49}$$

However, the liberation of O_2 from water requires a four-electron change.

$$2H_2O \longrightarrow O_2 + 4H^+ + 4e^- \tag{50}$$

If free-radical intermediates are to be avoided, it is necessary that the overall mechanism involves an intermediate species capable of charge accumulation. In other words, a catalyst capable of storing at least four oxidizing equivalents must be incorporated into the system if O_2 formation is to be observed, and prolonged irradiation of $ZnTMPyP^{4+}$ in the presence of a sacrificial electron acceptor, but in the absence of an added catalyst, certainly does not lead to the formation of O_2.

We have previously reported (*71*) that the quantum yield for the formation of O_2 from the $(bipy)_3Ru^{2+}$ photosensitized oxidation of water was markedly dependent on the type of catalyst used. Both RuO_2 and $CoSO_4$ were found to be effective O_2-producing catalysts, but the most efficient catalyst was RuO_2 supported on colloidal TiO_2. Using both the RuO_2–TiO_2 catalyst and $CoSO_4$, we were unable to observe O_2 formation from systems using $ZnTMPyP^{4+}$ as the photosensitizer. Thus irradiation of $ZnTMPyP^{4+}$ (1.0×10^{-5} *M*) in aqueous solution at pH 5 (acetate buffer) containing $Co(NH_3)_5Cl^{2+}$ (1.0×10^{-3} *M*) and various amounts of the RuO_2–TiO_2 catalyst gave no observable yield of O_2, as monitored with a membrane polarographic detector, and we estimate that the quantum yield for O_2 formation must be $<10^{-3}$. Similar negative results were observed for experiments performed at pH values between 2.0 and 6.7 and with $CoSO_4$ (5.0×10^{-3} *M*) as the catalyst. In no case was O_2 found as a reaction product.

Consequently, there appears to be some discrepancy between the findings of Borgarello *et al.* (*72*) and the work carried out in our laboratory. The most probable reason for this apparent lack of agreement lies with the catalysts used by the two research groups, and the development of active RuO_2–TiO_2 redox catalysts is now one of the most important goals in our work. It is essential that suitable catalysts be identified soon, and, similarly, that the mechanism for O_2 liberation on colloidal dispersions of RuO_2 be evaluated. Experiments are in progress, but the immediate future does not look too bright.

VII. Fuel Production

There has been a great surge of interest in the use of porphyrins and phthalocyanins for the photoreduction of water to H_2. In sacrificial, three-component systems, considerable success has been achieved and quantum efficiencies for H_2 production as high as 60% have been claimed. In such systems, a porphyrin or phthalocyanin is used to photoreduce MV^{2+} in slightly acidic solution (pH $<$ 6), and the oxidized form of the chromophore is reduced with an irreversible, secondary donor such as EDTA. Thus on steady-state irradiation the reduced form of the viologen builds up in solution, and this species, in the presence of a suitable catalyst, is capable of reducing water to H_2 (*50, 53*). Work in this area has

$$P^* + MV^{2+} \longrightarrow P^{\dagger} + MV^{\dagger} \quad (51)$$

$$P^{\dagger} + EDTA \longrightarrow P + EDTA^{\dagger} \quad (52)$$

$$EDTA^{\dagger} \longrightarrow \text{products} \quad (53)$$

$$2MV^{\dagger} + 2H^+ \overset{Pt}{\rightleftharpoons} 2MV^{2+} + H_2 \quad (54)$$

been reviewed (*74*), it need not be repeated here, and the most effective sensitizer was found to be $ZnTMPyP^{4+}$.

For the $ZnTMPyP^{4+}/MV^{2+}$/EDTA/Pt system, the optimum quantum efficiency for the formation of H_2 was about 60%, but as shown in Fig. 26, this decreased with prolonged irradiation time. This effect is owing, at least in part, to destruction of the viologen electron relay, which under such conditions is unstable with respect to oxidation and hydrogenation. The lack of longevity is the most serious problem associated with such photosystems, and it is essential that suitable replacements for the viologen be identified soon.

Several alternative cycles whereby a porphyrin is used to photoreduce water to H_2 have been considered (*74*) and reviewed. The problems associated with the poor stability of the viologen relay have been partially overcome by using reductive cycles in which the reduced form of the porphyrin is used to form H_2 directly. Such systems operate quite well,

$$P^* + EDTA \longrightarrow P^{\overline{\cdot}} + EDTA^{\dagger} \quad (55)$$

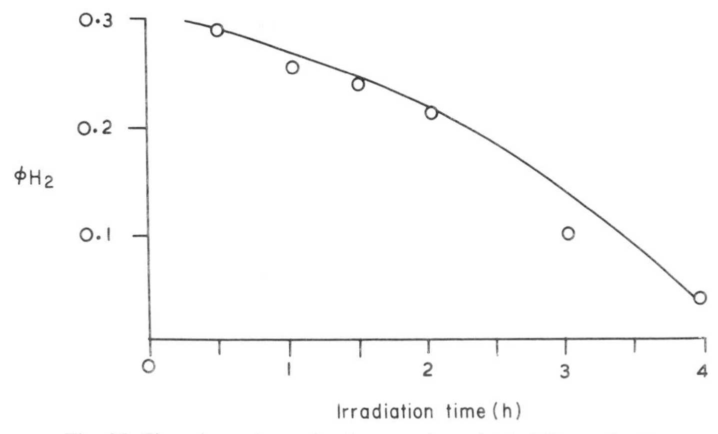

Fig. 26. Time dependence for the quantum yield of H_2 production.

$$\text{EDTA}^{+} \longrightarrow \text{products} \tag{56}$$

$$2\text{P}^{\overline{\cdot}} + 2\text{H}^{+} \xrightleftharpoons{\text{Pt}} 2\text{P} + \text{H}_2 \tag{57}$$

although the optimum efficiency for H_2 production is modest (~7%), and very high concentrations of Pt are required to inhibit undesirable side reactions that destroy the porphyrin.

All of the preceding systems have been based on production of H_2 at the expense of consuming some added sacrificial electron donor (e.g., EDTA). In the absence of such materials, no H_2 is formed, and, so far, it has not been possible to prepare colloidal Pt catalysts with sufficient selectivity to intercept reverse electron transfer between the photoredox ion products so that H_2 formation could occur. Work in this area is continuing, but preliminary results are not encouraging (*74*).

There have not been any reports describing investigations in which a metalloporphyrin has been used to photoreduce CO_2, although it is known that cobalt phthalocyanin can catalyze the electroreduction of CO_2 to formic acid.

VIII. Conclusions

Since the mid-1970s, considerable effort has been directed toward understanding and mimicking the natural photosynthetic processes of bacteria and green plants. Because of the complexity of the natural cells, interest has focused on small parts of the overall processes, and a great deal of progress has been made. We have a good (but not total) understanding of many of the individual steps in photosynthesis and are able to construct limited versions that mimic some of the essential features. This holds promise that, in the near future, we shall be able to construct a practical solar-energy storage device based on the photosynthetic unit of green plants. In this respect, it should be noted that the photoreduction of water to H_2 is not the bottleneck (many authors seem unaware of this point); the major problem remains to photooxidize water to O_2.

What must be done before we are in a position to build laboratory models capable of mimicking photosynthesis? At the moment, we are able to produce limited versions of bacterial photosynthesis in which red light is used to reduce water to H_2 at the expense of consuming some added electron donor. The efficiency of such model systems can be very high, but the longevity is poor. However, we have been unable to reproduce the essential features of PS-II, and metalloporphyrins have not proved to be good oxidants for water. From model systems, it seems that a strong oxidant is needed to liberate O_2 from water and an efficient catalyst must be used. Rather strangely, all successful model systems have used a high-valence metal ion as the oxidant, and there are no unambiguous examples in which an organic oxidant (such as a metalloporphyrin π radical-cation) has been used for O_2 evolution. Certainly, this is the area where further work is needed, and new sensitizers and catalysts are urgently required. Once such materials have been identified, it remains for the two photosystems (one producing H_2 and the other producing O_2) to be linked together, and for this purpose it will surely be necessary to use immobilized reactants.

Acknowledgments

I thank the SERC, the EEC, and GE (Schenectady) for financial support. Much of the work described in this review was carried out at the Royal Institution by J. R. Darwent, M. C. Richoux, P. Walters, and A. Wilowska, and their contributions are gratefully acknowledged.

References

1. Archer, M. D., and Porter, G., *ISR Interdiscip. Sci. Rev.* **1,** 119 (1976).
2. Porter, G., *Proc. R. Soc. London Ser. A* **362,** 281 (1978).
3. Harriman, A., and Barber, J., *Top. Photosynth.* **3,** 244 (1979).
4. Hill, R., and Bendall, F., *Nature (London)* **186,** 136 (1960).
5. Porter, G., Tredwell, C. J., Searle, G. F. W., and Barber, J., *Biochim. Biophys. Acta* **501,** 232 (1978).
6. Searle, G. F. W., Barber, J., Porter, G., and Tredwell, C. J., *Biochim. Biophys. Acta* **501,** 246 (1978).
7. Kelly, A. R., and Porter, G., *Proc. R. Soc. London Ser. A* **315,** 149 (1970).
8. Beddard, G. S., Carlin, S., and Porter, G., *Chem. Phys. Lett.* **43,** 27 (1976).
9. Costa, S. M., Froines, J. R., Harris, J. M., Leblanc, R. M., Orger, B. H., and Porter, G., *Proc. R. Soc. London Ser. A* **326,** 503 (1972).
10. Watson, W. F., and Livingston, R., *J. Chem. Phys.* **18,** 802 (1950).
11. Beddard, G. S., and Porter, G., *Nature (London)* **260,** 366 (1976).
12. Katz, J. J., Norris, J. R., and Shipman, L. L., *in* "Chlorophyll–Proteins, Reaction Centres and Photosynthetic Membranes" (J. M. Olsen, and G. Hind, eds.), p. 16. Brookhaven Natl. Labs., New York, 1976.
13. Jacobs, E. A., Holt, A. S., and Rabinowitch, E., *J. Chem. Phys.* **22,** 142 (1954).
14. Serlin, R., Chow, H. C., and Strouse, C. E., *J. Am. Chem. Soc.* **97,** 7237 (1975).
15. Shipman, L. L., Cotton, T. M., Norris, J. R., and Katz, J. J., *Proc. Natl. Acad. Sci. USA* **73,** 1971 (1976).
16. Boxer, S. G., and Bucks, R. R., *J. Am. Chem. Soc.* **101,** 1883 (1979).
17. Forster, T., *in* "Modern Quantum Chemistry" (O. Sinanoglu, ed.) Part 3, p. 93. Academic Press, New York, 1965.

18. Livingston, R., *Q. Rev. Chem. Soc.* **14,** 174 (1960).
19. Cotton, T. M., Ph.D. Thesis, Northwestern Univ., 1976.
20. Gouterman, M., Holten, D., and Lieberman, E., *Chem. Phys.* **25,** 139 (1977).
21. Anton, J. A., Kwong, J., and Loach, P. A., *J. Heterocycl. Chem.* **13,** 717 (1976).
22. Kagan, N. E., Mauzerall, D., and Merrifield, R. B., *J. Am. Chem. Soc.* **99,** 5484 (1977).
23. Boxer, S. G., and Closs, G. L., *J. Am. Chem. Soc.* **98,** 5406 (1976).
24. Harriman, A., *J. Chem. Soc. Faraday Trans. 1,* submitted.
25. Dirks, G., Moore, A. L., Moore, T. A., and Gust, D., *Photochem. Photobiol.* **32,** 277 (1980).
26. Moore, A. L., Dirks, G., Gust, D., and Moore, T. A., *Photochem. Photobiol.* **32,** 691 (1980).
27. Tsvirko, M. P., Solovev, K. N., and Sapunov, V. V., *Opt. Spektrosk.* **36,** 193 (1974).
28. Seely, G. R., *Photochem. Photobiol.* **27,** 639 (1978).
29. Whitten, D. G., *Acc. Chem. Res.* **13,** 83 (1980).
30. Kelly, J. M., and Porter, G., *Proc. R. Soc. London Ser. A* **319,** 319 (1970).
31. Hales, B. J., and Bolton, J. R., *J. Am. Chem. Soc.* **94,** 3314 (1972).
32. Holten, D., Gouterman, M., Parson, W. W., Windsor, M. W., and Rockley, M. G., *Photochem. Photobiol.* **25,** 85 (1977).
33. Gouterman, M., and Holten, D., *Photochem. Photobiol.* **23,** 415 (1976).
34. Rehm, D., and Weller, A., *Isr. J. Chem.* **8,** 259 (1970).
35. Beddard, G. S., Porter, G., and Weese, G. M., *Proc. R. Soc. London Ser. A* **342,** 317 (1975).
36. Kong, J., and Loach, P. A., *in* "Frontiers of Biological Energetics: From Electrons to Tissues" (P. L. Dutton, J. S. Leigh, and A. Scarpa, eds.), Vol. 1, p. 73. Academic Press, New York, 1981.
37. Ho, T. F., McIntosh, A. R., and Bolton, J. R., *Nature (London)* **286,** 254 (1980).
38. Kong, J., Spears, K. G., and Loach, P. A., *Photochem. Photobiol.* **35,** 545 (1982).
39. Migita, M., Okada, T., Mataga, N., Nishitami, S., Kurata, N., Sakata, Y., and Misumi, S., *Chem. Phys. Lett.* **84,** 263 (1981).
40. Dalton, J., and Milgrom, L. R., *J. Chem. Soc. Chem. Commun.* No. 378, p. 609 (1979).
41. Harriman, A., and Hosie, R. J., *J. Photochem.* **15,** 163 (1981).
42. Harriman, A., and Hosie, R. J., *J. Chem. Soc. Faraday Trans. 2* **77,** 1695 (1981).
43. Buck, R. R., Netzel, T. R., Fujita, I., and Boxer, S. G., *J. Phys. Chem.* **86,** 1947 (1982).
44. Fuoss, R., *J. Am. Chem. Soc.* **80,** 8059 (1958).
45. Gore, B. L., Harriman, A., and Richoux, M. C., *J. Photochem.* **19,** 209 (1982).
46. Harriman, A., Porter, G., and Searle, N., *J. Chem. Soc. Faraday Trans. 2* **75,** 1515 (1979).
47. Harriman, A., Porter, G., and Wilowska, A., *J. Chem. Soc. Faraday Trans. 1,* in press.
48. Harriman, A., *J. Chem. Soc. Faraday Trans. 2* **77,** 1281 (1981).
49. Harriman, A., *J. Chem. Soc. Faraday Trans. 1* **78,** 2727 (1982).
50. Harriman, A., and Richoux, M. C., *J. Chem. Soc. Faraday Trans. 2* **78,** 1873 (1982).
51. Richoux, M. C., *J. Photochem.* **22,** 1 (1983).
52. Holten, D., Windsor, M. W., Parson, W. W., and Gouterman, M., *Photochem. Photobiol.* **28,** 951 (1978).
53. Harriman, A., Porter, G., and Richoux, M. C., *J. Chem. Soc. Faraday Trans. 2* **77,** 1175 (1981).
54. Pasternack, R. F., and Spiro, E. G., *J. Am. Chem. Soc.* **100,** 968 (1978).
55. Chapman, R. D., and Fleischer, E. B., *J. Am. Chem. Soc.* **104,** 1575 (1982).
56. Walters, P., Ph.D. thesis, Univ. of London, 1983.
57. Neta, P., Scherz, A., and Levanon, H., *J. Am. Chem. Soc.* **101,** 3624 (1979).

58. Darwent, J. R., Ph.D. Thesis, Univ. of London, 1982.
59. Harriman, A., to be published.
60. Saier, H. D., and Strathmann, H., *Angew. Chem. Int. Ed. Engl.* **14,** 452 (1975).
61. Joliot, P., Barbieri, G., and Chabaud, R., *Photochem. Photobiol.* **10,** 309 (1969).
62. Kok, B., Forbush, B., and McGloin, M., *Photochem. Photobiol.* **11,** 457 (1970).
63. Joliot, P., and Kok, B., *in* "Bioenergetics of Photosynthesis" (Govindjee, ed.), p. 387. Academic Press, New York, 1975.
64. Radmer, R., and Cheniae, G. M., *Top. Photosynth.* **2,** 303 (1977).
65. Harriman, A., *Coord. Chem. Rev.* **28,** 147 (1979).
66. Carnieri, N., Harriman, A., and Porter, G., *J. Chem. Soc. Dalton Trans.*, p. 931 (1982).
67. Wang, J. H., *Proc. Natl. Acad. Sci. USA* **62,** 653 (1969).
68. Fong, F. K., and Galloway, L., *J. Am. Chem. Soc.* **100,** 3594 (1978).
69. Kiwi, J., and Grätzel, M., *Angew. Chem. Int. Ed. Engl.* **17,** 860 (1978).
70. Lehn, J. M., Sauvage, J. P., and Ziessel, R., *Nouv. J. Chim.* **4,** 81 (1980).
71. Harriman, A., Porter, G., and Walters, P., *J. Chem. Soc. Faraday Trans. 2* **78,** 2373 (1982).
72. Borgarello, E., Kalyanasundaram, K., Okuno, Y., and Grätzel, M., *Helv. Chim. Acta* **64,** 1937 (1981).
73. Neumann-Spallart, M., and Kalyanasundaram, K., *Z. Naturforsch.* **36b,** 596 (1981).
74. Darwent, J. R., Douglas, P., Harriman, A., Porter, G., and Richoux, M. C., *Coord. Chem. Rev.* **44,** 83 (1982).
75. Schwarz, F. D., Gouterman, M., Muljiani, Z., and Dolphin, D., *Bioinorg. Chem.* **2,** 1 (1972).
76. Ichimura, K., *Chem. Lett.*, p. 641 (1977).
77. Selensky, R., Holten, D., Windsor, M. W., Paine, J. B., Dolphin, D., Gouterman, M., and Thomas, J. C., *Chem. Phys.* **33,** 60 (1981).
78. Landrum, J. P., Reed, C. A., Hatano, K., and Scheidt, W. R., *J. Am. Chem. Soc.* **100,** 3232 (1978).
79. Chang, C. K., *J. Heterocycl. Chem.* **14,** 1285 (1977).
80. Battersby, A. R., Buckley, D. G., Hartley, S. G., and Turnbull, M. D., *J. Chem. Soc. Chem. Commun.* No. 916, p. 879 (1976).
81. Baldwin, J. E., Klose, T., and Peters, M., *J. Chem. Soc. Chem. Commun.* No. 917, p. 881 (1976).
82. Ogoshi, H., Sugimoto, H., and Yoshida, Z., *Tetrahedron Lett.*, p. 1515 (1977).
83. Diekmann, H., Chang, C. K., and Traylor, T. G., *J. Am. Chem. Soc.* **93,** 4058 (1971).
84. Chang, C. K., *J. Am. Chem. Soc.* **99,** 2819 (1977).
85. Kagan, N. E., *in* "Porphyrin Chemistry Advances" (F. R. Longo, ed.), p. 43. Ann Arbor Sci., Michigan, 1979.

7 Semiconductor Particulate Systems for Photocatalysis and Photosynthesis: An Overview

K. Kalyanasundaram

Institut de Chimie Physique
École Polytechnique Fédérale
Lausanne, Switzerland

ENERGY RESOURCES THROUGH
PHOTOCHEMISTRY AND CATALYSIS

ISBN 0-12-295720-2

I. Introduction

Studies of light-induced electron-transfer reactions have become an active area of research in photochemistry. Extensive studies with transition-metal complexes containing ligands such as bipyridyls and phenanthrolines, organic dyes, and metalloporphyrins have indicated the potential utility of these materials as *photosensitizers* in the photochemical conversion of solar energy (*3, 17, 21, 94, 138, 188*). These studies have also shown that the main factor that limits the efficiency in homogeneous solutions is the thermal reverse electron transfer between the redox products. Studies in organized molecular assemblies, such as micelles, vesicles, and microemulsions, show some promise in controlling to some extent the light-induced charge-separation process (*93, 105, 170*). An alternate approach to this problem is the utilization of semiconductor materials as light-absorbing units. Photoredox reactions between a semiconductor and redox species in solution occur in one direction and are, in general, not reversible. There has been a phenomenal growth and progress on photoelectrochemical cells with single-crystal or polycrystalline electrodes (*19, 50, 64, 132, 134*). The subject material for this chapter, however, is the study of the photoprocesses with finely divided semiconductor particulate systems. Their usage appears in three forms: as dry or wet powders for gas-phase reactions, as dispersions in aqueous or nonaqueous media, and as colloidal semiconductors. The utilization of "colloidal semiconductors" (particle dimensions often less than a few micrometers) and the introduction of catalysts to promote specific redox processes on semiconductor surfaces was developed in 1976. Colloidal semiconductors have the added advantage in providing transparent solutions for detailed mechanistic studies of photoredox processes by laser photolysis techniques.

Bard (*4, 5*) and Nozik (*133*) have proposed a classification of photochemical processes. Photocatalytic processes are reactions that are driven in a spontaneous direction ($\Delta G < 0$) (see Fig. 1) and the radiant

$$A + D \underset{}{\overset{\Delta G<0}{\rightleftharpoons}} A^- + D^+ \tag{1}$$

energy merely overcomes the energy of activation of the process. The term *heterogeneous photocatalysis* is often used in this context for studies with semiconductor powders and dispersions. In photosynthetic processes, the light is used to drive an overall reaction in a nonspontaneous direction so that the light energy is stored as chemical energy ($\Delta G > 0$) for reaction (1). We shall consider both types of processes. A major part of the chapter deals with reactions that occur on irradiation of semiconductors with light of energy equal to or greater than the band-gap energy, but for completeness we also include a discussion of studies in which the semiconductor powders are employed as inert carriers of catalysts or actively involved in dye sensitization and charge-injection studies. During the 1970s, several reviews on heterogeneous photocatalysis have appeared (*4, 5, 8, 49, 54, 60, 72, 74, 115, 136, 160, 174, 180–182, 184*). Our aim is to take an overview of the various types of photoprocesses that have been explored, with emphasis on more recent work. The companion chapters in this volume by Pelizzetti and Visca (Chapter 8), Kiwi (Chapter 9), Sakata and Kawai (Chapter 10), and Halmann (Chapter 15) examine in greater detail some of these reactions.

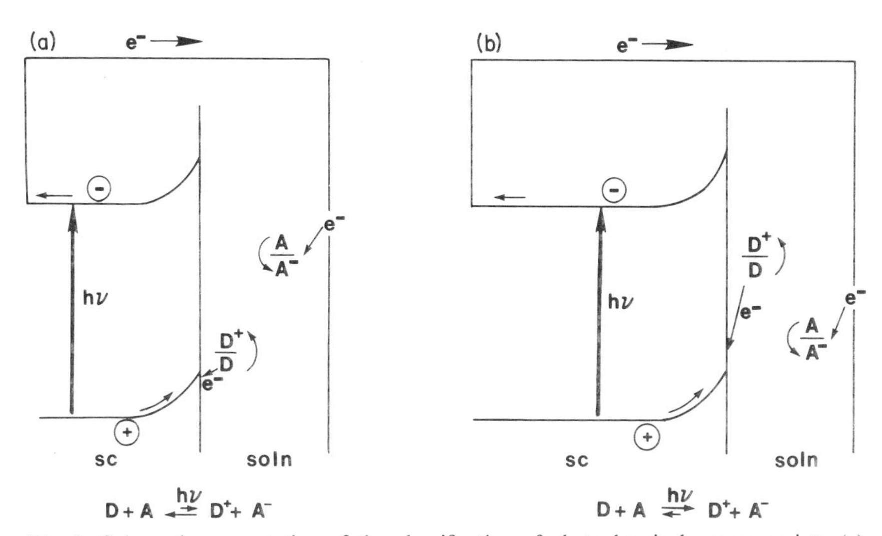

Fig. 1. Schematic presentation of the classification of photochemical processes into (a) photoelectrosynthetic ($\Delta G > 0$) and (b) photocatalytic ($\Delta G < 0$) systems.

Band Model and Redox Processes on Semiconductor Powders and Dispersions

The *band theory of solids* and *electronic theory of catalysis* have proved to be very useful in conceptual understanding and rationalization of various photoprocesses that have been studied with semiconductor electrode and powder systems. According to the band model (*124*), the occurrence and the efficiency of various photoredox processes are intimately related to the location of valence and conduction bands. In intrinsic semiconductors, the Fermi level for the electrons and holes are located midway between the valence and conduction bands, and depending on the nature and extent of doping, they move closer to either one of them (closer to valence band for *p*-type and to the conduction band for *n*-type). Absorption of light of energy $E > E_{\text{band gap}}$ leads to generation of electron–hole pairs, and these under the influence of the electric field move (within the band gap) into the conduction and valence bands, respectively. The resulting nonequilibrium distribution of electrons (e^-) and holes (h^+) gives rise to reduction or oxidation processes with adsorbed species, surface groups, or with the bulk semiconductor itself (Fig. 2). Often, because of the high absorption coefficients at the wavelength corresponding to the band gap ($\alpha_{\text{ZnO}}^{400\text{nm}} \approx 3 \times 10^7\ \text{m}^{-1}$) (direct band-gap materials), the incident photons are attenuated within a short distance from the surface, and hence surface properties of catalysts play a dominant role. The main advantage of semiconductor powders and dispersions is the very large surface area exposed to the reactants. A certain amount of band bending is necessary to provide the driving force for the chemical reactions. Ad-

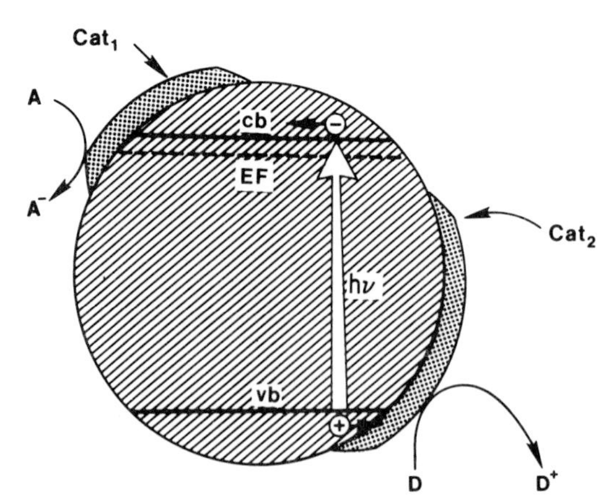

Fig. 2. Photogeneration of holes and electrons by band-gap excitation followed by redox processes, according to the band model.

sorption of gases such as O_2 on *n*-type or donor gases such as NH_3 or H_2 on *p*-type leads to electronic charge transfer from the solid to the adsorbate or vice versa (chemisorption), and as a consequence of movement of charges, a potential difference is created across the space-charge layer.

With the preceding band model, it is possible to correlate various types of photoprocesses that have been studied. Band-gap excitation of the semiconductor leads to generation of electron–hole pairs:

$$\mathrm{MO} \longrightarrow \mathrm{h}^+ + \mathrm{e}^- \qquad (2)$$

In the presence of adsorbed gases such as O_2, for example, there can be an electron transfer from the conduction band to O_2, leading to increased chemisorption (*photoadsorption of gases*). The superoxide radical-anions

$$\mathrm{O_2} + \mathrm{e}^- \longrightarrow \mathrm{O_2}^- \qquad (3)$$

(O_2^-) can subsequently protonate and give rise to production of H_2O_2 (*photosynthesis of* H_2O_2). Depending on the fate of the cogenerated h^+, the photoadsorption can be reversible or not. The process of photoadsorption and photodesorption can also be accompanied by *isotopic exchange*. In the absence of external additives, in semiconductors such as ZnO, the h^+ can react with the surface hydroxyl groups or with the bulk semiconductor (*photocorrosion*). In the presence of additives such as oxalates, formates, and aliphatic or aromatic compounds, the holes react with them giving rise to sustained *photooxidation of the organic substrates*. With substrates such as $Cr_2O_7^{2-}$, SO_4^{2-}, Cu^{2+}, $PtCl_4^{2-}$, etc., the reactions of holes and electrons can give rise to *photooxidation and photoreduction of inorganic substrates* and *photodeposition of metals*. In Section II, we review some of these studies with "naked" semiconductors.

There has been interest in the possible use of water itself as a hole reductant instead of organic or inorganic substrates. In the absence of

$$4\mathrm{h}^+ + 2\mathrm{H_2O} \longrightarrow 4\mathrm{H}^+ + \mathrm{O_2} \qquad (4)$$

oxygen, with semiconductors in which the conduction band is suitably located, [more negative with respect to $E_o(H^+/H_2)$], there can be direct

$$2\mathrm{e}^- + 2\mathrm{H_2O} \longrightarrow 2\mathrm{OH}^- + \mathrm{H_2} \qquad (5)$$

water reduction as well. Reactions (4) and (5) together would constitute sustained *photocleavage of water*. Owing to high overvoltages and the nature of multielectron transfer involved, in the absence of suitable redox catalysts these reactions either do not occur or occur very inefficiently. Photocorrosion often dominates. There have been several studies with metallized, catalyst-loaded semiconductor dispersions. Sustained water cleavage has been demonstrated for the first time with semiconductor dispersions such as Pt–TiO_2 and Rh–$SrTiO_3$ in water. Reactions such as photo-Kolbe and photo-Fenton are also being pursued vigorously. These topics are reviewed in Section III.

For a better understanding of heterogeneous photocatalysis it is a common practice to compare the results obtained with various specimens as well as with other semiconductor materials. Confusion often arises from lack of caution regarding such factors such as source of the semiconductor materials used (untreated, pretreatment of some kind or other, or addition of dopants at controlled level), reactions in the gas or liquid phase, and presence–absence of substrates (such as oxygen, organic additives, solvent system such as water, or even the possibility of the semiconductor itself acting as a substrate). Several physical techniques have been and are being developed for the characterization of the semiconductor dispersions employed and for the photoprocesses per se, and these are reviewed in Section IV.

II. Photoprocesses with "Naked" Semiconductor Powder Dispersions

A. Photoadsorption and Photodesorption of Gases

Adsorption and desorption of gases onto semiconductor powders and dispersions is probably the simplest form of photocatalysis, hence they have been studied rather extensively by various authors. According to the Morrison's model, gases such as O_2 initially *physisorb* on the surface of a semiconductor and subsequently transfer into *chemisorbed* species through charge exchange. The removal of electrons from the conduction band to the localized surface states can often be monitored by changes in conductivity (*82, 83*). By reversal of this process, desorption of chemisorbed species leads to increased conductivity of the sample (*123, 158*). At room temperature, physisorbed species are easily removed by applying a vacuum. Electronically, chemisorbed oxygen is envisaged to be present as O_2^-(ads) and also possibly as O^-(ads), as shown in the following scheme:

$$O_2(g) \rightleftharpoons O_2(ads) \tag{6}$$

$$O_2(ads) \overset{e^-}{\rightleftharpoons} O_2^-(ads) \tag{7}$$

$$O_2^-(ads) \rightleftharpoons O^-(ads) + O(ads) \tag{8}$$

Formation of electron-rich species such as O^- are limited in the dark to nonreduced *n*-oxides, as it requires surface sites capable of dissociating oxygen.

Photogeneration of electrons and holes on irradiation with band-gap light significantly alters the relative amounts of physi- and chemisorbed gases, and the effect is known as the *photoadsorptive effect*. The question of whether adsorption or desorption of chemisorbed gases occurs under illumination depends on the sign of the charge on the adsorbed species and on the existence of suitable trapping centers. Because of the charge transfer involved, the presence of "hole traps" often determine the reversibility of the processes. In general, photoadsorption is irreversible. Bickley (*8*), Lunsford (*117*), Steinbach (*160*), and Lichtman and Shapira (*115*) have summarized various studies of photoadsorption and desorption on semiconductor powders. An elegant semiquantitative analysis of the phenomenon on the basis of the electronic theory of catalysis has been provided by Wolkenstein (*181, 182*).

In ZnO, often there are excess interstitial zinc atoms as impurities, and this gives rise to donor levels located within the band gap at ~0.05 eV below the conduction band. Thermal excitations lead to placing some of the electrons in the conduction band. Thus in ZnO, in the dark, there are no holes; instead there are electrons in the conduction band. The interaction of these electrons with reactants can be studied in the dark. Dissociative electron attachment of gases such as N_2O, CH_3I, and CH_3Cl at the surfaces of vacuum-reduced ZnO has been observed by epr measure-

$$e^- + N_2O \longrightarrow N_2O^- \longrightarrow N_2 + O^-(ads) \tag{9}$$

$$e^- + CH_3X \longrightarrow CH_3X^- \longrightarrow CH_3 + X^- \tag{10}$$

ments (*28*). The excess Zn also acts as a trap for holes produced in the optical absorption process. Consistent with these are the observations of Barry and Stone (*7*) that photoadsorption of O_2 can be promoted on vacuum-reduced ZnO, whereas photodesorption can be induced from a specimen pretreated in O_2 at 873 K prior to illumination. Similarly, photoadsorption of O_2 increases on addition of Li^+ (acceptor) to ZnO and falls off on introduction of Ga or Al^{3+} (donor). Samples containing a stoichiometric deficiency of Zn show photoadsorption, and superstoichiometric levels of Zn in reduced ZnO lead to photodesorption.

The active centers for the photoadsorption have been postulated as being O^-(surface) according to

$$2O^-(ads) + O_2(g) \xrightarrow[\Delta]{h\nu} 2O_2^-(ads) \tag{11}$$

During illumination, the negatively adsorbed oxygen species (A^-) are driven to the surface by the electric field created, and they can react with holes to produce oxy radicals ($A\cdot$):

$$A^- + h^+ \longrightarrow A\cdot \tag{12}$$

The majority of powdered oxides are known to contain substantial surface coverage of hydroxyl groups (OH). If part of them carry negative charges, they can act as hole traps, producing radicals. In cases such as TiO_2 or ZnO, OH· radicals can dimerize to form adsorbed H_2O_2. In support, sustained photoproduction of H_2O_2 has been demonstrated with semiconductors such as ZnO, CdS, and phthlocyanins. In TiO_2, however, H_2O_2 is decomposed via Fenton-type reactions, involving Ti^{3+} ions that are produced by the e^- trapping at the Ti^{4+} centers. Studies by Boonstra and Munstears (*14*) on the stability of H_2O_2 on TiO_2 surfaces have shown that, whereas thermal decomposition of H_2O_2 starts at temperature 320 K and complete at 423 K, on uv photolysis, decomposition occurs even at room temperature (298 K). These authors (*15*) have also quantitatively correlated the amount of photoadsorbed oxygen with the surface hydroxyl group content of both anatase and rutile. Anatase was found to be considerably more active than rutile.

On TiO_2 samples that have been dehydrated by treatment *in vacuo* or in O_2 at elevated temperatures, the photoactivity associated with these surfaces is not associated with the presence of OH groups and is comparable with the majority of studies with ZnO. On careful outgassing, the photoadsorption activity of fully, oxidized rutile was found to be entirely destroyed, but the activity could be partially regenerated by treatment with water vapor. Liquid water was found to be much more effective in rehydroxylating the surface (*9*).

Munuera *et al.*, in a series of papers (*65, 66, 128–130*), have discussed the photoadsorption and photodesorption of oxygen on highly hydroxylated anatase (Degussa P-25) samples. Fully hydroxylated TiO_2 surfaces (with about 14–16 OH nm^{-2}) hold both OH groups and molecular water. The latter are easily removed under vacuum at temperatures less than 200°C. On these surfaces, fast photoadsorption of O_2 accompanied by slow photodesorption was observed. Similar to the results obtained by Boonstra and Munstears, a quantitative relation between the number of hydroxyl groups and their photoactivity has been observed. A change in mechanism from a first-order case to a diffusion-controlled process was also observed when the hydroxyl groups were progressively removed from the sample. The vanishing of photodesorption after outgassing of the samples at $T > 300$°C has been related to the formation of H_2O_2 via dimerization of OH· and $HO_2\cdot$ radicals and subsequent decomposition into O_2 and H_2O. Photodecomposition of H_2O_2 on dehydroxylated surfaces occurs following a diffusion law, whereas kinetics change to pseudo-zero-order at higher coverages of H_2O_2 on hydroxylated surfaces.

Photodesorption of CO_2 from ZnO surfaces has been examined by mass spectrometry (*156–158*). Studies at different temperatures have shown that the photodesorption occurs via neutralization of chemisorbed CO_2^- ion molecules by photogenerated holes (h^+). The rates of photodesorption and conductivity changes are initially fast and become slower with illumination time. This is owing to the buildup of an electron-accumulation layer, hence a barrier is established for hole flow to the surface. The photodesorption process is thermally activated with $E_{act} \approx 0.25$ eV. Among the gases O_2, CO, and CO_2, only O_2 was found to chemisorb from the gas phase on ZnO.

B. Photocatalytic Oxidation of CO, H_2, and NH_3

According to Tanaka and Blyholder (*166*), the photocatalytic reaction of CO with O_2 on ZnO is of half order in oxygen pressure and of zero order in CO pressure. With N_2O as the oxidant, the rate is given by

$$\text{rate} = kp_{N_2O}^{0.4}p_{CO}. \qquad (13)$$

The accompanying scheme represents the sequence of reactions believed to occur. Photocatalytic oxidation of CO has also been observed on other

With O_2:

$$CO + h^+ \longrightarrow CO^+ \qquad (14)$$

$$O_2 \longrightarrow 2O \qquad (15)$$

$$O + e^- \longrightarrow O^- \qquad (16)$$

$$O^- + CO^+ \longrightarrow CO_2 \qquad (17)$$

With N_2O:

$$N_2O + e^- \longrightarrow N_2O^- \qquad (18)$$

$$N_2O^- + CO^+ \longrightarrow CO_2 + N_2 \qquad (19)$$

oxides such as TiO_2, WO_3, and SnO_2 in the temperature range 100–230°C (*167*). Spectral dependence of the photooxidation on ZnO has shown the reaction to proceed with photons of energy greater than 3.0 eV and a quantum yield of about 40% (*189*). It is assumed that positive holes are trapped by oxygen lattice surface ions, giving O^- ions, which are agents for the oxidation (*173*). A similar mechanism is believed to be operative on TiO_2 (*54, 81*). The kinetics of photooxidation of CO on TiO_2 shows two simultaneous paths, one requiring oxygen from the lattice of TiO_2, the second involving dissociated adsorbed oxygen. At low CO pressures,

(<100 torr) "redox mechanisms" of the type shown in reactions (14)–(19) are operative. But at high CO pressures, the rate increases linearly with p_{CO}, suggesting the involvement of dissociated adsorbed oxygen atoms, as shown in the following scheme:

$$O_2(g) + e^- \longrightarrow O_2^-(ads) \quad (20)$$

$$O_2^-(ads) + h^+ \longrightarrow 2O(ads) \quad (21)$$

$$2O(ads) + 2CO(ads) \longrightarrow 2CO_2(g) \quad (22)$$

The participation of atomic oxygen species is in agreement with the isotope-exchange studies (*22*). Steinbach and Harborth (*161*) have also invoked formation of atomic adsorbed O on irradiated ZnO single crystals in the presence of O_2 and CO.

van Damme and Hall (*172*) have investigated possible photooxidation of H_2 and CO on ZnO, TiO_2, and various perovskites ($LaCoO_3$, $SrTiO_3$, $BaTiO_3$, and $Ba(Fe_{0.33}Ti_{0.67})O_{2.67}$). No effect of the light on the rate of H_2 oxidation was found, but a strong photocatalytic effect was observed for the CO oxidation on TiO_2, ZnO, $SrTiO_3$, and $BaTiO_3$. The effect of light is discussed in terms of band-to-band transitions followed by hole capture. Interactions of H_2 and CO with TiO_2 and ZnO surfaces have also been the subject of a study by Reymond *et al.* (*143*).

Photooxidation of NH_3 to N_2, N_2O, H_2O, and H_2 over semiconducting oxides TiO_2 and ZnO has been examined in some detail. Depending on the pretreatment, oxidation on ZnO yields N_2O, N_2, H_2, and H_2O (*39*). Higher yields of N_2 have been obtained on ZnO samples pretreated with H_2 and then heated in O_2 to about 300°C. Proposed mechanisms involve oxidation via adsorbed O_2^- radicals:

$$NH_3 + O_2^-(ads) \longrightarrow HNO + H_2O + e^-(ZnO) \quad (23)$$

$$HNO + HNO \longrightarrow N_2O + H_2O \quad (24)$$

The results obtained on TiO_2 (anatase) (in a dynamic differential flow photoreactor) with a conversion of 2 to 4% shows that two parallel pathways take place (*91*, *125*, *127*):

$$2NH_3 + 1.5O_2 \longrightarrow N_2 + 3H_2O \quad (25)$$

$$2NH_3 + 2O_2 \longrightarrow N_2O + 3H_2O \quad (26)$$

Because N_2O is not photodissociated under reaction conditions, N_2 is formed as a primary product. The selectivity toward formation of N_2 is of the order of 80%. Studies at various partial pressures of NH_3 and O_2 indicate that the formal kinetics corresponds to a model with a reaction between O_2 and NH_3 adsorbed (the former dissociatively) on different sites for N_2 formation and with another reaction between O_2 and a nitrogen containing intermediate (NH_3O or HNO?) for N_2O formation. ZnO,

SnO_2, and WO_3 also show photocatalytic activity for NH_3 oxidation with approximately 1–2 orders of magnitude smaller than that for titania.

C. Isotopic Exchanges on Semiconductor Surfaces

Study of exchanges of isotopes during adsorption and desorption of gases is often employed as a means of monitoring processes such as dissociative adsorption and interrelations between catalytic activity and lability of surface species. Tanaka (*164*) has studied oxygen isotope exchange on rutile at room temperature under irradiation with a medium-pressure Hg lamp. The isotopic exchange of a mixture of $^{18}O_2$ and $^{16}O_2$ (each at a pressure of 1 torr) reached an equilibrium within 3 min of irradiation, whereas in the dark the reaction was very slow. From an analysis of the isotopic composition of desorbed oxygen, the author concluded that there is homomolecular oxygen exchange via weakly bound O_3^- intermediates formed from O_2 and O^- because desorption of O_3^- is implausible at room temperature. For ZnO, similar results have been obtained (*31–35, 165*). Thus both dissociative ($O_2^- \rightleftharpoons 2O^-$) and associative ($O_2^- + O_2 \rightleftharpoons O_4^-$) mechanisms including O_2^- species are unfavorable under irradiation.

Oxygen isotopic exchange (OIE) of almost pure $^{18}O_2$ at low pressures (>1 torr) with surface O atoms on titania(anatase) during irradiation has been studied by Courbon *et al.* (*22*), and the results were interpreted as a process involving one surface atom and not two simultaneously. Similar

$$^{18}O_2(g) + {}^{16}O(\text{surface}) \longrightarrow {}^{16}O^{18}O(g) + {}^{18}O(\text{surface}) \qquad (27)$$

results have been obtained for other oxides (ZrO_2, ZnO, and SnO_2). Photoactivity for OIE is closely related to the photoactivity for isobutane oxidation ($TiO_2 > ZnO > ZrO_2 > SnO_2 > V_2O_5 = O$), hence the same dissociated surface O atoms have been invoked for both.

By flash irradiation and mass spectral studies on the interaction of alcohols with TiO_2 and ZnO, Cunningham *et al.* (*30*) have observed H–D exchange to occur simultaneously with the photodesorption of deuterated alcohols. Previously H_2–D_2 exchange on irradiation with near-uv light has been observed on MgO (*118*) and ZnO (*59*).

D. Photoproduction of H_2O_2 and Its Decomposition

Closely related to the photoadsorption of O_2 on surfaces of semiconductors is the photosynthetic production of H_2O_2 in the pressure of O_2.

With ZnO, it was one of the earliest and widely studied photoprocesses on semiconductor powders and dispersions. In the absence of any purposely added reducing agents, photolysis of aqueous, aerated–oxygenated solutions leads to the formation of H_2O_2, to a limiting concentration of $\sim 10^{-5}$ *M*. In the presence of reductants such as oxalate, formate, phenols, etc., the yields increase to 10^{-3} *M*, with CO_2 obtained as the oxidation product of organic additives. The quantum yield in the presence of additives approaches its maximum of 0.5. In both cases studied in aqueous media, it is known that the oxygen that is incorporated into H_2O_2 comes from dissolved oxygen (O_2) and that light of energy >3.2 eV (band-gap energy) is required. The formation of H_2O_2 shows very little specificity with respect to the reducing agents employed.

The mechanism by which H_2O_2 is formed in additive-free systems is not well understood. The reduction part can be understood via the accompanying scheme of reactions:

$$ZnO \longrightarrow h^+ + e^- \quad (28)$$

$$O_2 + e^- \longrightarrow O_2^-(ads) \quad (29)$$

$$O_2^-(ads) + H^+ \rightleftharpoons (HO_2\cdot)(ads) \quad (30)$$

$$(HO_2\cdot)(ads) + e^- \longrightarrow (HO_2)^-(ads) \quad (31)$$

$$(HO_2)^-(ads) + H^+ \rightleftharpoons (H_2O_2)(ads) \quad (32)$$

or

$$2(HO_2\cdot)(ads) \longrightarrow (H_2O_2)(ads) + O_2 \quad (33)$$

For the oxidation half of the synthesis, an oxidizable substance must be present. It has been suggested that impurities may be involved or that interstitial Zn may be an effective reductant. By spin-label studies, Harbour and Hair (*71*) have detected significant concentrations of OH· radicals ($\sim 6 \times 10^{-5}$ *M*) for them to play a major role. The source of OH· radicals can be surface hydroxyl groups or OH^-–H_2O or lattice oxygen oxidized by the hole. Thermodynamic considerations on various possible chemical reactions by Gerischer (*64*) suggest anodic dissolution of ZnO as

$$2ZnO + 4H_2O + 4h^+ \longrightarrow 2Zn(OH)_2 + O_2 + 4H^+ \quad (34)$$

a distinct possibility. In support of this, Rao *et al.* (*139*) have detected intense ir bands corresponding to OH^- groups on irradiated ZnO. (These bands are virtually absent in the spectra of starting materials and for materials kept in contact with water.) Along with H_2O_2, these authors have observed also the photosynthetic production of H_2.

To account for the limiting concentrations of H_2O_2 formed, a limiting reaction of the type

$$H_2O_2(ads) + OH\cdot \longrightarrow (HO_2\cdot)(ads) + H_2O \quad (35)$$

between the OH· radicals and adsorbed H_2O_2 has been proposed. Compared with closed systems, in open systems with continuous oxygen purging substantially higher (10^{-4} *M*) concentrations of H_2O_2 are obtained. With ZnS and CdS dispersions, esr spin-label studies have failed to indicate the production of OH· radicals.

In the presence of reductants such as formate, oxalate, EDTA, etc., the holes are scavenged via reactions (36) and (37), and H_2O_2 in the range of 10^{-3} *M* has been produced with ZnO, CdS, and *x*-phthalocyanin (*69, 70,*

$$h^+ + HCOO^- \longrightarrow CO_2^{-}\cdot + H^+ \tag{36}$$

$$h^+ + EDTA \longrightarrow CO_2 + CH_3CHO + \text{amines} \tag{37}$$

73, 111) dispersions. A wide variety of organic substrates, such as benzyl alcohol or toluene, have also been used.

Among the many semiconductors that have been examined (ZnO, TiO_2, CdS, ZnS, and H_2Pc), it is interesting to note that with ZnO, although it is the most efficient photoproducer of H_2O_2, no O_2^- has been detected. The lack of observation of O_2^- presumably is owing to rapid second-electron reduction, as manifested in the high quantum yield (*72*).

E. Photooxidation and Photoreduction of Inorganic Substrates

The strong oxidizing and catalytic properties exhibited by uv-irradiated aqueous suspensions of *n*-semiconductors at room temperature has been exploited to achieve photooxidation and photoreduction of various inorganic materials. In uv-irradiated aqueous suspensions of anatase (Degussa), iodide ions are oxidized in the presence of O_2 at a conversion ~80 times greater than that for Br^- ions, whereas Cl^- ions withstand oxidation (*80*). The quantum yield is about 2% with I^- ions. Although the nature of the final products depends on pH (halogens or hypohalites at low or high pH values, respectively), the kinetics of oxidation remain the same over the pH range. The reaction rate was found to be proportional to the surface coverage in halide ions.

Cyanide ions in aqueous oxygen-saturated solutions are photooxidized to OCN^- in the presence of TiO_2, CdS, and ZnO, whereas SO_3^{2-} ions are converted to SO_4^{2-} ions in the presence of TiO_2, CdS, ZnO, and Fe_2O_3 (*57, 58*):

$$CN^- + 2OH^- + 2h^+ \longrightarrow OCN^- + H_2O \tag{38}$$

$$SO_3^{2-} + 2H_2O + 2h^+ \longrightarrow SO_4^{2-} + 2H^+ \tag{39}$$

On TiO_2, the quantum yield for the SO_3^{2-} oxidation is higher (0.16) compared with CN^- oxidation (0.06) because of the oxidation of another molecule of SO_3^{2-} by the cogenerated product H_2O_2. The catalytic activity follows the order $Fe_2O_3 \approx ZnO \approx CdS > TiO_2$. Hydroquinones are

$$O_2 + 2e^- + 2H^+ \longrightarrow H_2O_2 \tag{40}$$

$$H_2O_2 + SO_3^{2-} \longrightarrow SO_4^{2-} + H_2O \tag{41}$$

also oxidized by white light in the presence of TiO_2 (anatase) and at longer wavelengths (>460 nm) with phthlocyanin-coated TiO_2 dispersions in aqueous Na_2SO_4 (*51*).

Finely divided CdS in aqueous, air-saturated phosphate-buffer suspensions sensitize photooxidation of cysteine, several different thiols, as well as inorganic sulfides (*159*). The photooxidation is most efficient with cysteine, although it is very slow with amino acids such as histidine, methionine, tryptophan, and tyrosine. The maximal quantum yield of 2.1% has been obtained at pH 9.5 with cysteine.

Photoproduction of O_2 from water by band-gap irradiation of aqueous semiconductor dispersions containing inorganic acceptor species has been reported by several authors. In *in vitro,* inorganic photosynthesis model systems, Krasnovskii and co-workers have demonstrated O_2 evolution from irradiated TiO_2, ZnO, and WO_3 in the presence of Fe^{3+}, $Fe(CN)_4^{3-}$ and *p*-benzoquinone (*53, 110*). Photocatalytic reduction of $Cr_2O_7^{2-}$ ions on TiO_2 (rutile and anatase), WO_3, Fe_2O_3, and $SrTiO_3$ has been reported by Yoneyama *et al.* (*186*). WO_3 was found to be most

$$2Cr_2O_7 + 16H^+ \longrightarrow 4Cr^{3+} + 8H_2O + 3O_2 \tag{42}$$

effective, with a quantum yield of 2% (at 400 nm). The relative activity of various oxides follow the order WO_3 > rutile > anatase > Fe_2O_3 > $SrTiO_3$.

Photocatalytic reduction of $S_2O_8^{2-}$ with concomitant oxidation of water to O_2 occurs efficiently on $SrTiO_3$–x-$LaCrO_3$ dispersions (mole fraction x = 0 to 1) (*168*). The rate of oxygen formation is dependent on the temperature ($E_{act} \approx 61$ kJ mol^{-1}) and is first order with respect to $S_2O_8^{2-}$ concentration. At 64°C and constant $[S_2O_8^{2-}]$, a maximum rate is observed for a sample with $x = 0.25$. Vonach and Getoff (*175*) have similarly used Ce^{4+} ions as acceptors for conduction-band electrons. In the absence of Ce^{4+}, in dilute H_2SO_4 solutions, these authors have observed H_2 and O_2 evolution from water in the ratio of 2 : 1.

Previously, Schrauzer and Guth (*155*) reported similar studies on the photodecomposition of water on pure and doped TiO_2 with band-gap irradiation. However, subsequent studies (*90, 147, 153, 171*) showed that the process is not catalytic and that H_2 production merely corresponds to the decomposition of surface OH groups under illumination. Yoneyama *et al.* (*185*), in a related study, reported photodecomposition of water to H_2

and O_2 over naked $SrTiO_3$ dispersions. Amorphous titania in the form of a porous glass has also been reported to photoproduce H_2 from water. (*112*).

Cunningham and Zainal (*29*) have determined quantum efficiencies in the range of 10^{-3} to 0.29 for reduction reactions by uv irradiation of aqueous ZnO suspensions containing $NaNO_3$, $KMnO_4$, indigo carmine, and *p*-nitrosodimethyl aniline. Band-gap irradiation of CdS suspensions in the presence of methylviologen (MV^{2+}) leads to production of the viologen radical via reduction with conduction-band electrons. ($MV^{2+} + e_{cb}^- \rightarrow MV^+$) (*144*). In the absence of other additives, photocorrosion of the semiconductor occurs by its reaction with the hole (h^+).

F. Photodeposition of Metals

Another photoreaction of wide practical utility is the photodeposition of metals onto semiconductor dispersions from their salt solutions. By xenon-lamp irradiation of TiO_2 and WO_3 dispersions, Bard and co-workers have demonstrated photodeposition of Pt, Cu, Pd, and Ag using their respective metal salt solutions (*107, 141*). The reaction occurs photocatalytically with over 80% conversion of the metal salt in the presence of hole reductants such as acetate. With Pt, the dark gray deposits on TiO_2 have been shown by esca to be Pt^0. Lehn *et al.* (*113*) have similarly deposited Rh and Pd on $SrTiO_3$ dispersions. Quantum yields in the range of 0.235 have been obtained for the reduction of Ag^+ by the 365-nm photolysis of ZnO suspensions containing $AgClO_4$ (*68*). The process is also sensitized in the visible (with much lower quantum yields, 0.003–0.015) by the addition of xanthene dyes such as uranine and rhodamine B. Photodeposition of Pd or Pt on TiO_2 single crystals has also been described (*187*). Bard's method thus appears to have a very general scope and is effective in the recovery of metals such as Cu, even from dilute solutions (<10 ppm) under oxygen-free conditions. Metallized semiconductors such as Pt–TiO_2 and Rh–$SrTiO_3$ are being used very extensively. With codeposited metal or metal oxides it is possible to catalyze specific redox reactions with reagents in solution, and these are reviewed in Section III.

G. Photoreduction of Carbon Dioxide and Nitrogen

Semiconductor dispersions have also been employed to achieve photoreduction of CO_2 and N_2. Photoreduction of carbon dioxide to formaldehyde, methanol, and methane has been shown to occur on a variety of semiconductors ($SrTiO_3$, WO_3, TiO_2, $CaTiO_3$, $BaTiO_3$, ZnO, SiC, HgS, Pb_3O_4, and Fe_2O_3) (*1, 84*). The efficiency of the process when carried out either as a gas–solid process (by passing CO_2 and H_2O vapor over illuminated semiconductor surfaces) or as a liquid–solid reaction (by illuminating aqueous dispersions through which CO_2 was bubbled) has also been examined (*1, 84*). The yields were better for the latter case and reduction products at various levels of reduction (HCHO, CH_3OH, and CH_4) were obtained. The best results were obtained with aqueous suspensions of $SrTiO_3$, WO_3, and TiO_2 with energy-conversion efficiencies of 6, 5.9, and 12%, respectively. Halmann (Chapter 15) has elaborated on this reaction.

In the presence of N_2, photolysis of chemisorbed water on incompletely outgassed TiO_2 has been reported to yield NH_3 and N_2H_4 (*155*) as

$$N_2 + 3H_2O \longrightarrow 2NH_3 + 1.5O_2 \tag{43}$$

$$N_2 + 2H_2O \longrightarrow N_2H_4 + O_2 \tag{44}$$

the photoproducts. Iron doping was found to enhance significantly the photocatalytic activity of rutile.

H. Photooxidation of Organic Materials

1. Reactions in the Gas Phase

Compared with thermal heterogeneous catalysis, which starts at ~400°C and results in free-radical-initiated total combustion of organic substrates, heterogeneous photocatalysis occurs at much lower temperatures and with much lower energy barriers involved and gives rise to various partial oxidation products. Photooxidation of various alkanes, alkenes, and alcohols with semiconductors has been investigated extensively by Teichner, Pichat, Bickley, Cunningham, and co-workers. Performed in a dynamic reactor, the photocatalytic oxidation can be a continuous process. With monochromatic radiation, the photocatalytic oxidation of isobutane oxidation increases with increasing photon flux up to saturation levels. The quantum yield based on calorimetric measurements was found to be close to unity, whereas when measured by actionometry it was only 0.1 (*40, 41*).

Alkanes. The mechanism of photooxidation of alkanes has been established through analysis of reaction products for various isomers of C_2 to C_8 alkanes (*55, 92, 126*). For given reaction conditions, the total conversion increases with the number of C atoms in the alkane family (*n*, neo, and iso), showing that each C atom reacts with oxygen. The reactivity of various C atoms has been found to follow the sequence $C_{tert} > C_{quart} > C_{sec} > C_{primary}$. The yield of ketones is in the range 40–60%, whereas that of aldehydes is 15–20%. In a few cases alcohols were found, and the rate of their photooxidation showed that they may be intermediates between alkenes and the products, ketones, and aldehydes. Because this initial formation of alcohols from paraffins is, in principle, possible with atomic oxygen species (neutral or ionic), all proposed mechanisms invoke these species as key intermediates.

Alcohols. Photooxidation of primary, secondary, and tertiary methylbutanols have been examined in detail under conditions similar to that of alkanes, and the accompanying scheme appears to hold well (*27, 31–34, 40, 177*).

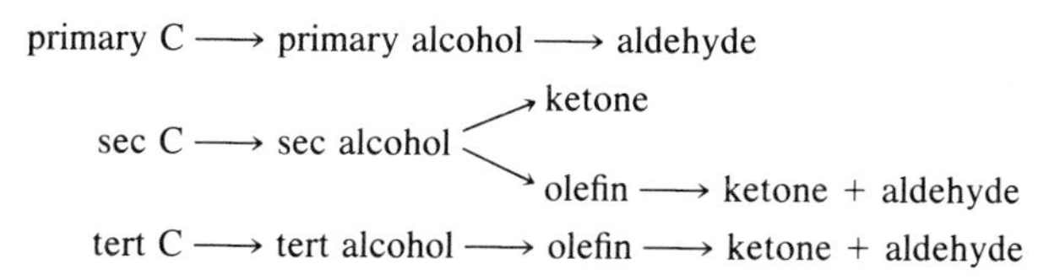

Alkenes. Pichat *et al.* (*137*) have examined the photocatalytic oxidation of propene at 320 K over a series of uvirradiated oxides (TiO_2, ZnO, V_2O_5, SnO_2, CeO_2, WO_3, and a Sn—O—Sb oxide). At this temperature, activity for the oxidation is observed only on band-gap irradiation, and the quantum yields varied widely with the metal oxides. Only V_2O_5 was found to be photoinactive. The selectivity greatly depended on the catalyst. For low and equivalent conversion levels, the total oxidation predominated for CeO_2, TiO_2, and, to a lesser extent, on ZnO and ZrO_2, whereas partial oxidation products only were obtained for SnO_2, WO_3, and Sn—O—Sb mixed oxide. For a given catalyst the product distribution was influenced by the activation mode (uv light or increase in temperature).

2. *Reactions in the Liquid Phase*

Photooxidation of alcohols in the liquid phase with semiconductor dispersions has also been explored. Cundall *et al.* (*25, 26*) have examined photocatalytic oxidation of liquid isopropanol to acetone using irradiated suspensions of rutile and anatase and by passing O_2, air, N_2O, or N_2 into the upper part of the vessel. In the presence of N_2, the reaction is not sustained. The quantum yield with O_2 for irradiation below the absorption edge of rutile is about 40%. The mechanism postulates trapping of holes by surface hydroxyls, giving OH· radicals and trapping of electrons by O_2 with the formation of adsorbed O_2^- species. Isopropanol photooxidation over other commercial pigments such as TiO_2, ZnO, $BaSO_4$, and $BaWO_4$ has also been examined (*48*). Photochemical oxidation of formic acid to CO_2 with oxygen in suspensions of TiO_2(anatase) has been studied in a photochemical slurry reactor (*10*).

Studies by Fujihara *et al.* (*61–63*) have shown that over uv-irradiated aqueous suspensions of TiO_2 (Aerosil P-25), aromatic compounds such as benzene, toluene, and acetophenone are hydroxylated to corresponding phenols. In addition, biphenyl from benzene, benzaldehyde from toluene, and phenol from acetophenone were also obtained. Both O_2 and illumination were found to be necessary for the oxidation. Addition of 0.05 *M* H_2SO_4 increased drastically the total yield of the oxidation products. The quantum yield is about 3% in additive-free systems and increases to 5% in 0.05 *M* H_2SO_4 and 0.05 *M* $CuSO_4$ mixtures. Formation of azo compounds on uv irradiation of ZnO in ethanolic solutions of anilines and toluidines has also been observed (*76, 77, 98*). Photocatalyzed oxidation of olefins (*56, 97*), lactams, and *N*-acyl amines (*135*) have also been reported to occur on TiO_2.

I. Photohydrogenation and Photodehydrogenation Reactions

Photohydrogenation of acetylene (to methane, ethane, ethylene, and propane) has been examined on TiO_2(anatase) powders containing OH groups (samples not degassed at 500°C) by Boonstra and Mutsaers (*16*). The reaction was found to occur; however, it was noncatalytic. On thoroughly degassed TiO_2 surfaces, no hydrogenation products were found. Formation of ethylene and acetaldehyde during photodesorption of ethanol (HCHO from CH_3OH) has been observed by Cunningham *et al.* (*32–34*). In the absence of O_2 in the gas phase, surface lattice oxygen and/or surface OH groups were involved in the oxidation. *tert*-Butanol gave acetone and isobutene, and isobutanol converted to isobutrylaldehyde. With $^{18}O_2$ incorporated in the system, acetone containing ^{18}O was found, indicating formation of this product via isobutene intermediacy.

Kinetics of conversion of 2-propanol and 2-butanol (to the corresponding ketones by photoassisted dehydrogenation, to lower aldehydes by photoassisted C_α—C_β bond cleavage, and to trace quantities of alkenes by photoassisted dehydration) on ZnO and TiO_2 at 348 K has been reported at O_2 partial pressures in the range 0–700 torr and at alcohol partial pressures of 0–60 torr (*35*). Two parallel pathways for the photodehydrogenation, one based on Langmuir–Hinshelwood and the other on Elex–Rideal, has been proposed.

III. Photoprocesses with "Metallized" Semiconductor Powder Dispersions

A. Photodecomposition of Water

Semiconductor powders coated with metals via photodeposition and other techniques are finding extensive applications in studies of photoreduction and oxidation of water. As in other photoreactions of semiconductors, band-gap excitation leads to generation of holes and electrons. With semiconducting oxides such as TiO_2, WO_3, and $SrTiO_3$, the photogenerated holes in the valence band have high oxidizing properties ($E_{vb} \geq$ 2.0 eV) that enable oxidation of water to O_2 [reaction (45)] occur unassisted by redox catalysts. It is not clear at present what the overvoltage

$$4h^+ + 2H_2O \longrightarrow O_2 + 4H^+ \tag{45}$$

requirements are for reaction (45) on various semiconductors, although the studies of Grätzel (*67*) and Sakata and Kawai (Chapter 10) have clearly demonstrated the catalytic effects of RuO_2 deposits on TiO_2 and CdS dispersions. Most of the semiconductors, however, are poor electrocatalysts for H_2 evolution, and one often needs metallic deposits such as Pt to reduce the overvoltage requirements for proton reduction to hydrogen.

Depending on the nature of the metal and the surface characteristics of the semiconductor, a metal–semiconductor contact may give rise to a Schottky barrier or just be an ohmic contact. Because a Schottky barrier would drive electrons away from the metal, for metal deposits to act as reduction centers for the photogenerated electrons the contact must be ohmic. Presumably, the Pt–TiO_2 contact appears to be so, as seen by its efficient performance in water-reduction reactions.

Efficient photosensitized H_2 evolution from water observed using platinized CdS dispersions and reductants such as cysteine or EDTA (*36, 38, 95, 96*) clearly demonstrates the advantages of redox catalysis in semiconductor photoreactions. The electron donors such as EDTA or cysteine are used to scavenge the holes rapidly, so that they do not undergo undesirable side reactions such as photocorrosion. A 20-fold increase in the relative rates of H_2 production indicates that in the absence of catalysts there is extensive recombination of the photogenerated holes and electrons in the bulk of the semiconductor. The formal quantum efficiency for H_2 production has been determined to be 0.04 for 436-nm light. Sulfides have also been used as efficient hole scavengers to generate H_2 with a quantum yield of 35% (*13*). Platinized Si, CdSe, and Cu phthalocyanins also evolve H_2 in similar experiments, albeit with low efficiency.

The earliest report on the water cleavage with metallized semiconductor dispersions has been that of Bulatov and Khidekel (*18*), who reported O_2 and H_2 production in the uv photolysis of platinized TiO_2 in 1 *N* H_2SO_4. Since then this system has been examined in detail in various laboratories. Wrighton *et al.* (*183*) described simultaneous H_2 and O_2 evolution from water on irradiation of platinized $SrTiO_3$, and $KTaO_3$ single crystals. (*183*).

Cleavage of water into H_2 and O_2 has been reported using metallized semiconductors such as Pt–TiO_2 (*11, 12, 147, 152, 153*) and Rh–$SrTiO_3$ (*113*). Pelizzetti and Visca (Chapter 8), Kiwi (Chapter 9), and Sakata and Kawai (Chapter 10) discuss in detail the various features, results, existing problems, and possible strategies of the photodecomposition of water with catalyst-loaded semiconductor dispersions.

Depending on the pretreatment, TiO_2 occurs in two crystalline modifications, anatase and rutile. The flat-band potential (electron Fermi level) of anatase is placed slightly more negative of H^+/H_2 redox potential (~300 mV more cathodic of rutile) and has a band gap of 3.22 V. (The band gap of rutile however is around 2.9–3.0 V.)

Photoinduced H_2 evolution from water using Pt–TiO_2 dispersions has been investigated under various conditions (*101, 147, 153*). TiO_2 by itself was found to have no activity for the photolysis of liquid- or gas-phase water, but platinized TiO_2 dispersions does (*147, 153*). Even with Pt–TiO_2, continuous photodecomposition of gas-phase water does not take place, apparently because of the thermal back reaction between the pho-

$$H_2 + \tfrac{1}{2}O_2 \xrightarrow{\text{Pt–TiO}_2} H_2O \tag{46}$$

toproducts catalyzed on Pt. The photocatalytic activity of Pt–TiO_2 for water decomposition is much improved by reducing TiO_2 with H_2. Hydrogen doping is known to produce oxygen vacancies in the TiO_2 lattice, so that the resistivity of TiO_2 is reduced. The higher activity presumably arises from this increased conductivity.

Sustained evolution of H_2 from the gas-phase water, however, is observed when the oxygen formed reacts with other substrates such as CO (*150*), hydrocarbons (*149*), active carbon (*151*), or lignite (*148*). Photolysis of NaOH-coated, platinized TiO_2 dispersions also show similar behavior (*152*). The quantum efficiency of H_2 and O_2 production reaches about 7%

$$C + O_2 \xrightarrow{\text{Pt–TiO}_2} CO_2 \tag{47}$$

at the beginning, but declines thereafter because of thermal back reactions of accumulated products on Pt. Wagner and Somorjai (*176*), in studies with NaOH-coated $SrTiO_3$ single crystals, have observed similar results. Ferrer and Somorjai (*52*) have provided evidence for photoproduction of

D_2 from D_2O vapor at a pressure of 10^{-7} torr on reduced $SrTiO_3$ single crystals (metal free) at 600 K. By electron energy-loss spectroscopy, a decrease in the surface concentration of Ti^{3+} during the photoproduction of D_2 was observed and is ascribed to the incorporation of oxygen into the crystal lattice.

Darwent and Mills (*37*) have examined photooxidation of water in aqueous WO_3 dispersions with Fe^{3+} as the electron acceptor. The formal quantum efficiency for O_2 production has been determined to be 3.1×10^{-3} at 405 nm. Oxygen production was found to be inhibited by Fe^{2+}, O_2, and high concentrations of Fe^{3+}. Deposits of RuO_2 on WO_3 catalyzed the rate of O_2 production, whereas those of Pt, Rh, and Ru were only inhibitive.

In studies with rhodium-deposited $SrTiO_3$ dispersions, high turnover numbers (~12,000 on Rh and ~50 on $SrTiO_3$) have been observed with no significant decomposition of the semiconductor (*113*). Temperature, pressure, and kinetic studies indicate that the splitting reaction occurs under conditions in which water is close to vaporization at a given temperature and pressure. Positive results have also been obtained for deposits of Ru, Ir, Pd, Pt, Os, Re, and Co on $SrTiO_3$ with relative efficiencies (with respect to Rh) in the order of 24, 13, 11, 13, 10, 16, and 4%, respectively. Essentially similar results have been obtained with $SrTiO_3$–$LaCrO_3$ dispersions (*168*). Although the role of Rh deposits in promoting the water-reduction reaction is clear, their role in catalyzing hole reaction cannot be ruled out.

B. Photooxidation of Halides and Cyanides

In oxygen-saturated solutions, Cl_2, Br_2, I_2 are easily photoproduced from their respective halide solutions using platinized TiO_2 dispersions (*142*). Although the photoproduction of Cl_2 and Br_2 via reaction (48) are

$$O_2 + X^- + 2H^+ \longrightarrow \tfrac{1}{2}X_2 + H_2O \qquad (48)$$
$$X = Cl, Br, I$$

photocatalytic processes, with I_2 it is photosynthetic. These studies are similar to that of Herrmann and Pichat (*80*), who used naked TiO_2 dispersions. With platinized TiO_2 the halide oxidations occur very efficiently, and O_2 is reduced to H_2O at the Pt centers. The rate of H_2 production follows the order $I_2 > Br_2 > Cl_2$, and quantum efficiencies of the order of 19, 0.9, and 1% have been calculated for I^-, Br^-, and Cl^-, respectively.

The photocatalytic oxidation of CN^- on platinized TiO_2, examined by Kago *et al.* (*102*), has shown that the rate doubles for ~50 monolayer thicknesses of Pt. The activity of the platinized catalyst seemed to be uninfluenced by the method of platinization. Ring-disk-electrode studies indicate that CN^- oxidation occurs in competition with water oxidation.

C. *Photooxidation of Carbonaceous Materials*

Light-induced H_2 evolution from water using platinized TiO_2 dispersions and various C compounds as substrates currently receives intense interest and activity. With water as the electron acceptor and CO as the hole reductant, the process corresponds to a *photoassisted water gas shift reaction*. Studies of kinetics of reaction (*49*) at 0–60°C have shown that on

$$CO + H_2O \xrightarrow[\text{Pt–TiO}_2]{h\nu} CO_2 + H_2 \tag{49}$$

Pt–TiO_2 dispersions the reaction is zero order both in CO and H_2O when p_{CO} is 0.3 torr, p_{H_2O} is 5 torr, and the activation energy ~7.5 kcal mol^{-1}. The quantum efficiency of the reaction was found to be 0.5% at 25°C. The photocatalytic activity of Pt–TiO_2 increases with increased NaOH coating of the powder (*152*).

The gas-phase reaction using active carbon or lignite (*151–153*) over illuminated, platinized TiO_2(anatase) leads to H_2, CO_2, and a small amount of oxygen:

$$H_2O(g) + \tfrac{1}{2}C(s) \longrightarrow H_2 + \tfrac{1}{2}CO_2 \tag{50}$$

The rate of the reaction declines with photolysis time owing to the accumulation of H_2 and resulting loss of good contact between the catalyst and C. With active carbon as the substrate, the reaction is zero order with respect to H_2O pressure, has an activation energy of ~5 kcal mol^{-1}, and a quantum efficiency of 2% in the early stages of photolysis. In liquid phase, oxidation of C is inhibited and water photodecomposition dominates. Using the same catalyst system, ethylene has been converted to ethane, CO_2, H_2, and traces of methane.

$$\tfrac{1}{2}C_2H_4(g) + 2H_2O(g) \longrightarrow CO_2(g) + 3H_2(g) \tag{51}$$

$$7.5C_2H_4(g) + 2H_2O(g) \longrightarrow CO_2(g) + 3C_2H_6(g) \tag{52}$$

$$\tfrac{1}{2}C_2H_6(g) + 2H_2O(g) \longrightarrow CO_2(g) + 3.5H_2(g) \tag{53}$$

Sakata and Kawai (*145, 146*) have observed photocatalytic production of H_2 and methane from ethanol and water mixtures using platinized TiO_2. The accompanying scheme has been proposed to account for the

$$C_2H_5OH \longrightarrow CH_3CHO + H_2 \tag{54}$$

$$H_2O + CH_3CHO \longrightarrow CH_3COOH + H_2 \tag{55}$$

$$CH_3COOH \longrightarrow CH_4 + CO_2 \tag{56}$$

observed products. The quantum efficiency (at 380 nm) for the hydrogen production has been determined to be 6.5, 13.6, 19.0, and 38.0% for TiO_2 dispersions loaded with Ni, Pd, Rh, and Pt, respectively. Photocatalytic production of methane from CO_2 and gaseous water has also been observed on $SrTiO_3$ crystals contacted with a Pt foil (*78, 79*). Kawai and Sakata have also demonstrated the possible use of various biomass sources such as wood, cotton, protein, and carbohydrate as substrates (*99–101, 145*).

Izumi *et al.* (*88*) have described heterogeneous photocatalytic oxidation of various hydrocarbons (benzene, hexane, cyclohexane, heptane, nonane, decane, and kerosene) in oxygen-containing solutions at platinized TiO_2. Along with CO_2, the formation of other products (phenols from benzene and alcohols from aliphatic hydrocarbons) has been explained in terms of reactions of photogenerated OH radicals at the TiO_2 surface.

D. Photo-Kolbe Reaction

A reaction that clearly demonstrates the features of photoprocesses on metallized semiconductor dispersions is the photo-Kolbe reaction studied extensively by Bard and co-workers (*89, 106, 108, 109*). Carried out in an electrochemical cell, at an illuminated *n*-TiO_2 electrode, decarboxylation of acetic acid–acetate yields CO_2 and ethane (*106*):

$$2CH_3COO^- \xrightarrow[\text{electrode}]{h\nu,\ TiO_2} CH_3CH_3 + 2CO_2 \tag{57}$$

However, the same reaction carried out in aqueous media with platinized TiO_2 dispersions yields CH_4 and CO_2 as major products (*106, 108*):

$$CH_3COOH \longrightarrow CH_4 + CO_2 \tag{58}$$

Detailed investigations have shown that the photo-Kolbe reaction is very general in scope and has no specificity either to the semiconductor (occurs also on Pt–WO_3) or to acetic acid. With Pt–TiO_2, several aliphatic carboxylic acids have been examined for their major photoproducts. The accompanying scheme accounts adequately for all the observed results.

$$h^+ + RCOO^- \longrightarrow R\cdot + CO_2 \tag{59}$$

$$e^- + RCOOH \longrightarrow H(ads) + RCOO \tag{60}$$

$$e^- + R\cdot + RCOOH \longrightarrow RH + RCOO^- \tag{61}$$

$$2R\cdot \longrightarrow R{-}R \tag{62}$$

$$2H(ads) \longrightarrow H{-}H \tag{63}$$

$$R\cdot + H(ads) \longrightarrow RH \tag{64}$$

In experiments with $Pt–TiO_2$ dispersions at reasonable light intensities, the large surface area of the catalysts results in low surface concentration of the radicals. This inhibits second-order reactions such as dimerization and disproportionation. Also, R· radicals are produced near the reducing sites (Pt) on the powder (facilitating reduction to RH), as contrasted to R· radicals produced at an illuminated TiO_2 electrochemical cell. This behavior would account for the major production of methane with dispersions.

With aromatic compounds such as benzoic acid, instead of the usual Kolbe products, major products are CO_2 and various phenolic resin polymers through the intermediate production of salicylic acid and phenol (*89*). In deaerated solutions, dicarboxylic acids such as adipic acid yield primarily CO_2 and butane, with lesser amounts of valeric acid and negligible quantities of oligomeric materials. A mechanism for the reaction based on photogeneration of OH· radicals and adsorption of intermediates onto the powder surface has been proposed.

Sato (*154*) has shown that photodecarboxylation of acetic acid with $Pt–TiO_2$ carried out in the gas phase with water vapor yields the normal Kolbe electrolysis product, namely, ethane in high yields.

E. Photosynthetic Production of Amino Acids

Xenon-lamp irradiation of platinized TiO_2 powders suspended in ammoniacal solutions saturated with methane leads to photosynthetic production of various amino acids (glycine, alanine, serine, aspartic acid, and glutamic acid) as well as methanol, ethanol, and methylamine (*43, 140*). Experiments with $^{15}NH_3$ has shown that the nitrogen in the amino acids originates with the ammonia rather than the contaminants. Nonplatinized semiconductor dispersions of TiO_2, Fe_2O_3, and $FeTiO_3$ showed no detectable activity. Because the same amino acids are also produced from $CH_4–NH_3–H_2O$ mixtures during the decomposition of H_2O_2 on a Pt foil, a mechanism based on free-radical reactions initiated by photogenerated OH· radicals has been proposed.

F. Other Photoprocesses

Photocatalytic isotopic exchange between cyclopentane and D_2 over $Pt–TiO_2$(anatase) at −10°C has been demonstrated by Courbon *et al.* (*191*). A selectivity in C_5H_9D of about 90% was found with a total quantum yield of 1.8%. The photocatalytic nature of the isotopic exchange was clearly demonstrated by (a) absence of any exchange in the dark or on replacement of TiO_2 by SiO_2, (b) a linear dependence of the exchange rate with illumination time, and (c) a turnover number of ~78.

IV. Photoprocesses in Semiconductor Dispersions Loaded with Oxides: Hole Transfer and Bifunctional Catalysis

Studies with metallized semiconductor dispersions, reviewed in Section III, have shown unambiguously that it is possible to catalyze various reactions of photogenerated electrons (in the conduction band) with solutes present in the solution. In an analogous manner, it should be possible to catalyze various oxidation reactions of the holes by depositing a thin layer of an electroactive catalyst such as an oxide. Studies using deposit of oxides of Ru, Rh, Ni, and Ir on semiconductor powders have shown some promising results in this context. For the oxide layer to act as a "hole-transfer catalyst," the catalyst–semiconductor junction should behave as a Schottky barrier. The height of the Schottky barrier will invariably determine the oxidizing power of the degenerated hole. If the catalytic redox reactions over these deposits are specific, then one can envisage "bifunctional redox catalysis" with combined deposits such as Pt and RuO_2 onto a semiconductor. Detailed electrochemical studies of such catalytic processes with semiconductor electrodes are not yet available.

The majority of studies has been with semiconductor dispersions coated with a layer of RuO_2. Because RuO_2 is also known to be a good electrocatalyst for H_2 evolution, interpretation of catalytic effects observed with these oxides on H_2 evolution are ambiguous as to what redox reaction they catalyze, that of holes or of electrons. For example, CdS dispersions loaded with RuO_2 catalysts efficiently catalyze H_2 evolution using visible light from aqueous solutions containing sulfide (S^{2-}) ions without undergoing significant photocorrosion (*13, 95, 96*). The proposed mechanisms are

$$S^{2-} + 2h^+ \xrightarrow{RuO_2} S \tag{65}$$

$$2e^- + 2H^+ \longrightarrow H_2 \tag{66}$$

Kawai and Sakata (*100*) have similarly shown that addition of RuO_2 to TiO_2 dispersions enables catalytic photodecomposition of water to H_2 and O_2 at a much higher rate than with TiO_2 alone. On NiO–$SrTiO_3$ and Co–$SrTiO_3$ powders, photodecomposition of water vapor proceeds steadily

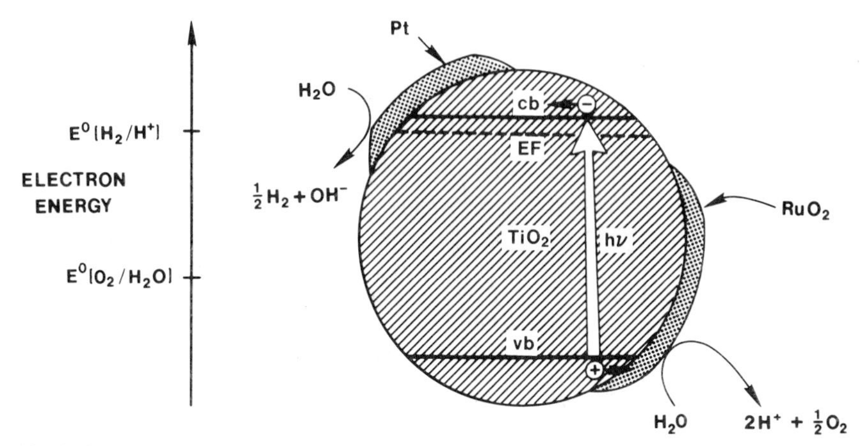

Fig. 3. Schematic representation of cyclic water cleavage in colloidal TiO_2 particles loaded with bifunctional redox catalysts.

for more than 100 h (*42*). In distinct contrast to the behavior with Pt–$SrTiO_3$, there is no synthesis of water from H_2 and O_2 in the dark.

Grätzel and co-workers have demonstrated cyclic water cleavage using colloidal TiO_2(amorphous and anatase) particles codeposited with Pt and RuO_2 (*11, 46, 67*). Figure 3 presents schematically the bifunctional redox catalysis with these particles. The model pictures photogeneration of electrons and holes followed by the migration of the former to Pt sites where H_2 is formed while the latter are channelled to RuO_2, which catalyzes oxygen evolution from water. The initial rate observed is high (about 2 to 3 ml H_2 per 25 ml solution per hour). As with experiments using Pt–TiO_2 dispersions, a photostationary state is soon reached, at which, if the irradiation is stopped, there is dark recombination of H_2 and O_2 over the Pt sites. The high rate, however, can be sustained by periodic flushing out of the products (H_2 and O_2) with argon or nitrogen. Doping of the semiconductor with metals such as Cr significantly improves the sensitivity of the photoprocess to the visible light (*105*). Bifunctional redox catalysis and related strategies are elaborated by Pelizzetti and Visca (Chapter 8) and Kiwi (Chapter 9).

V. Semiconductor Dispersions as "Carriers" of Catalysts and Photosensitizers

Photoprocesses with semiconductor dispersions described so far employed direct band-gap excitation of the semiconductor. Finely divided semiconductor particles are also finding increasing use as carriers of catalysts (''supported catalysts'') and photosensitizers. Finely divided metals such as Pt and Rh deposited on semiconducting powders such as TiO_2 or $SrTiO_3$ were found to exhibit the best catalytic activity for H_2 evolution from water in homogeneous photoredox systems such as $Ru(bipy)_3^{2+}$–methylviologen, (MV^{2+})–EDTA (*104*), and $Ru(bipy)_3^{2+}$–$Rh(bipy)_3^{3+}$–triethanolamine (*114*). In these systems, photogenerated reductants [MV^+ and $Rh(bipy)_3^{2+}$] directly reduce water to hydrogen at the catalytic sites provided by Pt or Rh.

$$2A^- + 2H_2O \xrightarrow{Pt\text{–}TiO_2} 2A + H_2 + 2OH^- \quad (67)$$

Visible-light photolysis of $Ru(bipy)_3^{2+}$, MV^{2+}, and a bifunctional redox catalyst $Pt\text{–}TiO_2\text{–}RuO_2$ has shown that it is also possible to achieve total water decomposition to H_2 and O_2 without EDTA (*12, 103, 104*). With surfactant derivatives of $Ru(bipy)_3^{2+}$, which adsorbs strongly onto the semiconductor, H_2 evolution has been observed even without MV^{2+} (*11, 12*). The photodecomposition of water in such systems is explained by direct charge injection from the excited state of $Ru(bipy)_3^{2+}$ to the conduc-

$$Ru(bipy)_3^{2+*} + TiO_2 \longrightarrow Ru(bipy)_3^{3+} + e_{cb}^-(TiO_2) \quad (68)$$

tion band of TiO_2 followed by water reduction at the Pt sites and RuO_2 catalyzed oxidation of water by the oxidized dye $Ru(bipy)_3^{3+}$ as illustrated

$$4Ru(bipy)_3^{3+} + 2H_2O \longrightarrow Ru(bipy)_3^{2+} + 4H^+ + O_2 \quad (69)$$

$$e_{cb}^-(TiO_2) + 2H_2O \longrightarrow H_2 + 2OH^- \quad (70)$$

in Fig. 4. In support of reaction (68), direct quenching of $Ru(bipy)_3^{2+*}$ by colloidal TiO_2 has been observed (*46*).

The preceding overall sequence of reactions amounts to ''sensitization'' of the semiconductor into the visible region by use of a dye. Such

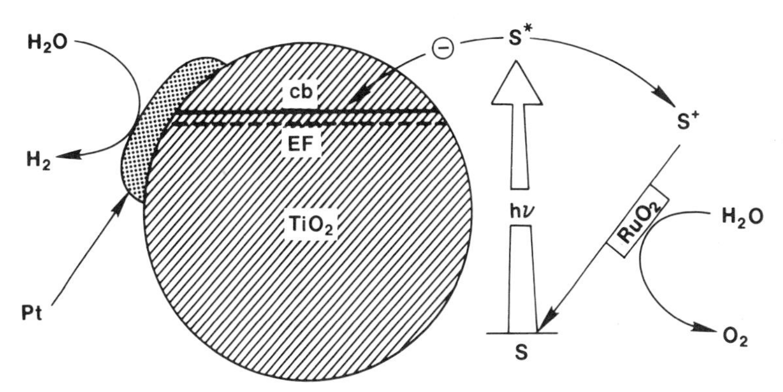

Fig. 4. Schematic illustration of dye-sensitized photocleavage of water in semiconductor dispersion systems.

"photocatalysis" and "photosynthesis" through excitation of adsorbed dyes has been explored by Honda *et al.*, who examined photochemical behaviour of dyes such as rhodamine B and methylene blue adsorbed onto CdS particles (*163, 179*). Although rhodamine B in aqueous solution is considerably stable to visible-light photoexcitation, efficient N-deethylation of the dye occurs, either by excitation of the adsorbed dye molecules or by excitation of CdS. Because excited rhodamine B is known to inject an e^- to the conduction band of the CdS electrode, the proposed mechanisms involved such charge injection to the conduction band of CdS particles [reaction (71)], followed by oxidative N-dealkylation of the radical-

$$\mathrm{RhB^*(ads)} + h\nu \longrightarrow \mathrm{(RhB)^+(ads)} + e_{cb}^-\mathrm{(CdS)} \tag{71}$$

cation of the dye. The absence of any N-dealkylation with methylene blue is consistent with its inability to sensitize the CdS electrode. Both dyes, however, undergo N-dealkylation on excitation of CdS with band-gap light through reactions of the photogenerated holes. Distinct changes in the electronic absorption spectra and catalytic activity observed with cobalttetraphenylporphyrin on its deposition onto TiO_2 powders (*122*) are explained by similar electronic interactions between the dye and the semiconductor.

VI. Physical Methods in the Study of Semiconductor Dispersions and Colloids

Quantitative analysis of various photoprocesses on semiconductor powders, dispersions, and colloids requires detailed knowledge on the particle characteristics (size, surface area, charge, etc.) and catalytic activity (product selectivity, kinetics, and rates). Establishment of mechanisms requires further information on the possible intermediates, their concentrations, and the sequence of their occurrence. An increasing number of physical methods are being brought into use, and Table I summarizes various methods available for the characterization of properties of semiconductor particulates and also photoreactions undergone by them. Here we briefly review some selected techniques as applied to semiconductor particulate systems.

A. Colloidal Semiconductors and Flash Photolysis

For a clear understanding of the redox processes across semiconductor–electrolyte interfaces, it is highly desirable to obtain detailed informa-

TABLE I

Physical Methods in the Characterization of Semiconductor Dispersions and in the Study of Their Photoreactions

Property, reaction	Technique
Characterization	
Size and polydispersity	Low-angle light scattering, x ray, electron microscopy
Concentration	Atomic absorption spectroscopy
Surface area, porosity	Gas adsorption such as BET
Charge	Micro- and photoelectrophoresis, slurry electrodes
Band gap	Photoacoustic, diffuse reflectance spectroscopy
Surface acidity or basicity	Infrared spectroscopy, thermogravimetry, temperature-programmed desorption (tpd)
Metal and metal oxide deposits	Electron spectroscopy (esca, Auger), transmission electron microscopy (tem, stem)
Reaction	
Adsorption and desorption of gases	Photoconductivity, electron spin resonance, dynamic mass spectrometry, Hall effect
Intermediates detection and mechanism	Flash photolysis, esr spin traps, product analysis in static or flow reactors, photoelectrochemical cells with macrosemiconductor electrodes

tion on the dynamics of the photoinduced electron transfer by fast kinetic spectroscopy. Hitherto this has been hampered for systems composed of solid electrodes and macrodispersions of semiconductor powders by various factors. Procedures have been and are being developed for the preparation of finely divided monodispersed semiconductor particles, such as those of TiO_2, CdS, and Fe_2O_3 (*119, 120*). The dimensions of these particles are small enough (often less than a few micrometers) to render scattering of light negligibly small. With such colloidal semiconductors, it is possible to directly monitor, by flash photolysis, the charge injection into or from the semiconductor to reagents present in solution. Used as carriers of catalysts, such systems also allow kinetic studies of redox catalysis. The first studies of this kind have been described by Grätzel and coworkers (*46, 47*).

Transparent sols of TiO_2 (anatase, particle radius 200 Å) can be produced via hydrolysis of titanium tetraisopropoxide in acidic aqueous solution. Kinetic analysis of visible-light-induced (excitation at 530 nm) electron-transfer reaction (72) in the presence of colloidal TiO_2 carrying either

$$Ru(bipy)_3^{2+} + MV^{2+} \rightleftharpoons Ru(bipy)_3^{3+} + MV^{+} \qquad (72)$$

Pt or RuO_2 clearly shows the effect of added catalysts on the life time of the redox products. The decay of MV^+ via reaction (73) is enhanced by the presence of Pt and that of $Ru(bipy)_3^{3+}$ by addition of RuO_2 [reaction (74)]. On excitation of TiO_2–methylviologen (MV^{2+}) solutions in the band

$$MV^+ + H^+ \xrightarrow{Pt} MV^{2+} + \tfrac{1}{2}H_2 \quad (73)$$

$$4Ru(bipy)_3^{3+} + 2H_2O \xrightarrow{RuO_2} 4Ru(bipy)_3^{2+} + 4H^+ + O_2 \quad (74)$$

gap (excitation at 347 nm; 15-ns laser pulses) of the semiconductor, rapid growth of absorption at 602 nm (because of the photoreduction of MV^{2+} by the conduction-band electrons) is readily observed [reactions (75) and (76)] (see Fig. 5). The MV^+ growth follows first-order kinetics with the

$$TiO_2 \xrightarrow{h\nu} h^+ + e_{cb}^- \quad (75)$$

$$e^- + MV^{2+} \longrightarrow MV^+ \quad (76)$$

rate constant in the range $2–30 \times 10^3\ s^{-1}$, dependent on the MV^{2+} concentration. From a quantitative analysis of the growth and decay of the MV^+ radical as a function of pH, MV^{2+}, etc., it is possible to comment on the lifetime of the $e^-–h^+$ pairs, the flat-band potentials, and on the kinetics of

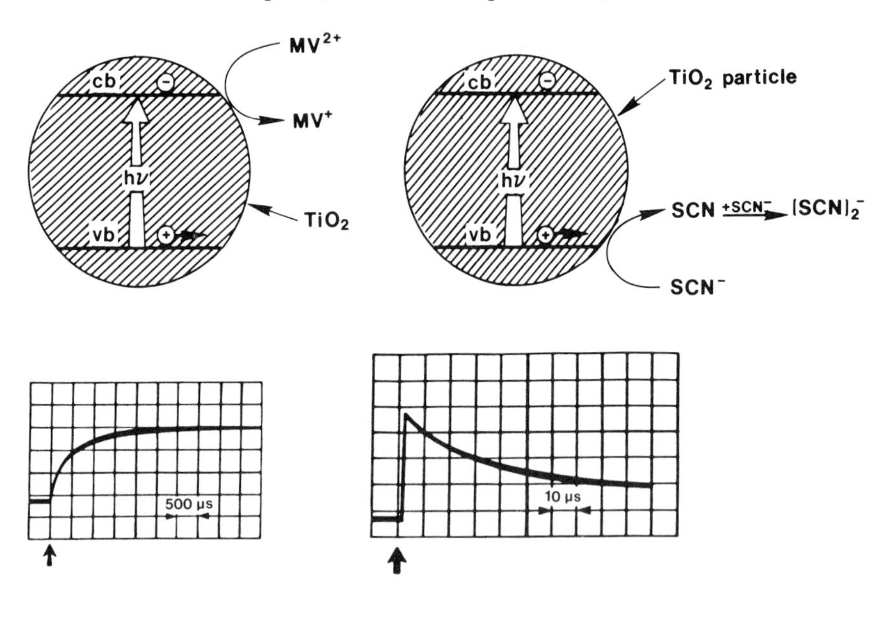

Fig. 5. Schematic representation with kinetic traces illustrating the time course of photoreduction of MV^{2+} by conduction-band electrons (a; λ_{max} = 602 nm) and of photooxidation of SCN^- by holes (b; λ_{max} = 470 nm). Arrows indicate laser pulses.

interfacial electron transfer. Similarly, with SCN^- as the hole reductant [reaction (77)], it is possible to follow hole-transfer kinetics. In the pres-

$$h^+ + SCN^- \longrightarrow SCN\cdot \tag{77}$$

ence of excess SCN^-, oxidized radicals ($SCN\cdot$) undergo complexation and the resulting $SCN_2^{\cdot -}$ radicals are easily monitored by their character-

$$SCN\cdot + SCN^- \longrightarrow (SCN)_2^- \tag{78}$$

istic optical absorption at 470 nm. In contrast to MV^+ growth (which occurs over a micro- to millisecond time domain), the growth of $SCN_2^{\cdot -}$ absorption is prompt, indicating transfer of electrons to adsorbed SCN^- species. Electron transfer from the conduction band to acceptors in solution and competitive trapping of e_{cb}^- by noble-metal deposits have also been investigated with colloidal CdS solutions. The technique thus appears very promising and is expected to yield some new results on interfacial electron-transfer processes.

B. Surface Charge and Electrophoresis

Hydrous oxide surfaces of semiconductor contain surface hydroxyl groups (OH) (about 4 to 15 OH groups per square nanometer), which exhibit amphoteric behavior. The surface chemistry of the oxide is determined by the dissociation equilibria [reaction (79)] and, depending on

$$M{-}OH_2^+ \overset{H^+}{\rightleftharpoons} M{-}OH \overset{H^+}{\rightleftharpoons} M{-}O^- \tag{79}$$

the solution pH, the surface may carry either a positive or a negative charge. (For metal oxide sols, H^+ and OH^- ions are thus known as potential-determining ions.) Consequently, for dispersions composed of particles of large dimensions, there exists an electrical double layer with an excess of counterions in the vicinity of the particle. The potential at the shearing plane associated with the particle–solution interface is defined as the zeta potential ζ. The *isoelectric point* (iep) refers to the pH at which the particle carries no net charge (or zero zeta potential). Electrophoresis is the technique of measuring the "mobility" for the movement of charged particles in a suspension under the influence of an applied external electric field. Processes such as surface doping, mode of preparation, and specific adsorption of ions can significantly influence the iep. *Photoelectrophoresis* is a means of monitoring changes in the surface charges under illumination (*169*). If photogenerated holes can react with reductants in solution, but the electrons do not react, a net negative charge can be built up on the particle during illumination.

The iep of colloidal TiO_2 (anatase), prepared by hydrolysis of titanium tetraisopropoxide, has been determined to be 3.2 (*46*). For anatase and rutile dispersions, values in the range 5.0–5.5 have been reported (*44, 162, 192*). Loading of colloidal TiO_2 by Pt or RuO_2 decreases the iep by approximately 0.3 units. Studies with "naked" anatase dispersions under band-gap illumination show that there is a significant buildup of negative charges on the particle surface (iep shifts to lower values). Platinized dispersions do not exhibit this effect. The results are easily understood in terms of the band model. Band-gap irradiation produces e^-–h^+ pairs. The holes are energetic enough to oxidize water, and the electrons accumulate on the surface. On platinized TiO_2 dispersions there is no such effect, owing to rapid reduction of protons by electrons to yield H_2.

Sugikara and co-workers have shown that rutile(TiO_2) particles suspended in a highly insulating liquid alter the sign of their ζ potential on exposure to visible light, and this phenomenon of photoelectrophoresis can be used in direct image-storage and -display devices (*162, 169*). Possible utility of titania particles stabilized with various polymeric materials in "electrophoretic display" devices has also been explored (*23, 24*).

Closely related to photoelectrophoresis is the phenomenon of *memory effect* observed with oxides such as ZnO, and it is used extensively in *electrophotography*. In the dark-adapted state, O_2 is chemisorbed at the surface of ZnO grains (in the electrophotographic layer), and the interior of the grains are depleted of electrons. On exposure to light, desorption of the gas occurs, and as mentioned in Section II, this is accompanied by an increase in the conductivity of the sample (adsorbed O_2^- ions annihilate the photogenerated holes, and the electrons freely move from one grain to

$$O_2^-(\text{ads}) + e_{cb}^- \xrightarrow[\text{desorption}]{} O_2 \tag{80}$$

another). This increased conductivity is preserved even after the irradiation is turned off. This charged photoconductor is now exposed to produce the latent image. Weigl (*180*) has elaborated on semiconductor-dispersion applications in electrophotography.

C. Electrochemical Methods and Slurry Electrodes

A unifying hypothesis for all the photoprocesses induced by semiconductor powders and dispersions is that they are all the consequences of photogeneration of holes and electrons (according to the band model) and subsequent reactions of these with species adsorbed or present in solution. The capture of holes by the reductants and of electrons by the oxidants follow the same thermodynamic picture proposed by Gerischer (*64*) for reactions on semiconductor electrodes in electrochemical cells. It is also becoming clear that the existing numerous studies of photoredox processes on electrodes can be adapted for semiconductor dispersions and colloids and, conversely, that the mechanisms of novel reactions observed with particulate systems are easily probed by the corresponding "macro" electrodes. Thus experiments of photodecomposition of water with Pt–TiO_2 and Rh–$SrTiO_3$ dispersions are nothing but miniature (short-circuited) versions of corresponding photoelectrochemical cells. However, few methods currently exist for the direct probing of the detailed mechanisms or the catalyst properties *in situ*.

In addition to the photoelectrophoresis, *slurry electrodes* composed of semiconductor dispersions and metal electrodes have been shown to be operational, demonstrating the light-induced charge-separation process. Early electrochemical studies on suspensions of semiconductors have been those of Loutfy and Sharp (*116*), who demonstrated voltammetric oxidations and reductions of α, β, and x forms of metal-free phthalocyanins dispersed in CH_2Cl_2.

Figure 6 presents diagrammatically one such slurry electrode cell composed of an illuminated compartment (containing TiO_2 dispersions in 0.02 *M* HCl and a Pt-gauze collector electrode) and a dark compartment (another Pt electrode immersed in oxygenated 0.02 *M* HCl). Illumination of the TiO_2 dispersions with band-gap light leads to a large initial cathodic current (20 μA) followed by a steady-state anodic current (8–10 μA) (*45*,

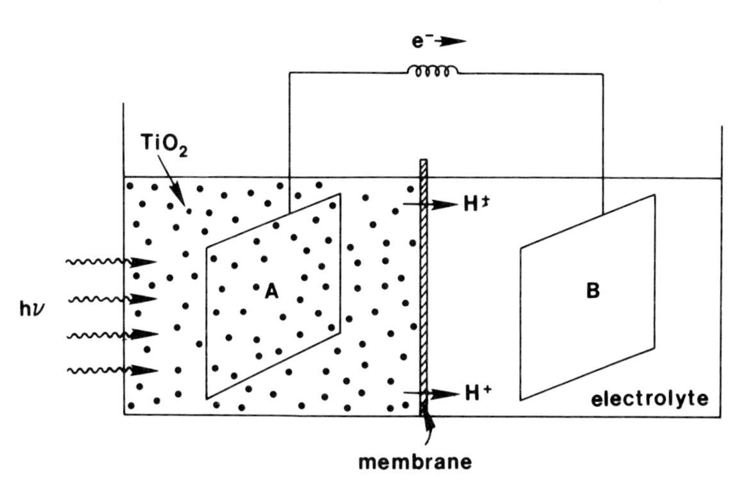

Fig. 6. Photoelectrochemical cell involving semiconductor particles as photo- and electroactive species (slurry electrodes). A, Collector electrode; B, counterelectrode.

44). The observation of significant photocurrents (under both the two-electrode-cell configuration and for the slurry electrode under potentiostated conditions) coupled with its action spectrum (consistent with a current onset at 3.1 eV) clearly show that the photogenerated charges on the particles can be collected at an electrode. Chojnowski *et al.* (*20*) have also examined a similar cell composed of TiO_2 (Degussa P-25) slurry and observed cyclic water cleavage (photocurrents ~20 μA). With easily oxidizable substrates such as acetate (instead of water), much higher photocurrents (550 μA) are obtainable. Addition of Cu^{2+}, Fe^{3+}, and methylviologen as the electron scavenger also produce significant enhancement of photocurrents (*6, 178, 190*). Slurry electrodes made up of CdS dispersions also appear to be feasible (*6*).

D. Spectral Sensitivity by Photoacoustic and Diffuse Reflectance Spectroscopy

In studies of "naked" semiconductors and in dye sensitization processes, action spectrum for the spectral sensitivity of photoreactions on the semiconducting dispersions is a useful guide in probing the origin of the photoprocesses. Owing to the severe light-scattering problems, optical measurements of action spectra are difficult to carry out, and results are grossly inaccurate. Photoacoustic spectroscopy and diffuse reflectance spectroscopy has been shown to be very useful in this context.

Absorption of light by a sample is converted to heat by radiationless processes, and for samples placed in a close volume this gives rise to pressure changes within the cell. In photoacoustic spectroscopy (pas), these photoinduced pressure changes are detected by a microphone or by piezoelectric detection and can, therefore, be further processed. That photoacoustic spectroscopy is a useful technique for the determination of the onset of spectral response (onset is closely related to the band gap) in semiconductor dispersions has been demonstrated by Iwasaki *et al.* (*85–87*). Photoacoustic spectra have been reported for ZnO powder and for 1,1′-diethyl-2,2′-cyanin chloride rhodamine B dyes adsorbed on ZnO powders (*85*). For the cyanin dye, prior to photolysis distinct spectral maxima at 525 and 575 nm (corresponding to molecular *J* aggregates) are observed; these decrease significantly on photolysis, confirming the model of charge injection from the excited state of the dye to the conduction band followed by the decomposition of the oxidized dye. Similar

$$\mathrm{RhB^* + ZnO \longrightarrow (RhB)^+ + e_{cb}^-(ZnO)} \tag{81}$$

behavior was observed only with other semiconducting powders such as TiO_2, CdS, AgCl, and AgBr, but not on nonsemiconducting Al_2O_3. In subsequent studies (*87*), photoacoustic signals corresponding to the transient intermediates (radical-cations) of the dye eosin have also been recorded.

Halmann and co-workers have similarly demonstrated the utility of diffuse reflectance spectroscopy to determine the band gap and spectral sensitivity of semiconducting dispersions active in the photoreduction of CO_2 (*1*).

E. Radical Intermediates and Utility of esr Spin Traps

Electron spin resonance (esr) is very powerful technique for the study of radical intermediates and is employed extensively in photochemistry and radiation chemistry. In early esr studies of semiconducting particulate systems, the emphasis was in the identification of the radical intermediates involved in the photoadsorption and photodesorption of gases such as O_2, NO, and CO. In these studies, semiconducting powders were photolyzed in the presence of adsorbing–adsorbed gases in the cavity of an esr spectrometer and the resulting signals analysed *in situ*. Lunsford (*117*), for example, has elaborated on various studies of adsorbed oxygen species on semiconducting oxides by esr: O^-, O_2^-, O_3^-, O_2^+, and O_4^-. Progress in this area has been outlined by Meriaudeau and Vedrine (*121*), Bickley (*8*), and Gonzales-Elipe *et al.* (*65, 66*).

A variation of the esr technique for the study of homogeneous solutions and of dispersions is *spin-trapping*. Here, short-lived radicals are transformed into long-lived spin adducts with suitable spin traps. Commonly

$$\text{radical} + \text{spin trap} \longrightarrow \text{spin adduct}$$

used spin traps have been DMPO (5,5-dimethyl,1 : pyrrolinyl-1-oxy), PBN (α-phenyl-*N*-*tert*-butylnitrone), and OBN (α-4-pyridyl 1-oxide-*N*-tetrabutylnitrone) (*2*). Harbour and Hair (*72*) have provided a summary of their spin-trap studies of semiconductor particulate systems.

F. Characterization of Catalyst Deposits By Electron Spectroscopy

An elegant way of monitoring the chemical nature of the metal and metal oxide deposits on the semiconductor powders (prior to as well as after the photoreactions) is by electron spectroscopy. Both x-ray photoelectron spectroscopy (esca, as it is often called) and Auger spectroscopy can be employed, as has been illustrated (*193–195*). Thus deposition of Rh onto $SrTiO_3$, either by photodeposition or by thermal air oxidation methods using aqueous solutions of $RhCl_3 \cdot 3H_2O$, leads mainly to oxides of Rh (*193*). As monitored by esca, the catalytic activity of rhodium correlates well with the oxidation state of the rhodium species, with increasing efficiency observed on progressive oxidation of the deposited metal to surface Rh(III) species. In analogous studies, Bard *et al.* (*194*) have used combined esca and scanning Auger spectroscopy to demonstrate the uniform coverage of Pt on TiO_2 (present as zerovalent Pt^0) when the metal is deposited from an aqueous solution of the metal salt and acetate by a photodeposition process.

VII. Addendum

Since the submission of the original manuscript, nearly 50 publications have appeared on the topic under review, reflecting the intense research activity in this area. For lack of space, we review them only briefly here.

Participation of surface bonds in the process of dark and photoadsorption of O_2 and CO has been described on ZnO (*196*) and $SrTiO_3$ (*197*) powders as well as on $SrTiO_3$ single crystals (*198, 199*). The mechanism of oxidation of organic substrates such as alcohols (*200–203*), amines (*204, 205*), and aromatic hydrocarbons (*206*) on TiO_2 continue to receive close scrutiny. Novel chemistry photoassisted by semiconductor dispersions that has been reported include alkene isomerization (*207*) and photocycloaddition (*208*). Several inorganic materials, such as $S_2O_8^{2-}$ (*209*), S (*210*), Cl^- (*211*), halates (*212*), Ag^+ (*213*), and N_2H_4 (*214*) have been successfully employed as "sacrificial agents" in semiconductor particulate systems designed for the overall evolution of either H_2 or O_2 from water.

There have been several reports on the preparation and catalytic activity of metal- and metal-oxide-coated semiconductor dispersions. According to Mills (*215*), precipitation of a Pt sol using an inert electrolyte in the presence of a semiconductor powder appears to be the best means of platinization. With Pt–TiO_2, relatively high yields have been obtained during the photodehydrogenation of alcohols (quantum yields ~0.45 for CH_3OH, ~0.2 for C_2H_5OH, and ~0.1 for C_3 and C_4 alcohols; *216*). An optimal initial rate of H_2 production was found for Pt content in the 0.1–1 weight % range, and the maximum reaction rate has been attributed to an optimum attraction of free electrons of titania by Pt crystallites (*217*). On both oxidized and reduced Pt–TiO_2, adsorption of CO_2 occurs dissociatively, forming chemisorbed oxygen atoms and CO (*218*).

The rate of photoassisted water-gas shift reaction [reaction (49)] on platinized titania has been found (*219*) to be (a) independent of the method of Pt deposition, (b) first order in light intensity, (c) independent of the chemical state or Pt loading above 2 weight %, (d) dependent on reduction of titania, and (e) strongly dependent on the surface concentration of NaOH. A photochemical diode model has been elaborated (*220*) to explain the photocatalytic properties of Pt–TiO_2. Cyclic photodecomposition of water into H_2 and O_2 both in the vapor phase and in liquid water continues to be examined carefully on various catalyst-loaded semiconductor dispersions (*221–225*). It appears that reaction conditions of reduced pressure and elevated temperatures are necessary for simultaneous production of the two gases H_2 and O_2 (*223*). ZnS also acts a photosensitizer for heterogeneous photoreduction of water (*226*). Considerable progress has been made on the photoreduction of CO_2 as well as HCHO to methanol and other C_1–C_3 compounds using rare-earth-doped or -deposited titanates (*227, 228*).

Studies with colloidal semiconductors are becoming widespread, with procedures being made available for the preparation of monodispersed colloidal semiconductors [e.g., CdS (*229*) and hemitite (*230*)]. Dynamics of electron transfer from the conduction band or to the valence band using various redox couples has been studied on colloidal CdS (*231–235*), TiO_2 (*236, 237*) and Fe_2O_3 (*238*) systems. A kinetic model to describe the interfacial electron transfer has also been proposed (*239*). As mentioned earlier, depending on the nature of the preparation, a photocatalyst activity can often vary by several orders of magnitude, and caution needs to be exercized in the comparison of different systems. Childs and Ollis (*240*) have elaborated on the information that is needed to assess whether a reported photocatalysis is indeed catalytic.

Acknowledgments

It is a pleasure to acknowledge the fruitful collaboration and discussions on the materials covered in this review with Professor M. Grätzel. Our thanks are also due to the Swiss National Funds for Scientific Research for financial support.

References

1. Aurian-Blajeni, B., Halmann, M., and Manassen, J., *Sol. Energy* **25,** 165 (1980).
2. Aurian-Blajeni, B., Halmann, M., and Manassen, J., *Photochem. Photobiol.* **35,** 157 (1982).

3. Balzani, V., Boletta, F., Gandolfi, M. T., and Maestri, M., *Top. Curr. Chem.* **75,** 1 (1978).
4. Bard, A. J., *J. Photochem.* **10,** 59 (1979).
5. Bard, A. J., *Science (Washington, D.C.)* **207,** 139 (1980).
6. Bard, A. J., Pruiksma. R., White, J. R., Dunn, W., and West, M., *Proc. Electrochem. Soc. 160th,* **82–83,** 381 (1982).
7. Barry, T. I., and Stone, F. S., *Proc. R. Soc. London Ser. A* **255,** 124 (1960).
8. Bickley, R. I., *Spec. Period. Rep. Chem. Phys. Solids* **7,** 118 (1978).
9. Bickley, R. I., and Jayanthi, R. K. M. *Discuss. Faraday Soc.* **58,** 194 (1974).
10. Bideau, M., Claudel, B., and Otterben, M., *J. Photochem.* **14,** 291 (1980).
11. Borgarello, E., Kiwi, J., Pelizzetti, E., Visca, M., and Grätzel, M., *J. Am. Chem. Soc.* **103,** 6324 (1981).
12. Borgarello, E., Kiwi, J., Visca, M., and Grätzel, M., *Nature (London)* **289,** 158 (1981).
13. Borgarello, E., Kalyanasundaram, K., Grätzel, M., and Pelizzetti, E., *Helv. Chim. Acta* **65,** 243 (1982).
14. Boonstra, A. H., and Mutsaers, C. A. H. A., *J. Phys. Chem.* **79,** 1940 (1975).
15. Boonstra, A. H., and Mutsaers, C. A. H. A., *J. Phys. Chem.* **79,** 1694 (1975).
16. Boonstra, A. H., and Mutsaers, C. A. H. A., *J. Phys. Chem.* **79,** 2025 (1975).
17. Bolton, J. R. (ed.), "Solar Power and Fuels." Academic Press, New York, 1977.
18. Bulatov, A. V., and Khidekel, M. L., *Izv. Akad. Nauk SSSR Ser Khim.* 1902 (1976).
19. Cardon, F., Gomes, W., DeKeyser, W. (eds.), "Photovoltaic and Photoelectrochemical Solar Energy Conversion." Plenum, New York, 1981.
20. Chojnowski, F., Clechet, P., Martin, J.-R., Herrmann, J. M., and Pichat, P., *Chem. Phys. Lett.* **84,** 555 (1981).
21. Connolloy, J. S. (ed.), "Photochemical Conversion and Storage of Solar Energy." Academic Press, New York, 1981.
22. Courbon, H., Formenti, M., and Pichat, P., *J. Phys. Chem.* **81,** 550 (1977).
23. Croucher, M. D., and Hair, M. L., *Ind. Eng. Chem. Prod. Res. Dev.* **20,** 324 (1981).
24. Croucher, M. D., Harbour, J., Hopper, M., and Hair, M. L., *Photog. Sci. Eng.* **25,** 80 (1981).
25. Cundall, R. B., Hulme, B., and Salim, M. S., *J. Chem. Soc. Faraday Trans. 1* **72,** 1642 (1976).
26. Cundall, R. B., Hulme, B., Rudham, R., and Salim, S., *J. Oil Colour Chem. Assoc.* **61,** 351 (1978).
27. Cunningham, J., and Hodnett, B. K., *J. Chem. Soc. Faraday Trans. 1* **77,** 2777 (1981).
28. Cunningham, J., and Penny, A. L., *J. Phys. Chem.* **78,** 870 (1974).
29. Cunningham, J., and Zinal, H., *J. Phys. Chem.* **76,** 2362 (1972).
30. Cunningham, J., Finn, E., and Samman, N., *Discuss. Faraday Soc.* **58,** 160 (1974).
31. Cunningham, J., Doyle, B., and Samman, N., *J. Chem. Soc. Faraday Trans. 1* **72,** 1495 (1976).
32. Cunningham, J., Doyle, B., and Leahy, E. M., *J. Chem. Soc. Faraday Trans. 1* **75,** 2000 (1979).
33. Cunningham, J., Morrissey, D. J., and Goold, E. L., *J. Catal.* **53,** 68 (1978).
34. Cunningham, J., Goold, E. L., and Leahy, E. M., *J. Chem. Soc. Faraday Trans. 1* **75,** 305 (1979).
35. Cunningham, J., and Fierro, J. L. G., *J. Chem. Soc. Faraday Trans. 1* **78,** 785 (1982).
36. Darwent, J. R., *J. Chem. Soc. Faraday Trans. 2* **77,** 1703 (1981).
37. Darwent, J. R., and Mills, A., *J. Chem. Soc. Faraday Trans. 2* **78,** 359 (1982).
38. Darwent, J. R., and Porter, G., *J. Chem. Soc. Chem. Commun.* No. 4, 145 (1981).
39. Den Besten, I. E., and Qasim, M., *J. Catal.* **3,** 387 (1964).

40. Djeghri, N., Formenti, M., Juillet, F., and Teichner, S. J., *Discuss. Faraday Soc.* **58,** 185 (1974).
41. Djeghri, N., and Teichner, S. J., *J. Catal.* **62,** 99 (1980).
42. Domen, K., Naito, S., Soma, M., Onishi, T., and Tamaru, K., *J. Chem. Soc. Chem. Commun.* No. 12, 543 (1980).
43. Dunn, W. W., Aikawa, Y., and Bard, A. J., *J. Am. Chem. Soc.* **103,** 6893 (1981).
44. Dunn, W. W., Aikawa, Y., and Bard, A. J., *J. Am. Chem. Soc.* **103,** 3456 (1981).
45. Dunn, W. W., Aikawa, Y., and Bard, A. J., *J. Electrochem. Soc.* **128,** 222 (1981).
46. Duonghong, D., Borgarello, E., and Grätzel, M., *J. Am. Chem. Soc.* **103,** 4685 (1981).
47. Duonghong, D., Ramsden, J., and Grätzel, M., *J. Am. Chem. Soc.* **104,** 2977 (1982).
48. Egerton, T. A., and King, C. J., *J. Oil. Colour Chem. Assoc.* **62,** 45 (1979).
49. "Photoeffects in Adsorbed Species," *Discuss. Faraday Soc.* **58,** (1974).
50. "Photoelectrochemistry," *Discuss. Faraday Soc.* **70,** (1980).
51. Fan, F.-R. F., and Bard, A. J., *J. Am. Chem. Soc.* **101,** 6139 (1979).
52. Ferrer, S., and Somorjai, G. A., *J. Phys. Chem.* **85,** 1464 (1981).
53. Fomin, G. V., Brin, G. P., Genkin, M. V., Lyubimova, A. K., Blyumenfeld, L. A., and Krasnovskii, A. A., *Dokl. Akad. Nauk SSSR Ser. Khim.* **212,** 424 (1973).
54. Formenti, M., and Teichner, S. J., *Spec. Period. Rep. Catal.* **2,** 87 (1978).
55. Formenti, M., Juillet, F., Meriaudeau, P., and Teichner, S. J., *Chem. Technol.* **1,** 680 (1971).
56. Fox, M. A., and Chen, C. C., *J. Am. Chem. Soc.* **103,** 6757 (1981).
57. Frank, S. N., and Bard, A. J., *J. Phys. Chem.* **81,** 1484 (1977).
58. Frank, S. N., and Bard, A. J., *J. Am. Chem. Soc.* **99,** 4667 (1977).
59. Freund, T., *J. Catal.* **3,** 289 (1964).
60. Freund, T., and Gomes, W. P., *Catal. Rev.* **3,** 1 (1969).
61. Fujihara, M., Satoh, Y., and Osa, T., *Nature (London)* **293,** 206 (1981).
62. Fujihara, M., Satoh, Y., and Osa, T., *Chem. Lett.,* 1053 (1981).
63. Fujihara, M., Satoh, Y., and Osa, T., *J. Electroanal. Chem.* **126,** 277 (1981).
64. Gerischer, H., *Top. Appl. Phys.* **31,** 115 (1979).
65. Gonzales-Elipe, A. R., Munuera, G., and Soria, J., *J. Chem. Soc. Faraday Trans. 1,* **75,** 748 (1979).
66. Gonzales-Elipe, A. R., Soria, J., and Munuera, G., *Z. Phys. Chem. (Wiesbaden)* **126,** 251 (1981).
67. Grätzel, M., *Acc. Chem. Res.* **14,** 376 (1981).
68. Hada, H., Tanemura, H., and Yonezawa, Y., *Bull. Chem. Soc. Jpn.* **51,** 3154 (1978).
69. Harbour, J. R., and Hair, M. L., *J. Phys. Chem.* **82,** 1397 (1978).
70. Harbour, J. R., and Hair, M. L., *Photochem. Photobiol.* **28,** 721 (1978).
71. Harbour, J. R., and Hair, M. L., *J. Phys. Chem.* **83,** 652 (1979).
72. Harbour, J. R., and Hair, M. L., *NATO Adv. Study Inst. Ser C* **61,** 431 (1980).
73. Harbour, J. R., Tromp, J., and Hair, M. L., *J. Am. Chem. Soc.* **102,** 1874 (1980).
74. Hauffe, K., and Wolkenstein, Th. (eds.), "Electronic Phenomena in Chemisorption and Catalysis." de Gruyter, Berlin, 1969.
75. Hauffe, K., and Bode, U., *Discuss. Faraday Soc.* **58,** 281 (1974).
76. Hema, M. A., Ramakrishnan, V., and Kuriakose, J. C., *Indian J. Chem. Sect. B* **15,** 947 (1977).
77. Hema, M. A., Ramakrishnan, V., and Kuriakose, J. C., *Indian J. Chem. Sect. B* **16,** 619 (1978).
78. Hemminger, J. C., Carr, R., and Somorjai, G. A., *Chem. Phys. Lett.* **57,** 100 (1978).
79. Hemminger, J. C., Carr, R., Lo, W. J., and Somorjai, G. A., *Adv. Chem. Ser.* No. 184, p. 13 (1980).

80. Herrmann, J. M., and Pichat, P., *J. Chem. Soc. Faraday Trans. 1* **76,** 1138 (1980).
81. Herrmann, J. M., Vergnon, P., and Teichner, S. J., *J. Catal.* **37,** 57 (1975).
82. Herrmann, J. M., Disdier, J., Mozzanega, M. N., and Pichat, P., *J. Catal.* **60,** 369 (1979).
83. Herrmann, J. M., Disdier, J., and Pichat, P., *J. Chem. Soc. Faraday Trans. 1* **77,** 2815 (1981).
84. Inoue, T., Fujishima, A., Konishi, S., and Honda, K., *Nature (London)* **277,** 637 (1979).
85. Iwasaki, T., Oda, S., Kamada, M., and Honda, K., *J. Phys. Chem.* **84,** 1060 (1980).
86. Iwasaki, T., Sawada, T., Kamada, H., Fujishima, A., and Honda, K., *J. Phys. Chem.* **83,** 2142 (1979).
87. Iwasaki, T., Oda, S., Sawada, T., and Honda, K., *J. Phys. Chem.* **84,** 2800 (1980).
88. Izumi, I., Dunn, W. W., Wilbourn, K. P., Fan, F. R. F., and Bard, A. J., *J. Phys. Chem.* **84,** 3207 (1980).
89. Izumi, I., Fan, F. R. F., and Bard, A. J., *J. Phys. Chem.* **85,** 218 (1981).
90. Jaeger, C. D., and Bard, A. J., *J. Phys. Chem.* **83,** 3146 (1979).
91. Juillet, F., Lecomte, F., Mozzenega, H., Teichner, S. J., and Vergnon, P., *Symp. Faraday Soc.* **7,** 57 (1973).
92. Kaliaguine, S. L., Shelimov, B. N., and Kazansky, V. B., *J. Catal.* **55,** 384 (1978).
93. Kalyanasundaram, K., *Chem. Soc. Rev.* **7,** 453 (1978).
94. Kalyanasundaram, K., *Coord. Chem. Rev.* **46,** 159 (1982).
95. Kalyanasundaram, K., Borgarello, E., and Grätzel, M., *Helv. Chim. Acta* **64,** 362 (1981).
96. Kalyanasundaram, K., Borgarello, E., Duonghong, D., and Grätzel, M., *Angew. Chem. Int. Ed. Engl.* **20,** 987 (1981).
97. Kanno, T., Oguchi, T., Sakuragi, H., and Tokumaru, K., *Tetrahedron Lett.* **21,** 467 (1980).
98. Kasturirangan, H., Ramakrishnan, V., and Kuriakose, J. C., *J. Catal.* **69,** 216 (1981).
99. Kawai, T., and Sakata, T., *Nature (London)* **286,** 474 (1980).
100. Kawai, T., and Sakata, T., *Chem. Phys. Lett.* **72,** 87 (1980).
101. Kawai, T., and Sakata, T., *Chem. Lett.* p. 81 (1981).
102. Kago, K., Yoneyama, H., and Tamura, H., *J. Phys. Chem.* **84,** 1705 (1980).
103. Kiwi, J., *Chem. Phys. Lett.* **83,** 594 (1981).
104. Kiwi, J., Borgarello, E., Pelizzetti, E., Visca, M., and Grätzel, M., *Angew. Chem. Int. Ed. Engl.* **19,** 646 (1980).
105. Kiwi, J., Kalyanasundaram, K., and Grätzel, M., *Struct. Bonding (Berlin)* **49,** 37 (1982).
106. Kraeutler, B., and Bard, A. J., *J. Am. Chem. Soc.* **100,** 2239 (1978).
107. Kraeutler, B., and Bard, A. J., *J. Am. Chem. Soc.* **100,** 4317 (1978).
108. Kraeutler, B., and Bard, A. J., *J. Am. Chem. Soc.* **100,** 5985 (1978).
109. Kraeutler, B., Jaeger, C. D., and Bard, A. J., *J. Am. Chem. Soc.* **100,** 4903 (1978).
110. Krasnovskii, A. A., and Brin, G. P., *Dokl. Akad. Nauk SSSR Ser. Khim.* **147,** 656 (1962).
111. Krasnovskii, A. A., Brin, G. P., Luganskaya, A. N., and Nikandrov, V. V., *Dokl. Akad. Nauk SSSR Ser. Khim.* **249,** 896 (1979).
112. Kruczynski, L., Gesser, H. D., Turner, C. W., and Speers, E. A., *Nature (London)* **291,** 399 (1981).
113. Lehn, J. M., Sauvage, J. P., and Ziessel, R., *Nouv. J. Chim.* **4,** 623 (1980).
114. Lehn, J. M., Sauvage, J. P., and Ziessel, R., *Nouv. J. Chim.* **5,** 291 (1981).
115. Lichtman, D., and Shapira, Y., *CRC Crit. Rev. Solid State Sci.* **8,** 93 (1978).

116. Loutfy, R. O., and Sharp, J. H., *J. Appl. Electrochem.* **7,** 315 (1977).
117. Lunsford, J. H., *Catal. Rev.* **8,** 135 (1973).
118. Lunsford, J. H., and Leland, T. W., *J. Phys. Chem.* **66,** 2591 (1962).
119. Matijevic, E., *Acc. Chem. Res.* **14,** 22 (1981).
120. Matijevic, E., *Pure Appl. Chem.* **50,** 1193 (1978).
121. Meriaudeau, P., and Vedrine, J. C., *J. Chem. Soc. Faraday Trans. 2* **72,** 472 (1976).
122. Mochida, T., Tsuji, K., Suetsugu, K., Fujitsu, H., and Takeshita, K., *J. Phys. Chem.* **84,** 3159 (1980).
123. Morgan, P., Onnsgaard, J. H., and Tongand, S., *J. Appl. Phys.* **47,** 5094 (1976).
124. Morrison, S. R., "Electrochemistry at Metal and Semiconductor Electrodes." Plenum, New York, 1980.
125. Mozzanega, H., Ph.D. Thesis, Lyon, France, 1975.
126. Mozzanega, M. N., Herrmann, J. M., and Pichat, P., *Tetrahedron Lett.* **34,** 2965 (1977).
127. Mozzanega, H., Herrmann, J. M., and Pichat, P., *J. Phys. Chem.* **83,** 2251 (1979).
128. Munuera, G., Rives-Arnau, V., and Saucedo, A., *J. Chem. Soc. Faraday Trans. 1,* **75,** 736 (1979).
129. Munuera, G., Gonzales-Elipe, A. R., Soria, J., and Sanz, J., *J. Chem. Soc. Faraday Trans. 1,* **76,** 1535 (1980).
130. Munuera, G., Navio, A., and Rives-Arnau, V., *J. Chem. Soc. Faraday Trans. 1,* **77,** 2747 (1981).
131. Nagarjunan, T. S., and Calvert, J. G., *J. Phys. Chem.* **68,** 7 (1964).
132. Nozik, A. J., *Annu. Rev. Phys. Chem.* **29,** 189 (1978).
133. Nozik, A. J., *Phil. Trans. R. Soc. London Ser. A* **295,** 453 (1980).
134. Nozik, A. J. (ed.), "Photoeffects at Semiconductor Electrolyte Interfaces." *ACS Symp. Ser.* No. 146 (1981).
135. Pavlik, J. W., and Tantayanon, S., *J. Am. Chem. Soc.* **103,** 6755 (1981).
136. Peshev, O., Malakhov, V., and Wolkenstein, Th., *Prog. Surf. Sci.* **6,** 63 (1975).
137. Pichat, P., Herrmann, J. M., Disdier, J., and Mozzenega, M. N., *J. Phys. Chem.* **83,** 3122 (1979).
138. Porter, G., and Archer, M. D., *Interdiscip. Sci. Rev.* **1,** 119 (1976).
139. Rao, M. V., Rajeswar, K., Pai Vernekar, V. R., and DuBow, J., *J. Phys. Chem.* **84,** 1987 (1980).
140. Reiche, H., and Bard, A. J., *J. Am. Chem. Soc.* **101,** 3127 (1979).
141. Reiche, H., Dunn, W. W., and Bard, A. J., *J. Phys. Chem.* **83,** 2248 (1979).
142. Reichman, B., and Byvik, C. E., *J. Phys. Chem.* **85,** 2255 (1981).
143. Reymond, J. P., Vergnon, P., Gravelle, P. C., and Teichner, S. J., *Nouv. J. Chim.* **1,** 197 (1977).
144. Saeva, F. S., Olin, G. R., and Harbour, J. R., *J. Chem. Soc. Chem. Commun.,* 401 (1980).
145. Sakata, T., and Kawai, T., *Chem. Phys. Lett.* **80,** 341 (1981).
146. Sakata, T., and Kawai, T., *Nouv. J. Chim.* **5,** 279 (1981).
147. Sato, S., and White, J. M., *Chem. Phys. Lett.* **72,** 83 (1980).
148. Sato, S., and White, J. M., *Ind. Eng. Chem. Prod. Res. Dev.* **19,** 542 (1980).
149. Sato, S., and White, J. M., *Chem. Phys. Lett.* **70,** 131 (1980).
150. Sato, S., and White, J. M., *J. Am. Chem. Soc.* **102,** 7206 (1980).
151. Sato, S., and White, J. M., *J. Phys. Chem.* **85,** 336 (1981).
152. Sato, S., and White, J. M., *J. Catal.* **69,** 128 (1981).
153. Sato, S., and White, J. M., *J. Phys. Chem.* **85,** 592 (1981).
154. Sato, S., *J. Chem. Soc. Chem. Commun.,* 26 (1982).

155. Schrauzer, G. N., and Guth, T. D., *J. Am. Chem. Soc.* **99,** 7189 (1977).
156. Shapira, Y., Cox, S. M., and Lichtman, D., *Surf. Sci.* **50,** 503 (1975).
157. Shapira, Y., Cox, S. M., and Lichtman, D., *Surf. Sci.* **54,** 43 (1976).
158. Shapira, Y., McQuiston, R. B., and Lichtman, D., *Phys. Rev. B: Condens. Matter* **15,** 2163 (1977).
159. Spikes, J. D., *Photochem. Photobiol.* **34,** 549 (1981).
160. Steinbach, F., *Top. Curr. Chem.* **25,** 117 (1972).
161. Steinbach, F., and Harborth, R., *Discuss. Faraday Soc.* **58,** 143 (1974).
162. Takahashi, A., Aikawa, Y., Toyoshima, Y., and Sukigara, M., *J. Phys. Chem.* **83,** 2854 (1979).
163. Takizawa, T., Watanabe, T., and Honda, K., *J. Phys. Chem.* **82,** 1391 (1978).
164. Tanaka, K., *J. Phys. Chem.* **78,** 555 (1974).
165. Tanaka, K., and Miyahara, K., *J. Phys. Chem.* **78,** 2303 (1974).
166. Tanaka, K., and Blyholder, G., *J. Phys. Chem.* **76,** 1807 (1972).
167. Thevenet, A., Juillet, F., and Teichner, S. J., *Jpn. J. Appl. Phys.* **2,** *(Suppl.),* 529 (1974).
168. Thewissen, D. H. W. M., Eenwhorst-Reinten, M., Timmer, K., Tinnemans, A. H. A., and Mackor, A., *in* "Photochemical, Photoelectrochemical Processes" (D. O. Hall, and E. Palz, eds.), p. 56. Reidel, Amsterdam, 1982.
169. Toyoshima, Y., Takahashi, A., Nozaki, H., Iida, T., and Sukigara, M., *Photogr. Sci. Eng.* **21,** 29 (1977).
170. Turro, N. J., Braun, A. M., and Grätzel, M., *Angew. Chem. Int. Ed. Engl.* **19,** 675 (1980).
171. van Damme, H., and Hall, W. K., *J. Am. Chem. Soc.* **101,** 4373 (1979).
172. van Damme, H., and Hall, W. K., *J. Catal.* **69,** 371 (1981).
173. Volodin, A. M., and Cherkashin, A. E., *Kinet. Katal.* **22,** 1227 (1981).
174. Völz, H. G., Kaempf, G., Fitzky, H. G., and Klaeren, A., *ACS Symp. Ser.* No. 151, p. 163 (1981).
175. Vonach, W., and Getoff, N., *Z. Naturforsch. A* **36,** 876 (1981).
176. Wagner, F. T., and Somorjai, G. A., *J. Am. Chem. Soc.* **102,** 5494 (1980).
177. Walker, A., Formenti, M., Meriaudeau, P., and Teichner, S. J., *J. Catal.* **50,** 237 (1977).
178. Ward, M. D., and Bard, A. J., *J. Phys. Chem.* **86,** 3599 (1982).
179. Watanabe, T., Takizawa, T., and Honda, K., *J. Phys. Chem.* **81,** 1845 (1977).
180. Weigl, T. W., *Angew. Chem. Int. Ed. Engl.* **16,** 374 (1977).
181. Wolkenstein, Th., *Adv. Catal.* **23,** 157 (1973).
182. Wolkenstein, Th., *Prog. Surf. Sci.* **6,** 213 (1975).
183. Wrighton, M. S., Wolczanski, P. T., and Ellis, A. B., *J. Solid State Chem.* **22,** 17 (1977).
184. Yesodharan, E. P., Ramakrishnan, V., and Kuriakose, J. C., *J. Sci. Industr. Res.* **35,** 712 (1976).
185. Yoneyama, H., Koizumi, M., and Tamura, H., *Bull. Chem. Soc. Jpn.* **52,** 3449 (1979).
186. Yoneyama, H., Yamashita, Y., and Tamura, H., *Nature (London)* **282,** 817 (1979).
187. Yoneyama, H., Nishimura, N., and Tamura, H., *J. Phys. Chem.* **85,** 268 (1981).
188. Zamareav, K. I., and Parmon, V. N., *Catal. Rev.* **22,** 261 (1980).
189. Zahkarenko, V. S., Cherkashin, A. E., Keier, N. P., and Koshcheev, S. V., *Kinet. Katal.* **16,** 182 (1975).
190. Ward, M. D., White, J. R., and Bard, A. J., *J. Am. Chem. Soc.* **105,** 27 (1983).
191. Courbon, H., Herrmann, J. M., and Pichat, P., *J. Catal.* **72,** 129 (1981).
192. Parfitt, G. D., *Progr. in Surface & Membrane Sci.* **11,** 181 (1976).

193. Lehn, J.-M., Sauvage, J.-P., Ziessel, R., and Hilaire, L., *Isr. J. Chem.* **22,** 168 (1982).
194. Dunn, W. W., and Bard, A. J., *Nouv. J. Chim.* **5,** 651 (1981).
195. Chen, B.-H., White, J. M., *J. Phys. Chem.* **86,** 3534 (1982).
196. Volodin, A. M., and Cherkashin, A. E., *Kinet. Katal.* **22,** 598 (1981).
197. Hieu, N. V., and Lichtman, D., *J. Catal.* **73,** 329 (1982).
198. Ferrar, S., and Somorjai, G., *Surf. Sci.* **94,** 41 (1980).
199. Ferrar, S., and Somorjai, G., *Surf. Sci.* **97,** 1304 (1980).
200. Cunningham, J., Hodnett, B. K., Ilyas, M., Leahy, E. M., and Tobin, J. P., *J. Chem. Soc. Faraday Trans. 1,* **78,** 3297 (1982).
201. Childs, L. P., and Ollis, D. F., *J. Catal.* **67,** 35 (1981).
202. Teratani, S., Nakamuchi, J., Taya, K., and Tanaka, K., *Bull. Chem. Soc. Jpn.* **55,** 1688 (1982).
203. Herrmann, J.-M., Disdier, J., and Pichat, P., *In* "Metal-Support and Metal-Additive Effects in Catalysis" (B. Imelik, ed.), p. 27. Elsevier, Amsterdam, 1982.
204. Hubesch, B., and Malieu, B., *Inorg. Chim. Acta. Lett.* **65,** L65 (1982).
205. Manassen, J., *Isr. J. Chem.* **22,** 190 (1982).
206. Fujihara, M., Satoh, Y., and Osa, T., *Bull. Chem. Soc. Jpn.* **55,** 666 (1982).
207. Kodama, S., Yabuta, M., and Kubokawa, Y., *Chem. Lett.* **21,** 1671 (1982).
208. Barber, R. A., de Mayo, P., and Okada, K., *J. Chem. Soc. Chem. Commun.* No. 11, 1073 (1982).
209. Thewissen, D. H. M. W., Timmer, K., Eeuwhorst-Reinten, M., Tinnemans, A. H. A., and Mackor, A., *Isr. J. Chem.* **22,** 173 (1982).
210. Matsumoto, Y., Nagai, H., and Sato, E., *J. Phys. Chem.* **86,** 4664 (1982).
211. Kiwi, J., and Grätzel, M., *J. Chem. Soc. Faraday Trans. 2,* **78,** 931 (1982).
212. Oosawa, Y., *Chem. Lett.* **8,** 423 (1982).
213. Hada, H., Yonezawa, Y., and Saikawa, M., *Bull. Chem. Soc. Jpn.* **55,** 2010 (1982).
214. Oosawa, Y., *J. Chem. Soc. Chem. Commun.* No. 2, 221 (1982).
215. Mills, A., *J. Chem. Soc. Chem. Commun.* No. 3, 367 (1982).
216. Pichat, P., Hermann, J.-M., Disdier, J., Courbon, H., and Mozzanega, M.-N., *Nouv. J. Chim.* **5,** 627 (1981).
217. Pichat, P., Mozzanega, M.-N., Disdier, J., and Hermann, J.-M. *Nouv. J. Chim.* **6,** 559 (1982).
218. Tanaka, K., and White, J. M., *J. Phys. Chem.* **86,** 3977 (1982).
219. Fang, S., Chen, B.-H. and White, J. M., *J. Phys. Chem.* **86,** 3126 (1982).
220. Sakata, T., Kawai, T., and Hashimoto, K., *Chem. Phys. Lett.* **88,** 50 (1982).
221. Domen, K., Naito, S., Onishi, T., Tamaru, K., and Soma, M., *J. Phys. Chem.* **86,** 3657 (1982).
222. Domen, K., Naito, S., Onishi, T., and Tamaru, K., *Chem. Phys. Lett.* **92,** 433 (1982).
223. Mills, A., and Porter, G., *J. Chem. Soc. Faraday Trans. 1,* **78,** 3659 (1982).
224. Borgarello, E., Kiwi, J., Grätzel, M., Pelizetti, E., and Visca, M., *J. Am. Chem. Soc.* **104,** 2996 (1982).
225. Stevens, C. G., Wessman, N. J., Bowman, J. E., and Ramsey, W. J., *Chem. Phys. Lett.* **91,** 335 (1982).
226. Buecheler, J., Zeng, N., and Kirsch, H., *Angew. Chem. Int. Ed., Engl.* **21,** 783 (1982).
227. Ulman, M., Aurien-Blajeni, B., and Halmann, M., *Isr. J. Chem.* **22,** 177 (1982).
228. Tinnemans, A. H. A., Koster, T. P. M., Thewissen, D. H. M. W., and Mackor, A., *Nouv. J. Chim.* **6,** 373 (1982).
229. Matijevic, E., and Wilhemy, D. M., *J. Coll. Int. Sci.* **86,** 476 (1982).
230. Hamada, S., and Matijevic, E., *J. Chem. Soc. Faraday Trans. 1,* **78,** 2147 (1982).
231. Kuczynski, J., and Thomas, J. K., *Chem. Phys. Lett.* **85,** 445 (1982).

232. Henglein, A., *Ber. Bunsenges. Phys. Chem.* **86,** 301 (1982).
233. Henglein, A., *J. Phys. Chem.* **86,** 2291 (1982).
234. Alfassi, Z., Bahnemann, D., and Henglein, A., *J. Phys. Chem.* **86,** 4656 (1982).
235. Rossetti, R., and Brus, L., *J. Phys. Chem.* **86,** 4470 (1982).
236. Henglein, A., *Ber. Bunsenges. Phys. Chem.* **86,** 241 (1982).
237. Fox, M. A., Lindig, B., and Chen, C. C., *J. Am. Chem. Soc.* **104,** 5828 (1982).
238. Moser, J., and Grätzel, M., *Helv. Chim. Acta* **65,** 1436 (1982).
239. Grätzel, M., and Frank, A. J., *J. Phys. Chem.* **86,** 2964 (1982).
240. Childs, L. P., and Ollis, D. F., *J. Catal.* **66,** 383 (1980).

8 Bifunctional Redox Catalysis: Synthesis and Operation in Water-Cleavage Reactions

Ezio Pelizzetti

Istituto di Chimica Analitica
Università di Torino
Torino, Italy

Mario Visca

Centro Ricerche SIBIT (Montedison)
Spinetta Marengo, Italy

I. Introduction

The possibility of converting solar radiation into chemical or electrical energy has received ever increasing attention (*1–10*). Chemical energy

ENERGY RESOURCES THROUGH
PHOTOCHEMISTRY AND CATALYSIS

ISBN 0-12-295720-2

can be easily stored and transported; consequently, considerable interest is currently focused on the possibility of using visible light to break the hydrogen–oxygen bonds of water to obtain molecular hydrogen and oxygen. These gases can be stored and, during recombination, release energy to give heat and electricity.

Some photochemical systems, based on the semiconductor–liquid junction, can produce both electrical and chemical energy, and considerable work has been done since 1970 on the basis of this idea (*11–18*). The principles of photoelectrochemical cells, combined with suitable cycles of photoredox reactions and extensive studies on hydrogen and oxygen evolution through catalysis by colloidal particles, can be extended to design particulate systems for achieving water cleavage. (*19–21*).

The particulate systems present some advantages compared with photoelectrochemical cells: fewer constraints on the type of material (in photoelectrochemical devices single-crystal or at least polycrystalline electrodes must be selected), the efficiency can be high, the catalytic surface is large, and particulate systems are simpler and less expensive to assemble and use.

II. Required Properties for Efficient Colloidal Semiconductors

The principles of photoelectrochemical cells have been extensively reviewed and, as discussed in Section I, can be a good starting point to select a proper particulate system (*11–21*). When fuel formation is the goal, the desired energetic situation for semiconductor material is that the conduction band (E_C) and the valence band (E_V) straddle the two redox potentials of the solution couples ($E^\circ_{H^+/H_2}$ and $E^\circ_{O_2/H_2O}$ in the case of water cleavage). In addition, other criteria are required, such as a band gap somewhat greater than the energy for decomposing water to ensure fast rates, but small enough to absorb an appreciable amount of solar radiation; stability against corrosion; and good properties to act as supporting material for the proper catalysts for H_2 and O_2 evolution. Figure 1 illustrates the positions of the bands for various semiconductors in solution at pH 1, in comparison with the standard potentials of some redox couples (*12*).

Efficient solar conversion should require materials with band gap of 1.3 ± 0.3 eV—which means absorption of light at 1000 ± 250 nm—but, taking into account the overpotentials for driving the electron-transfer reactions,

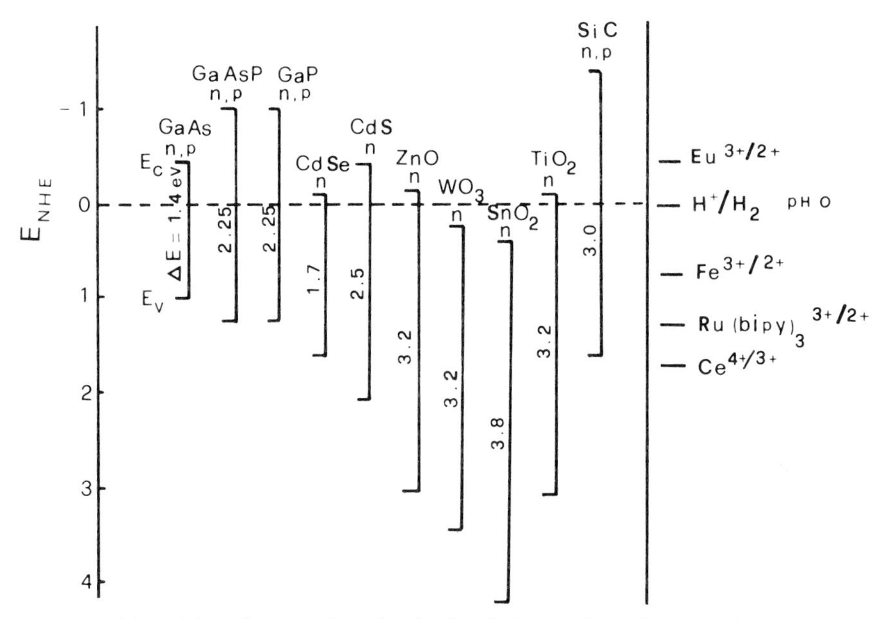

Fig. 1. Position of the valence- and conduction-band edges at the surface of various semiconductors in contact with aqueous electrolyte at pH 1. From Memming (*12*).

a value around 2 eV is suggested (*7, 20*). Generally, when the band gap is narrow the stability is unsatisfactory. Only semiconductors with large band gaps fulfill the condition that the anodic process under illumination is oxygen evolution instead of chemical dissolution, but these materials show very little absorption in the solar spectrum.

Since the work of Fujishima and Honda appeared (*22*), extensive studies have been stimulated concerning TiO_2 as a suitable material for photoelectrochemical (*11–18*) and photochemical (*23–26*) water cleavage.

TiO_2 can exist in two different crystalline forms, rutile and anatase. As illustrated in Fig. 2, the semiconductor band energies are slightly different for anatase and rutile (*27*). The anatase form has a slightly larger band gap than that of rutile (3.23 and 3.02 eV, respectively), probably shifting the flat-band potential (which for heavily doped materials is very near to the conduction-band level) negatively with the result that the conduction-band level becomes more negative than the H^+–H_2 level. The position and the change of the flat-band potential of various samples of TiO_2 varies with pH as shown in Fig. 3.

Moreover, TiO_2 shows, as do most of the oxides, very good stability in the photoelectrochemical processes. It can be noted from Fig. 2 that the redox potential for the oxidation of water to O_2 is more negative than that

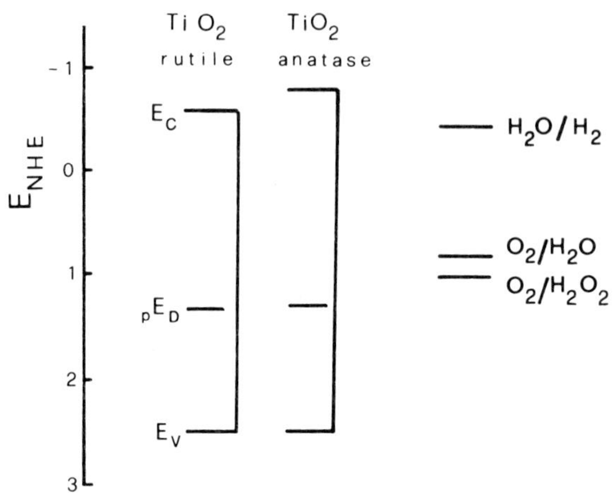

Fig. 2. Position of the energy bands for rutile and anatase. The position of the H^+/H_2, O_2/H_2O, and O_2/H_2O_2 redox couples are indicated at the right. The band positions and the redox levels are referred at pH 7. Decomposition Fermi energies are also reported. From Rao *et al.* (*25*). Copyright 1980 American Chemical Society.

for the oxidation of TiO_2 to O_2; consequently, the water-oxidation reaction is thermodynamically favored with respect to TiO_2 oxidation. Although kinetic factors must also be considered, the experimental data show that the water-oxidation process almost exclusively predominates. Nevertheless, TiO_2 can be slowly corroded in 1 *M* acids such as H_2SO_4 and $HClO_4$ (*29*), although it appears to be stable in alkaline media (some evidence of photoanodic corrosion in 5–7 *M* NaOH at very high light intensity has been reported) (*30*).

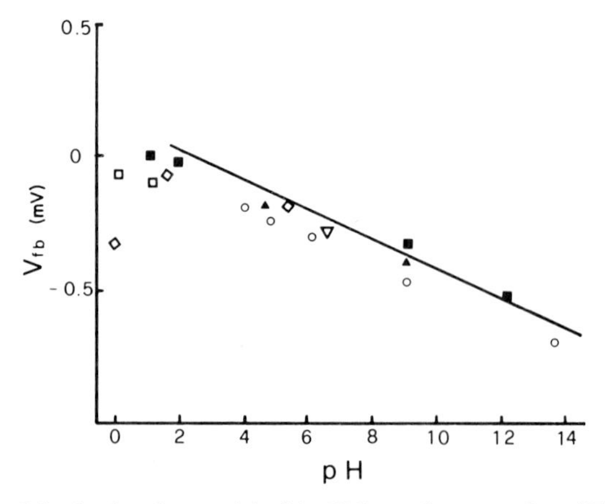

Fig. 3. Variation of the flat-band potential with pH for various samples of TiO_2. From Dutoit *et al.* (*28*).

The other relevant properties of TiO_2 and colloidal TiO_2 are extensively reported in other chapters.

III. Preparation and Characteristics of Colloidal Titanium Dioxide

A. Preparation of the Colloids

Titanium dioxide in hydrous and crystalline form can be prepared by various techniques from different starting compounds. The most widely used procedures are (a) hydrolysis of aqueous Ti(IV) solutions (*31–33*), (b) vapor- or aerosol-phase hydrolysis (*34, 35*), (c) thermal decomposition of Ti(IV) alkoxides or coordination compounds (*36*), and (d) high-temperature oxidation of $TiCl_4$ (*37*).

The industrial preparation of TiO_2 from $TiCl_4$ has been the subject of a large number of patents. Some of them are reported, for example, under United States patents 3,062,621, 3,078,148, 3,403,001, 3,663,283, and 3,914,396.

Model systems consisting of uniform spherical TiO_2 particles have been prepared by the hydrolysis of aqueous solutions of $TiCl_4$ in 6 *M* HCl, in the presence of sulfate ions (*32*), and by reacting aerosols of hydrolyzable liquid Ti(IV) compounds, such as $TiCl_4$ and Ti(IV) alkoxides (*34*), with water vapor. Figure 4a shows an electron micrograph of TiO_2 particles prepared by the aerosol method.

The procedures commonly used in the industrial manufacture of titanium dioxide are vapor-phase oxidation of $TiCl_4$ and hydrolysis of Ti(IV) sulfate solutions. In the first process, vaporized $TiCl_4$ is reacted in a flame with oxygen. Carrying out reaction (1) in specially designed burners leads

$$TiCl_4 + O_2 \longrightarrow TiO_2 + 2Cl_2 \quad (1)$$

to the formation of TiO_2 particles in the pigmentary size range ($\sim 0.2\ \mu m$). Particle size, uniformity, and crystal structure can be controlled by modifying the oxidation conditions (temperature, flow conditions, and contact times) and by adding to the $TiCl_4$ feed small amounts of conditioning agents like $AlCl_3$ and $SiCl_4$.

In the sulfate process, a titaniferous material such as ilmenite (natural $FeTiO_3$) is digested at high temperature with excess H_2SO_4, yielding a solid, porous mass consisting of Ti(IV) and Fe(II) sulfates containing absorbed H_2SO_4. The digested ore is dissolved in diluted H_2SO_4 under reducing conditions to avoid oxidation of divalent iron. The black solution obtained typically has a composition as shown in Table I and contains

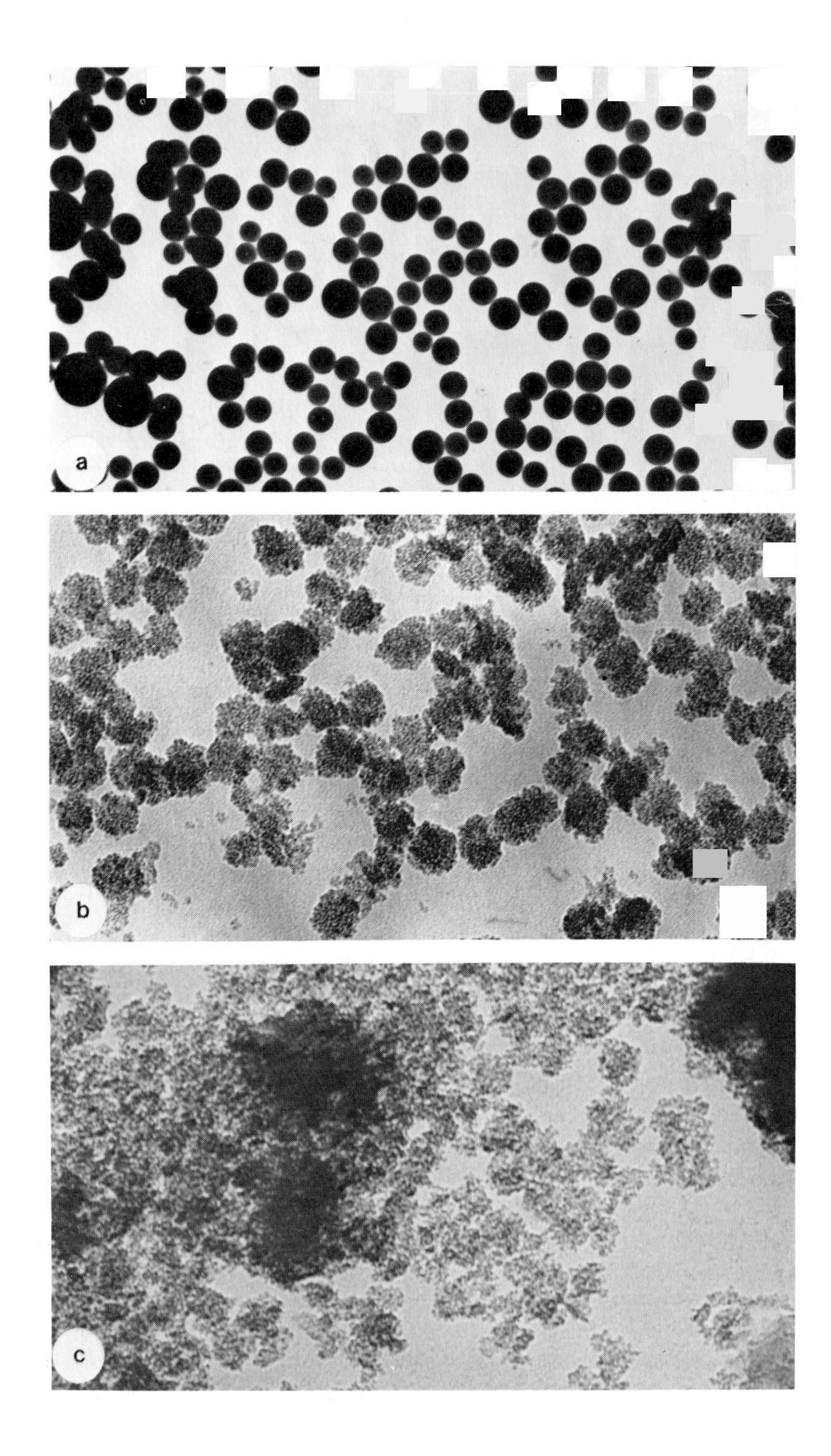
a
b
c

TABLE I

Composition of Black Sulfuric Liquor from the Digestion of Australian Ilmenite

Component	Concentration (mol/liter)
Ti(IV)	3
Ti(III)	0.05
Fe	1.4
H_2SO_4 (free)	2

hydroxosulfato complexes of Ti(IV), Ti(III), and Fe(II) (*32, 38–45*). After the elimination of undissolved material by clarification and filtration, the solution is heated to about 90–100°C and diluted with water to hydrolyze the Ti(IV) ions (*31*). Hydrolysis nuclei, necessary to increase the hydrolysis rate and to control the particle size of the final product can be generated *in situ* or added during the process.

In the Blumenfeld procedure (*31*), the black sulfuric liquor kept at ~95°C is added under stirring to a batch of hot water. The sudden dilution of the solution in the very first stage of the addition results in the precipation of hydrolyzed material, which then undergoes peptization because of the increasing acidity of the solution and acts as nucleating material.

Following the Macklemburg method (*31*), hydrolysis nuclei are prepared in a separate batch by thermal aging at pH 2.5–3.5 of $TiOSO_4$ solutions.

The hydrolysis products obtained by the just-described procedures appear as aggregates of primary nuclei having a diameter of about 50–100 Å. Figure 4b shows typical hydrolysis products, and Table II shows the average composition of the hydrous titania gel after filtration and thorough washing. X-Ray analysis of the dried solid shows the particles to be essentially amorphous, with anatase-oriented microcrystalline structure. The dimension of the microcrystals is only slightly dependent on the preparation conditions, whereas the secondary aggregate shape and size is a function of the hydrolysis conditions and of the physicochemical properties of the black sulfuric liquor (*46–51*). Hydrous titania gel can be converted to crystalline anatase or rutile particles in the submicronic size range by calcination at about 1000°C. Hydrated rutile-oriented particles

Fig. 4. Electron micrograph of (a) spherical TiO_2 obtained by the hydrolysis of Ti(IV) ethoxide aerosol (magnification 9400×), (b) hydrous titania obtained by the Macklemburg procedure (150,000×), and (c) hydrous titania obtained by the Blumenfeld procedure (190,000×).

TABLE II

Composition of Hydrous Titania Gel

	Filtered cake (wt%)	Dried solid (wt%)
TiO_2	36	76
SO_3	4	9
H_2O	60	15

can be prepared by the thermal hydrolysis of $TiCl_4$ solutions or by thermal aging of sodium titanate dispersions in the presence of excess HCl (*31, 52–54*).

B. Surface Properties

The surface of titanium dioxide has been extensively studied (*55*) because of its important applications in fields such as pigments (*56*), catalysis (*57–63*), photocatalysis (*37, 52, 64–68*), and as a model oxide system for surface electrochemical studies (*55, 69–76*). The reported results and surface characteristics, however, are often different because the surface properties of the product strongly depend on the method of preparation, crystal structure, surface area, and presence or absence of impurities. For example, isoelectric points of TiO_2 have been reported ranging from 2.5 to 8 (*55, 77–83*), although the most probable values quoted are 6.1 for anatase and 4.8–5.6 for rutile (*55, 83*). The present section is concerned only with the product from the hydrolysis of sulfate solutions because of its applications in the photoassisted decomposition of water.

As is apparent from the data reported in Table II, titanium dioxide obtained by the thermal hydrolysis of sulfuric solutions is highly hydrated and contains appreciable amounts of sulfate ions. Sulfate impurities are strongly coordinated to the gel, probably as basic titanium sulfates, and cannot be easily eliminated by simple washing. Continuous washing of a hydrous titania gel dispersion using the serum-replacement technique (*84*) indicated an excess sulfate concentration in the ultrafiltrate resulting from slow desorption of the anion. Sulfate impurities strongly affect the surface properties of the gel, both *in vacuo* and in aqueous solutions (*55, 85*).

Elimination of the sulfate contaminant could be achieved either by calcination above 700°C (*55, 85*) or by neutralization with a base such as NH_4OH, followed by thorough washing. Calcination of the gel also has a tremendous effect on the surface properties of the product. The main changes occurring during this process are crystal growth, particle growth, elimination of coordinated and adsorbed water and of surface hydroxyls, general decrease in the surface reactivity, and migration of ionic impurities from the bulk to the surface (*55, 85, 86*).

Infrared studies (*55, 86*) have shown that in sulfate-based products the elimination of physically adsorbed water occurs at ~150°C and strong dehydroxylation is observed at temperatures above 600°C. For product calcined above 800°C, the dehydration of the surface shows some irreversibility. In Fig. 5 (*87*) the results of the titration with Hammet indicators (*88–90*) of the surface acidic sites are reported as a function of the calcination temperature; the elimination of hydroxyls and surface sulfates leads to a decrease in the total acid amount and strength of the acidic sites. Also, the nature of the sites is changed, because surface sulfates have been reported to generate Brönsted-type acidity (*85*) such as OH and incompletely coordinated Ti(IV) ions (*55, 86*).

The change in surface composition is shown in Fig. 6, in which the xps analyses data for samples calcined at different temperatures are reported (*87*).

The surface segregation of impurities such as Zn, K, and S with ionic radii incompatible with the TiO_2 lattice structure is apparent, whereas elements such as Nb(V), having an ionic radius similar to that of Ti(IV), are evenly distributed throughout the particle (*91*). It is noticeable that the sharp rise in the surface concentration of impurities coincides with the crystal rearrangement preceding the anatase–rutile phase transition starting at ~800°C. The drop in surface sulfur is observed at ~700°C resulting from the elimination of SO_3.

The variation of surface area of the TiO_2 particles is shown in Fig. 7 and is a result of particle growth and loss of surface porosity (*87*).

C. Optical Properties

Titanium dioxide, in both the anatase and the rutile crystal modifications, is colorless and appears white in powdered form. Indeed, as shown in Fig. 8, no absorption bands are present in the visible region for crystalline anatase, rutile, and hydrous titania. The strong absorption band lying in the near uv corresponds to the vb–cb charge transfer, with energy ≥3 eV.

The refractive index of crystalline TiO_2, as shown in Table III (*32, 92*), is very high and makes this product the most widely used white pigment, primarily in applications where vehicles with high refractive indexes are used. Hydrous titania has a somewhat lower refractive index, but no

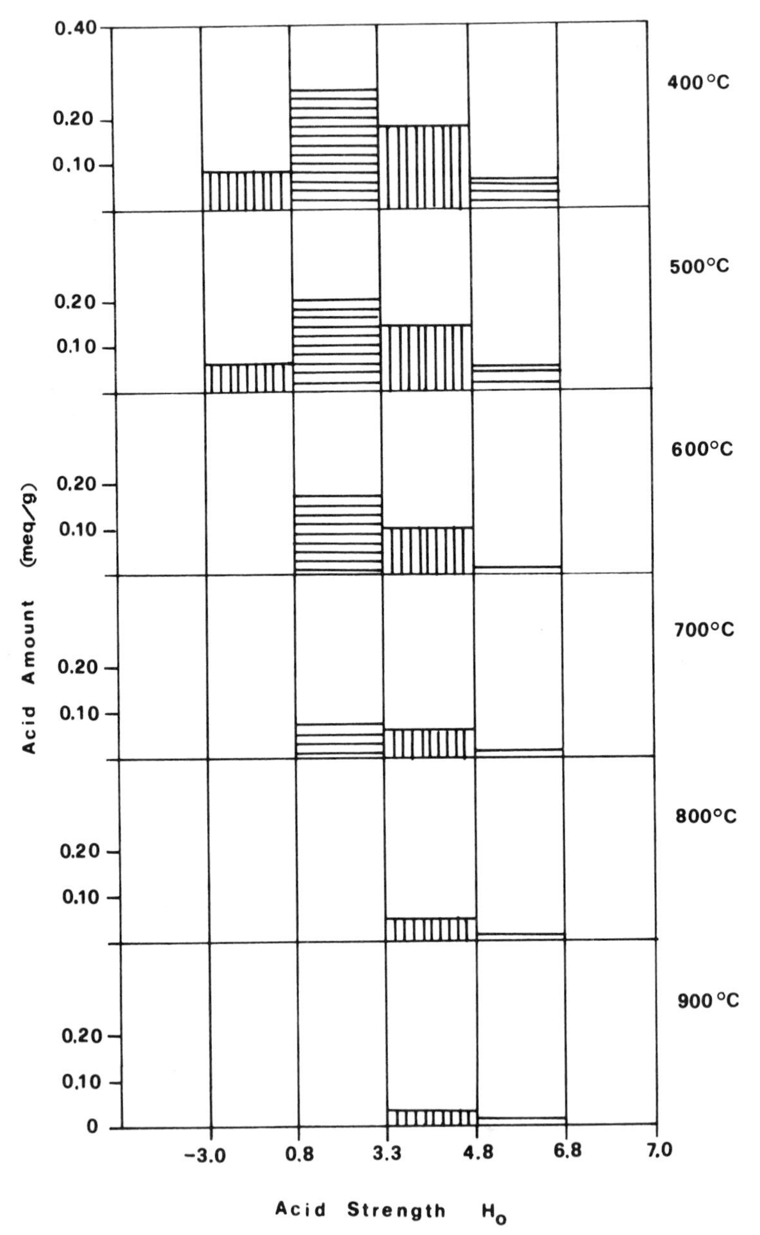

Fig. 5. Amount and strength of surface acidic sites of a TiO_2 sample as a function of the calcination temperature.

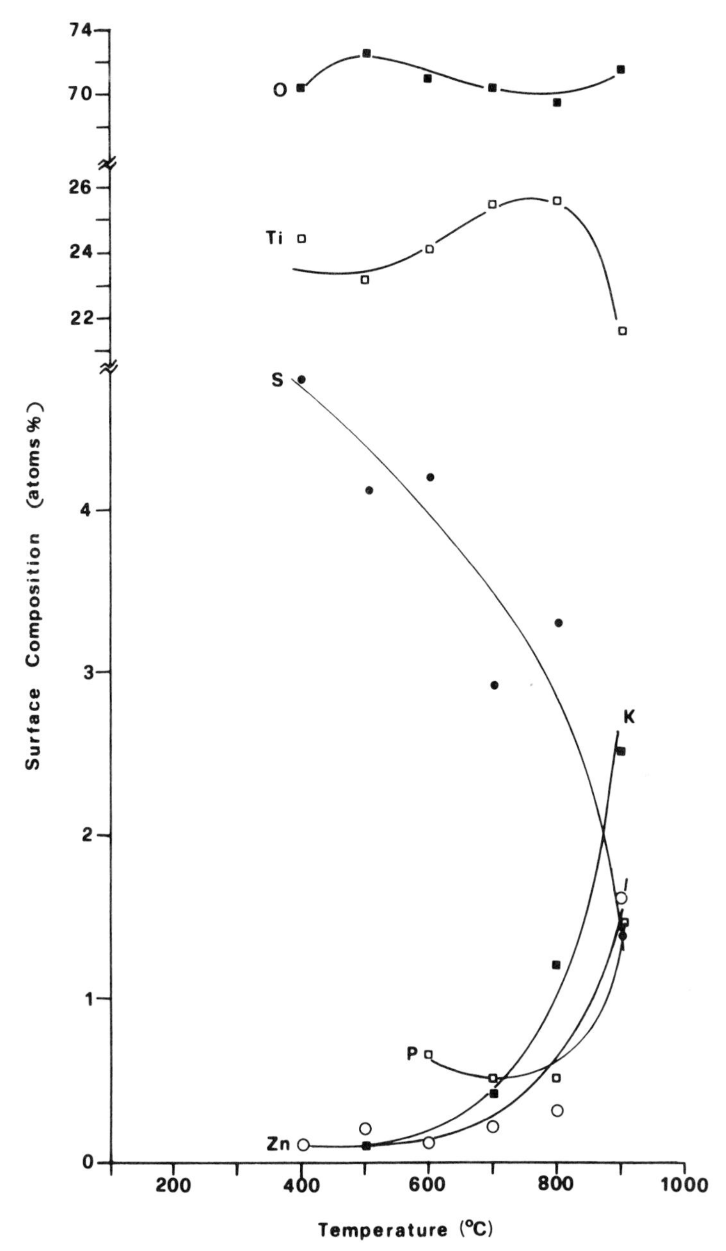

Fig. 6. Surface composition of a TiO_2 sample as a function of the calcination temperature. Lower-temperature composition approaches bulk values.

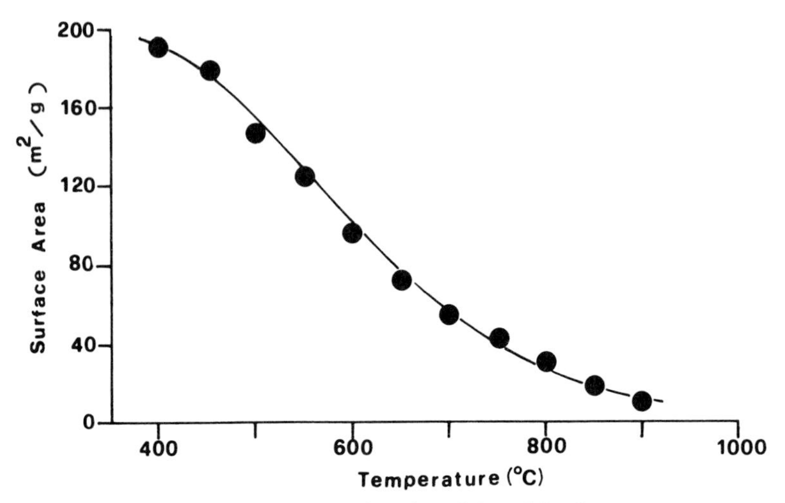

Fig. 7. Surface area of TiO_2 as a function of the calcination temperature.

reliable data are available concerning this material. Light-scattering measurements on uniform spherical titania particles prepared by the aerosol technique (*34*) and dispersed in liquids with different refractive indexes (*93*) resulted in the determination of a refractive index of ~2 (*94–97*). The scattering efficiency of the TiO_2 particles dispersed in water is, as a consequence, very high, even in hydrous form, and depends on particle size (*93, 98*). Particles having a diameter of ~0.2 μm fall within the range of maximum scattering efficiency; indeed, the great scattering of the incident light was responsible for the difficulties that arose in the determination of the quantum yield in the photoassisted water-cleavage experiments. A possible way to overcome such a problem would be the preparation of particles in the Rayleigh size range (*93*), because very small particles show much lower scattering efficiencies. As is well known by TiO_2-pigment manufacturers (*92*), the color of TiO_2 powders is strongly affected

TABLE III

Refractive Indexes of Anatase and Rutile

	λ (436 nm)		λ (546 nm)		λ (589 nm)	
Orientation	Ordinary	Extraordinary	Ordinary	Extraordinary	Ordinary	Extraordinary
Anatase	2.76	2.67	2.59	2.52	2.56	2.49
Rutile	2.85	3.20	2.65	2.94	2.61	2.90

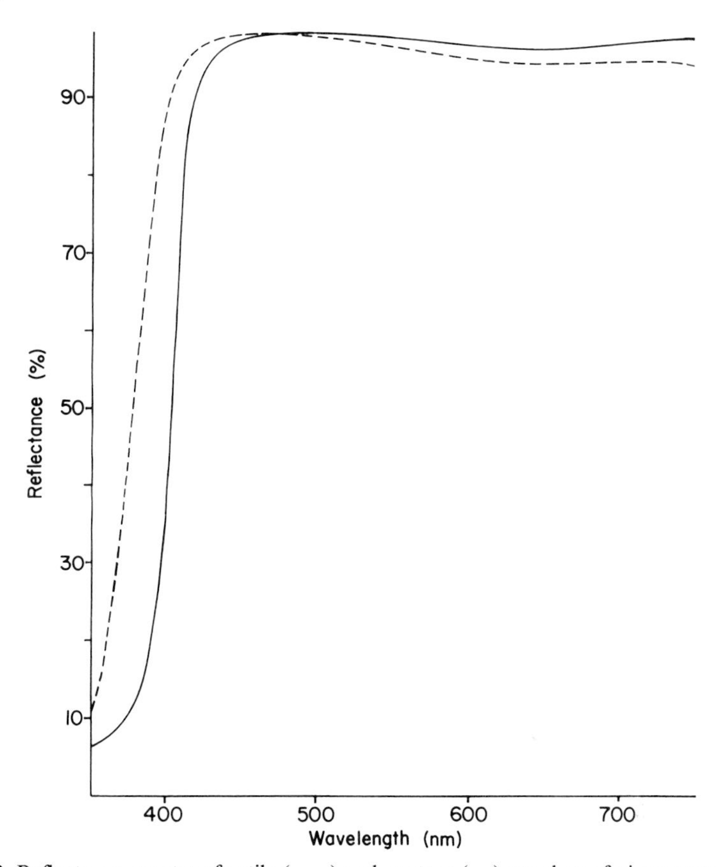

Fig. 8. Reflectance spectra of rutile (——) and anatase (---) powders of pigmentary size.

by oxygen deficiencies in the lattice or by doping with colored transition metals. Oxygen can be reversibly removed from the TiO_2 lattice by evacuation at high temperatures (*55*) or by evacuation under uv illumination (*55, 94–97*). In the absence of electromagnetic radiation, the process is enhanced by the presence of anionic impurities such as SO_4^{2-}, Cl^-, or by organic contamination (*55*). Loss of oxygen is associated with the generation of surface Ti(III) sites, which are responsible for the blue-gray color assumed by the product (*55*).

Transition-metal doping has been used to extend the spectral response of single crystals of TiO_2 and $SrTiO_3$ (*99–108*) and of TiO_2 particle dispersions toward visible light. Indeed, as shown in Fig. 9, the reflectance spectra of TiO_2 are modified by doping with Fe(III), Cr(III), and V(V).

The shape of the reflectance spectra indicates that doping has no influence on the band gap of TiO_2 because the charge-transfer band edge

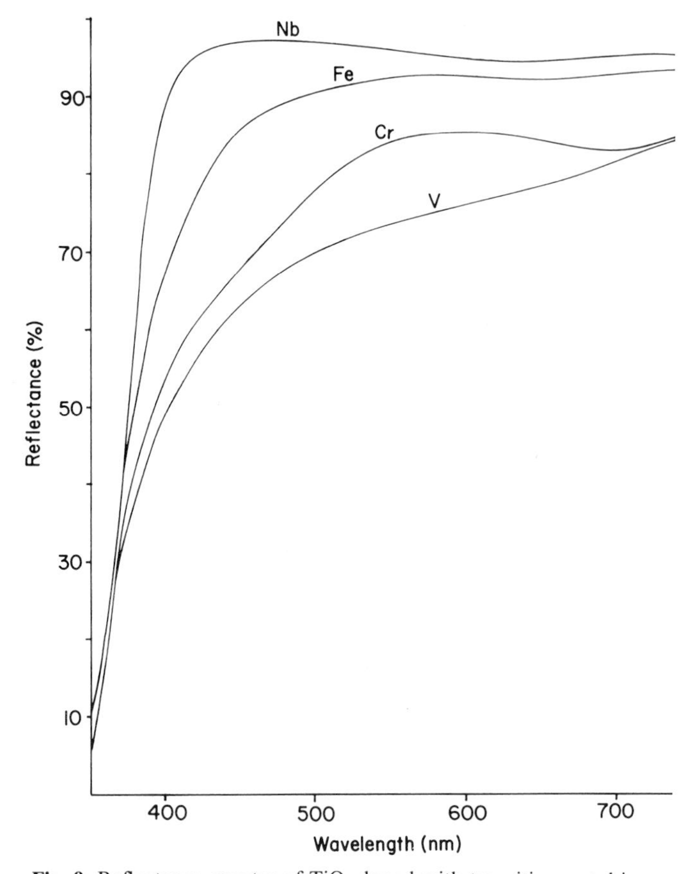

Fig. 9. Reflectance spectra of TiO_2 doped with transition-metal ions.

remains almost unchanged. The absorption in the visible-light range is attributed to electron transitions from *d* levels of the dopants to the TiO_2 conduction band (*99–110*).

IV. Photoinduced Redox Reactions

These reactions represent the basis of almost all the wireless processes of conversion of solar energy to chemical fuels actually under investigation (*111–115*). The thermodynamics of a light-driven reaction can be represented as

$$S + R \underset{\Delta}{\overset{h\nu}{\rightleftharpoons}} S^{+} + R^{-} \qquad (2)$$

The electron transfer does not occur when the reactants are in the ground state, but only when the species S (called the sensitizer) is excited by the light. R is an electron-acceptor species (often called the electron relay). Suitable sensitizer–relay systems have been found (good absorption spectrum, long-lived excited state, high quantum yield and efficiency in the electron-transfer process, redox properties—that is, $E°(S^+/S)$ and $E°(R/R^-)$—in agreement with the desired fuel formation), primarily by using transition-metal complexes such as $Ru(bipy)_3^{2+}$ (*116, 117*) or porphyrins (*118*) as sensitizers and 1,1′-dialkyl-4,4′-bipyridine (viologens, MV^{2+}) (*119, 120*), $Rh(bipy)_3^{2+}$ (*121, 122*), and others (*123*) as electron relays.

The other two main problems to be solved concern the prevention of the backward reaction (2), which is thermodynamically favored, and the acceleration of the reactions giving rise to fuel formation, that is (in the case of water cleavage),

$$2R^- + 2H_2O \longrightarrow 2R + 2OH^- + H_2 \quad (3)$$

$$4S^+ + 2H_2O \longrightarrow 4S + 4H^+ + O_2 \quad (4)$$

Organized molecular assemblies (*124*) have been extensively investigated as a means of minimizing the back reaction rate. Many different structures, such as micelles (*125–129*), microemulsions (*130, 132*), polyelectrolytes (*133–135*), vesicles (*136–137*), monolayers (*138*), membranes (*139*), and inorganic colloids (*140, 141*), have been exploited to control the kinetics of electron-transfer reactions. Considerable success has been obtained with these multiphase systems, but this aspect represents only the first step of the entire process of water cleavage.

V. Colloidal Redox Catalysts

It can easily be recognized that the crucial aspect in the whole water-cleavage process is represented by the acceleration of reactions (3) and (4). Consequently, significant efforts have been devoted to find proper redox catalysts.

Reactions (3) and (4) proceed, in the absence of catalysts, through the intermediate formation of radicals (H· in water reduction and OH· in water oxidation), which requires that a relevant energy barrier must be overcome. As a means of lowering these barriers, thus allowing these processes to compete with back reaction (2), colloidal catalysts have been investigated (Fig. 10) (*142*).

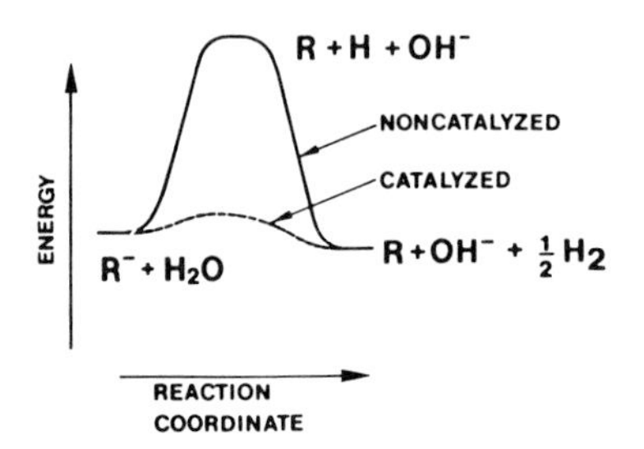

Fig. 10. Representation of the noncatalyzed and catalyzed hydrogen evolution from water. From Kiwi *et al.* (*21*).

Based on consideration derived from electrocatalytic reagents used on macroelectrodes, colloidal metals have been proposed. Thus the reaction

$$2R^- + 2H_2O \xrightarrow{\text{catalyst}} 2R + 2OH^- + H_2 \tag{5}$$

has been examined in detail from mechanistic and kinetic points of view; the roles of the method of preparation, dimensions of the particles, and nature of the protective agents have been thoroughly investigated, particularly for Au (*143–145*), Ag (*146–149*), and Pt (*150–155*) colloids, the latter appearing to be the most effective. From mechanistic aspects, in the Ag and Au colloids (prepared by reduction and then polymer stabilized), the initial step in the transformation of reducing equivalents to hydrogen is electron transfer from the reducing radical to the metallic particle (*143–149*). A large number of electrons can be stored on each particle, and the capacitance of such microelectrodes can be measured.

The reaction

$$nR^- + (M)_c \longrightarrow nR + (M)_c^{n-} \tag{6}$$

where $(M)_c$ represents the metal colloid, is diffusion controlled, whereas the protonation of the charged particle is much slower and very strongly pH dependent:

$$(M)_c^{n-} + H^+ \longrightarrow (M)_c^{(n-1)-} + H_{ads} \tag{7}$$

Therefore, the protonation step or a later step such as desorption, be it chemical or electrochemical, represents the rate-determining step.

$$2H_{ads} \longrightarrow H_2 \tag{8}$$

$$(M)_c^{(n-1)-} + H^+ + H_{ads} \longrightarrow (M)_c^{(n-2)-} + H_2 \tag{9}$$

In the case of Pt (prepared from H_2PtCl_6 by reduction and then stabilized with polyvinylalcohol or other polymers) (*150–155*), a combined pulse radiolytic and conductometric technique showed that the proton discharge immediately follows the electron-transfer reaction from the radical to the particle (Fig. 11) (*156*), because of the lower overpotential and higher exchange-current density. The Pt particles can, therefore, be

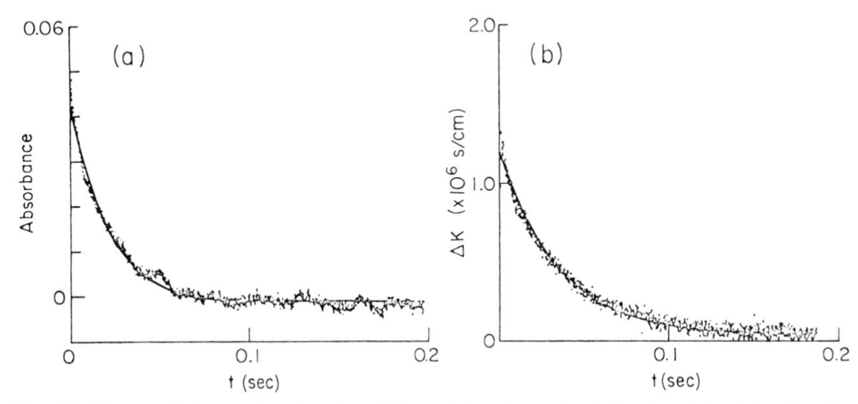

Fig. 11. Decay of absorption signal at 605 nm (a) and conductivity signal (b) following the pulse radiolytic production of MV^{+} in Pt-containing solutions. Both are deaerated and contain $[MV^{2+}] = 5 \times 10^{-4}$ *M*, [Pt] $= 7.5 \times 10^{-6}$ *M*, $[H^+] = 5 \times 10^{-4}$ *M*, 1% propan-2-ol, and 1% acetone. Solid lines represent the nonlinear mean squares best fit to a first-order decay-rate law. From Matheson *et al.* (*156*). Copyright 1983 American Chemical Society.

viewed as storage pools of hydrogen atoms rather than electrons. Hydrogen desorption from the particle appears to be the slowest step in the process.

It is worthwhile to mention that a quantitative electrochemical approach has been developed for the electrodic behavior of these metallic particles (*157, 158*).

A surprising observation, still unexplained, is the dependence of the rate of radical decay on the colloidal Pt concentration, which is greater than first order (*150–155, 157, 158*). A particle–particle interaction could explain this feature (*156*), although a diffusion-controlled reaction at multiple spherical electrodes could partially account for the phenomenon (*157, 158*). Another way to achieve high efficiency, by depositing ultrafine Pt on inorganic colloids, has been attempted (*159–161*). For example, the Pt–TiO_2 sol (deposit of Pt on TiO_2 particles with hydrodynamic radii of 200 Å) strongly enhances the rate of MV^{+} reaction (with corresponding H_2 evolution). The observed astonishingly high rate can probably be explained if the TiO_2 particles are considered to act as electron acceptors. The electrons transferred to the conduction band of the particle are then channeled to the Pt sites, where hydrogen evolution occurs (*21*).

The difficulty in the generation of O_2 lies in the coupling of a one-electron oxidant with the four-electron requirement for oxygen formation from water. Because the one-electron water oxidation requires too strong an oxidant and involves the formation of the OH· radical, two-, and three- and, particularly, four-electron oxidations are highly desirable. In the latter case, because four oxidants and two water molecules are involved, the presence of a redox catalyst is crucial.

Noble-metal oxides such as PtO_2 (*162*), IrO_2 (*162*), RuO_2 (*163–166*), and also MnO_2 (*167*) in powdered or colloidal form have been proven to be effective catalysts for oxygen evolution. Oxidants such as Ce(IV), $IrCl_6^{2-}$, and ML_3^{3+} (where M = Ru, Fe, or Os and L = bipyridine-like ligands; these can also be generated photochemically in a proper sacrificial system) have been examined.

The oxygen yields are pH dependent, and the optimal pH values are influenced by the oxidant as well as by the catalyst. Considerable attention has been devoted to RuO_2-based catalysts, which have been shown to be highly effective; in particular, colloidal RuO_2 stabilized by styrene–maleic anhydride copolymer or other protective agents exhibits a high catalytic activity (Fig. 12) (*165, 168*).

The kinetics of the oxidant disappearance in the presence of redox catalysts follows the first-order dependence in the case of Ce(IV) or Ru-$(bipy)_3^{3+}$ reduction with powdered catalysts (PtO_2, RuO_2) (*163*).

With colloidal RuO_2, the kinetics of the reaction

$$4Fe(bipy)_3^{3+} + 4OH^- \longrightarrow 4Fe(bipy)_3^{2+} + 2H_2O + O_2 \quad (10)$$

is similar to that reported in noncatalyzed media. As expected for the heterogeneous mechanism for the oxygen generation, the reaction rate increases linearly with RuO_2 concentration (up to 3.5 mg/liter), hence with catalyst surface area (*168*). Similar results have been reported using MnO_2 as catalyst (*167*).

The exact mechanism of O_2 formation through redox catalysis is not completely understood. If it is similar to the mechanism inferred in homogeneous O_2 evolution, the formation of ionic sites at the surface of the catalyst [e.g., Ru(VI) in RuO_2 or Mn(V) in MnO_2] could be responsible for the reaction with water. The method of preparation and stabilization of the colloids is a determining factor in the catalyst activity; the stabilizer is

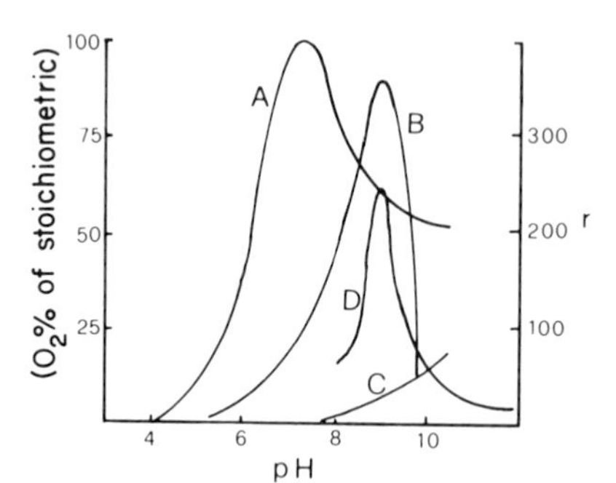

Fig. 12. (Left scale) Oxygen yields as a function of pH from $Fe(bipy)_3^{3+}$: (A) RuO_2 powder (300 mg/liter), (B) colloidal RuO_2 (30 mg/liter), (C) catalyst-free solution. (Right scale) (D) Ratio of the observed rate constant in the presence of colloidal RuO_2 (3.5 mg/liter) with respect to the rate constant in the catalyst-free solution. From Kalyanasundaram *et al.* (*165*). See also Pramauro and Pelizzetti (*168*).

particularly important because it can influence the charge, availability of sites, and stability of the particles.

Also, for this process the possibility of using inorganic particles as carriers of ultrafine deposits of RuO_2 opens a very promising area. Owing to the similarity of the Ru^{4+} and Ti^{4+} ionic radii, it is possible to prepare efficient TiO_2–RuO_2 catalysts. Colloidal TiO_2 particles (200–500 Å) loaded with RuO_2 are extremely active catalysts for oxygen evolution (*169, 170*).

From the mechanistic point of view, a study combining flash photolysis and conductance techniques showed that in the oxidation of water by $Ru(bipy)_3^{3+}$, decay of the oxidant and increase in conductivity (owing to proton formation) occur simultaneously; thus the particle acts as a local element, and the positive charges transferred from the oxidant to the RuO_2 deposited on TiO_2 are immediately transferred to water, giving O_2 and protons. The rate seems limited by the diffusion of the oxidant toward the particles (*170*).

The dependence of the reaction rate on the catalyst concentration is noteworthy (Fig. 13). For up to 150 mg/liter of TiO_2 (loaded with 2%

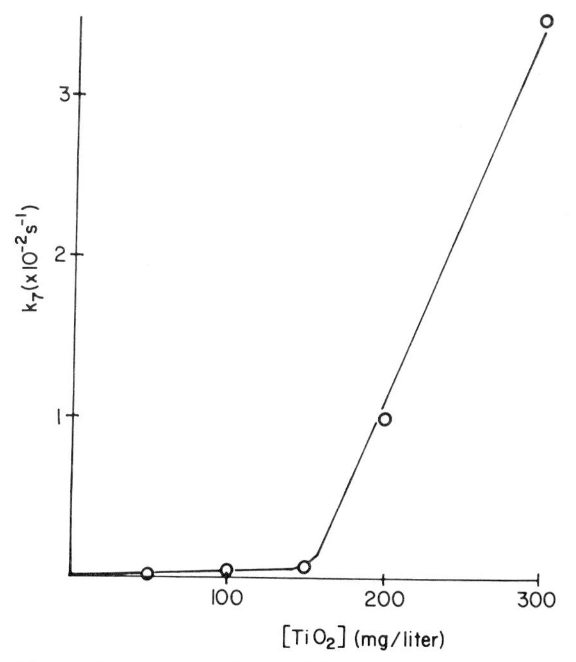

Fig. 13. Effect of the catalyst concentration (colloidal TiO_2 charged with 2% RuO_2) on the observed rate constant for water oxidation by $Ru(bipy)_3^{3+}$ at pH 5. From Humphrey-Baker *et al.* (*170*). Copyright 1982 American Chemical Society.

RuO_2), there is a linear dependence that could be a result of the filling of the active sites or a partial coagulation of the colloid in these conditions. Above this catalyst concentration there is a significant deviation from simple pseudo-first-order behavior, as observed in the previously reported H_2-evolution studies in the presence of Au and Pt colloids. Detailed studies on these arguments as well as further improvement of the catalyst are presently being pursued to increase the efficiency of the O_2-evolution process.

VI. Cyclic Water Cleavage with Bifunctional Redox Catalysts

Because it has been shown that $Ru(bipy)_3^{3+}$ and $MV^{\dagger}$ can generate oxygen and hydrogen when proper catalysts are present, a cyclic system in which both processes take place simultaneously has been checked (*171*).

$$Ru(bipy)_3^{2+} + MV^{2+} \xrightarrow{h\nu} Ru(bipy)_3^{3+} + MV^{\dagger} \quad (11)$$

In this case, to compete efficiently with the back reaction the catalysis must be extremely efficient and highly specific. Selectivity and yield can be improved with proper selection of the hydrophobic–hydrophilic properties of the sensitizer–relay system and of the protective agents. In the light of the results presented previously on the high catalytic efficiency of the ultrafine Pt and RuO_2 deposits on colloidal TiO_2, a catalyst has been prepared that can act simultaneously for water-oxidation and -reduction processes.

A. *Preparation and Properties of the Bifunctional Catalyst*

The catalysts are based on hydrous titania obtained by thermal hydrolysis of sulfuric Ti(IV) solutions (anatase oriented) or by thermal aging of sodium titanate dispersions in the presence of excess HCl (rutile oriented). The particles were doped with Nb and coated with the redox catalysts RuO_2 and Pt. Doping with Nb was found to be of crucial importance for rutile-oriented catalyst particles, although it increased the reactivity of hydrous anatase-oriented TiO_2. Owing to the special characteristics of the product, the most common doping procedure, that is, high-temperature diffusion of the dopant (*172*), could not be used; instead, Nb_2O_5 was coprecipitated with TiO_2. The required amount of a soluble Nb(V) compound, such as freshly precipitated Nb_2O_5, $NbCl_5$, or potassium niobate, was dissolved in the sulfuric Ti(IV) solution, and the obtained liquor was subjected to thermal hydrolysis, as described previously. Because of the instability of the sulfuric solutions of Nb(V), the hydrolyzed product contained the whole amount of added Nb. This method allowed hydrous TiO_2 containing up to 0.6% Nb_2O_5 to be prepared, owing to the low solubility of Nb(V) in acidic solutions.

Higher levels of doping could be obtained by adding an alkaline potassium niobate solution to the Ti(IV) solution during the hydrolysis process, but this required a calcination step to obtain active products.

In the case of 0.4% Nb_2O_5 doping, the hydrous TiO_2 consists of particles having an average diameter of ~0.2 μm; these particles are aggregates of primary nuclei of ~100 Å in diameter. They have a BET surface area of 200 m^2/g and are porous (*52*). The isoelectric point (iep) of the particles, as determined by electrophoresis, is 3.5–4, indicating the presence of sulfate surface impurities.

The redox catalysts RuO_2 and Pt must be deposited onto a TiO_2 surface to obtain a highly active catalyst. Ruthenium dioxide was precipitated as the inner coating from homogeneous $RuCl_3$ solutions. The mechanism of surface precipitation of hydrous oxides onto dispersed systems has been extensively studied (*173–176*) and involves either the adsorption of hydrolyzed complexes at the solid–solution interface or the heterocoagulation of finely divided hydrolysis products with the dispersed phase.

In our case, $RuCl_3$ solution was added to a TiO_2 dispersion in distilled water under sonication, and the resulting slurry was neutralized at pH 6 with KOH. Under these conditions, TiO_2 particles should act as precipitation nuclei for hydrolyzed ruthenium. The obtained product was filtered and dried overnight in air under reduced pressure at 100°C. The amount of RuO_2 precipitated on TiO_2 ranged between 0.1 and 3% by weight. An esca analysis performed on the dried product showed (F. Garbassi, personal communication) the presence of the $3D_{5/2}$ transition of Ru(IV), with a binding energy of 280.6 eV.

Pt loading of RuO_2-coated TiO_2 was achieved either by surface precipitation from homogeneous solutions or by adsorption of an ultrafine Pt sol onto the surface of the particles. Reduction of H_2PtCl_6 or H_2PtCl_4 solutions in the presence of the dispersed phase was obtained either by the photoplatinization method (*177, 178*) or by refluxing the dispersion for several hours in the presence of excess formaldehdye. The solution was then evaporated to a small volume several times to eliminate the unreacted organic compound. The photoplatinized product showed higher reactivity, probably because of a more controlled surface reduction of Pt ions, leading to more highly dispersed surface Pt sites.

The other method for loading the catalyst with platinum consisted of preparing a finely divided Pt sol by reducing a homogeneous chloroplatinate solution with citrate ions at 90°C for some hours (*172, 179*).

After a careful purification procedure over ion-exchange resins, the calculated amount of the obtained Pt sol was added to a TiO_2–RuO_2 dispersion in water at pH 3.5–4.5. After stirring, complete adsorption of the Pt sol onto TiO_2–RuO_2 particles was observed. This can be qualitatively explained by taking into account that the two colloidal systems have oppositely charged surfaces between pH 2.5 (iep of Pt sol) and pH 4.5 (iep of TiO_2–RuO_2) and are of greatly different sizes (*180–182*). However, the activity of the obtained product showed some irreproducibity because of changes in the effectiveness of the heteroagulation process, possibly caused by the presence of surfactant-like impurities coming from the ion-exchange resins (*183, 184*).

B. Cyclic Water Cleavage

Water splitting induced by visible light (cutoff filter 400 nm) on a system consisting of $Ru(bipy)_3^{2+}$ and MV^{2+} has been achieved in the presence of the bifunctional catalyst prepared as described previously. The hydrogen-evolution rate remains constant for 40 h, as shown in Fig. 14. It is noteworthy that in such a dispersion, a quantum yield of one-fifth that obtained with a sacrificial system is obtained (*172*).

These results imply that the participation of the electronic states of the semiconductor and/or the adsorbed reactants in the process occurs. Moreover, the presence of both redox catalysts (Pt and RuO_2) has been shown to be beneficial, because the activity of the bifunctional catalyst far exceeds the sum of the activities of the two individual components (Pt–TiO_2 and RuO_2–TiO_2), thus implying a synergistic effect. The colloidal Pt–TiO_2–RuO_2 particles showed a very high efficiency in water cleavage in the absence of sensitizer and electron relay when irradiated without a cutoff filter; the direct band-gap excitation gave hydrogen and oxygen much more abundantly than previously reported with other semiconductor powders. These results confirm the efficient electron–hole separation in this colloidal catalyst (*178, 185*) (see Fig. 15).

The participation of adsorbed species can be inferred because hydrogen evolution is also observed, although at a very low rate, in the absence of electron relay. Similarly, the shift in the iep of TiO_2 particles in the presence of $Ru(bipy)_3^{2+}$ and the adsorption isotherm suggest the participation of adsorbed sensitizer in the process (*52*). These results prompted the design of a relay-free system with a sensitizer that can easily be adsorbed at the catalyst surface. Sensitizers having a long alkyl chain lead to a

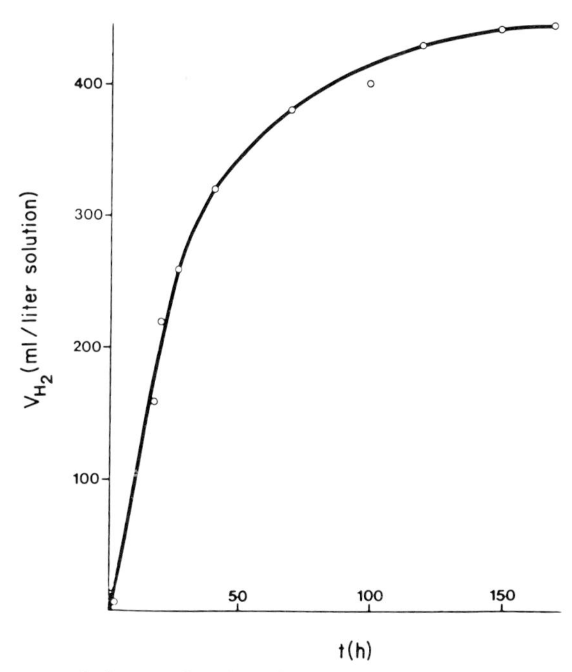

Fig. 14. Hydrogen evolution as a function of time for visible-light photolysis of a solution at pH 4.7 containing $[Ru(bipy)_3^{2+}] = 1 \times 10^{-4}\ M$, $[MV^{2+}] = 5 \times 10^{-3}\ M$, TiO_2 particles (500 mg/liter) loaded with 0.2% RuO_2, and 40 mg/liter of Pt. From Kiwi *et al.* (*21*).

significant improvement in the H_2-evolution rate and yield. Among the different derivatives, the mono-$C_{12}H_{25}$ substituted $Ru(bipy)_3^{2+}$ exhibited the highest efficiency for H_2 generation (Fig. 16) (*186*).

These experimental results can be interpreted with the aid of the scheme shown in Fig. 17. Excitation of the adsorbed sensitizer is followed by charge injection into the conduction band of the semiconductor, and the electron is channeled to the Pt sites where H_2 evolution takes place;

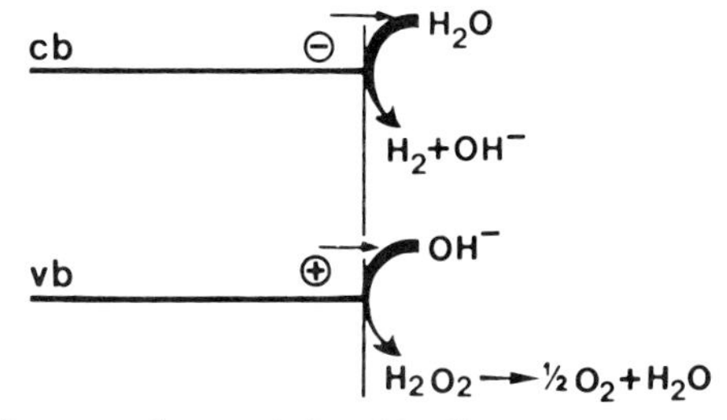

Fig. 15. Band model for water cleavage induced by direct band-gap excitation. From Kiwi *et al.* (*21*).

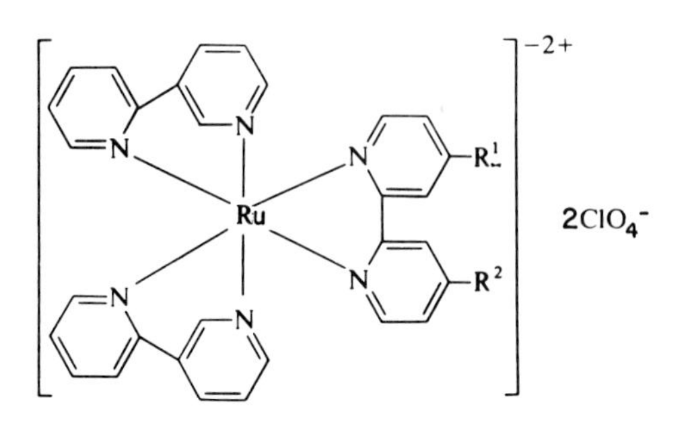

(a)

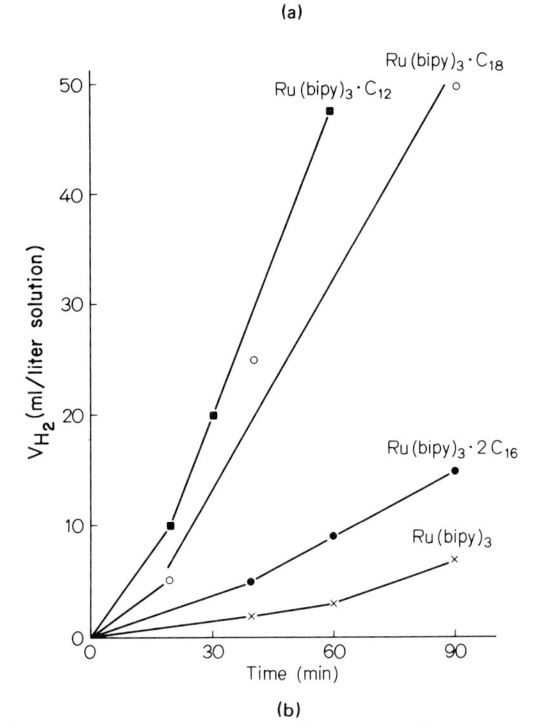

(b)

Fig. 16. (a) Surfactant derivatives of $Ru(bipy)_3^{2+}$. $Ru(bipy)_3^{2+}$, $R^1 = R^2 = H$; $Ru(bipy)_3^{2+} \cdot 2C_{16}$, $R^1 = R^2 =$ *n*-hexadecyl; $Ru(bipy)_3^{2+} \cdot C_{18}$, $R^1 = CH_3$, $R^2 =$ *n*-octadecyl; $Ru(bipy)_3^{2+} \cdot C_{12}$, $R^1 = CH_3$, $R^2 =$ *n*-dodecyl. (b) Hydrogen yields from visible-light photolysis of solutions at pH 4.5, containing 500 mg/liter of TiO_2 (loaded with 0.1% RuO_2 and 40 mg/liter Pt) and the sensitizer concentrations: $[Ru(bipy)_3^{2+}] = 2 \times 10^{-4}$ *M*, $[Ru(bipy)_3^{2+} \cdot 2C_{16}] = [Ru(bipy)_3^{2+} \cdot C_{18}] = 7 \times 10^{-5}$ *M*, and $[Ru(bipy)_3^{2+} \cdot C_{12}] = 5 \times 10^{-5}$ *M*. From Borgarello *et al.* (*186*). From *Nature* (*London*). Copyright 1980 Macmillan Journals Ltd.

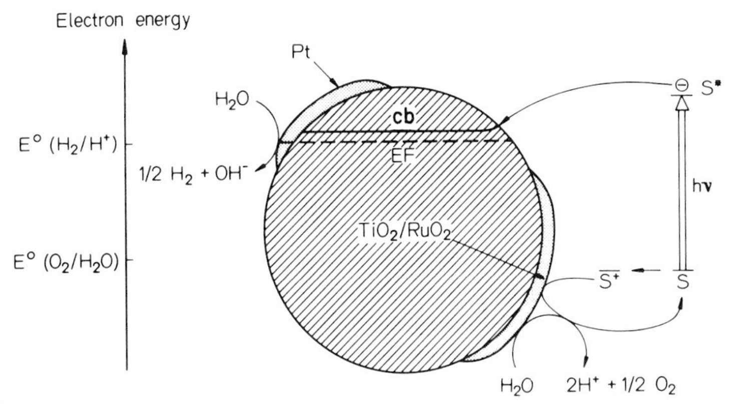

Fig. 17. Representation of the processes involved in the visible-light photolysis of water in an electron-relay free system. cb, Conduction band; EF, energy of Fermi level. From Kiwi *et al.* (*21*).

the back conversion of the oxidized sensitizer into the original form is mediated by RuO_2 with concomitant O_2 evolution. The role of RuO_2 is to accelerate the hole transfer from the valence band of TiO_2 to the aqueous solution. The low overvoltage characteristic of O_2 evolution on RuO_2 renders hole capture by water particularly effective and hence inhibits electron–hole recombination. This aspect shows another interesting feature of semiconductor particles: the band bending at the semiconductor–liquid interface allows an efficient charge separation owing to the mobility of charges in the conductor and valence bands. Thus differing with respect to the homogeneous photoredox systems, it should be possible to reach very high conversion efficiencies.

Another relevant feature of the system is the drastic increase in H_2 yield with increasing temperature. In a 50°C range the H_2 yield increases by a factor of 50 (*185*).

VII. Increasing the Efficiency and the Sunlight Response

The reported results stimulated interest in developing systems with higher efficiency, better absorption of solar radiation, good stability, and reduced cost.

Several different approaches are now being pursued to overcome these problems. These include dye sensitization, impurity sensitization, new small-band-gap semiconductors, heterostructure design, and new redox catalysts and sensitizers.

A. Dye Sensitization

As described previously, large-band-gap semiconductors such as TiO_2 are relatively stable and the use of a sensitizer layer (*187–190*) [e.g., surfactant derivatives of $Ru(bipy)_3^{2+}$] (*186*) improves the response to visible light.

Several processes can take place when an electron is raised to the excited state of a dye molecule, such as an electron transfer to the conduction band (the charge changes by 1+ in the molecule), an electron transfer from the semiconductor to the ground state of the dye (charge variation 1−), and also the simple deexcitation of the molecule to its ground state. In particular, the electron transfer from the excited state of the sensitizer to the conduction band of a *n*-type semiconductor implies that the lowest vacant level of the sensitizer is situated above the conduction band of the semiconductor (Fig. 18).

A large number of dyes (absorbing in the visible light with high absorptivity) are capable of performing this process; among these there are inorganic complexes [such as $Ru(bipy)_3^{2+}$], organic dyes (rhodamine, cyanins, and phenothiazines), and metal phthalocyanin.

In this line, the use of zinc tetramethylpyridyl porphine as a sensitizer

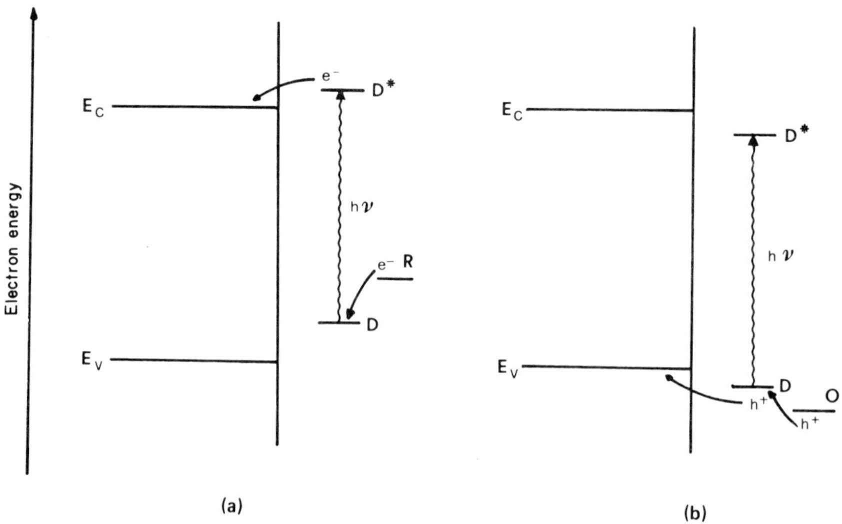

Fig. 18. Mechanism of dye sensitization in a semiconductor. The excited dye molecule can act as an electron donor ($D^* \rightarrow D^+ + e^-$; a) or as an electron acceptor ($D^* \rightarrow D^- + h^+$; b). From Memming (*12*).

with the bifunctional Pt–TiO_2–RuO_2 catalyst showed H_2 and O_2 evolution (*191*). Obviously, not only the spectral and adsorbing properties are to be considered, the redox potentials must also be selected to improve the performance of the systems (*192*).

Other important aspects of dye sensitization are the possibility of desorption of the sensitizer in the electrolyte solution and the degradation of response owing to the bleaching of the dye.

B. Impurity Sensitization

It appears that a good way to overcome many problems might be the alteration of the properties of the materials through the introduction of impurities (*99–108*). Several effects may be induced by the presence of a dopant: (a) protection against corrosion, (b) change in the resistance of the bulk semiconductor and in the thickness of the space-charge layer with a consequent higher conversion efficiency, (c) variation in the bandgap, (d) shift of the flat-band potential, and (e) formation of new band by the interaction of an interstitial dopant with the semiconductor lattice.

The reported effect of Nb on the TiO_2 colloids revealed that the doping is critical when rutile-like particles are used. Nb^{5+} has a radius very close to that of Ti^{4+} and, therefore, isomorphically replaces titanium ions in the TiO_2 lattice. As an *n*-dopant it produces a Schottky barrier at the particle–solution interface that assists electron injection from the reduced electron relay. Nb_2O_5 also shifts the flat-band potential cathodically (*193*).

However, the most interesting possibilities are the shift of the band gap to lower energies or the introduction of donor levels that can be excited into the semiconductor band.

Although investigation of the effect of several impurities on the band gap of semiconductor electrode materials showed that often no relevant change occurs as a consequence of doping, the photoresponse is increased by the introduction of transition metals (*194*). In the light of this observation, colloidal TiO_2 doped with Cr and loaded with RuO_2 and Pt was investigated in the cyclic water-cleavage system (*195*). Hydrogen evolution was observed in the absence of any sensitizer and electron relay when a cutoff filter (415 nm) was present (Fig. 19). This observation may be explained in terms of excitation of electrons from the Cr^{3+} center to the conduction band of TiO_2 by visible light. These electrons are then available for water reduction (mediated by Pt), and holes (Cr^{4+}) are ready for oxidation of water (mediated by RuO_2). The method of preparation of these doped colloids suggests that Cr^{3+} is present primarily in the surface region of the particles. Increasing the Cr doping much beyond its solubility limit in TiO_2 ($\sim$0.4%) is likely to produce a chromic oxide surface layer

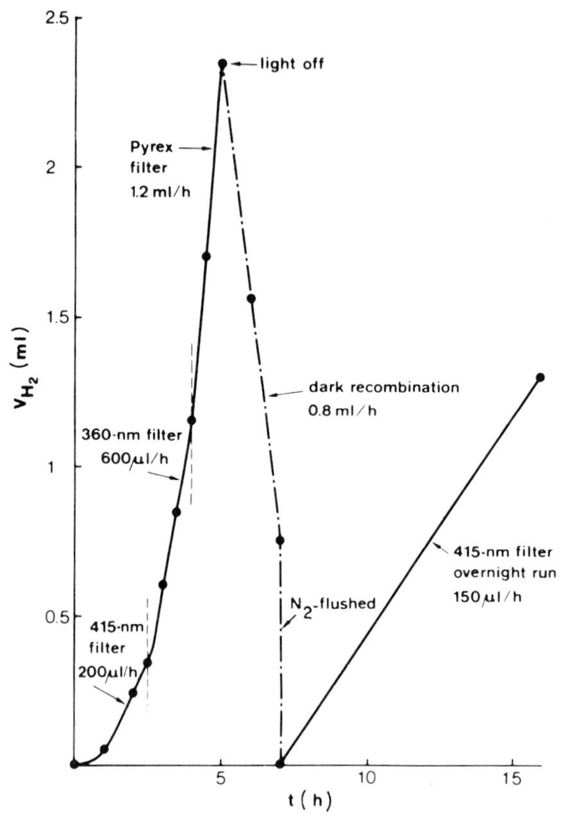

Fig. 19. Light-induced hydrogen evolution in TiO_2–Cr dispersions (Cr doping 570 ppm; annealed at 400°C), loaded with 0.2% RuO_2 and 1% Pt. Filters used as indicated. From Borgarello *et al.* (*195*). Copyright 1982 American Chemical Society.

that is photochemically inactive and, through its insulating and light-absorbing properties, reduces the photoactivity of the TiO_2 particles. It is noteworthy that the optimal doping level and annealing temperature for the particles are similar to those reported for polycrystalline TiO_2 electrodes (*196*). Many reports are available on the locations and position of impurity levels relative to the band gap in semiconductors (*197, 198*).

C. Other Semiconductor Particles

Many semiconductor materials can be prepared and investigated for the just described purposes. Particular interest arises from compounds having a band gap around 2 eV for the reasons outlined previously.

The more common *n*-type semiconductors in this range, such as CdS, CdSe, GaAs, and Si, are known to be unstable (*11–18*). Some of these can be stabilized with appropriate solutions containing S_n^{2-}–S^{2-} and Se_n^{2-}–Se^{2-} (for CdS, CdSe) (*199–201*), through surface modification (*202*), or by covering the surface with a passivating or polymer layer (*203*).

A strong decrease in the photocorrosion of CdS particles has been observed when RuO_2 loading is performed. This catalyst (Pt–CdS–RuO_2) promotes water cleavage (although much less efficiently than the TiO_2 dispersions) through direct band-gap illumination in the visible (*204, 205*); no degradation of the photocatalyst was observed after 60 h of irradiation.

D. Heterostructure Design

This method is an attempt to combine the corrosion resistance of wide-band-gap semiconductors with the good sunlight response of small-band-gap materials. Overcoating with metal oxides (TiO_2, SnO_2, etc.) onto CdS, GaP, InP, Si, etc. has not given encouraging results in macroelectrodes because localized corrosion has been observed (*206, 207*). Nevertheless, if it becomes possible to prepare very thin and regular films of large-band-gap semiconductors on suitable small-band-gap materials, this approach could have an interesting future.

VIII. Outlook

Assuming that the desired goal is the conversion of solar energy to chemical fuels by using properly designed particulate systems, future research work will be directed toward the improvement of the efficiency of the systems and to finding new photocatalysts and sensitizers as well as new and cheaper redox catalysts. Also, the design of devices for separate evolution of the gaseous products must be developed.

Moreover, other oxidation reactions can be taken into account. Among these are the photooxidation of halides, that is, $2Cl^- \rightarrow Cl_2 + 2e^-$ (*208, 209*) and hydrogen sulfide cleavage $H_2S \rightarrow H_2 + S$ (*210*). Processes that use organic materials derived from biomass or available as wastes such as carbohydrates (*211*), ethanol (*212, 213*), and acetic acid (*214, 215*) could also become important. Another interesting possibility is the reduction of CO_2 to obtain methanol and other reduced species (*216, 217*).

The new field concerned with catalysis by polyfunctional semiconductor particles requires new techniques for the characterization of the colloids and aggregates (*218–220*) and for the investigation of the photoprocesses in solution and at the interface (*221, 222*).

It appears that this approach of conversion of light to chemical products is extremely promising in the solution of energy problems as well as in several related fields.

References

1. Marcus, R. J., *Science (Washington, D.C.)* **123,** 399 (1956).
2. Balzani, V., Moggi, L., Manfrin, M. F., Bolletta, F., and Gleria, M., *Science (Washington, D.C.)* **189,** 852 (1975).
3. Calvin, M., *Photochem. Photobiol.* **23,** 425 (1976).
4. Porter, G., and Archer, M., *Interdisc. Sci. Rev.* **1,** 119 (1976).
5. Claesson, S. (ed.), "Photochemical Conversion and Storage of Solar Energy." Swedish Energy Board, Stockholm, 1977.
6. Bolton, J. R. (ed.), "Solar Power and Fuels." Academic Press, New York, 1977.
7. Connolly, J. S. (ed.), "Photochemical Conversion and Storage of Solar Energy." Academic Press, New York, 1981.
8. Grätzel, M., *Acc. Chem. Res.* **14,** 376 (1981).
9. Wrighton, M., *Chem. Eng. News* **57,** 29 (1979).
10. Goodenough, J. B., *Proc. Indian Acad. Sci.* **88,** 69 (1979).
11. Gerischer, H., *Pure Appl. Chem.* **52,** 2649 (1980).
12. Memming, R., *Electrochim. Acta* **25,** 77 (1980).
13. Nozik, A. J., *Annu. Rev. Phys. Chem.* **29,** 189 (1978).
14. Butler, M. A., and Ginley, D. S., *J. Mater. Sci.* **15,** 1 (1980).
15. Rajeswar, K., Singh, P., and DuBow, J., *Electrochim. Acta* **23,** 1117 (1978).
16. Wrighton, M., *Acc. Chem. Res.* **12,** 303 (1979).
17. Heller, A., *Acc. Chem. Res.* **14,** 154 (1981).
18. Maruska, H. P., and Ghosh, A. K., *Sol. Energy* **20,** 443 (1978).
19. Bard, A. J., *J. Photochem.* **10,** 59 (1979).
20. Bard, A. J., *J. Phys. Chem.* **86,** 172 (1982).
21. Kiwi, J., Kalyanasundaram, K., and Grätzel, M., *Struct. Bonding (Berlin)* **49,** 39 (1982).
22. Fujishima, A., and Honda, K., *Nature (London)* **238,** 37 (1972).
23. Schrauzer, G. N., and Guth, T. O., *J. Am. Chem. Soc.* **99,** 7189 (1977).
24. Van Damme, H., and Hall, W. K., *J. Am. Chem. Soc.* **101,** 4373 (1979).
25. Rao, M. V., Rajeshwar, K., Pal Verneker, V. R., and DuBow, J., *J. Phys. Chem.* **84,** 1987 (1980).
26. Sato, S., and White, J. M., *J. Phys. Chem.* **85,** 592 (1981).
27. Kraeutler, B., and Bard, A. J., *J. Am. Chem. Soc.* **100,** 5985 (1978).
28. Dutoit, E. C., Cardon, F., and Gomes, W. P., *Ber. Bunsenges, Phys. Chem.* **80,** 477 (1976).
29. Harris, L. A., Cross, D. R., and Geruster, M. E., *J. Electrochem. Soc.* **124,** 839 (1977).
30. Bocarsly, A. B., Bolts, J. M., Cummings, P. G., and Wrighton, M. S., *Appl. Phys. Lett.* **31,** 568 (1977).
31. Barksdale, J., "Titanium." Ronald Press, New York, 1966.
32. Matijevic, E., Budnik, M., and Maites, L., *J. Colloid Interface Sci.* **61,** 302 (1977).

33. U.K. Patent No. 1, 085,724.
34. Visca, M., and Matijevic, E., *J. Colloid Interface Sci.* **68,** 308 (1979).
35. U.S. Patent No. 1,842,620 and 1,931,380.
36. Bradley, D.C., and Hillyer, M. J., *Trans. Faraday Soc.* **62,** 2367, 2374, 2382 (1966).
37. George, A. B., Murley, R. D., and Place, E. R., *Symp. Faraday Soc.* **7,** 63 (1973).
38. Dolmatov, Y., *J. Appl. Chem. USSR* (*Engl. Transl. of Zh. Prikl. Khim.* [*Leningrad*]) **42,** 1627 (1969).
39. Bobyrenko, Y., and Dolmatov, Y., *J. Appl. Chem. USSR* (*Engl. Transl. of Zh. Prikl. Khim.* [*Leningrad*]) **44,** 1006 (1971).
40. Babko, A., Mazurenko, E., and Nabivanets, I., *Russ. J. Inorg. Chem.* (*Engl. Transl. of Zh. Neorg. Khim.*) **14,** 1091 (1969).
41. Nazarenko, V., Antonivich, V., and Nevskaya, E., *Russ. J. Inorg. Chem.* (*Engl. Transl. of Zh Neorg. Khim.*) **16,** 530 (1971).
42. Benkepkamp, J., and Herrington, K. D., *J. Am. Chem. Soc.* **82,** 3025 (1960).
43. Hixon, A. W., and Fredrickson, R. E. C., *Ind. Eng. Chem.* **37,** 678 (1945).
44. Reeves, R. E., and Blouin, F. A., *J. Am. Chem. Soc.* **76,** 5233 (1954).
45. Jerman, Z., *Colloid Czech. Chem. Commun.* **31,** 2481 (1966).
46. Dolmatov, Y., and Sheinkman, I., *J. Appl. Chem. USSR* (*Engl. Transl. of Zh. Prikl. Khim.* [*Leningrad*]) **43,** 257 (1970).
47. Dolmatov, Y., and Antonova, N., *J. Appl. Chem. USSR* (*Engl. Transl. of Zh. Prikl. Khim.* [*Leningrad*]) **50,** 2328 (1977).
48. Bekkerman, L., Dobrovl'skii, I., and Ivakin, A., *Russ. J. Inorg. Chem.* (*Engl. Transl. of Zh. Neorg. Khim.*) **21,** 223 (1976).
49. Sohnel, O., and Maracek, J., *Proc. Symp. Ind. Crystallogr.* 6th, p. 155 (1975).
50. Duncan, J., and Richards, R., *N.Z. J. Sci.* **19,** 185 (1976).
51. Becker, H., Kein, E., and Rechman, H., *Farbe & Lack* **79,** 779 (1964).
52. Pelizzetti, E., Visca, M., Borgarello, E., Pramauro, E., and Palmas, A., *Chim. Ind.* (*Milan*) **63,** 805 (1981).
53. Bérubé, Y. G., and De Bruyn, P. L., *J. Colloid Interface Sci.* **27,** 305 (1968).
54. Parfitt, G. D., Ramsbotham, J., and Rochester, C. H., *J. Chem. Soc. Faraday Trans.* **67,** 3100 (1971); **68,** 17 (1972).
55. Parfitt, G. D., *Prog. Surf. Membr. Sci.* **11,** 181 (1976).
56. Wiseman, T. J., *in* "Characterization of Powder Surfaces" (G. D. Parfitt, and K. S. W. Sind, eds.), p. 159. Academic Press, New York, 1976.
57. Rives-Arnau, V., and Sheppard, N., *J. Chem. Soc. Faraday Trans. 1* **76,** 394 (1980); **77,** 953 (1981).
58. Boonstra, A. H., and Mutsaers, C. A. H. A., *J. Phys. Chem.* **79,** 2025 (1975).
59. Onishi, Y., and Hamamura, T., *Bull. Chem. Soc. Jpn.* **43,** 996 (1970).
60. Onishi, Y., *Bull. Chem. Soc. Jpn.* **45,** 922 (1972).
61. Gorbunova, E. V., Deev, Y. S. and Ryabov, E. A., *Vysokomol. Soedin, Ser. A* **23,** 811 (1981).
62. Greene, H. L., and Kapoor, A., *Chem. Eng. Commun.* **7,** 169 (1980).
63. Engels, S., Freitag, B., Moerke, W., Roschke, W., and Wilde, M., *Z. Anorg. Allg. Chem.* **474,** 209 (1981).
64. Jaeger, C. D., and Bard, A. J., *J. Phys. Chem.* **83,** 3146 (1979).
65. Dunn, W. W., Aikawa, Y., and Bard, A. J., *J. Am. Chem. Soc.* **103,** 3456 (1981).
66. Kawai, T., and Sakata, T., *Nature* (*London*) **282,** 283 (1979).
67. Sato, S., and White, J. M., *J. Phys. Chem.* **85,** 336 (1981).
68. Kruczynski, L., Gesser, H. D., Turner, C. W., and Speers, E. A., *Nature* (*London*) **291,** 399 (1981).

69. Bérubé, Y. G., and De Bruyn, P. L., *J. Colloid Interface Sci.* **28,** 92 (1969).
70. Ahmed, S. M., and Maksimov, D., *J. Colloid Interface Sci.* **29,** 97 (1969).
71. Bobyrenko, Y. Y., and Guzairova, A. A., *Zh. Prikl. Khim.* (*Leningrad*) **45,** 2265 (1972).
72. Schindler, P. W., and Gamsjager, H., *Kolloid Z. Z. Polym.* **250,** 759 (1972).
73. Yates, D. E., James, R. O., and Healy, T. W., *J. Chem. Soc. Faraday Trans. 1* **76,** 1 (1980).
74. Yates, D. E., and Healy, T. W., *J. Chem. Soc. Faraday Trans. 1* **76,** 9 (1980).
75. Tschapek, M., Wasowski, C., and Torres Sanchez, R. M., *J. Electroanal. Chem.* **74,** 167 (1967).
76. Ashida, M., Sasaki, M., Kan, H., Yasunaga, T., Hachiya, K., and Inoue, T., *J. Colloid Interface Sci.* **67,** 219 (1978).
77. Parks, G. A., *Chem. Rev.* **65,** 177 (1965).
78. Parfitt, G. D., Ramsbotham, J., and Rochester, C. H., *J. Colloid Interface Sci.* **41,** 437 (1972).
79. James, R. O., and Healy, T. W., *J. Colloid Interface Sci.* **40,** 53 (1972).
80. Wiese, G. R., and Healy, T. W., *J. Colloid Interface Sci.* **51,** 427 (1975).
81. Tewari, P. H., and Lee, W., *J. Colloid Interface Sci.* **52,** 77 (1975).
82. Cornell, R. H., Posner, A. M., and Quirk, J. P., *J. Colloid Interface Sci.* **53,** 6 (1975).
83. Furlong, D. N., and Parfitt, G. D., *J. Colloid Interface Sci.* **65,** 548 (1978).
84. Ahmed, S. M., El Aasser, M. S., Pauli, G. H., Poehlein, G. W., and Vanderhoff, J. W., *J. Colloid Interface Sci.* **73,** 388 (1980).
85. Morterra, C., Ghiotti, G., Garrone, E., and Fisicaro, E., *J. Chem. Soc. Faraday Trans. 1* **76,** 2102 (1980).
86. Morterra, C., Chiorino, A., Zecchina, A., and Fisicaro, E., *Gazz. Chim. Ital.* **109,** 691 (1979).
87. Garbassi, F., Mello Ceresa, E., and Visca, M., *J. Phys. Chem. Solids,* submitted.
88. Hammet, L. P., and Deyrup, A. J., *J. Am. Chem. Soc.* **54,** 2721 (1932).
89. Walling, C., *J. Am. Chem. Soc.* **72,** 1164 (1950).
90. Tanabe, K., "Solids Acids and Bases." Academic Press, New York, 1970.
91. Wadsley, A. D., Acta Crystallogr. **14,** 660 (1961).
92. Kronos Titanium Co., "Kronos Guide." Kölnische Verlag, Cologne, 1968.
93. Kerker, M., "The Scattering of Light and Other Electromagnetic Radiations." Academic Press, New York, 1969.
94. Cox, S. M., and Lichtman, D., *Surf. Sci.* **54,** 675 (1976).
95. Bickley, R. I., and Stone, F. S., *J. Catal.* **31,** 389 (1973).
96. Cunningham, J., Goold, E. L., and Healy, E. M., *J. Chem. Soc. Faraday Trans. 1* **75,** 305 (1979).
97. Courbon, H., Formenti, M., and Pichat, P., *J. Phys. Chem.* **81,** 550 (1977).
98. Mitton, P. B., *in* "Pigment Handbook" (T. C. Patton, ed.), Vol. 3, p. 289. Wiley, New York, 1973.
99. Memming, R., and Schroppel, F., *Chem. Phys. Lett.* **62,** 207 (1979).
100. Fleischauer, P. D., and Allen, J. K., *J. Phys. Chem.* **82,** 432 (1978).
101. Hammett, A., Dare-Edwards, M. P., Wright, R. D., Seddon, K. R., and Goodenough, J. B., *J. Phys. Chem.* **83,** 3200 (1979).
102. Rauh, R. D., Buzby, J. M., Reise, T. F., and Alkaitis, S. A., *J. Phys. Chem.* **83,** 2221 (1979).
103. Maruska, H. P., and Ghosh, A. K., *Sol. Energy Mater.* **1,** 237 (1979).
104. Guruswami, V., and Bockris, J. O. M., *Sol. Energy Mater.* **1,** 441 (1979).

105. Matsumoto, Y., Kurimoto, J., Amagasaki, Y., and Sato, E., *J. Electrochem. Soc.* **127,** 2148 (1980).
106. Mackor, A., and Blasse, G., *Chem. Phys. Lett.* **77,** 6 (1981).
107. Blasse, G., and Dirksen, G. J., *Chem. Phys. Lett.* **77,** 9 (1981).
108. Lam, R. U. E.'t, deHaart, L. G. J., Wiesma, A. W., Blasse, G., Tinnemans, A. H. A., and Mackor, A., *Mater. Res. Bull.* **16,** 1593 (1980).
109. Goodenough, J. B., *Adv. Chem. Ser.,* No. 186, p. 113 (1980).
110. Campet, G., Verniolle, J., Doumerc, J. P., and Claverie, J., *Mater. Res. Bull.* **15,** 1135 (1980).
111. Porter, G., *Proc. R. Soc. London Ser. A* **362,** 281 (1978).
112. Balzani, V., Bolletta, F., Gandolfi, M. T., and Maestri, M., *Top. Curr. Chem.* **75,** 1 (1978).
113. Calvin, M., *Acc. Chem. Res.* **11,** 369 (1978).
114. Sutin, N., *J. Photochem.* **10,** 19 (1979).
115. Whitten, D., *Acc. Chem. Res.* **13,** 83 (1980).
116. Adamson, A. W., and Gafney, H. D., *J. Am. Chem. Soc.* **94,** 8238 (1972).
117. Bock, C. R., Meyer, T. J., and Whitten, D. G., *J. Am. Chem. Soc.* **96,** 4710 (1974).
118. Kalyanasundaram, K., and Grätzel, M., *Helv. Chim. Acta* **63,** 478 (1980).
119. Moradpour, A., Amouyal, E., Keller, P., and Kagan, H., *Nouv. J. Chim.* **2,** 547 (1978).
120. Kalyanasundaram, K., Kiwi, J., and Grätzel, M., *Helv. Chim. Acta* **61,** 2720 (1978).
121. Lehn, J. M., and Sauvage, J. P., *Nouv. J. Chim.* **1,** 449 (1977).
122. Chan, S. F., Chou, M., Creutz, C., Matsubara, T., and Sutin, N., *J. Am. Chem. Soc.* **103,** 369 (1981).
123. Koryakin, B. V., Dzhabier, T. S., and Shilov, A. E., *Dokl. Akad. Nauk SSSR Ser. Khim.* **298,** 620 (1977).
124. Thomas, J. K., *Chem. Rev.* **80,** 283 (1980).
125. Grätzel, M., *in* "Micellization and Microemulsions" (K. L. Mittal, ed.), Vol. 2, p. 531. Plenum, New York, 1977.
126. Pelizzetti, E., *in* "Energy Storage" (J. Silverman, ed.), p. 441. Pergamon, New York, 1980.
127. Pelizzetti, E., and Pramauro, E., *Inorg. Chem.* **19,** 1407 (1980).
128. Lachish, U., Ottolenghi, M., and Rabani, J., *J. Am. Chem. Soc.* **99,** 8062 (1977).
129. Meisel, D., Matheson, M., and Rabani, J., *J. Am. Chem. Soc.* **100,** 117 (1978).
130. Kiwi, J., and Grätzel, M., *J. Am. Chem. Soc.* **100,** 6314 (1978).
131. Willner, I., Ford, W. E., Otvos, J. W., and Calvin, M., *Nature (London)* **280,** 823 (1979).
132. Jones, C. A., Weaner, L. E., and Mackay, R. A., *J. Phys. Chem.* **84,** 1495 (1980).
133. Meisel, D., and Matheson, M. S., *J. Am. Chem. Soc.* **99,** 6577 (1977).
134. Meyerstein, D., Rabani, J., Matheson, M. S., and Meisel, D., *J. Phys. Chem.* **82,** 1879 (1978).
135. Kelder, S., and Rabani, J., *J. Phys. Chem.* **85,** 1637 (1981).
136. Calvin, M., *Int. J. Energy Res.* **3,** 73 (1979).
137. Fendler, J., *J. Phys. Chem.* **84,** 1485 (1980).
138. Kuhn, H., *J. Photochem.* **10,** 111 (1979).
139. Tien, H. Ti, and Karvaly, B., *in* "Solar Power and Fuels" (J. R. Bolton, ed.), p. 167. Academic Press, New York, 1977.
140. Willner, I., Yang, J. M., Leane, C., Otvos, J. W., and Calvin, M., *J. Phys. Chem.* **85,** 3277 (1981).
141. Leane, C., Willner, I., Otvos, J. W., and Calvin, M., *Proc. Natl. Acad. Sci. USA* **78,** 5928 (1981).

142. Wagner, C., and Traud, W., *Z. Elektrochem.* **44,** 397 (1938).
143. Henglein, A., *J. Phys. Chem.* **83,** 2200, 2958 (1979); **84,** 3461 (1980).
144. Henglein, A., and Lilie, J., *J. Am. Chem. Soc.* **103,** 1059 (1981).
145. Lee, P. C., and Meisel, D., *J. Catal.* **70,** 160 (1981).
146. Meisel, D., *J. Am. Chem. Soc.* **101,** 6133 (1979).
147. Kopple, K., Meyerstein, D., and Meisel, D., *J. Phys. Chem.* **84,** 870 (1980).
148. Meisel, D., Mulac, W. A., and Matheson, M. S., *J. Phys. Chem.* **85,** 179 (1981).
149. Westerhausen, J., Henglein, A., and Lilie, J., *Ber. Bunsenges. Phys. Chem.* **85,** 182 (1981).
150. Kalyanasundaram, K., Kiwi, J., and Grätzel, M., *Helv. Chim. Acta* **61,** 2720 (1978).
151. Moradpour, A., Amouyal, E., Keller, P., and Kagan, H., *Nouv. J. Chim.* **2,** 547 (1978).
152. Kirsch, M., Lehn, J. M., and Sauvage, J. P., *Helv. Chim. Acta* **62,** 1345 (1979).
153. Kiwi, J., and Grätzel, M., *J. Am. Chem. Soc.* **101,** 7214 (1979).
154. Brugger, P. A., Cuendet, P., and Grätzel, M. *J. Am. Chem. Soc.* **103,** 2923 (1981).
155. Keller, P., and Moradpour, A., *J. Am. Chem. Soc.* **102,** 7193 (1980).
156. Matheson, M. S., Lee, P. C., Meisel, D., and Pelizzetti, E., *J. Phys. Chem.* **87,** 394 (1983).
157. Miller, D., Bard, A. J., McLendon, G., and Ferguson, J., *J. Am. Chem. Soc.* **103,** 5336 (1981).
158. Miller, D., and McLendon, G., *J. Am. Chem. Soc.* **103,** 6791 (1981).
159. Stol, R. J., Von Helden, A. K., and De Bruyn, P. L., *J. Colloid Interface Sci.* **57,** 115 (1976).
160. Dousma, M., and De Bruyn, P. L., *J. Colloid Interface Sci.* **72,** 314 (1979).
161. Hanson, F. V., and Boudart, M., *J. Catal.* **53,** 56 (1978).
162. Kiwi, J., and Grätzel, M., *Angew. Chem. Int. Ed. Engl.* **17,** 860 (1978).
163. Kiwi, J., and Grätzel, M., *Angew. Chem. Int. Ed. Engl.* **18,** 624 (1979).
164. Lehn, J. M., Sauvage, J. P., and Ziessel, R., *Nouv. J. Chim.* **3,** 423 (1979).
165. Kalyanasundaram, K., Micic, O., Pramauro, E., and Grätzel, M., *Helv. Chim. Acta* **62,** 2432 (1979).
166. Harriman, A., Porter, G., and Walters, P., *J. Chem. Soc. Faraday Trans. 2* **77,** 2373 (1981).
167. Shafirovich, V. Ya., Khannanov, N. K., and Shilov, A. E., *J. Inorg. Biochem.* **115,** 113 (1981).
168. Pramauro, E., and Pelizzetti, E., *Inorg. Chim. Acta Lett.* **45,** 131 (1980).
169. Minero, C., Lorenzi, E., Pramauro, E., and Pelizzetti, E., *Inorg. Chim. Acta,* submitted.
170. Humphrey-Baker, R., Lilie, J., and Grätzel, M., *J. Am. Chem. Soc.* **104,** 422 (1982).
171. Kalyanasundaram, K., and Grätzel, M., *Angew. Chem. Int. Ed. Engl.* **18,** 701 (1979).
172. Kiwi, J., Borgarello, E., Pelizzetti, E., Visca, M., and Grätzel, M., *Angew. Chem. Int. Ed. Engl.* **19,** 646 (1980).
173. James, R. O., and Healy, T. W., *J. Colloid Interface Sci.* **40,** 42; 53, and 65 (1972).
174. Matijevic, E., *J. Colloid Interface Sci.* **43,** 217 (1973); **58,** 374 (1977).
175. Howard, P. B., and Parfitt, G. D., *Croat. Chem. Acta* **50,** 15 (1977).
176. Furlong, D. N., Sing, K. S. W., and Parfitt, G. D., *J. Colloid Interface Sci.* **69,** 409 (1979).
177. Krautler, B., and Bard, A. J., *J. Am. Chem. Soc.* **100,** 4318 (1978).
178. Duonghong, D., Borgarello, E., and Grätzel, M., *J. Am. Chem. Soc.* **103,** 4685 (1981).
179. Turkevich, J., Aika, K., Ban, L. L., Okura, I., and Namba, S., *J. Res. Inst. Catal. Hokkaido Univ.* **24,** 54 (1976).

180. Kolakowski, J. E., and Matijevic, E., *J. Chem. Soc. Faraday Trans. 1* **75,** 65 (1979).
181. Kuo, R., and Matijevic, E., *J. Chem. Soc. Faraday Trans. 1* **75,** 2014 (1979).
182. Hausen, F. K., and Matijevic, E., *J. Chem. Soc. Faraday Trans. 1* **76,** 1240 (1980).
183. Shenkel, J. H., and Kitchener, J. A., *Nature (London)* **182,** 131 (1958).
184. Van den Hul, H. J., and Vanderhoff, J. W., *J. Electroanal. Chem.* **37,** 161 (1972).
185. Borgarello, E., Kiwi, J., Pelizzetti, E., Visca, M., and Grätzel, M., *J. Am. Chem. Soc.* **103,** 6324 (1981).
186. Borgarello, E., Kiwi, J., Pelizzetti, E., Visca, M., and Grätzel, M., *Nature (London)* **289,** 158 (1980).
187. McLeod, G. L., *Photogr. Sci. Eng.* **13,** 93 (1969).
188. Clark, W. D. K., and Sutin, N., *J. Am. Chem. Soc.* **99,** 4676 (1977).
189. Jaeger, C. D., Fan, F. R. F., and Bard, A. J., *J. Am. Chem. Soc.* **102,** 2592 (1980).
190. Tinnemans, A. H. A., and Mackor, A., *Rec. Trav. Chim. Pays-Bas* **100,** 295 (1981).
191. Borgarello, E., Kalyanasundaram, K., Okuno, Y., and Grätzel, M., *Helv. Chim. Acta* **64,** 1937 (1981).
192. Minero, C., Pelizzetti, E., Barni, E., and Savazino, P., unpublished data.
193. Salvador, P., *Sol. Energy Mater.* **2,** 413 (1980).
194. Houlihan, J. F., Armitage, D. B., Hoovler, T., Bonaquist, D., Madacsi, D. P., and Mulay, L. N., *Mater. Res. Bull.* **13,** 1205 (1978).
195. Borgarello, E., Kiwi, J., Grätzel, M., Pelizzetti, E., and Visca, M., *J. Am. Chem. Soc.* **104,** 2996 (1982).
196. Mannier, A., and Augustynski, J., *J. Electrochem. Soc.* **127,** 1576 (1980).
197. Ghosh, A. K., Lauer, R. B., and Addiss, R. R., *J. Phys. Rev. B* **8,** 4842 (1973).
198. Stokowski, S. E., and Schawlow, A. L., *Phys. Rev.* **178,** 457 (1969).
199. Ellis, A. B., Kaiser, S. W., and Wrighton, M. S., *J. Am. Chem. Soc.* **98,** 1635 (1976).
200. Hodes, G., Manassen, J., and Cahen, D., *Nature (London)* **261,** 403 (1976).
201. Miller, B., and Heller, A., *Nature (London)* **262,** 680 (1976).
202. Bolts, J. M., Bocarsly, A. B., Palazzotto, M. C., Walton, E. G., Lewis, N. S., and Wrighton, M. S., *J. Am. Chem. Soc.* **101,** 1378 (1979).
203. Cooper, G., Noufi, R., Frank, A. J., and Nozik, A. J., *Nature (London)* **295,** 578 (1982).
204. Kalyanasundaram, K., Borgarello, E., and Grätzel, M., *Helv. Chim. Acta* **64,** 362 (1981).
205. Borgarello, E., Grätzel, M., Balducci, L., Visca, M., and Pelizzetti, E., unpublished data.
206. Tomkiewicz, M., and Woodall, J. M., *J. Electrochem. Soc.* **124,** 1436 (1977).
207. Kohl, P. A., Frank, S. N., and Bard, A. J., *J. Electrochem. Soc.* **124,** 225 (1977).
208. Kiwi, J., and Grätzel, M., *Chem. Phys. Lett.* **78,** 241 (1981).
209. Reichman, B., and Byvik, C. E., *J. Phys. Chem.* **85,** 2255 (1981).
210. Borgarello, E., Kalyanasundaram, K., Grätzel, M., and Pelizzetti, E., *Helv. Chim. Acta* **65,** 243 (1982).
211. Kawai, T., and Sakata, T., *Nature (London)* **286,** 474 (1980).
212. Sakata, T., and Kawai, T., *Chem. Phys. Lett.* **80,** 341 (1981).
213. Borgarello, E., and Pelizzetti, E., *Chim. Ind. (Milan)*, submitted.
214. Krautler, B., and Bard, A. J., *J. Am. Chem. Soc.* **99,** 7729 (1977).
215. Borgarello, E., Minero, C., Grätzel, M., Pelizzetti, E., and Visca, M., unpublished data.
216. Inoue, T., Fujishima, A., Konishi, S., and Honda, K., *Nature (London)* **277,** 637 (1979).

217. Halmann, M., *Nature* (*London*) **275,** 115 (1978).
218. Dunn, W. W., Aikawa, Y., and Bard, A. J., *J. Am. Chem. Soc.* **103,** 3456 (1981).
219. De Pauw, E., and Marien, J., *J. Phys. Chem.* **85,** 3550 (1981).
220. Huizinga, T., and Prins, R., *J. Phys. Chem.* **85,** 2156 (1981).
221. Duonghong, D., Ramsden, J., and Grätzel, M., *J. Am. Chem. Soc.* **104,** 2977 (1982).
222. Grätzel, M., and Frank, A. J., *J. Phys. Chem.* **86,** 2964 (1982).

9 Examples for Photogeneration of Hydrogen and Oxygen from Water

J. Kiwi

Institut de Chimie Physique
École Polytechnique Fédérale
Lausanne, Switzerland

I. Evolution of H_2 Induced by Visible Light in Sacrificial Systems

Considerable effort has been directed toward finding chemical processes that are capable of quantum storage of light energy. It has been shown through a number of studies (*39, 70, 77*) that certain noble-metal dispersions are suitable in mediating light-induced hydrogen and oxygen evolution from water. If solar energy is stored in the form of an activated chromophore S* that absorbs light in the visible, then processes could be started leading to endoergic storage of visible energy as shown previously (*53, 63, 64*).

With a sensitizer S such as $Ru(bipy)_3^{2+}$ (ruthenium tris-bipyridyl) and a relay MV^{2+} (1,1-dimethyl-4,4′-bipyridinium^{2+}), one can obtain the separation of charge ($S^+ + R^-$) if the inverse electron transfer between S^+ and R^-

ENERGY RESOURCES THROUGH PHOTOCHEMISTRY AND CATALYSIS

ISBN 0-12-295720-2

can be suppressed. The stabilization of the redox intermediates that can readily give H_2 and O_2 via reactions (1) and (2) could be achieved by multiphase systems, such as micelles, microemulsions, and vesicles (*61, 66, 117*). Metal or metal oxide dispersions on polymer or inorganic materials could be used as catalysts. For the metal colloid to be effective as

$$2S^+ + H_2O \xrightarrow{\text{cat 1}} \tfrac{1}{2}O_2 + 2H^+ + 2S \quad (1)$$

$$2R^- + 2H_2O \xrightarrow{\text{cat 2}} H_2 + 2OH^- + 2R \quad (2)$$

catalyst it must overcome the following difficulties: (a) the rapid nature of reverse electron transfer between S^+ and R^-, (b) the destructive reactions of the intermediate radicals produced during the reaction, and (c) the high selectivity required for two-electron transfer because two protons are reduced simultaneously to form a molecule of H_2. Also, reaction (2) should not be poisoned over many cycles and should be quantitative. The actual form of the redox catalyst can be homogeneous (as finely dispersed colloidal Pt particles) or heterogeneous in the form of pellets on a solid support. Colloidal particles have the advantage of providing large surface area for catalysis. A centrifuged colloidal Pt catalyst stabilized by polyvinyl alcohol (PVA) showed exceptionally high activity in promoting hydrogen evolution from water via

$$2MV^+ + H_2O \xrightarrow{\text{Pt}} H_2 + 2OH^- + 2MV^{2+} \quad (3)$$

where MV^+ stands for reduced methylviologen.

Among the various acceptor relays that have been examined, methylviologen has been found to be suitable because of its redox potential, high solubility in aqueous solutions, and the ease of reduction. This process can be catalyzed by Pt dispersions (*40*). This is an example of the process described by reaction (2). The experimental results are shown in Fig. 1; many colloidal dispersions have been prepared with different protective agents. The particular polyvinyl alcohol–Pt dispersion was prepared following Nord's procedure (*29*). By varying the time and speed of centrifugation, Pt–PVA colloids of different particle sizes were prepared and are presented in Fig. 2. To compare the efficiency of these preparations, the same Pt concentration in solution had to be maintained. The losses incurred by centrifugation were compensated for by increasing the amount of the starting material. The particle size of the kinetic unit Pt polymer as measured by quasi-elastic scattering (qes) techniques has a definite influence on the activity of the catalyst. This is illustrated in Fig. 2, which shows the relationship between the radius of Pt–PVA aggregates and their activity in producing H_2 under irradiation of the solutions. It is seen that the decrease of the radius from 500 to 100 Å leads to a drastic increase in

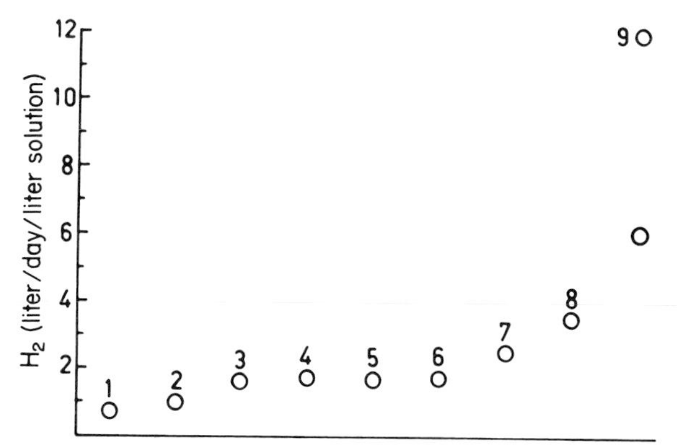

Fig. 1. (a) Evolution of H_2 of a solution of 4×10^{-5} *M* $Ru(bipy)_3^{2+}$, 2×10^3 *M* MV^{2+}, and 3×10^{-2} *M* EDTA at pH 5, under irradiation with light ($\lambda > 400$ nm) in the presence of various catalysts as indicated. For preparation of catalyst 9, the higher point contains 3.5 times the amount of Pt as the lower point. Catalysts used: 1, Pt–citric acid; 2, Pt–polyvinyl pyrrolydon; 3, Pt–gelatin; 4, Pt–TiO_2; 5, Pt–phosphorus; 6, Pt reduced by $NH_4 \cdot HSO_4$; 7, Pt–cetyltrimethylammonium chloride centrifugated; 8, Pt–polyacrylic acid hydrazide; 9, Pt–polyvinyl alcohol 60,000.

the hydrogen-evolution rate, which can be as high as 12 liter/day/liter of solution. The size reported here refers to the kinetic unit of macromolecular dimensions, and the present observations are related to the diffusion constant of random aggregates that have been evaluated by the Einstein–Stokes relation (*19*). The size effects are explained by the fact that the Pt

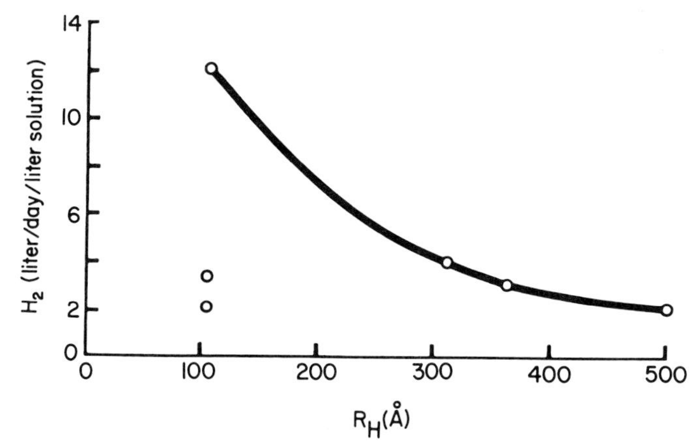

Fig. 2. Correlation for the observed yields for H_2 in the system 4×10^{-5} *M* $Ru(bipy)_3^{2+}$, 10^{-3} *M* MV^{2+}, and 3×10^{-2} *M* EDTA at pH 5 and the radius of the catalysts employed. The lower two points at 110 Å represent 0.2 mg of Pt and 0.3 mg of Pt per 25 ml of solution, respectively. The catalyst used was Pt–polyvinyl alcohol 60,000.

particles intervene as microelectrodes in the hydrogen-evolution reaction. The role of these microelectrodes is to couple the anodic oxidation of the reduced relay MV^+ as shown by reaction (3) with H_2 generation from water.

Figure 3 shows how the electron transfer from MV^+ to the Pt particles leads subsequently to stoichiometric formation of H_2 as expressed in Eq.

$$MV^+ + Pt \longrightarrow MV^{2+} + Pt^- \tag{4}$$

(2); it also shows the increase of the MV^{2+} concentration in the absence of catalyst as a function of time of irradiation. The incident light was restricted in wavelength to 420–480 nm by means of a K-45 Balzers interference filter. The actinometry carried out here under the same conditions of illumination gave a value of 0.60×10^{17} quanta/min. The linear portion in Fig. 3 gives a value $\phi(MV^+) = 0.30$ at irradiation times of less than 4 min. This compares favorably with $\phi(MV^+) = 0.15$–0.30 derived from laser photolysis experiments (*53*). The H_2 yields show an induction period of up to 1 min. This may be because sufficient MV^+ is not yet formed to subsequently generate H_2. Also, initially at very low levels of H_2 produced, the latter is retained in the liquid. After this period, the H_2 quantum yield remains at $\phi(H_2) = 0.13 \pm 10\%$ until the EDTA is consumed. We conclude that, in the presence of suitable redox catalyst, hydrogen formation from water can occur rapidly and quantitatively, even if the driving force of the reaction, as in the case for MV^+, amounts to only several millivolts.

The choice of the catalytic material is based on the same considerations that apply to electrocatalytic reagents used on macroelectrodes: the exchange current densities for electron transfer must be high. (*126;* Chapter

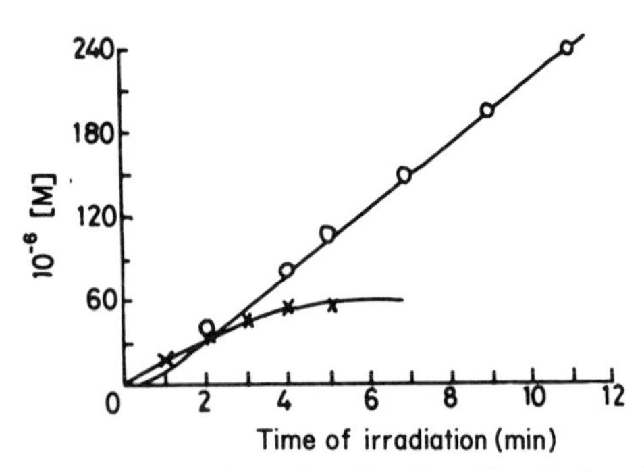

Fig. 3. Dependence of maximal MV^+ formed on duration of irradiation, using an interference filter at 450 nm. Concomitant dependence of H_2 formed under the same conditions of illumination in the system 4×10^{-5} *M* $Ru(bipy)_3^{2+}$, 2×10^{-3} *M* MV^{2+}, and 3×10^{-2} *M* EDTA at pH 5. ○, H_2 (Pt–PVA catalyst); X, MV^+ (no catalyst).

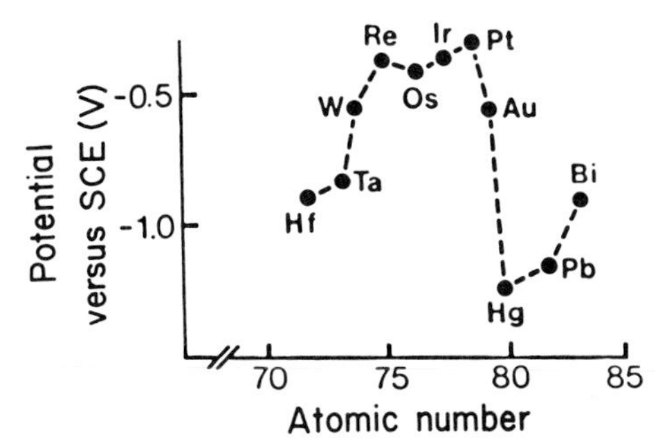

Fig. 4. (a) Periodic dependence of overpotential at a current density of 2 mA/cm² for hydrogen evolution from 0.1 *N* H_2SO_4 at 80°C on atomic number.

4, this volume). Colloidal platinum appears to be a suitable candidate to mediate reaction (3). Figure 4 shows the periodic dependence of overpotential at current density of 2 mA/cm². Hydrogen evolution takes place in 0.1 *N* H_2SO_4. The lowest overpotential is shown to exist for Pt metal (*59*). The very high catalytic activity of these Pt particles enables one to catalyze H_2 evolution from reactive free radicals in competition with side reactions such as disproportionation in many other systems (*4, 38, 54*). A further advantage of these finely divided platinum dispersions is that the solutions remain completely transparent even at high catalyst concentration. Good dispersion of the catalyst is essential for good catalytic activity. For optimal use of Pt, a preferred size of Pt cluster can be predicted. For catalytic activity of exposed atoms, this would take place around point C in Fig. 5 (*116*). Above ~75 atoms the electronic structure of the cluster is identical to the bulk metal (*2*).

Figure 6 presents the hydrogen yields obtained when a flask with 25 cm³ of solution was irradiated with a 450-W Xe lamp in the visible region under the conditions described by Brugger *et al.* (*10*). The pH of the solution was 4.7 (buffered with tampon potassium phthalate 10^{-2} *M*). A

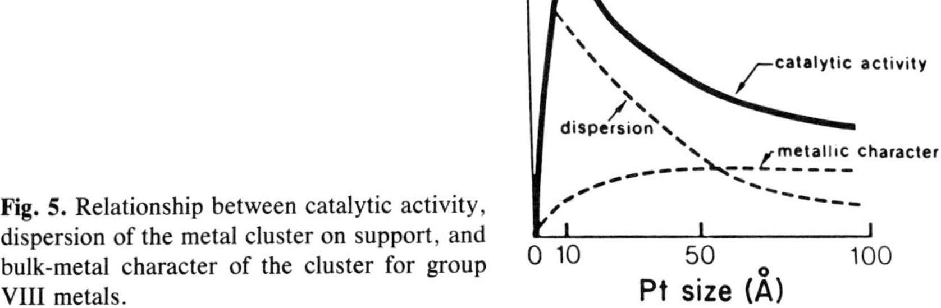

Fig. 5. Relationship between catalytic activity, dispersion of the metal cluster on support, and bulk-metal character of the cluster for group VIII metals.

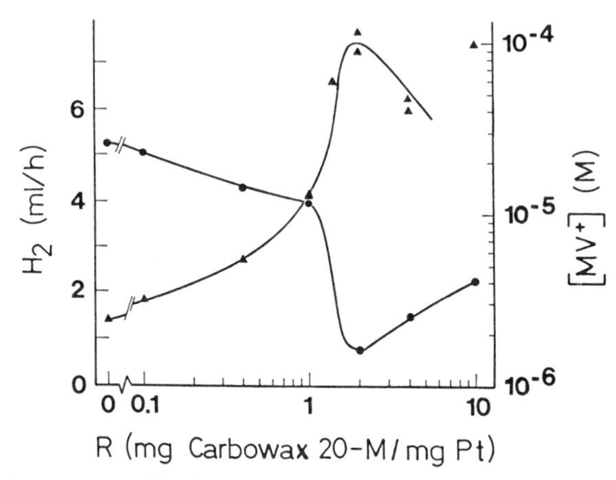

Fig. 6. Effect of mass ratio of Carbowax-20M–platinum on (a) the rate of H_2 generation and (b) the photostationary concentration of MV^+ measured 30 s after the beginning of photolysis. [Pt] = 6 mg/liter; other conditions as in Fig. 1.

change in pH would move the potential for H^+ reduction to less favorable values for such a process. When Carbowax was used as protective polymer, it showed a hydrogen output of 8 ml/h for 25 cm^3 of irradiated solution. This rate of H_2 generation can be sustained over a long time, and the effect of Carbowax concentration on the rate of light-induced H_2 evolution is shown in Fig. 6. On increasing the mass ratio of polymer to platinum (R) from 0 to 2, the hydrogen output augments by a factor of more than 4. At the same time, the MV^+ level present under photostationary conditions decreases. This effect is particularly pronounced between R values of 1 and 2, at which the MV^+ concentration diminishes abruptly by a factor of 8. At R values above 2, one notes a decrease in the hydrogen output concomitant with a rise in the MV^+ level present in the photostationary state. Apparently, at very high Carbowax concentrations, the coating of the Pt surface by the polymer blocks the access of the electron relay to the active sites. At R values below 1, the Pt sol is unstable as was shown previously. Larger aggregates are formed, which leads to a decreased activity under these conditions. With the Carbowax–Pt ratio (R = 2.7) kept constant, the effect of catalyst concentration on the hydrogen evolution rate was examined, and the results are presented in Fig. 7. The hydrogen output under photostationary conditions rises steeply with Pt concentration up to 1.4 mg of Pt/liter, and from there on further augmentation is relatively slow. At higher Pt levels the H_2 evolution rate decreases again. This effect is caused by absorption of light by the Pt particles. The extinction coefficient of the Pt solution is $2.3 \times 10^3/M$/cm at 450 nm. At the break point, the rate is already astonishingly high in view of

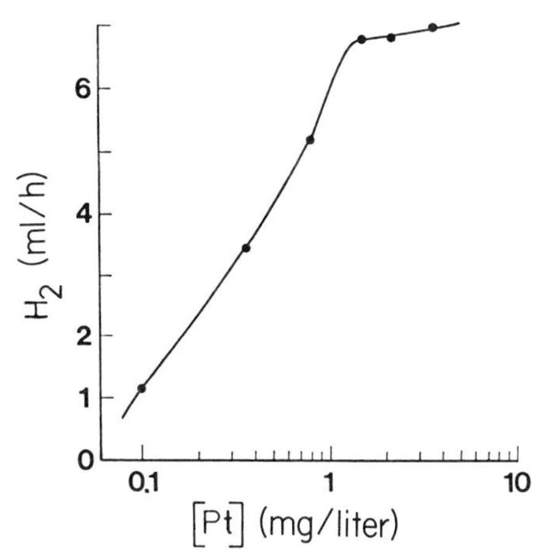

Fig. 7. Effect of Pt–Carbowax concentration on the H_2-generation rate (R = 2.7), $Ru(bipy)_3^{2+} = 4 \times 10^{-5}$ M, $MV^{2-} = 2 \times 10^{-3}$ M, EDTA = 10^{-2} M, potassium hydrogen phthalate buffer = 5×10^{-3} M at pH 4.5.

the very small Pt concentration ($\sim 7 \times 10^{-6}$ M) present in solution. In fact, this figure can even be further improved because at the $Ru(bipy)_3^{2+}$ concentration employed, only a fraction of the incident light (the maximum percentage is 86 for λ = 452 nm) is absorbed by the solution This condition was selected intentionally for our kinetic studies to avoid inhomogeneities resulting from complete light absorption over a small path length. Given these facts, the hydrogen evolution obtained with 1.4 mg of Pt–Carbowax per liter becomes comparable with that observed for Pt–PVA at ~100 mg of Pt/liter. Figure 6 shows that heterogeneous electron transfer from reduced methylviologen (MV^+) to the Pt particle may be enhanced by the protective agent at certain defined concentrations. In the most favorable situation leading to high H_2 evolution, the concentration of MV^+ is very small, indicating that the reoxidation through electron transfer to the Pt particle occurs very rapidly. The process of H_2 generation in sacrificial systems would then involve scavenging of MV^+ by the Pt particles through hydrophobic interactions with the protective agent and subsequent charge transfer and water reduction on the Pt particles. The 32-Å Pt particles, used in these experiments, have been produced in aqueous solutions by citrate reduction of H_2PtCl_6 (*120*). A particle size of 32 Å is favorable to catalytic processes such as the one under study because it ensures high dispersion, approximately 50% (*5, 8*), and the necessary metallic character (>15 Å) for electron transfer through the catalyst. The process of H_2 generation slows down once the solution pH rises as a result of exhaustion of the buffer. The quantum yield for H_2 production is about 0.08 in the sacrificial system shown in Table I.

TABLE I

Scheme for H_2 Evolution Induced by Visible Light in Sacrificial Systems

$$Ru(bipy)_3^{2+} \xrightarrow{h\nu} Ru(bipy)_3^{2+*}$$
$$Ru(bipy)_3^{2+*} \longrightarrow Ru(bipy)^{2+} + h\nu(Fl) + \text{radiationless deactivation}$$
$$Ru(bipy)_3^{2+} + MV^{++} \longrightarrow Ru(bipy)_3^{3+} + MV^{+}$$
$$Ru(bipy)^{3+} + EDTA \longrightarrow Ru(bipy)_3^{2+} + EDTA^{+}$$
$$MV^{+} + H_2O \xrightarrow[h\nu \geq 400\ nm]{Pt\ cat.} \tfrac{1}{2}H_2 + OH^{-} + MV^{2+}$$

Structure of Polymer-Protected Platinum Particles

We shall discuss in the following section the mode of protection of Pt particles by Carbowax-20M. Carbowax-20M consists of two units of $(HC_2—CH_2O)_{600}$ with OH terminals that are connected through a hydrophobic portion. The stabilization of the Pt dispersion is thought to be a result of two main effects (*18*): (a) good adsorption by anchoring of the hydrophobic portion of the polymer to a predominantly hydrophobic adsorbate such as Pt and (b) strong repulsion as a result of the two large hydrophilic foils extending into the medium. Steric stabilization seems to be very effective for metal polymers in this case because coagulation with 0.24% aqueous NaCl solution of a dispersion containing 30 mg/liter Pt and 100 mg/liter of polymer did not exceed 1% after 1.5 days and 5% after 15 days (*132*). Small-angle neutron-scattering experiments have been carried out on different compositions of Carbowax 20M–Pt, and the results are shown in Table II (*69*).

Small-angle neutron-diffraction scattering measurements have been carried out in the D-11 detector of the neutron beam of a high-flux reactor. A neutron wavelength of 6 Å from the cold source of this reactor and a detector distance of 10 meters were used. The angular resolution attained in this way for the observed signals in 0.1-cm optical cells is about 10%. The storage of the data was performed on magnetic tape and transferred to the PDP-11 data-handling system. That enough scattering intensity was available from the dilute samples under study was visualized from the intensities on the display unit attached to the data-acquisition system. The obtained scattering was always normalized in transmission (D_2O) and background (K1) by the use of a SPOLLY program. The numerical value for R_g was extrapolated from the obtained radial distribution of the neutron-scattered intensities and the GUIMG program.

TABLE II

Radius of Gyration R_g of the Pt–Carbowax-20M Particles Determined by Low-Angle Neutron Scattering

Sample	Carbowax-20M/Pt ppm ratio	R_g–D_2O (Å)	Range QR_g	R_g (Å)
K1	3/30	199	0.8–2.4	—
K2	10/30	156	0.6–1.9	152
K3	20/30	190	0.8–2.3	190
K6	200/30	159	0.6–1.9	152
K8	1000/30	162	0.6–1.9	160
K18	40/100	204	0.8–2.4	205
K19	200/100	193	0.8–2.3	193
K20	500/100	171	0.7–2.0	169

The ranges of QR_g are shown in the fourth column of Table II. For an exact Guinier approximation for the scattered intensity the values of QR_g should be well below 1, which is not the case in this example. The implicit model for the Pt polymer particle is a particle having a metal core with the coherent scattering length of 6.29 and layers of Carbowax-20M (CH_2CH_2O) having a coherent scattering length of 0.45. The H_2O and D_2O have values of −0.56 and 0.41, respectively, for this parameter. By the use of the contrast-variation method when the scattering length of Pt and the sum of H_2O plus D_2O are matched, a solution 98.2% in D_2O is required to provide the contrast necessary to compensate for the polymer layer. The concentration of platinum has little effect on R_g (compare the results obtained using 30 and 100 mg Pt). The values obtained for R_g have been corrected for the D_2O background (as shown in the third column of Table II) and Carboxwax-20M 3/30 background (as shown in the fifth column of the table).

In spite of a narrow distribution in the molecular weight of Carbowax-20M, the heavier polymer constituents of this distribution will be adsorbed more readily by the 32-Å Pt core. Nevertheless, the lighter particles remaining in solution will not significantly affect the total scattering statistics observed because of the low weight of its contribution. In the Pt polymer dispersions prepared some of the polymers will stay in solution. This supernatant polymer has been taken as the background for each different concentration of polymer. To show that the concentration of platinum has no influence on R_g, we made the three subsequent runs (K18 to K20) at 100-ppm platinum concentration obtaining similar values for R_g. From the measured probes in Table II, it is seen that all particles have a radius $152 \leq R_g \leq 205$ Å, indicating that an association of Carbowax-

20M and platinum takes place. The number of particles has been determined per cubic centimeter: $10^{14}/cm^3$ of Pt and $10^{15}/cm^3$ for Carbowax-20M units for the K18 sample in Table II. Assuming that a 32-Å Pt particle has ~1000 atoms (coarse approximation in spherical symmetry), then the weight of the platinum per particle is about 200,000. Because Carbowax-20M is 10 times more abundant per particle, it will contribute 200,000 to the total molecular weight, giving a final value of 400,000 for the aggregate. That very narrow MW distribution has been verified for the Carbowax family of polymers should help in making this value valid in a narrow distribution region.

By quasi-elastic light-scattering techniques (*20*) the same radius was obtained (~180 Å) for Pt–Carbowax-20M particles at similar concentrations, and they show consistency for the size of this aggregate studied by two different techniques. The conformation of platinum–polymer colloidal systems cannot be determined by either of the scattering techniques employed in the present study. The metal polymers are not discrete aggregates of the type for which a relation of mass and geometrical size may lead to conclusions about their shape (*131*). Hydroxyl groups on the metal form an ether linkage to Carbowax-20M, but most remain available for catalysis. The hydroxylic channel between water and metal (*89*) is supposed to be the point of contact between these two components.

Table III summarizes the results of our systematic investigation on copolymers used as dispersion agents. These copolymers contain a component having low solubility in water and consequently a good affinity for metal surfaces and a second component that is highly soluble in water providing the steric barrier. Vigorous agitation of the protective agent with Pt–citrate solution is necessary to ensure that the adsorbent surface (which is irregular) becomes accessible to the polymer. Table III shows that, of the polymers investigated, the ones with more hydrophobic groups give higher rates of hydrogen evolution. This may be related to the better protective action of such polymers. Moreover, the table shows that the higher the molecular weight of the polymer, the higher the rate of evolution of hydrogen.

The polymer concentration used was sufficient for complete coverage of the metal particles. The variation of hydrogen evolution with molecular weight for polyoxyethylene polymers as well as the maleic anhydride styrene copolymers is illustrated in Fig. 8. The latter polymer solutions have been prepared by prolonged hydrolysis with NaOH at pH 8. In part B of Table III, UCON random copolymers have been used containing 75, 50, and 25% polyethyleneoxide, the remainder being polypropyleneoxide (from BP Suisse). It is readily seen that for the UCON series the polymer with the largest hydrophobic portion (UCON 25) and the highest molecu-

TABLE III

Sacrificial System $[Ru(bipy)_3^{2+}] = 10^{-4}$ *M*, $[MV^{2+}] = 5 \times 10^{-3}$ *M*, [EDTA] $= 3 \times 10^{-2}$ *M*, pH 4.7, 450-W Xe Lamp[a]

Protective agent	MW	H_2 (ml/liter solution/h)
(A) Polyoxyethylene	1,600	20
Carbowax	4,000	50
Carbowax	6,000	80
Polyoxyethylene	14,000	130
Carbowax-20M	20,000	180
(B) UCON 75 H	4,000	72
UCON 50 H	10,000	120
UCON 25 H	20,000	185
(C) Maleic anhydride–styrene (50 : 50%)	10,000	90
Maleic anhydride–styrene	40,000	175
Maleic anhydride–styrene	50,000	185
(D) Polyacrylic-acid–styrene	20,000	95
Pluronic P-75	4,200	135
Maleic anhydride–vinyl acetate	10,000	170

[a] All irradiation at 80 mg/liter of protective agent, 40 mg/liter Pt.

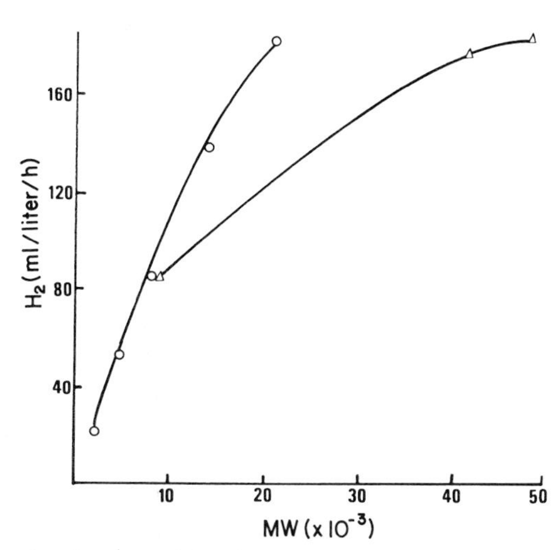

Fig. 8. Influence of molecular weight of protective agent on hydrogen yields for solutions irradiated under the experimental conditions cited in Table III. ○, Polyoxyethylene copolymer series; △, maleic anhydride–styrene copolymer series.

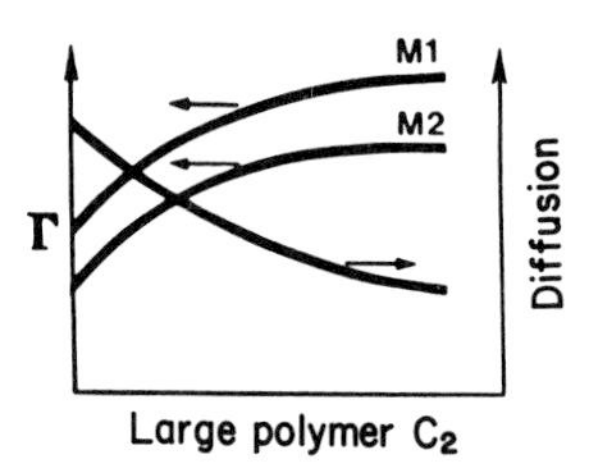

Fig. 9. Schematic relationship of adsorption isotherms for polymers of different molecular weight and their diffusion toward the substrate as a function of concentration of the polymer.

lar weight provides the best-dispersing system, as reflected by the observed H_2 yields. The amount of polypropyleneoxide in contact with the surface is, therefore, important for protection. This is also shown with Pluronic, which also gives high hydrogen yield, reflecting its good stabilizing action for hydrophobic particles. Pluronic P-75 is a Wyandotte Chemical Co. block copolymer with composition similar to the random copolymer UCON 75 H.

If diffusion of the polymer toward the interface takes place in processes in which irreversible adsorption occurs, these processes are governed by the molar mass of the polymer involved (*114*). This is schematically shown in Fig. 9, which shows that diffusion decreases with increase of polymer concentration and also shows the increase of adsorption with increase of molecular weight. Thus, although diffusion is faster for the low-molecular-weight polymers, the higher-molecular-weight polymers give higher adsorption values ($M_2 > M_1$ in Fig. 9), and this is reflected in better stabilization of the Pt catalyst.

II. Evolution of O_2 in Dark- and Light-Induced Processes in Sacrificial Systems

The idea of cyclic water decomposition has generated interest for some years. Heidt (*47*) proposed production of hydrogen by photochemical processes using uv light absorbed by cerous ions, thereby oxidizing them to ceric ions. Oxygen would, as a second step, be produced by the part of the light absorbed by the ceric ions, reducing them to cerous ions and oxidizing water. The net result of the two reactions would be the photochemical decomposition of water into its elements, resulting in a gain in total chemical energy at the expense of the light energy. As for other transition-metal ions, this process also proceeds using uv light, but the yields are low ($\phi < 0.1$). Oxyanions such as VO_2^+, MnO_4^-, and $S_2O_8^{2-}$ have produced O_2 on photolysis, but the source of oxygen has never been mechanistically elucidated, and this reaction is not going to be further elaborated here.

In a publication (*60*) from our laboratory, the first evidence was presented that the production of oxygen via the reaction

$$4Ce^{4+} + 2H_2O \longrightarrow 4Ce^{3+} + 4H^+ + O_2 \quad (5)$$

is mediated by a redox catalyst such as PtO_2 and IrO_2. Although this reaction is thermodynamically favored by a few millivolts, it needs a redox catalyst to be feasible as a dark reaction. In H_2SO_4, Ce^{4+} ($E° = 1.44$ eV) should be capable of spontaneous oxygen evolution from water. Nevertheless, oxygen evolutions occurs concomitantly with the reduction of Ce^{4+} ion only when this reaction is promoted efficiently by a suitable redox catalyst. The redox catalyst would act as a defect electron storage system; through electron transfer from the catalyst to Ce^{4+}, an anodic potential is imposed on these microelectrodes that is sufficiently positive to induce water oxidation. In a subsequent publication (*62*) we reported on the use of RuO_2 powder of ~20 m^2/gm as redox catalyst. The very low overvoltage for O_2 evolution on RuO_2 (*72, 112*) allows the redox process to occur quantitatively at a higher rate than with PtO_2.

The initial rate of Ce^{4+} reduction of O_2 was found to increase linearly with Ce^{4+} concentration. It is also a function of the amount of catalyst present and the rate of stirring, that is, of the efficiency of contact between the catalyst and the solution. The method for detection of O_2 is selective, the principle being based on the response of the potential of a $Zr–ZrO_2$ (solid electrolyte) potential to O_2. At 20°C the Ce^{4+} reduction rate is 1.63×10^{-3} *M*/l/h, which is 16 times higher than that observed with PtO_2 under identical conditions. In a semilogarithmic plot the decrease of $[Ce^{4+}]$ as the conversion to Ce^{3+} proceeds is linear, with a rate constant of 0.3/h; the process is first order with respect to $[Ce^{4+}]$. This would indicate that a step in the process takes place at the surface of the catalyst where contact between reagents determines the rate of the reaction.

At 50°C the Ce^{4+} reduction rate is drastically increased, the rate constant being 2/h. This result is interesting because similar temperatures prevail in solar-energy devices. In the preceding experiments 5×10^{-3} *M* RuO_2 was used. If the redox catalyst RuO_2 is employed at a concentration of 1.2×10^{-2} *M*, the reaction becomes 6 times faster. Evidently, the increase in RuO_2 concentration increases the area of contact of the catalyst with the aqueous solution. In the course of our work with RuO_2 it was found that water oxidation can also be achieved by oxidants other than Ce^{4+}. As an example, $Ru(bipy)_3{}^{3+}$ at pH 1 is reduced by RuO_2 to $Ru(bipy)_3{}^{2+}$, with simultaneous liberation of oxygen. This result is potentially important with regard to the eventual use of $Ru(bipy)_3{}^{2+}$ as a sensitizer in light-energy conversion. For the oxygen-evolution reaction, the catalytic activity in acid solution is found to be in the order Ir ≈ Ru > Pd

> Rh > Pt > Au. The work of Srinivasan and Salzano (*112*) has established that oxygen evolution always takes place on oxide-covered surfaces, whereas oxygen reduction can occur on bare metallic as well as on oxide-covered surfaces (*100*).

Sacrificial systems for light-induced O_2 evolution have been developed (*54*) whereby a photosensitizer, an electron acceptor, and a redox catalyst operate in the following way: photolysis of $Ru(bipy)_3^{2+}$ as sensitizer takes place with $[Co(NH_3)_5Cl]^{2+}$ complex as electron acceptor and RuO_2 as redox catalyst. This is shown in Fig. 10. In the presence of a suitable catalyst, $Ru(bipy)_3^{3+}$ efficiently oxidizes H_2O to O_2 according to

$$4Ru(bipy)_3^{3+} + 2H_2O \xrightarrow[\text{cat}]{h\nu} 4Ru(bipy)_3^{2+} + O_2 + 4H^+ \tag{6}$$

$Fe(bipy)_3^{3+}$ is capable of oxidizing water in dark processes with stoichiometric yields in the presence of RuO_2 powder (*55*) and colloidal dispersions of RuO_2. The oxygen yield depends strongly on the pH, reaching a maximum between pH 7 and 8 and corresponds stoichiometrically to

$$4Fe(bipy)_3^{3+} + 2H_2O \xrightarrow{RuO_2} 4Fe(bipy)_3^{2+} + 4H^+ + O_2 \tag{7}$$

The yield is stoichiometric only within the relatively narrow pH region between 7 and 7.5 and decreases as the pH moves toward the acidic or basic side. When colloidal RuO_2 is used, the amount of RuO_2 needed is 10-fold smaller than in the case of the powder. The remarkable effect introduced by the RuO_2 catalyst is that it reduces the pH requirement for O_2 evolution by 6 units and renders this process quantitative. In fact, from the standard potentials of the two couples involved in the redox reaction, $E°(O_2/H_2O) = 820$ mV and $E°[Fe(bipy)_3^{3+}/Fe(bipy)_3^{2+}] = 980$ mV, the driving force of the oxygen generation at pH 7 (maximum efficiency of the RuO_2 powder) is only 160 mV/electron transferred. The low overvoltage for water oxidation on RuO_2 explains the low driving force required for reaction (7). Because Ru exists in valences from 3+ to 8+, intermediate valence states of this ion can be formed that live long enough to react with solution species, by which the original Ru^{4+} state is restored (*27, 87*). The presence of variable valence states facilitating charge transfer in solution seems to be a valid concept (*13*) and has been invoked in charge transfer in photoelectrochemistry with oxides of Ta, W, Ti, Fe, and Ru.

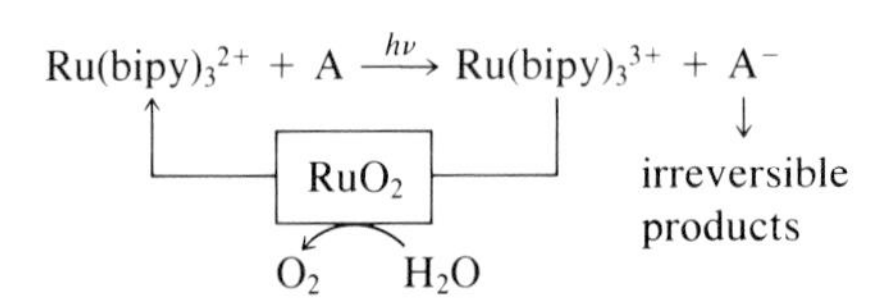

Fig. 10. Scheme for water oxidation induced by visible light where $Ru(bipy)_3^{2+}$ is the sensitizer and $A = Co(NH_3)_5Cl^{2+}$ the electron acceptor.

In Fig. 6 it is assumed that surface hydroxyl radicals are involved in the catalytic reaction stated in Eq. (5). Oxidized ruthenium ions on the surface react with H_2O molecules through the OH group attached to the surface. The surface oxide is hydrated (as done in this work), and exact thermodynamic assessment of the individual steps of the scheme proposed in Fig. 11 is, therefore, difficult (*80, 89*). The preparation of colloidal RuO_2 and its electrostatic stabilization via charge-protective agents such as sodium lauryl sulfate or polyvinyl pyridine has been discussed (*68*). Depending on local conditions the hydrous oxide will adsorb a protective agent, and the stability of these finely divided systems toward aggregation will depend on charge, hydration, and the composition of the liquid environment. Hydrolysis of ruthenate salts has been employed in preparing the hydrous oxide sol.

Figure 12 shows the amount of O_2 produced (as percentage of stoichiometric yield) when $Fe(bipy)_3^{3+}$ perchlorate is added to the solution as a function of time. In all cases 250 mg of this salt have been dissolved. The reaction was run at pH 7 for RuO_2 powder and at pH 9 for NaLS–RuO_2 colloid, obtaining stoichiometric yields of O_2 when 0.5 mg RuO_2 (content of RuO_2 in the colloid) was used. The higher efficiency of the RuO_2 colloid as compared to the powder is related to the available surface of RuO_2 per gram of catalyst used in the form of colloid as compared to the powder, which has a 2360-Å particle size. The quantity of O_2 evolved increases as more catalyst is used. It is possible that under the experimental conditions used here the coulombic attraction may bring the $Fe(bipy)_3^{3+}$ close to the interface, affecting this reaction favorably. The major difficulty associated with the oxidation of water is that the formation of oxygen requires a

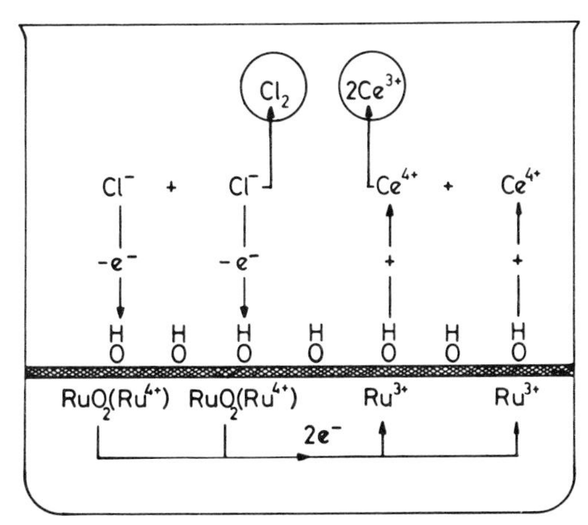

Fig. 11. Mechanism for oxidation of water in sacrificial system catalyzed by RuO_2. Ce^{4+} is the oxidant. The scheme shows the possible Ru^{4+}/Ru^{3+} couple at the catalyst surface that is active in catalysis.

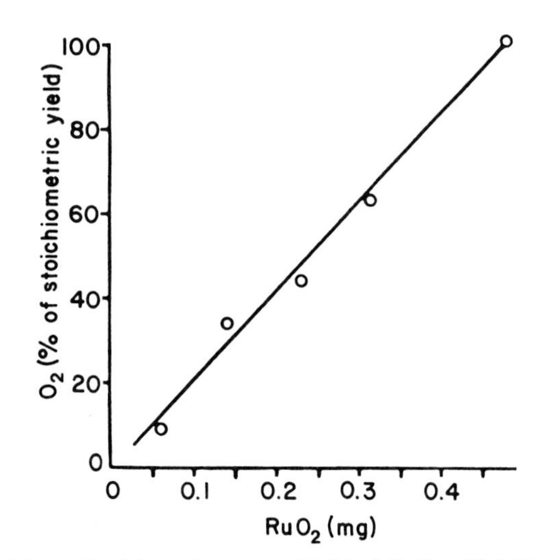

Fig. 12. Amount of O_2 evolved from the system $Fe(bipy)_3^{2+}$ (2×10^{-3} *M*) used as oxidizing agent at pH 9 as a function of RuO_2 content in RuO_2–NaLS in 60 cm^3 of solution after 50 min.

concerted four-electron transfer as shown in reaction (7). Production of free-radical intermediates is undesirable because it involves unnecessary expenditure of energy.

Figure 12 shows the amount of O_2 evolved by reaction (7) as function of concentration of catalyst used. It is readily seen that the higher the concentration of catalyst the higher the yields of O_2 obtained. The process goes to 100% stoichiometric yield of O_2 in 50 min in the case of NaLS for 0.5 mg of RuO_2 in 60 cm^3 of solution. The lower concentrations showed a lower evolution rate after 50 min of stirring. Figure 13 shows the O_2 production for NaLS and PVP–RuO_2 colloids as a function of pH. The maximum O_2 production takes place at pH 9, at which the yield is stoichiometric for the NaLS colloid. The observed variation of O_2 yield as a function of the pH at which the reaction is carried out can be explained by the changes in chemical composition in the electrolyte medium. It appears that colloidal RuO_2 intervenes in reaction (7) more efficiently than the powder because stoichiometric O_2 yield is obtained at pH 9 using 40-fold less material than when powder is used under optimal conditions. Negatively charged hydrophilic colloids are good protective agents for negative sols, showing that electrostatic interaction is not the determining factor in the stabilization of the colloid. Other transition-metal oxides such as cobalt or nickel oxide also produce oxygen in the presence of a suitable oxidizing agent, but the yield is far lower than stoichiometric (*106*). Shafirovich and Shilov (*105–107*) have investigated oxygen evolution from water in dark reactions of strong oxidants, such as MnO_4^-, IO_4^-, PbO_2,

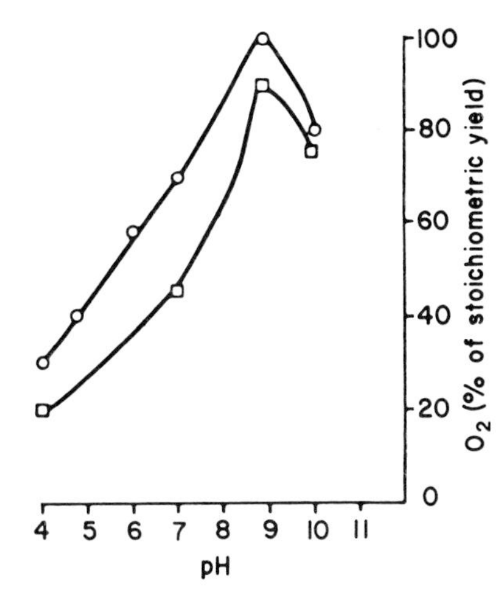

Fig. 13. Dependence of the yield of oxygen on the pH from 2×10^{-3} *M* $Fe(bipy)_3^{3+}$ using colloidal RuO_2 (0.5 mg RuO_2 per 60 cm^3 of solution) ○, NaLS–RuO_2; □, PVP–RuO_2.

etc., catalyzed by homogeneous catalysts, such as Co^{2+} and Mn^{2+}, that are generated in the course of the reaction. The mechanism proposed for the observed O_2 has yet to be elucidated. Water oxidation mediated by homogeneous redox catalysts such as Co^{2+} are strongly pH dependent, as reported by these authors. Oxidants derived from bipyridyl phenanthroline complexes of such ions as Fe^{3+}, Ru^{3+}, and Os^{3+} have been suggested for oxidizing water in alkaline media, and a possible mechanism for these reactions has been postulated (*21, 91, 93, 113*).

$Ru(bipy)_3^{3+}$ has been photogenerated via irreversible redox reactions (*78*) as shown in Fig. 5. The O_2 generation in acidic pH involves an excited $Ru(bipy)_3^{2+}$ with subsequent oxidation to $Ru(bipy)_3^{3+}$ by a Co^{3+} complex. The $Ru(bipy)_3^{3+}$ oxidizes water with O_2 evolution in the presence of RuO_2, regenerating the initial $Ru(bipy)_3^{2+}$. Lehn and Ziessel (*79*) have examined photoproduction of O_2 with Y-zeolite-supported IrO_2 and RuO_2, showing high rates for O_2 evolution on these catalysts.

Homogeneous catalysts for O_2 evolution from water by $Ru(bipy)_3^{3+}$ have been studied (*30*) in the presence of Co, Fe, and Cu complexes. As complexing agents, ethylenediamine and oxyacids were used in alkaline and weak acidic media. Phosphate and acetate buffers, when used, inhibited O_2 evolution. The homogeneous catalysis was postulated to proceed via partially hydrolyzed, unsaturated forms of the metal complexes used. $Ru(bipy)_3^{3+}$ has also been shown to be active in water-oxidation processes through the use of Cr, Mn, Ni, Al, Cu, Zn, Fe, and Co metallophthalocyanins (*31*). The most active catalysts were the Co complexes. Metal-free phthalocyanin were inactive when used in these processes. The mechanism of the catalytic action by metalloporphyrin is still unclear at this stage, and further work is necessary in this direction.

III. Hydrogen Evolution Induced by Light-Catalyzed Colloidal TiO_2-Loaded Systems

Until now, heterogeneous photocatalysis on TiO_2 has primarily been concerned with oxidations (*3, 22, 33*) or with oxygen surface reactions (*84, 90*) that have been carried out on diverse pure forms of this material. A wider field of catalytic reactions is becoming available because of the use of metals and oxides deposited on semiconductor oxides (1). These semiconductor powders, such as TiO_2 and $SrTiO_3$ (*127*), assist uphill reactions such as water splitting under uv or visible light. Water splitting induced by visible light on TiO_2 dispersions, in which $Ru(bipy)_3^{2+}$ was used as sensitizer and methylviologen (MV^{2+}) as relay, was first achieved in our laboratory in 1980 (*65*). The photoinduced reduction of methylviologen (*N,N'*-dimethylbipyridine dication, MV^{2+}) by a suitable sensitizer such as $Ru(bipy)_3^{2+}$, can be exploited to achieve cleavage of water into hydrogen and oxygen. The photoredox process is performed in the presence of two redox catalysts, one of which, for example, colloidal platinum, effects reduction of water, and the other, for example, RuO_2, oxidation of water. This is shown in Scheme 1.

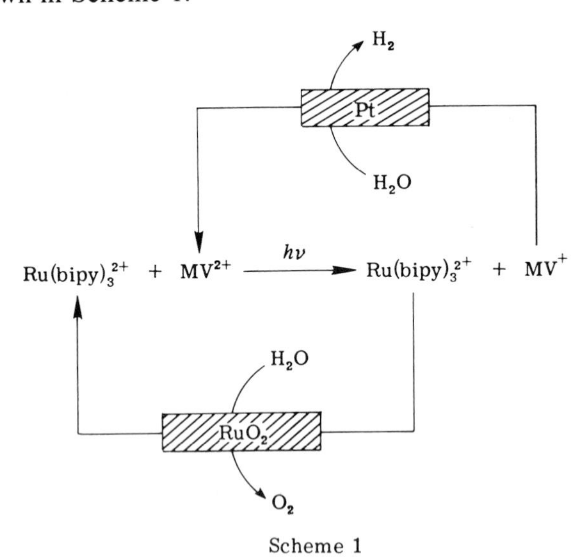

Scheme 1

The observed evolution of H_2 proceeds after a reaction period of up to 5 min. This induction period occurs because an H-atom layer first forms at the surface over the catalyst. When using 100 mg/liter $Pt–TiO_2–RuO_2$

protective agent instead of 500 mg/liter, the observed H_2 evolution after 10 min of irradiation was higher, and only after 15 min did the situation clearly invert in favor of the most concentrated catalyst. The measured yields of O_2 are about 5% of stoichiometric because of the known high adsorption of O_2 on the TiO_2 surface.

Platinum supported on titania is used for many catalytic processes because Pt is one of the best catalysts for dissociating or recombining hydrogen or hydrogen compounds (H_2O in this case). The electron affinity of this material for water decomposition seems adequate. In contrast, TiO_2 has been widely used in catalysis by metals as a nonconventional support that induces strong interactions with noble metals such as Ru (*122*), Pt (*115*), and also Ni (*123*). Work in our laboratory has characterized efficient catalysts capable of sustaining cleavage of water induced by visible light and direct band-gap irradiation (*5*). The bifunctional redox catalyst composed of Pt and RuO_2 supported on colloidal TiO_2 (anatase) had a hydrodynamic radius of 470 Å as determined by quasi-elastic scattering. The effect of TiO_2 structure, *n*-doping, and RuO_2 loading of the catalyst on the efficiency of water cleavage is shown in Fig. 14. Anatase doped with 0.4% Nb_2O_5 and loaded with 0.1% RuO_2 exhibits optimum activity. $Ru(bipy)_3^{2+}$ was invariably employed as a sensitizer and MV^{2+} as relay, and 40 mg of Pt deposited on 500 mg of anatase were used. In all cases, the presence of RuO_2 was required in anatase-based catalysts to obtain efficient water splitting. The activity of anatase is by far superior to that of rutile-based redox catalysts. Thus, at a loading of 0.1% RuO_2, the rate of hydrogen generation obtained with the former is at least four times higher than that observed with the latter modification.

The difference in activity observed between rutile- and anatase-based catalysts may be accounted for by the location of the conduction band in the semiconductors. The flat-bound potential of rutile coincides almost exactly with the NHE potential, whereas that of anatase is shifted cathodically by ~200 mV (*97*). Hence, only in the latter case is a driving force for water reduction available.

To explain the effect of Nb_2O_5, we draw attention to the fact that Nb^{5+} has a radius very close to that of Ti^{4+} (*108*) and, therefore, replaces isomorphically titanium ions in the TiO_2 lattice (*71*). As an *n*-dopant it produces a Schottky barrier at the particle–solution interface, which assists electron injection from the reduced methylviologen. Moreover, Nb_2O_5 shifts the flat-band potential of rutile cathodically (*103*). The effect of sensitizer and TiO_2 concentration as well as pH have also been examined (*6*).

Niobium added to TiO_2 affects the Ti^{4+}–Ti^{3+} ratio. For every Nb^{5+} substitutionally incorporated in TiO_2, we create a Ti^{3+}. Because the elec-

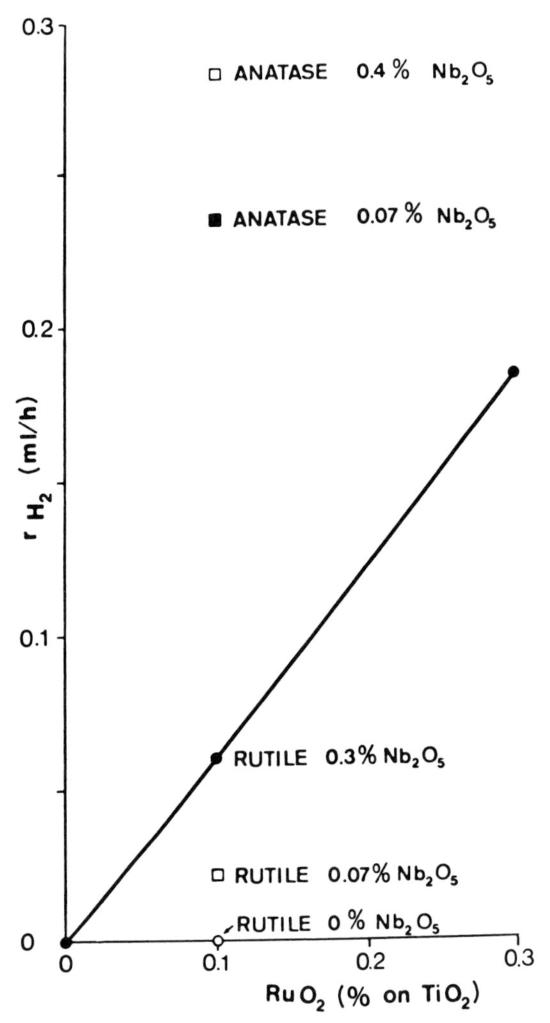

Fig. 14. Influence of catalyst and composition on the visible-light- ($\lambda > 400$ nm)-induced dissociation of water. Conditions: 500 mg TiO_2 per liter of solution loaded with 40 mg Pt, Nb_2O_5 doping, and RuO_2 loading as shown, pH 4.7, 25°C, $[Ru(bipy)_3^{2+}] = 10^{-4}$ *M*, and $[MV^{2+}] = 5 \times 10^{-3}$ *M*.

tron affinity of Ti^{4+} is greater than that of Nb^{5+}, charge neutrality is preserved by the creation of an Nb^{5+}–Ti^{3+} pair. The Nb^{5+}–Ti^{3+}-bound pair has a donor level at ~0.04 eV below the conduction band (*49*). Invoking band-bending arguments for the Nb action, the following could be stated. When an electrolyte is put into contact with an *n*-type semiconductor, band bending occurs because the Fermi level of the semiconductor must equal the Fermi level associated with the redox couple in the electrolyte. When Nb is added, the depletion layer becomes narrower. Therefore, band bending would take place on a narrower width through the depletion layer and become more pronounced. The increased band bending would allow electron tunneling to proceed more easily at the interface. In this case, the electron injection is facilitated.

In another project performed in our laboratory (*26*), transparent TiO_2 sols of colloidal dimensions (~200 Å) were produced via hydrolysis of titanium tetraisopropoxide in acidic aqueous solution. Because scattering effects are small, these solutions can be subjected to a detailed photochemical analysis both by conventional and flash-photolysis techniques. When loaded simultaneously with ultrafine Pt and RuO_2 deposits, these particles are also active as water-decomposition catalysts. After a short induction period, the H_2-generation rate in the case of the bifunctional redox catalyst establishes itself at 2.8 ml/h. The process was stopped after ~7 h when significant pressure had built up in the reaction vessel, and the gas produced was flushed out with N_2. On reillumination, H_2 generation resumes at the initial rate. This cycle can be repeated many times. To measure the quantum yield for H_2 production, we repeated the experiment by illuminating through the Balzers RUV 308 interference filter. The efficient water splitting obtained under illumination by these colloidal particles may be ascribed to the high state of hydroxylation of the TiO_2 surface.

Sensitization of semiconductors involves transfer of electrons from a dye in contact with a semiconductor. For good sensitizing activity, the dye must be strongly adsorbed at the surface of the powder (*86*). Results from photoelectrochemistry (*37*) have revealed that for $Ru(bipy)_3^{2+}$ the observed photocurrent is caused by the dye adsorbed on the electrode, not from that dissolved in the liquid phase. This result also holds for other dyes (*83*). Efficient electron injection via surfactant $Ru(bipy)_3^{2+}$ has led to water splitting. In this system only two components have been used, an amphiphilic sensitizer that adsorbs well to the microelectrode surface by hydrophobic interactions. This hydrophobic effect is also helped because we have worked at pH 4.7, not far from the iep of TiO_2 (4.7–6), at which the oxide is not highly dissociated and can better adsorb the amphiphilic dye. The mono-C_{12}-substituted $Ru(bipy)_3^{2+}$ exhibited optimal efficiency for H_2 generation. The electron transfer from $Ru(bipy)_3^{2+}$ to TiO_2 implies that the lowest excited state of $Ru(bipy)_3^{2+}$ is situated above the conduction band of TiO_2. The regeneration of the dye involves water oxidation assisted by the RuO_2 catalyst. In this context, a useful review on redox potentials of dyes has been published (*15*). The role of RuO_2 in our water-splitting system is to accelerate the hole transfer from the dye to the aqueous solution. The low overvoltage characteristic for oxygen evolution on RuO_2 renders hole capture by water particularly effective and hence inhibits electron–hole recombination.

Many reducing agents or dyes are able to inject electrons into semiconductors such as TiO_2. A study was undertaken to clarify certain aspects of redox–dye–semiconductor interactions with improved response to visible light. Using laser flash photolysis, it was established that $Ru(bipy)_3^{2+}$ in the excited state injects electrons into TiO_2 and that this injection is temperature dependent. Molecules adsorbed on TiO_2 or inside the diffusion layer will be active in electron injection. The width of the diffusion layer is $(2D\tau)^{1/2}$, where D is the diffusion coefficient and τ the excited-state lifetime. TiO_2, as well as the Pt on its surface, are active in mediating electron injection from $Ru(bipy)_3^{2+}$ under light irradiation. In this same study, experiments were conducted as shown in Fig. 15 for uv and visible-light irradiation of this system. Figure 15 shows that H_2 evolution is proportional to the first power of the light intensity and, because the points are satisfactorily aligned, one arrives at the conclusion that H_2 is formed in a monophotonic reaction and that one electron is involved in H^+ reduction (*67*).

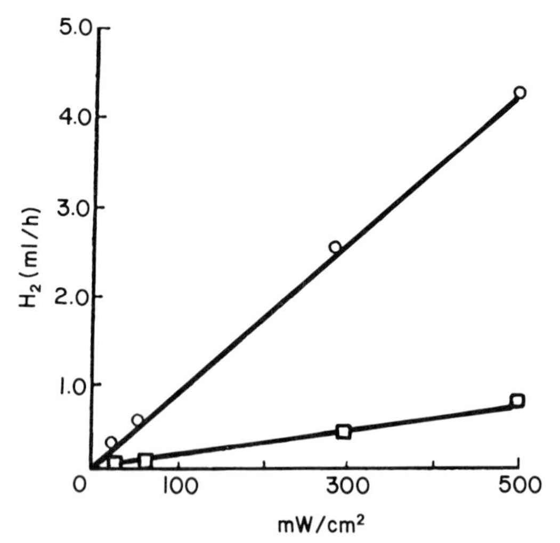

Fig. 15. Effect of radiation intensity on the amount of hydrogen evolved in solutions of volume 25 ml: (○) TiO_2 500 mg/liter loaded with 0.1% RuO_2, 0.42% Nb_2O_5, and 60 mg Pt/liter, with pH 4.3, 75°C, and $\lambda > 250$ nm. (□) Same conditions as in (○), but irradiation was carried out with $\lambda > 400$ nm and $[Ru(bipy)_3^{2+}] = 2 \times 10^{-4}$ M.

IV. Design of Spinel- and Perovskite-Type Semiconductors Active in H_2 Evolution Induced by Visible Light

A. Preliminary Considerations

Ultrafine powder or colloidal particles represent a special case: the size of the particles is so small that their surface properties begin to dominate over the bulk characteristic of the material (*24*). In the past, some interest has been given to the study of powders either by themselves or as units of micrometer-size aggregates (*73*). An important problem that arises in particulate catalysts is that once the product formed at oxidizing sites (O_2) builds up, it is preferentially reduced at reducing sites (Pt on TiO_2). This hinders the long-term activity of a system of the type $Ru(bipy)_3^{2+}$ on TiO_2–RuO_2–Pt catalyst as discussed previously and greatly reduces the efficiency of the water splitting process. The reaction

$$Pt^- + O_2 \longrightarrow Pt + O_2^- \tag{8}$$

would depolarize the microcathode whenever a sizable amount of O_2 built up in the system. Separation of the products could overcome this problem by using inorganic membranes as shown in the last part of this section.

Surface aspects are important in the materials that are prepared for water-splitting processes. Somorjai (*111*) has shown that the surface of solids contain terraces, kinks, vacancies, and atoms having different oxidation states that act differently in catalysis. As an example, Yao (*129*) found that the catalytic activity of perovskite ($LaCoO_3$), used for CO and C_2H_4 oxidation, depends strongly on the departing substances and preparation method. (*96*). The difference in catalytic activity in perovskites is related to the different degrees of heterogenity in their surface. It is interesting to point out that $LaCoO_3$, which was prepared at higher temperature and showed lower surface area, exhibited the highest specific activity (*121*). This is related to the formation of a lower number of pores having larger diameters on the catalyst surface (*128*). Because water splitting is effectively accelerated by surface reactions for either the production of H_2 or O_2, it is expected that the number of charges reaching the surface (induced by light or by a dye in the catalyst) and their mobility are the controlling factors that render a catalyst kinetically efficient in water-splitting processes. The surface recombination velocity is an extremely sensitive function of surface constitution (doping and ions adsorbed) and physical parameters (BET area and pore size distribution). The development and maintenance of both appropriate catalytic specificity and high area in porous perovskite powders are the major aims in catalysis leading to water-splitting processes via powdered semiconductors.

Water must be chemisorbed on the semiconductor material used for water splitting (*46*). In all of our work, we regard the surface of TiO_2 available for interaction with water and the Pt and RuO_2 catalytic sites on TiO_2 as the points where the charges generated in TiO_2 are channelled, producing on these points H_2 and O_2, respectively. Therefore, the stoichiometry of the chemisorption reaction cannot be stated because active centers on the catalyst surface that form the chemisorbed complex in the interfacial layer are unknown (*25*). At present, a standard ordered scheme does not exist for the electronic levels of the oxides of the transition metals, which, like RuO_2, are important in O_2 evolution from water by light processes. Because of the partially filled *d* levels, metallic conduction is observed in RuO_2 and Pt. No matter what the mechanism of catalytic action of RuO_2 or Pt loaded on TiO_2, the fact that only small clusters of Pt ($r \approx 15$ Å) deposited on TiO_2 give efficient H_2 generation in water splitting is revealing.

Semiconducting materials that are used as photoanodes must be stable under conditions of prolonged oxygen evolution (*99*). The compounds most likely to have this property are metal oxides, in which the metal ion is in its highest stable oxidation state. No *n*-type nonoxide semiconductor has yet been found that is stable under conditions of photoelectrochemical oxygen evolution, their anodic dissolution always being favored, both thermodynamically and kinetically [e.g., GaAs, CdS (X = S, Se, or Te)] (*50*). In sulfides corrosion takes place at ~0 V. Arsenites and antimonides corrode at potentials more negative than 0 V. Only oxides are thermodynamically expected to undergo chemical corrosion at potentials around 1 V (*81*). It is important in this context to recall the nature of available oxide semiconductors that, because of their behavior, could be classified into three groups (*35*): (a) reductives such as ZnO, CdO, and TiO_2 that, because of nonstoichiometric oxygen defects, induce excess electrons in the crystals; (b) oxidatives such as Cu_2O, NiO, Mn_2O_3, Co_3O_4, and UO_3 that, because of nonstoichiometric excess metal ions, induce oxygen absorption and show *p*-type defect electron behavior; and (c) superoxidatives such as CaO and BaO that, having one available valence in the metal, can absorb O_2 beyond their given stoichiometry and show *p* behavior.

If there are ions in the powder that can exist in more than one stable valence state, the catalytic reaction may proceed through the formation of an intermediate metastable state that lives long enough to react with the electrolyte, restoring the original oxidation state. Catalytic reactions involve ions that undergo oxidation–reduction processes that can account for charge transfer. This process, involving several stable valence states, is common in oxides (*110*) of Ti, W, Ta, and Fe, and evidence for their stability in photoelectrochemical reactions has been given. It is known that substances that are catalytically active in oxidation–reduction reactions are sometimes colored. A study (*16*) of the effect of various oxides on the decomposition of the unstable oxygen-containing compounds $KMnO_4$, $KClO_3$, HgO, and Ag_2O showed that the most active oxides (NiO, Co_3O_4, MnO_2, and CuO) are intensely colored materials in which black and brown tones predominate. These last materials have also been reported as being the best oxide-type oxidation catalysts for organic reactions (*17, 34*). Then come the less active substances, which have red, orange, and yellow colors (SnO_2, PbO, ZnO, CdO, and others). The slightly active and inactive substances are white (Al_2O_3, MgO, TiO_2, and SnO_2).

A further complication in the use of powder or colloidal materials is that the position of the conduction (Fermi) band is a function of the subdivision of the material used. This assumption has been experimentally verified by Shapiro *et al.* (*109*) when comparing the photon-energy desorption of CO_2 on ZnO. Another complication in irradiated powders is that light penetration is not easily determined. Formenti *et al.* (*32*) have determined the acetone-production rate on TiO_2 powdered catalyst as a function of mass of TiO_2 (anatase) used. The reaction rate increases linearly as function of the catalyst mass up to a limiting value. The results obtained point to a penetration depth of 2 μm. Measurements of the diffuse reflection of powders can be converted to absorption spectra and used to find the electron transitions taking place from the valence band to the conduction band and are widely used in optical acoustic spectroscopy (oas). Cahen (*12*) has estimated the depth of penetration of light to assess the width of the active carrier layer in irradiated semiconductor materials. More studies of the field created by light may be helpful in the design of new semiconducting materials that are usually obtained first in the powder form.

In this context it is desirable to investigate the possibility of developing *d* band semiconductor powders through the introduction of d^n metal ions into the lattice of metal oxides. The goal is to produce photoactive transitions between a *d* band (produced by the d^n cation sublattice) and the oxide conduction band. The new low-lying interband transitions should be possible without affecting the position of the oxide conduction band. The formation of "*d* bands" can occur only if the *d* orbitals of the incorporated transition-metal ions show significant overlap. This overlap might be achieved, for example, if these ions were ordered in a given plane. Otherwise, the d^n additives can form localized states lying within the band gap. Carriers in the localized levels have extremely low mobilities. Hence, excitation of an electron from the localized level to the conduction band in the bulk (depending on the material) would not lead to efficient separation of the electron and hole. Only charges produced at the surface would be active in producing photocurrent via reaction with the solution species, in this case, H_2O.

The strategy employed in designing new materials will involve substances with a band gap of ~2 eV for good absorption in the visible, such as ferrates, titanates, chromites, manganites, tungstates, and niobiates. Also, in this way two metals coexisting with different valence states may lead to materials that have more than one center transitions. These materials must combine good spectral characteristics (absorption in the visible), adequate band gap, as described, and have a conduction band negative enough to allow hydrogen evolution to proceed at adequate rates.

B. Design and Fabrication of New Titanates and Perovskite-Powder Semiconductors

Perovskite-type oxides have been found to be of great importance in catalysis. Among the most important reactions in which these compounds have been used as catalysts are oxidation of CO and hydrocarbons, reduction of NO_2 (*124*), oil hydrotreating (*23*), SO_2 reduction (*51*), hydrogenation of hydrocarbons (*52*), and oxidation of NH_3 (*125*).

New semiconductors to be formed may be derived from perovskite oxides (ABO_3) and from titania (TiO_2). They should have band gaps around 2 eV, low electron affinities, and be stable under irradiation in water-splitting processes. Sr^{2+} or La^{3+} could be used as A (in ABO_3) because they do not significantly influence the electronic structure of the crystal (*14*). Titanates can be efficiently doped if the radius of the dopant does not differ markedly from the Ti^{4+} ionic radius. This is the case for V^{3+}, Cr^{3+}, Mn^{3+}, and Fe^{3+} used as dopants to extend the titanates' response in the visible. The radius of these cations is similar to that of Ti^{4+}, which is 0.68 Å ± 8%. Good photoresponse and photocurrent onset, negative of the H_2-evolution potential, have been obtained with $SrTiO_3$, $BaTiO_3$, and $CaTiO_3$. The design of metal-ion-doped perovskite compounds is based on the excitation of well-localized d^n orbitals of the metal ion by visible light (instead of the well-known π electron of O^{2-}) and the transition A–B to the σ^*, π^* antibonding orbitals of the conduction band of ABO_3. This has been shown with La- and Sr-doped perovskites (*14*) and in our laboratory using Cr-doped powders (*7*). Visible-light-induced oxygen evolution has been observed in these materials used as electrodes

(*36*) because valence bands derived from MO-theory orbitals lie deep in almost all oxides because of the high electronegativity of oxygen.

In designing semiconductors suitable for the splitting of water, one must allow for kinetic overpotentials and band bending. Because of these factors, a value of ~1.8 eV is an optimistic lower value for the band gap necessary to accomplish efficient water cleavage. Myamlin and Pleskov (*92*) have estimated the following losses for semiconductors useful in water splitting: 0.2 eV for losses at a cathode for efficient electron transfer, 0.2 eV band bending for efficient charge separation at the cathode, 0.1 eV for H_2 overvoltage at the cathode, 0.5 eV for O_2 overvoltage at the anode, and 0.2 eV for efficient hold transfer at the anode. Because the total loss for a semiconductor electrode is ~1.2 eV, the band gap of such a material would be ~2.4 eV to be efficient in water decomposition. A band gap of this order would be a practical upper limit for powder semiconductors.

Five standard procedures are currently employed to prepare perovskites:

1. Standard ceramic technique. Oxides or carbonates are mixed to the desired proportions, milled in H_2O or ethanol, dried and heated to effect the solid diffusion (*85, 98*) of the reactants.
2. Wet techniques. Solutions of nitrates, oxalates, acetates, or alcoxides are mixed in the desired proportions, and from the mixture the metals are coprecipitated by adding ammonia, etc. Then a lower heat is necessary to produce the perovskite structure because these preparations show higher reactivity (*118*).
3. Cryochemical methods. The use of this method eliminates the possibility of segregation resulting from the differences in solubility products of the precipitates. A solution of salts is frozen rapidly by spraying the solution into liquid N_2, and in this way the degree of dispersion of the original solution is preserved (*119*).
4. Forced hydrolysis. Salts are digested in an autoclave for several hours under controlled concentration, pH, and temperature conditions to effect hydrolysis (*82*).
5. Volatile halide contact salt. Two oxides are brought together and a good contact is established through a relatively volatile halide. This material is volatilized and by further increasing the temperature in Ar or air, the desired perovskite is formed (*11*).

In the new semiconductor materials, the efficiency of carrier separation at the surface will depend on the recombination of the electrons generated together with holes. In the case of partial transitions, they will increase the intensity of the main transition induced by visible light. Overall *p*-type semiconductors, as stated previously, will allow the electron to diffuse in a more favorable setup because no energy is involved in this step. It has been observed that the typical *p*-type semiconductors are more active than *n*-type primarily when oxidation reactions are involved [e.g., Li-doped ZnO versus ZnO for organic dehydrogenations, as studied by Hauffe (*45*)]. However, such a simple classification is not always valid for more complex perovskite systems (*104*). Doping of semiconductor oxides and perovskites (*28*) to 10^{19} to 10^{20} cm^{-3} (large density of doping) may reduce the thickness of the space charge down to 10 Å, as assumed by Möllers and Memming (*88*) and Heller (*48*), in semiconductor electrodes. The space-charge layer becomes sufficiently thin for tunnelling of the electrons (*76*).

Figure 16 shows that the optical band gap and flat-band potential are related and that smaller band gaps are accompanied by more positive flat-band potentials (*44, 74*). A compilation of available values is plotted in Fig. 10. The slope of this line is 1.4. The linear relationship is explained by

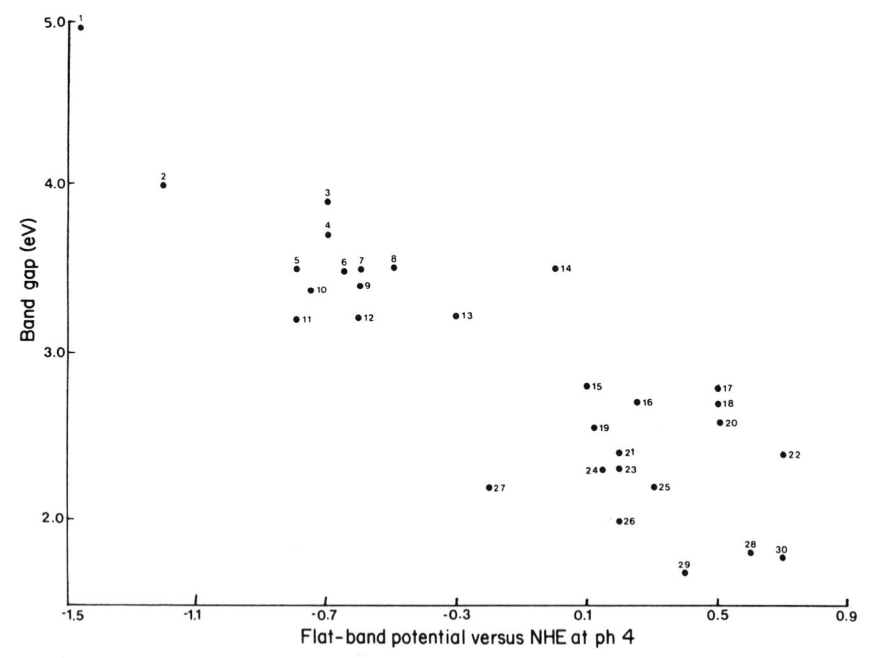

Fig. 16. Band gap of diverse semiconductor materials plotted versus flat-band potential at pH 4 (NHE): 1, ZrO_2; 2, Ta_2O_5; 3, $SrNb_2O_7$; 4, $SrNb_2O_6$; 5, $KTaO_3$; 6, $CaTiO_3$; 7, Nb_2O_5; 8, NbO_2; 9, $Ba_{0.5}Sr_{0.5}Nb_2O_6$; 10, $SrTiO_3$; 11, ZnO; 12, TiO_2; 13, $BaTiO_3$; 14, SnO_2; 15, PbO; 16, CdO; 17, V_2O_5; 18, $FeTiO_3$; 19, $YFeO_3$; 20, In_2O_3; 21, WO_3; 22, Cd_2SnO_4; 23, $CdFe_2O_4$; 24, $LaCrO_3$; 25, Fe_2O_3; 26, $FeNbO_4$; 27, $LuRhO_3$; 28, $Hg_2Nb_2O_7$; 29, $Hg_2Nb_2O_7$; 30, $CdSnO_3$.

the fact that the oxides (or double oxides) have similar potentials for the valence band because the valence band is made up of O^{-2} π electrons. The change in band gap then stems from a change in the conduction band and, therefore, as seen in Fig. 16, a decreased band gap means a more positive flat-band potential.

In perovskites the band gap of the entire system moves between the limits attained by the individual components of the system. $Fe_{2-x}Cr_xO_3$ will have a band gap between the 1.68 eV of Cr_2O_3 and the 2.2. eV of Fe_2O_3. Bockris (*41*) has utilized $LaCrO_3$ and $LaCrO_4$ as electrodes to accomplish efficient photoelectrolysis of H_2O because the energy gap of these materials ranges from 1.68 (the band gap of Cr_2O_3) to 2.5 eV (the band gap of La_2O_3) and the absorption of solar light is superior to TiO_2. More interesting, Roy and Nag (*102*) have shown that $LaCrO_4$–TiO_2 has three times the efficiency for the same type of processes than TiO_2 because of d^n interband transitions.

If various photoactive centers would act independently in a perovskite, this might allow for the required negative flat potentials needed to split H_2O as well as an adequate low band gap in a stable matrix. Whenever these numbers become available they should fill the lower left part of Fig. 16. Only oxides and perovskites comprising double oxides have been listed for reasons of photoelectrochemcal stability (*130*). Even unstable electrode oxides [when generating photocurrents under illumination (*42, 43*)] could be stabilized in powder form (*57*). The behavior of niobate and titanate systems in Fig. 16 is consistent with the view that the A site (in ABO_3) is responsible for variation in flat-band potential and that the structure (BO_3) is responsible for the magnitude of the observed band gap.

C. Porous Material Loading for H_2 Production from Water

The distance between the Pt and RuO_2 deposited on pellets of different dimensions (or porous glass) is a subject of investigation in our laboratory and should be a controlling factor in the efficiency of water splitting. Until now, the codepositation of Pt and RuO_2 on 500-Å TiO_2 powder has not allowed a detailed investigation of this problem. Van Olphen (*94*) has reported metallic-island deposits on clays and has characterized them thoroughly by electron microscopy. These clay–metal sols are important as catalysts in the petrochemical industry. When simultaneously producing H_2 and O_2 photochemically, porous inorganic membranes of TiO_2 may play an important role in separating the two gases (*75*). Pellets of TiO_2 of the desired dimensions are routinely pressed based on the cohesive properties of TiO_2 (*95*).

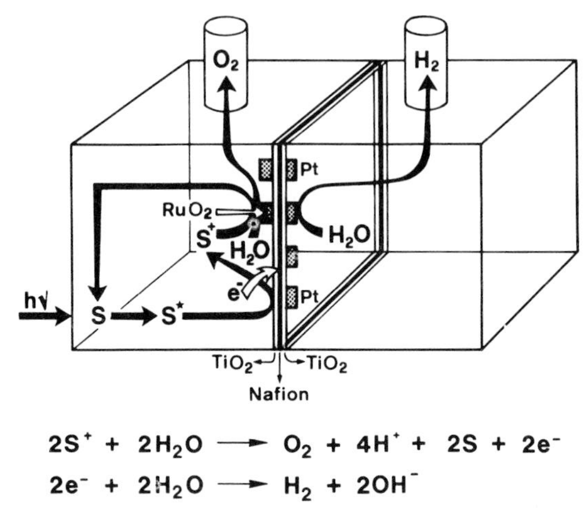

$$2S^+ + 2H_2O \longrightarrow O_2 + 4H^+ + 2S + 2e^-$$

$$2e^- + 2H_2O \longrightarrow H_2 + 2OH^-$$

Fig. 17. Ideal operation for a two-cell reactor in which a porous TiO_2 plate loaded with RuO_2 and Pt catalytic sites is used to transport the charges and separate the O_2 and H_2 produced in photolysis.

The deposition of metals on TiO_2 films or plates is the basis of the PD-R method of manufacturing printed circuit boards (*101*), and the nucleation of metals on TiO_2 has been reported (*58*). Figure 17 shows a hypothetical version of a small-scale solar-energy reactor based on a porous plate of TiO_2 that separates the chamber evolving O_2 to the left from the one evolving H_2 at the right-hand side. Incident light activates the absorbing dye S, and the S* (excited state) produced on contact with the TiO_2 oxidizes to S^+. In the TiO_2 plate, the holes generated in this way evolve O_2 and the RuO_2 sites. The photogenerated electrons in the hole pair diffuse to the Pt catalytic site, and the Pt^- so formed discharges through H_2O reduction. In this way, H_2 is produced in this system.

References

1. Bard, J. A., *J. Photochem.* **10,** 59 (1979).
2. Bond, G., *Platinum Met. Rev.* **19,** 126 (1975).
3. Boehm, H. P., *J. Catal.* **22,** 347 (1971).
4. Borgarello, E., Kalyanasundaram, K., Pelizzetti, E., and Grätzel, M., *Helv. Chim. Acta* **64,** 1937 (1981).
5. Borgarello, E., Kiwi, J., Pelizzetti, E., Visca, M., and Grätzel, M., *Nature (London)* **289,** 5794 (1981).
6. Borgarello, E., Kiwi, J., Pelizzetti, E., Visca, M., and Grätzel, M., *J. Am. Chem. Soc.* **103,** 6234 (1981).

7. Borgarello, E., Kiwi, J., Pelizzetti, E., Visca, M., and Grätzel, M., *J. Am. Chem. Soc.* **104,** 2996 (1982).
8. Boudart, M., Aldag, A., Ptak, L., and Benson, J., *J. Catal.* **11,** 35 (1968).
9. Breckenbridge, R., and Hosler, W., *Phys. Rev.* **91,** 793 (1953).
10. Brugger, P.-A., Cuendet, P., and Grätzel, M., *J. Am. Chem. Soc.* **103,** 2923 (1981).
11. Burton, J., and Garten, R., "Advanced Materials in Catalysis." Academic Press, New York, 1977.
12. Cahen, D., *Appl. Phys. Lett.* **33,** 810 (1978).
13. Cahen, D., Manassen, J., and Hodes, G., *Sol. Energy Mater.* **1,** 343 (1979).
14. Campet, G., Dare-Edwards, P., and Goodenough, J., *Nouv. J. Chim.* **4,** 501 (1980).
15. Chan, M. S., and Bolton, J., *Sol. Energy* **24,** 561 (1980).
16. Clark, A., *Ind. Eng. Chem.* **45,** 1476 (1953).
17. Cimino, A., Cordischi, D., Guarino, G., and Micheli, A., *J. Chem. Soc. Faraday Trans.* **67,** 1776 (1971).
18. Couper, A., and Eley, D., *J. Polym. Sci.* **3,** 345 (1968).
19. Chu, B., "Laser Light Scattering" Academic Press, New York, 1974.
20. Chromatix, Application Note LS8, 560 Oakmead Parkway, Sunnyvale, California (1978).
21. Creutz, C., and Sutin, M., *Proc. Natl. Acad. Sci. USA* **72,** 2858 (1975).
22. Cundall, R., Rudham, R., and Salim, M., *J. Chem. Soc. Faraday Trans. 1* **72,** 1642 (1976).
23. Cusumano, S., Dalla Beta, R., and Levy, R., "Catalysis in Coal Conversion." Academic Press, New York, 1978.
24. Davies, J., and Rideal, E., "Interfacial Phenomena." Academic Press, New York, 1963.
25. Dowden, D., "Surface Science," Vol. 2. Int. Atomic Energy Agency, Vienna, 1975.
26. Duonghong, D., Borgarello, E., and Grätzel, M., *J. Am. Chem. Soc.* **103,** 4685 (1981).
27. Durrant, P. J., and Durrant, B. *in* "Advanced Inorganic Chemistry" p. 1072. Wiley (Interscience), New York, 1962.
28. Dunn, W., Aikawa, Y., and Bard, A. J., *J. Electrochem. Soc.* **128,** 222 (1981).
29. Dunworth, W., and Nord, F., *Adv. Catal.* **6,** 125 (1954).
30. Elizarova, G., Matvienko, L., Lozhkina, N., Parmon, V., and Zamaraev, K., *React. Kinet. Catal. Lett.* **16,** 191 (1981).
31. Elizarova, G., Matvienko, G., Maizlich, V., and Parmon, V., *React. Kinet. Catal. Lett.* **16,** 285 (1981).
32. Formenti, M., Juillet, F., Meriaudeau, P., and Teichner, S., *Chem. Technol.* **1,** 680 (1971).
33. Formenti, M., and Teichner, S., Catalysis, Specialist Periodical Report, Vol. 2. Chem. Soc., London, 1979.
34. Germain, J. E., *Intra Sci. Chem. Rep.* **6,** 101 (1972).
35. Gerisher, H., *Pure Appl. Chem.* **52,** 2649 (1980).
36. Ghosh, A., and Maruska, P., *J. Electrochem. Soc.* **124,** 1516 (1977).
37. Ghosh, P., and Spiro, T., *J. Am. Chem. Soc.* **102,** 5543 (1980).
38. Grätzel, C., and Grätzel, M., *J. Am. Chem. Soc.* **101,** 7741 (1979).
39. Grätzel, M., *Discuss. Faraday Soc.* **70,** 359 (1980).
40. Green, D., and Strickland, L., *Biochem. J.* **28,** 898 (1934).
41. Guruswamy, V., Keillor, P., Campbell, G., and Bockris, J.O'M., *Sol. Energy Mater.* **4,** 11 (1980).
42. Hardee, K., and Bard, A. J., *J. Electrochem. Soc.* **123,** 1024 (1976).
43. Hardee, K., and Bard, A. J., *J. Electrochem. Soc.* **124,** 215 (1977).

44. Harris, W., and Wilson, R., *Annu. Rev. Mater. Sci.* **8,** 99 (1978).
45. Hauffe, K., *DECHEMA Monogr.* **26,** 222 (1956).
46. Hauffe, K., Raveling, H., and Rein, D., *Naturwissenschaften* **64,** 91 (1977).
47. Heidt, L., and Smith, M., *J. Am. Chem. Soc.* **70,** 2476 (1948).
48. Heller, A., *Acc. Chem. Res.* **14,** 154 (1981).
49. Hensch, L. L., and Dove, D. B., "Physics of Electron Ceramics." Dekker, New York, 1977.
50. Heller, A. (ed.), "Semiconductor Liquid-Junction Solar Cells." Electrochem. Soc., Princeton, New Jersey, 1977.
51. Hibbert, D., and Tseung, A. C., *J. Chem. Technol. Biotechnol.* **29,** 713 (1979).
52. Ichimura, K., Inoue, Y., Kojima, I., Miyasaki, E., and Yasumori, Y., *Proc. Int. Congr. Catal. 7th,* Paper No. B44.
53. Kalyanasundaram, K., Kiwi, J., and Grätzel, M., *Helv. Chim. Acta* **61,** 2720 (1978).
54. Kalyanasundaram, K., and Grätzel, M., *Angew. Chem. Int. Ed. Engl.* **18,** 701 (1979).
55. Kalyanasundaram, K., Mićić, O., Promauro, E., and Grätzel, M., *Helv. Chim. Acta* **62,** 2432 (1979).
56. Nord, G., and Wernberg, J. *J. Chem. Soc. Dalton,* p. 845 (1975).
57. Kalyanasundaram, K., Borgarello, E., and Grätzel, M., *Helv. Chim. Acta* **64,** 362 (1981).
58. Kelly, J., and Vandeling, J., *J. Electrochem. Soc.* **122,** 1104 (1975).
59. Kita, H., *J. Electrochem. Soc.* **113,** 1095 (1966).
60. Kiwi, J., and Grätzel, M., *Angew. Chem. Int. Ed. Engl.* **17,** 860 (1978).
61. Kiwi, J., and Grätzel, M., *J. Am. Chem. Soc.* **100,** 6314 (1978).
62. Kiwi, J., and Grätzel, M., *Chimia* **33,** 289 (1979).
63. Kiwi, J., and Grätzel, M., *J. Am. Chem. Soc.* **101,** 7214 (1979).
64. Kiwi, J., and Grätzel, M., *Nature* (*London*) **281,** 657 (1979).
65. Kiwi, J., Borgarello, E., Pelizzetti, E., Visca, M., and Grätzel, M., *Angew. Chem. Int. Ed. Engl.* **19,** 647 (1980).
66. Kiwi, J., and Grätzel, M., *J. Am. Chem. Soc.* **84,** 1503 (1980).
67. Kiwi, J., *Chem. Phys. Lett.* **83,** 594 (1981).
68. Kiwi, J., *J. Chem. Soc. Faraday Trans.* **78,** 339 (1982).
69. Kiwi, J., and Hässlin, Hans-W., *Colloids Surfaces* (1983).
70. Kiwi, J., Kalyanasundaram, K., and Grätzel, M., *Struct. Bonding,* (*Berlin*) **49,** 37 (1982).
71. Kofstad, P., "Nonstoichiometry, Diffusion and Electrical Conductivity in Binary Metal Oxides." Wiley (Interscience), New York, 1972.
72. Kuhn, A., and Mortimer, C., *J. Electrochem. Soc.* **120,** 231 (1973).
73. Kuhn, W. (ed.), "Ultrafine Particles." Wiley, New York, 1962.
74. Kung, H., Jarrett, H., Sleight, A., and Ferreti, A., *J. Phys. D* **48,** 2463 (1977).
75. Krucyinsky, L., Gesser, H., Turner, C., and Speers, A., *Nature* (*London*) **291,** 399 (1981).
76. Laitinen, H., Vincent, C., and Bednarski, T., *J. Electrochem. Soc.* **117,** 1343 (1970).
77. Lehn, J.-M., and Sauvage, J., *Nouv. J. Chim.* **1,** 449 (1977).
78. Lehn, J.-M., Sauvage, J. P., and Ziessel, R., *Nouv. J. Chim.* **3,** 423 (1979).
79. Lehn, J.-M., and Ziessel, R., *Nouv. J. Chim.* **4,** 355 (1980).
80. Lu, P., and Srinivasan, S., "BNL Rep. 24914" Upton, New York, 1979.
81. Maruska, P., and Ghosh, A., *Sol. Energy Mater.* **1,** 237 (1979).
82. Matijević, E., *Pure Appl. Chem.* **52,** 1179 (1980).
83. Matsumara, M., Nomura, Y., and Tsubomura, H., *Bull. Chem. Soc. Jpn.* **50,** 2533 (1977).

84. Bickley, R., and Stone, F., *J. Catal.* **31,** 389 (1973).
85. Meadowcroft, D., *Br. J. Appl. Phys.* **2,** 1225 (1969).
86. Meier, H., "Spectral Sensitization." Focal Press, London, 1968.
87. Mills, A., and Zeeman, M. L., *J. Chem. Soc. Chem. Commun.* No. 713, p. 948 (1981).
88. Möllers, F., and Memming, R., *Ber. Bunsenges. Phys. Chem.* **76,** 469 (1972).
89. Mott, N. F., and Jones, H., "Theory and Properties of Metals and Alloys." Oxford Univ. Press, London and New York, 1936.
90. Munuera, G., Felipe, R. G., Soria, J., and Sanz, J., *J. Chem. Soc. Faraday Trans. 1* **76,** 1535 (1980).
91. Murray, W., *Acc. Chem. Res.* **13,** 135 (1980).
92. Myamlin, V., and Pleskov, Y., "Electrochemistry of Semiconductors." Plenum, New York, 1967.
93. Nord, G., and Wernberg, O., *J. Chem. Soc. Dalton Trans.,* 866 (1972).
94. van Olphen, H., "Clay Colloid Chemistry." Wiley, New York, 1965.
95. Parfitt, G., *Prog. Surf. Membrane Sci.* **11,** 181 (1976).
96. Quinn, K., Nasby, R., and Baughman, R., *Mater. Res. Bull.* **11,** 1011 (1976).
97. Rao, M., Rajeshwar, K., Verneker, V., and Du Bow, J., *J. Phys. Chem.* **84,** 1980 (1978).
98. Rao, S., Wanklin, B., and Rao, C., *J. Phys. Chem. Solids* **32,** 345 (1971).
99. Rauh, R., Buzby, J., Reise, T., and Alkaitis, S., *J. Phys. Chem.* **83,** 2221 (1979).
100. Reymond, P., Vergnon, P., Gravelle, P., and Teichner, S., *Nouv. J. Chim.* **1,** 197 (1977).
101. Roberts, E., *Philips Tech. Rundsch.* **35,** 72 (1975).
102. Roy, A., and Nag, K., *J. Inorg. Nucl. Chem.* **40,** 1501 (1978).
103. Salvador, P., *Sol. Energy Mater.* **2,** 413 (1980).
104. Sato, T., Sugihara, M., and Saito, M., *Rev. Electr. Commun. Lab.* **11,** 26 (1963).
105. Shafirovich, V., Moravskii, A., Dzhabiev, T., and Shilov, A., *Kinet. Katal.* **18,** 509 (1977).
106. Shafirovich, V. Ya., and Shilov, A. E., *Kinet. Katal.* **20,** 1156 (1979).
107. Shafirovich, V. Ya., Khannanov, M. K., and Strelets, V. V., *Nouv. J. Chim.* **4,** 81 (1980).
108. Shanon, R. D., *Acta Crystallogr Sect. A* **32,** 751 (1976).
109. Shapiro, Y., Cox, A., and Lichtman, D., *Surf. Sci.* **54,** 43 (1976).
110. Smith, R. A., "Semiconductors." Cambridge Univ. Press, Cambridge, 1978.
111. Somorjai, G. A., *Science (Washington, D.C.)* **201,** 489 (1978).
112. Srinivasan, S., and Salzano, F., *Int. J. Hydrogen Energy* **2,** 53 (1977).
113. Sutin, M., *J. Photochem.* **10,** 9 (1979).
114. Tadros, Th. F., *Adv. Colloid Interface Sci.* **12,** 141 (1980).
115. Tauster, S., Fung, S., and Garten, R., *J. Am. Chem. Soc.* **100,** 170 (1980).
116. Toshima, N., Kuriyama, M., Yamada, Y., and Hirai, H., *Chem. Lett.,* 793 (1981).
117. Tricot, I., Kiwi, J., Niederberger, N., and Grätzel, M., *J. Phys. Chem.* **85,** 862 (1981).
118. Trim, D., "Design of Industrial Catalysts." Elsevier, New York, 1980.
119. Tseung, A. C., and Bevan, H., *J. Mater. Sci.* **5,** 604 (1970).
120. Turkevich, J., Aika, K., Ban, L., Okura, I., and Namba, S., *J. Res. Inst. Catal. Hokkaido Univ.* **24,** 54 (1976).
121. Turnock, A., *J. Am. Chem. Soc.* **49,** 177 (1966).
122. Vannice, M., and Garten, R., *J. Catal.* **56,** 236 (1980).
123. Vannice, M., and Garten, R., *J. Catal.* **63,** 255 (1980).
124. Voorhoeve, R., Johnson, D., Remeika, P., and Gallagher, P., *Science (Washington, D.C.)* **195,** 827 (1977).

125. Vrieland, E., *J. Catal.* **32,** 415 (1974).
126. Wagner, C., and Traud, W., *Z. Electrochem.* **44,** 397 (1938).
127. Wagner, F., and Somorjai, G. A., *J. Am. Chem. Soc.* **102,** 5494 (1980).
128. Wheeler, A., *in* "Reaction Rates and Selectivity in Catalyst Pores in Catalysis" (P. H. Emmett, ed.), Vol. 2. Reinhold, New York, 1955.
129. Yao, Y., *J. Catal.* **36,** 266 (1975).
130. Yoneyama, H., Ohkubo, T., and Tamura, H., *Bull. Chem. Soc. Jpn.* **54,** 404 (1981).
131. Young, C., Meissel, P., Mazer, N., Benedek, G., and Carey, M., *J. Phys. Chem.* **82,** 1375 (1978).
132. Zsigmondy, R., *Z. Anal. Chem.* **40,** 197 (1901).

10 Photosynthesis and Photocatalysis with Semiconductor Powders

T. Sakata
T. Kawai

Institute for Molecular Science
Okazaki, Japan

ISBN 0-12-295720-2

I. Introduction

Total photosynthetic production on the Earth is estimated to be 2×10^{11} t/yr, expressed as carbon (*1*). A huge amount of energy, 6–10 times as much as the total energy demand of mankind, is stored every year through the photosynthetic activity of green plants and some bacteria. Its utilization to produce fuels has been proposed by a number of people (*2, 3*). The production of methane or ethanol by fermentation of biomass or the extraction of hydrocarbons from some special plants such as *Hevea* and *Euphorbia* have been considered. When it is taken into consideration that photosynthetic production corresponds to only 0.03% of the solar energy reaching the Earth, it seems highly desirable to promote this new type of agriculture to produce fuels. Biomass utilization can be called a classical and biological method for solar-energy conversion. In contrast, many attempts at chemical conversion are being carried out, aimed at "artificial photosynthesis." Concerning artificial photosynthesis, it may be said that there are three important problems to be solved: (a) decomposition of water, (b) reduction of carbon dioxide, and (c) nitrogen fixation. if we succeed in splitting water efficiently with sunlight, we can obtain hydrogen from water, which exists inexhaustibly on the Earth. Hydrogen production from water and the sun would constitute a recycling and ecologically sound energy system. The establishment of a method for carbon dioxide reduction would enable us to synthesize various organic compounds such as textiles and drugs from carbon dioxide, which is the final and stable oxidation product of carbon compounds. Nitrogen fixation under mild conditions, which is already achieved by root-nodule bacteria, would be an important means of producing nitrogen-containing com-

TABLE I

Typical Photocatalytic Reactions for Artificial Photosynthesis

Reaction	Description	$\Delta G°$/transferred electron (eV)	Reference
(1) $H_2O \longrightarrow H_2 + \frac{1}{2}O_2$	Water splitting	1.23	*6*
(2) $CO_2 + 2H_2O \longrightarrow CH_3OH + \frac{3}{2}O_2$	Carbon dioxide reduction	1.21	*37, 38*
(3) $CO_2 + 2H_2O \longrightarrow CH_4 + 2O_2$	Same as Eq. (2)	1.06	*36*
(4) $N_2 + 3H_2O \longrightarrow 2NH_3 + \frac{3}{2}O_2$	Nitrogen reduction	1.17	*8*
(5) $N_2 + 2H_2O \longrightarrow N_2H_4 + O_2$	Same as Eq. (4)	1.10	*8*
(6) $CO_2 + H_2O \longrightarrow \frac{1}{6}C_6H_{12}O_6 + O_2$	Photosynthetic reaction	1.25	

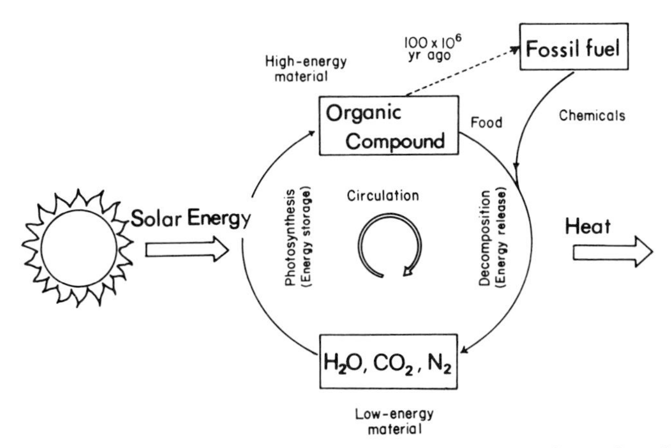

Fig. 1. Circulation of energy and matter on the Earth. Stable molecules such as H_2O, CO_2, and N_2 are converted into various organic compounds by the photosynthesis of green plants, through which solar energy is stored as chemical energy. These high-energy materials, some of which are utilized to synthesize useful materials for human beings, are burned (oxidized) or decomposed. This produces a steady flow of energy and a circulation of materials on the earth.

pounds. In nature, three molecules (H_2O, CO_2, and N_2) are activated to produce energy-rich compounds through photosynthetic processes (Fig. 1). The energy contained in these photosynthetic products is released again to maintain life. The circulation of energy and materials forms a world of life. Table I shows some representative reactions of artificial photosynthesis together with the photosynthetic reaction of green plants. As can be seen in this table, water splitting is the most fundamental of all the reactions. In each reaction, water is involved as a reducing agent, and, in the reduction of carbon dioxide or in nitrogen fixation, hydrogen from water is captured by CO_2 or N_2.

II. Photocatalytic Effect of Semiconductors

To induce these redox reactions a photocatalyst or photosensitizer is used, because the photon energy of sunlight is too small to excite these simple molecules electronically. The role of the photocatalyst is essentially the same as the role chlorophyll plays in the photosynthetic system. Various kind of photocatalysts have been exploited. They are classified into two groups: (a) dye molecules in homogeneous systems such as $Ru(bipy)_3^{2+}$ and metal porphyrin and (b) semiconductors in a heterogeneous system. Sometimes a semiconductor is used, combined with a dye.

In addition, a photocatalyst–thermal catalyst such as Pt or RuO_2 is used to promote charge separation and to catalyze the reactions. The semiconductor photocatalyst has essentially the same properties as the molecular photocatalyst (dye). However, it has several properties characteristic of the solid:

1. Electron orbitals dispersed throughout the solid compose a band at each state. For this reason, the semiconductor photocatalyst functions as a pool of electrons or holes, which is advantageous for multielectron-transfer processes such as H_2 and O_2 evolution.
2. A space-charge layer is formed at the interface between the semiconductor and a liquid (or gas). At this layer electrons and holes, which are generated under illumination, are separated efficiently.
3. The surface often becomes a place on which catalytic reactions proceed. Usually, a catalyst such as Pt and RuO_2 is added on the surface to make an active catalytic surface.

The second characteristic is the origin of the photovoltaic effect of the semiconductor, which is utilized in a solar cell. A photoelectrochemical cell composed of a semiconductor electrode and an electrolyte solution is closely connected with the photocatalytic action of the semiconductor (*4, 5*). When a semiconductor absorbs photons, electrons are excited to the conduction band and holes are generated in the valence band. These electrons and holes are used for reduction and oxidation reactions, respectively. The energy level of the bottom of the conduction band can be considered to be a measure of the reduction strength of the photoexcited electrons, whereas that of the upper edge of the valence band is a measure of the oxidation strength of the holes. Figure 2 shows the energy levels of several semiconductors in an aqueous solution of pH 0 together with the redox potential of hydrogen evolution (H^+/H_2) and oxygen evolution (O_2/H_2O). The potential of these semiconductors for oxidation and reduction can be classified into four groups from the view point of the water-splitting reaction:

1. OR type. The oxidation and reduction power is strong enough to enable (in principle) hydrogen and oxygen to evolve. Examples are $SrTiO_3$, TiO_2, and CdS. O is the abbreviation for oxidation and R for reduction. OR indicates a strong ability for both oxidation and reduction.

2. R type. Only the reduction power is strong enough to reduce water (H_2 evolution). The oxidation power is too weak to oxidize water. Examples are CdTe, CdSe, and Si.

3. O type. The valence band is located deeper than the O_2/H_2O level so that the oxidation power is strong enough to oxidize water but the

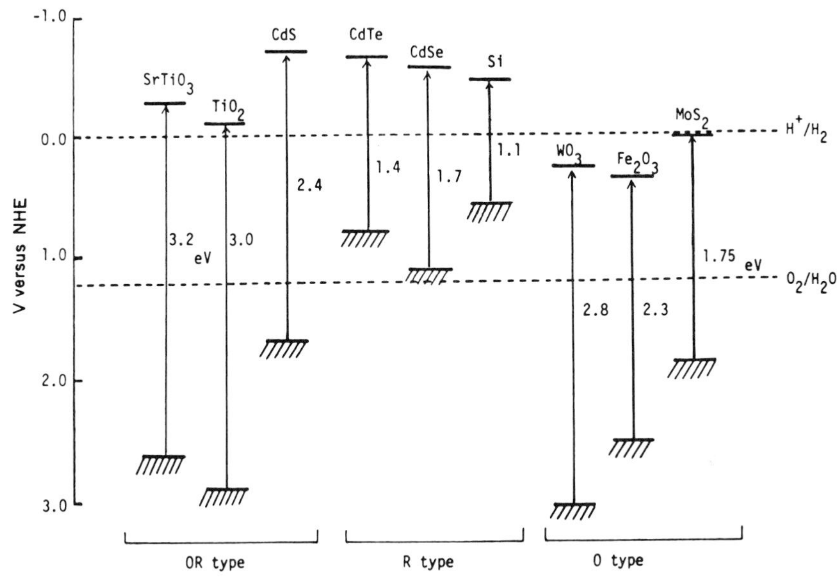

Fig. 2. Energy-level diagram indicating the energy positions of the conduction and valence bands for various semiconductors in aqueous solution at pH 0. The location of the valence band is a measure of the oxidation power of the photogenerated holes, and that of the conduction band is a measure of the reduction power of electrons photoexcited to that band.

reduction power not strong enough to reduce water. Examples are WO_3, Fe_2O_3, MoS_2, and Bi_2O_3.

4. X type. The conduction and valence bands are located between the H^+/H_2 and O_2/H_2O levels. Therefore, both the oxidation and reduction powers are so weak and neither oxygen nor hydrogen can be evolved.

Among these types the OR-type semiconductor is needed for the complete decomposition of water. For half-decomposition, the R type can be applied to hydrogen production using a sacrificial reducing agent and the O type to oxygen evolution using an electron acceptor. As well as the location of energy levels, the photocatalytic properties of a semiconductor depend on the lifetime of electrons or holes (minority carriers), their mobility, and the catalytic activity of the surface.

III. Photoassisted Decomposition of Water with Powdered Semiconductors

TiO_2, which is typical of OR-type semiconductors, can decompose water, as was shown for the first time by Fujishima and Honda (*6, 6a*). The quantum yield is reported to be 10^{-3}–10^{-4} for a single-crystal photoelectrode in a photoelectrochemical cell (*7*). The reason for this low quantum yield is ascribed to the difficulty of hydrogen evolution because of the small negative value of the flat-band potential (−0.11 V versus NHE). For $SrTiO_3$ (E_g = 3.2 eV) the efficiency is improved, because the conduction band is at a level more negative than that of TiO_2.

Photoassisted water decomposition by powdered semiconductors such as TiO_2 (*8*) and $SrTiO_3$ (*9*) has also been reported, although there is some doubt about the catalytic nature (*10*). Interestingly, photocatalytic activity is increased remarkably by supporting a metal or metal oxide such as Pt, Rh, NiO, and RuO_2 on the semiconductor surface (*11–16*). Hydrogen and oxygen evolve continuously with the photocatalyst. Figure 3 shows an example of the use of a RuO_2–TiO_2–Pt photocatalyst. Table II shows the effect of Pt and RuO_2. As can be seen in this table, oxygen could hardly be detected for TiO_2 alone as Damme and Hall reported (*10*). The hydrogen-evolution rate for the RuO_2–TiO_2–Pt photocatalyst is about 10^2–10^3 times larger than that for TiO_2 alone, and oxygen evolves stoichiometrically and continuously for this surface-modified photocatalyst. Lehn *et al.* also reported the effect of depositing various metals on the $SrTiO_3$

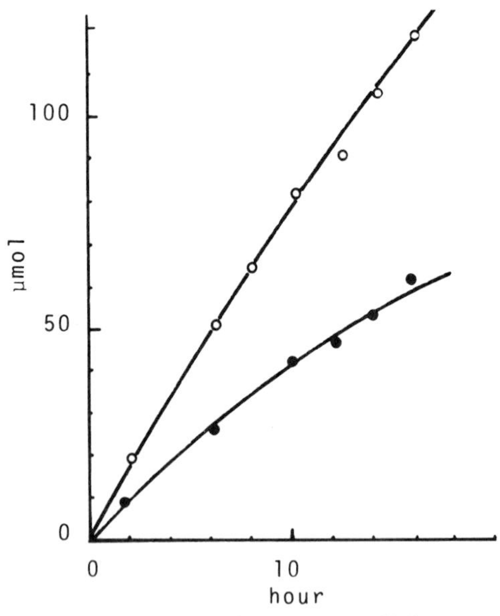

Fig. 3. Continuous evolution of hydrogen (○) and oxygen (●) from water versus the irradiation time. Experimental conditions: water (20 ml), RuO_2–TiO_2–Pt (300 mg; weight ratio 5 : 100 : 10), 500-W high-pressure Hg lamp (Philips Co.).

TABLE II

Simultaneous Hydrogen and Oxygen Evolution from Water by Use of Various TiO_2 Photocatalysts[a]

Photocatalyst	Reactant	H_2 (μmol/10 h)	O_2 (μmol/10 h)
TiO_2	H_2O(g)	0.004	0
TiO_2	H_2O(l)	0.1	0
TiO_2	6 *N* NaOH	0.3	0
TiO_2–Pt	H_2O(l)	19	12
RuO_2–TiO_2–Pt[b]	H_2(l)	82	38

[a] For the reaction $H_2O \rightarrow H_2 + \frac{1}{2}O_2$, hydrogen and oxygen were detected by a quadrupole mass spectrometer. A 500-W Hg lamp (Philips Co.) was used as a radiation source; photocatalyst (300 mg), water (20 ml).
[b] RuO_2–TiO_2–Pt (weight ratio 5 : 100 : 10), TiO_2–Pt (100 : 10).

surface on the water-splitting reaction (*12*). For Rh–$SrTiO_3$, the hydrogen-evolution rate increased 60–3000 times over that for $SrTiO_3$ alone. Oxygen evolves stoichiometrically against hydrogen for these metal-loaded $SrTiO_3$ catalysts, whereas no oxygen could be detected for $SrTiO_3$ alone (as was also observed for TiO_2 alone). Duonghong *et al.* reported very efficient water splitting (quantum yield of H_2 evolution, 30%) by using transparent TiO_2 sol (particle radius 200 Å), on the surface of which ultrafine RuO_2 and Pt were deposited simultaneously (*17*). Grätzel's group also reported water splitting with visible light, using RuO_2–CdS–Pt (E_g = 2.4 eV) (*18*). They reported that this photocatalyst is stable under long-time irradiation and that both RuO_2 and Pt are indispensable for the activity. Although CdS on which only Pt or RuO_2 is supported does not show any photocatalytic activity for water splitting (nor does CdS alone), simultaneous addition both of Pt and RuO_2 brings about the activity even though the efficiency is not good. This observation is worthy of notice, because the CdS electrode easily undergoes photocorrosion, which is very hard to prevent without addition of strong reducing agents such as S^{2-} and Sn^{2-} (*19, 19b, 20*). In the preceding active photocatalysts, RuO_2–TiO_2 (or CdS–Pt), RuO_2 has been considered to catalyze oxygen evolution because of its high activity for oxygen evolution. However, energetically it would also be possible for RuO_2 on an *n*-type semiconductor to function as a reducing site, because it shows a metallic conductivity. Actually, powdered RuO_2 is a good catalyst for H_2 evoution (*21*), but its role is not clear.

As shown in the preceding examples, research on photoassisted water decomposition is still at an early stage and far from practical application.

However, it is desirable to develop the research because the success of the reaction would bring an incalculable benefit to mankind.

IV. Hydrogen Production from the Photocatalytic Reaction of Water and Organic Compounds

In the decomposition of water, oxygen and hydrogen evolve simultaneously, and the mixture is explosive in a closed system. When a molecular photocatalyst such as $Ru(bipy)_3^{2+}$ or a semiconductor photocatalyst is used, the oxidation products such as $OH\cdot$, $O_2H\cdot$, H_2O_2, and O_2 can be separated from the reduction products such as $H\cdot$ and H_2 only with great difficulty. Therefore, back reactions take place easily. This is one of the most important reasons for the poor decomposition efficiency in a closed system. However, the hydrogen-evolution efficiency was found to increase by a factor of $10–10^3$ with addition of organic compounds. In this reaction, the organic compounds are oxidized and water is reduced to produce hydrogen. The principle is shown schematically in Fig. 4. With TiO_2 photocatalysts, organic compounds are finally oxidized to CO_2, owing to the strong oxidation power of TiO_2 (Fig. 2).

A. Photoinduced Water-Gas Reaction and Hydrogen Evolution from Fossil Fuels and Water

$$C + 2H_2O \xrightarrow[h\nu]{\text{photcat}} 2H_2O + CO_2 \qquad \Delta G^\circ = +63 \quad \text{kJ/mol} \tag{1}$$

$$C + H_2O \xrightarrow[h\nu]{\text{photcat}} H_2 + CO \qquad \Delta G^\circ = +92 \quad \text{kJ/mol} \tag{2}$$

Solid carbon reacts with water at room temperature to produce hydrogen (*22, 23*). This reaction is well known as the water-gas reaction and proceeds thermally at a temperature higher than 800°C. By use of the TiO_2 photocatalyst and light energy, it is possible to drive this reaction at room temperature. This reaction is interesting because it describes the gasification of coal at room temperature. Actually, hydrogen evolves from an aqueous suspension of coal. Table III shows its application to various fossil fuels and simple molecules like benzene as model compounds for coal (*24*).

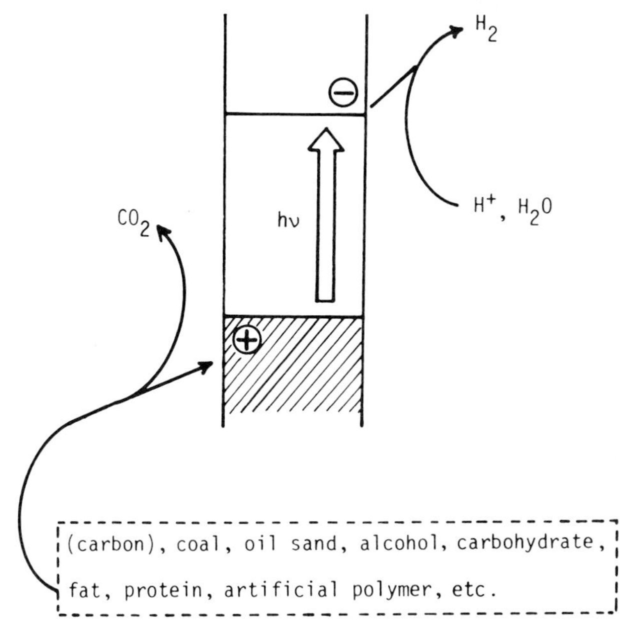

Fig. 4. Schematic illustration of the principle of photocatalytic hydrogen evolution from water and various organic materials. Organic materials function as reducing agents and reduce water with the assistance of photoexcitation.

B. Efficient Hydrogen Evolution from Alcohols and Water

Methanol (*25*)

$$CH_3OH + H_2O \longrightarrow 3H_2 + CO_2 \qquad (3)$$
$$\Delta G^\circ = 9.0 \quad \text{kJ/mol}$$

Ethanol (*26*)

$$C_2H_5OH + H_2O \longrightarrow 2H_2 + CH_4 + CO_2 \qquad (4)$$
$$\Delta G^\circ = -33.9 \quad \text{kJ/mol}$$

Alcohols react photocatalytically with water and produce hydrogen vigorously (*25, 26*). In the case of ethanol, methane is produced as well as hydrogen, and there seem to be side reactions in addition to reaction (4). Table IV shows the quantum yield of hydrogen production from a water–methanol mixture and a water–ethanol mixture under monochromatic irradiation at 380 nm. The quantum yield is defined by the equation

$$\text{Quantum Yield} = \frac{2 \times \text{the number of } H_2 \text{ molecules}}{\text{number of incident photons}}. \qquad (5)$$

TABLE III

Hydrogen Evolution from Water and Fossil Fuels or Their Model Compounds Using the TiO_2–Pt Photocatalyst

Reactant	H_2 evolution (μmol)[a]	
	Neutral	Alkaline (5 *N*)
Fossil fuel		
Coal	13	60–90[b]
Oil sand	5	70–100[b]
Pitch	10	90–100[b]
Aliphatic hydrocarbon		
n-Pentane	44	84
n-Heptane	66	110
Isooctane	24	94
n-Paraffin	12	23
Polyethylene	10	39
Aromatic compound		
Benzene	40	360
Phenol	41	330
Pyridine	27	310

[a] 500-W Xe lamp; irradiated for 10 h. TiO_2–Pt (300 mg).
[b] Hydrogen-evolution rate at the early stage of the reaction. The evolution rate decreases gradually with irradiation time.

As can be seen in Table IV, the quantum yield is very high for RuO_2–TiO_2–Pt. In Fig. 5 is shown the dependence of the hydrogen-evolution rate from an ethanol–water mixture (1 : 1 in volume) on the kind of metal that is loaded on the TiO_2 surface. Among the various metals that have

TABLE IV

Quantum Yields of Hydrogen Production from Water–Alcohol Mixtures under Monochromatic Irradiation at 380 nm.

Alcohol	Photocatalyst	Quantum yield (%)
Methanol	TiO_2	3
	RuO_2–TiO_2–Pt	55
Ethanol	TiO_2	0.9
	TiO_2–Pt	38

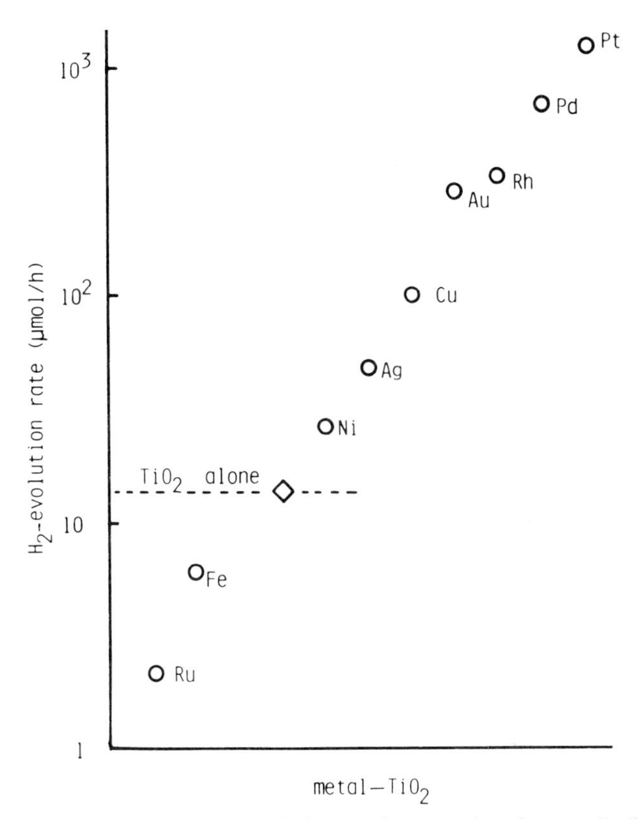

Fig. 5. Dependence of the hydrogen-evolution rate from an ethanol–water (1 : 1) mixture on the kind of metal that is supported on TiO_2 photochemically (molar ratio of metal : TiO_2 = 3.0%).

been loaded on TiO_2, Pt, Pd, and Rh have a remarkable effect on the hydrogen-evolution rate. Because these metals are known to be good hydrogenation catalysts, this result seems to indicate the importance of the catalytic effect of metals on hydrogen evolution. The photocatalytic activity of hydrogen production from alcohols was maintained during long-time irradiation.

C. Hydrogen Production from Photosynthetic Products and Water

Carbohydrates, fat, and protein are oxidized to produce hydrogen, reducing water at the same time.

1. Carbohydrates

Glucose, sugar, starch, and cellulose can react with water to produce hydrogen and carbon dioxide (*27*). This reaction is most interesting when combined with the photosynthetic reaction of green plants. Taking glucose as an example:

$$6CO_2 + 6H_2O \xrightarrow{\text{photosynthesis}} C_6H_{12}O_6 + 6O_2 \quad (6)$$
$$\Delta G^\circ = 2870 \quad \text{kJ/mol}$$

$$C_6H_{12}O_6 + 6H_2O \xrightarrow[\text{reaction}]{\text{photocatalytic}} 6CO_2 + 12H_2 \quad (7)$$
$$\Delta G^\circ = -32 \quad \text{kJ/mol}$$

$$12H_2O \longrightarrow 6O_2 + 12H_2 \quad (8)$$
$$\Delta G^\circ = 2838 \quad \text{kJ/mol}$$

In nature, glucose is formed by photosynthetic reaction (6), through which light energy is stored. In photocatalytic reaction (7), glucose reacts with water to produce hydrogen, through which a small amount of free energy is lost. However, the loss is very small, only 1.1% of the energy stored in reaction (6). Therefore, photocatalytic reaction (7) is close to isoenergetic. The total reaction (8), which is the sum of reactions (6) and (7), is the water-splitting reaction.

Figure 6a shows the evolution of H_2 and CO_2 from sugar plotted against irradiation time. In a neutral, aqueous solution, the ratio of H_2 to CO_2 is 2.0, as expected from reaction (7). Interestingly, sugar reacts completely with water after long-time irradiation, producing H_2 and CO_2 in amounts that agree well with those calculated theoretically (*27*). This indicates the strong oxidizing power of TiO_2 under illumination, which is in a marked contrast to the mild oxidation process of fermentation by microorganisms.

When D_2O was used instead of H_2O for the decomposition of cellulose (cotton), D_2 made up 92% of the evolved hydrogen gas, DH 6%, and H_2 2% (*28*). Similar results were obtained for other biomasses. This finding supports the proposed mechanism in Fig. 4 in which water reacts with organic compounds and is reduced on the photocatalyst to produce hydrogen.

2. Amino Acids, Protein, and Fat

Amino acids and proteins are decomposed by reaction with water in the presence of TiO_2–Pt catalyst. NH_3 as well as H_2 is produced in these nitrogen-containing compounds. Reaction (9) is offered as one of the main

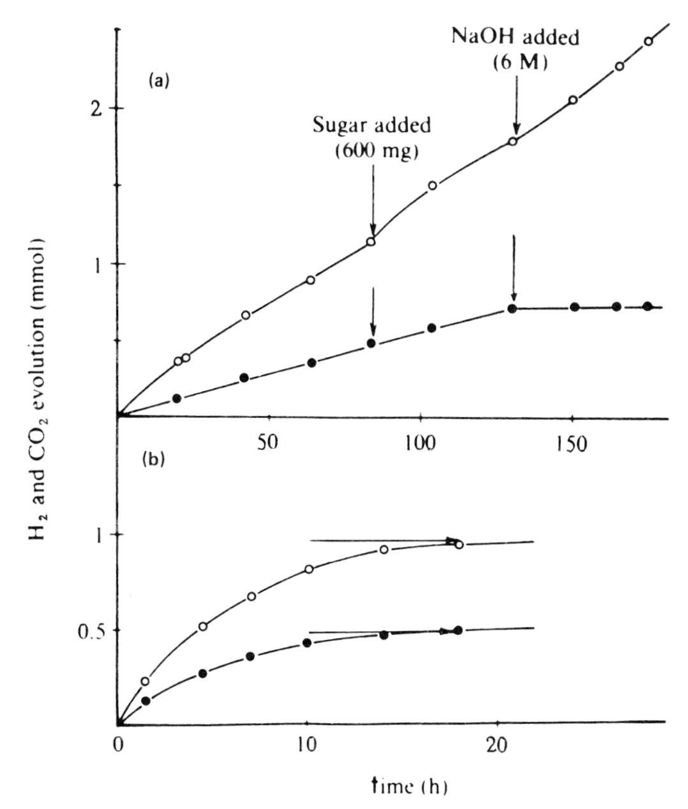

Fig. 6. (a) H_2 (○) and CO_2 (●) evolution from sugar in water versus the irradiation time. Experimental conditions: sugar (600 mg), H_2 (40 ml), RuO_2–TiO_2–Pt (300 mg), 500-W Xe irradiation. Additional sugar (600 mg) and NaOH (9.6 g) were added to the solution at 85 and 130 h after the start of the reaction, respectively. (b) H_2 (○) and CO_2 (●) production from a small amount of sugar versus the irradiation time. Experimental conditions: 500-W ultra-high-pressure Hg lamp, sugar (13 mg), H_2O (40 ml), RuO_2–TiO_2–Pt (150 mg). Arrows indicate the estimated amounts of H_2 and CO_2 that are produced from 13 mg of sugar.

paths for the decomposition of glycine (*29*). Table V shows the yield of hydrogen production from the reaction of water with carbohydrates, proteins, and organic acids.

$$NH_2CH_2COOH + 2H_2O \longrightarrow 3H_2 + NH_3 + 2CO_2 \quad (9)$$

3. Various Biomasses

In Table VI the rate of hydrogen production for various biomasses in an aqueous solution is shown. In the fraction trapped at −197°C, products

TABLE V

Hydrogen Evolution from Water and Carbohydrates, Amino Acids, or Fatty Acids Using a TiO_2–Pt Photocatalyst

Reactant	Formula	H_2 (μmol/10 h; neutral H_2O)
Carbohydrate		
Glucose	$C_6H_{12}O_6$	1130
Sugar	$C_{12}H_{22}O_{11}$	920
Starch	$(C_6H_{10}O_5)_n \doteqdot 100$	240
Cellulose (filter paper)	$(C_6H_{10}O_5)_n \doteqdot 1000$	40
Protein (amino acid)		
Glycine	NH_2CH_2COOH	220
Glutamic acid	$HOOC(CH_2)_2CHNH_2HCOOH$	126
Proline	$C_5H_9NO_2$	130
Gelatin	Glycine, proline, and glutamic acid	71
Fatty acid		
Stearic acid	$CH_3(CH_2)_{16}COOH$	88
Olive oil	—	32
Pyruvic acid	$CH_3COCOOH$	323
Fumaric acid	$C_2H_2(COOH)_2$	25
Bee wax	—	42

A 500-W Xe lamp (Ushio CO. Ltd.) was used as the radiation source.

such as C_2H_6, CH_3OH, C_2H_5OH, $(CH_3)_2CO$, and NH_3 were observed, as shown in Table V (*29*). The quantity and kind of products varied, depending on the kind of biomass. These observations show that simple organic molecules are produced by the decomposition of starch, cellulose, lignin, proteins, and fatty acids, of which biomass is composed. Because much CO_2 is produced in a neutral aqueous medium, the components of biomass seem to be oxidized finally into CO_2 by water through photocatalytic reactions, as demonstrated in the cases of alcohols and carbohydrates.

In Fig. 7 is shown a schematic illustration of this method combined with the photosynthesis of green plants (*28*). As can be seen, water is oxidized by the photosynthetic reactions of green plants and is reduced by the photocatalytic reaction of semiconductors. Consequently, the total reaction is the decomposition of water into H_2 and O_2, where both biomass and CO_2 are considered as intermediates. The same principle was applied to hydrogen production from artificial polymers such as polyethylene, nylon, polyvinylchloride, and urea resin (*29*).

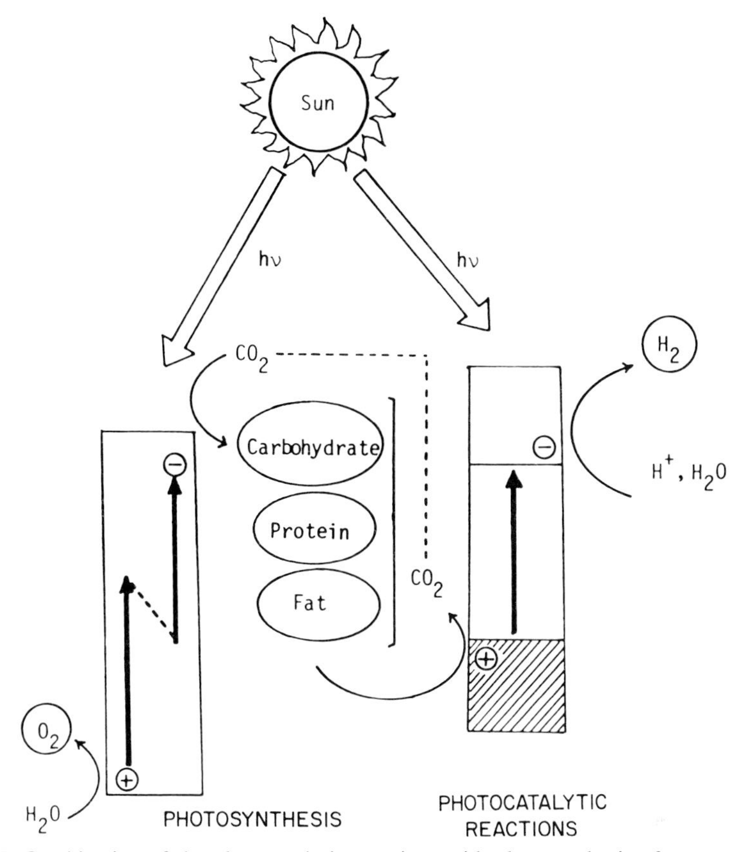

Fig. 7. Combination of the photocatalytic reactions with photosynthesis of green plants. Solar energy is stored in the photosynthetic products, which are decomposed to produce hydrogen in the photocatalytic reaction. The total reaction is water splitting. The whole system is semiartificial in the sense that the photosynthetic products are indispensable.

V. Hydrogen Production by Visible Light

TiO_2 acts under uv irradiation, because its band gap is 3.0 eV. There exist a variety of semiconductors that absorb visible light. The requirement of the photocatalyst for H_2 production is that the conduction band should be equal to or higher than the redox potential of H^+/H_2. Furthermore, the deeper the valence-band position, the better the oxidation of organic compounds. Table VII shows H_2-production rates from water with various platinized semiconductor powders and with EDTA as a donor. TiO_2, CdS, and GaP, the conduction bands of which are higher than the redox potential of H^+/H_2, produced much hydrogen, and WO_3 and Fe_2O_3, the

TABLE VI

Hydrogen Production from Water and Various Biomasses Using the TiO_2–Pt Photocatalyst

Biomass	H_2 (μmol; neutral)[a]	H_2 (μmol; 5 *M* NaOH)[a]	Quantum yield (%) at 380 nm (5 *M* NaOH)	Other products
Natural products				
Glucose	1130	490	8.5[b]	CH_3CHO, C_2H_5OH, and $(CH_3)_2CO$
Ethanol	5080	520	38[b]	C_2H_6, CH_3CHO, and CH_3COOH
Cellulose (cotton)	21	200	1.5	C_2H_6, C_2H_5OH, and $(CH_3)_2CO$
Lignin	12	77	0.6	—[c]
Pyruvic acid	323	206	2.4[b]	C_2H_5OH
Glycine	220	—[c]	1.6[b]	NH_3
Food				
Sweet potato	39	378	2.8	CH_3OH and $(CH_3)_2CO$
Fatty oil	72	212	1.6	C_2H_6
Herbs and Wood				
Cherry wood	—[c]	148	1.1	C_2H_6, CH_3OH, and $(CH_3)_2CO$
White Dutch clover	54	142	1.1	CH_4, CH_3OH, and C_2H_5OH, NH_3
Goldenrod	12	55	0.4	CH_3OH and NH_3
Water hyacinth	25	130	0.9	NH_3
Rice plant	23	175	1.3	—[c]
Green algae and other seaweeds				
Chlorella	73	270	2.0	NH_3
Seaweed (wakame)	74	166	1.2	NH_3
Laver	0	332	3.3	—[c]
Dead animals and excrement				
Cockroach	—[c]	86	0.6	NH_3
Human urine	18	228	1.7	NH_3 and $(CH_3)_2CO$
Cow dung	12	198	1.5	—[c]

[a] 10 h of irradiation by a 500-W Xe lamp.
[b] Quantum yield (%) in a neutral aqueous solution.
[c] not determined.

conduction-band positions of which are lower than H_2/H_2O, produced only a very small amount of H_2, even though EDTA could be easily oxidized. CdS and GaP act under visible light, in contrast to the TiO_2 photocatalyst. Because CdS had an excellent performance, we have attempted H_2 production from water and various organic materials using CdS (Table VIII). Visible light can be used effectively for H_2 production

TABLE VII

Hydrogen Evolution with EDTA as an Electron Donor and a Platinized Semiconductor Photocatalyst[a]

Platinized semiconductor	H_2-evolution rate (μmol/10 h)	
	Xe full spectrum	$\lambda > 430$ nm
TiO_2	3320	24
CdS	3260	1152
GaP	768	432
WO_3	12	0
Fe_2O_3	25	0

[a] 500-W Xe lamp was used as a radiation source.

TABLE VIII

Hydrogen Evolution from Various Organic Materials and Water with Platinized CdS as Photocatalyst[a]

Reactant	Formula	Xe full spectrum (nm)	$\lambda > 430$ nm
Methanol	CH_3OH	1200	480
Ethanol	C_2H_5OH	8400	4000
Isopropanol	C_2H_6CHOH	7100	3000
n-Propanol	C_3H_7OH	2700	1350
n-Butanol	C_4O_9OH	2500	1600
Glucose[b]	$C_6H_{12}O_6$	1400–1100	440
Sugar[b]	$C_{12}H_{22}O_{11}$	830–560	370
Cellulose[b]	$(C_6H_{10}O_5)_n$	20	17–7
Polyethylene glycol	$(CH_2CH_2O)_n$	1400	540
Glycine[b]	NH_2CH_2COOH	2200–1500	1100–780
Proline[b]	NHC_4H_7COOH	1200–560	430
Urea resin[b]	$(CH_2NHCONH)_n$	1200–400	300–200
Acetaldehyde[b]	CH_3CHO	840	420
Acetic acid	CH_3COOH	0	0
Polyvinylchloride	$(CH_2CHCl)_n$	0	0
Polyethylene	$(CH_2CH_2)_n$	0	0
None	—	0	0

[a] 500-W Xe lamp (H_2 : μmol/10 h) was used as the radiation source.
[b] Basic solution (5 *N* NaOH).

from various organic compounds, such as alcohols, carbohydrates, or amino acids. Even urea resin, a synthetic polymer, could be decomposed by this photocatalyst. The reactivity increased more than 40 times by depositing Pt. One of the characteristics of this photocatalyst is that its valence-band position is shallower than that of TiO_2. Accordingly, the hole created in CdS shows a weaker oxidation power than that of TiO_2. Acetic acid, for example, can be decomposed to produce CH_4, C_2H_6, and CO_2 on TiO_2–Pt, but this reaction does not occur on CdS–Pt. Another example is the oxidation of aldehyde. An alcohol is oxidized to form an aldehyde, and the aldehyde is then oxidized to form the carboxylic acid, with the simultaneous production of H_2. The oxidation of acetaldehyde is about 5 times slower on CdS–Pt than on TiO_2–Pt. When a CdS–Pt–ethanol–water mixture is used, acetaldehyde can be accumulated in the solu-

TABLE IX

Hydrogen Production from Water–Ethanol Mixture by Use of Various Platinized Semiconductors[a]

Semiconductor	H_2-production rate (μmol/10 h)	
	Neutral	Alkaline
SiC	120	420
SiC ($\lambda > 430$ nm)	66	290
GaP	42	28
Si	69[b]	—
CdSe	170[b]	1,300
CdS	8300	12,500
TiO_2	5800	600
$MoSe_2$	380[b]	170[b]
MoS_2	360	1,100
$MoTe_2$	110	96
Fe_3O_4	92	180
Fe_2O_3	30	150
WO_3	10	—
CdTe	102	24
WS_2	92	530
WSe_2	480	25
GaAs	48[b]	740[b]
InP	26	25

[a] 500-W Xe lamp, platinized photocatalyst (300 mg), ethanol–water (~1 : 1).

[b] Photocatalytic activity decreased with time.

tion. Polyethylene and polyvinylchloride, which could be decomposed on TiO_2–Pt, cannot be decomposed on CdS–Pt. Accordingly, for CdS we emphasize a milder oxidizing power than that of TiO_2 as well as the utilization of visible light.

Because ethanol is an excellent reducing agent, by using various semiconductor powders we have studied H_2 production from a H_2O–ethanol mixture, as is shown in Table IX. Photocatalytic H_2 production was found to be possible for several semiconductor powders. The highest reactivity is shown by TiO_2 and CdS. Fe_2O_3 and WO_3, the conduction-band positions of which are below the redox potential of H^+/H_2, are poor photocatalysts in spite of their deep valence-band positions. Si, GaP, and CdTe have conduction bands high enough for H_2 production, but their valence-band positions are not deep. For this reason, they are also poor photocatalysts for H_2 production from the ethanol–H_2O mixture. CdS, TiO_2, and MoS_2 have conduction-band positions close to the redox potential of H^+/H_2 and, furthermore, have moderate valence-band positions. They are good photocatalysts for H_2 production when used with visible and uv light. For the CdS–Pt–ethanol–water system, a 200-ml bulb is filled with H_2 at 1 atm within only 10 h by the use of visible light (500-W lamp, Ushio Co. Ltd.). The H_2-production rate on Si or CdSe powders decreases with time. This is because these semiconductor surfaces are not stable in aqueous solution under irradiation.

VI. Energy Structure of TiO_2–Pt Particle and Its Photocatalytic Activity

A. "Photochemical Diode" Model of Powdered Semiconductor Photocatalysts and Its Electronic Structure

Nozik, considering the photochemical properties of a metal–semiconductor diode, designated it a "photochemical diode" (*30*). Although Nozik was referring to a wafer of the semiconductor ohmically contacted with a metal layer at the back, a small, metal-loaded semiconductor particle can be considered to have similar properties in some points (*31*). Figure 8 shows the electronic structure of a TiO_2–Pt particle in a neutral aqueous solution. The energy structure under illumination is shown by the dotted lines. It can be seen that H_2 evolution becomes possible owing to the upper shift of the Fermi level of Pt under illumination, which was determined by a photovoltage measurement of the slurry electrode. The quantum yield of hydrogen evolution was found to depend on the wave-

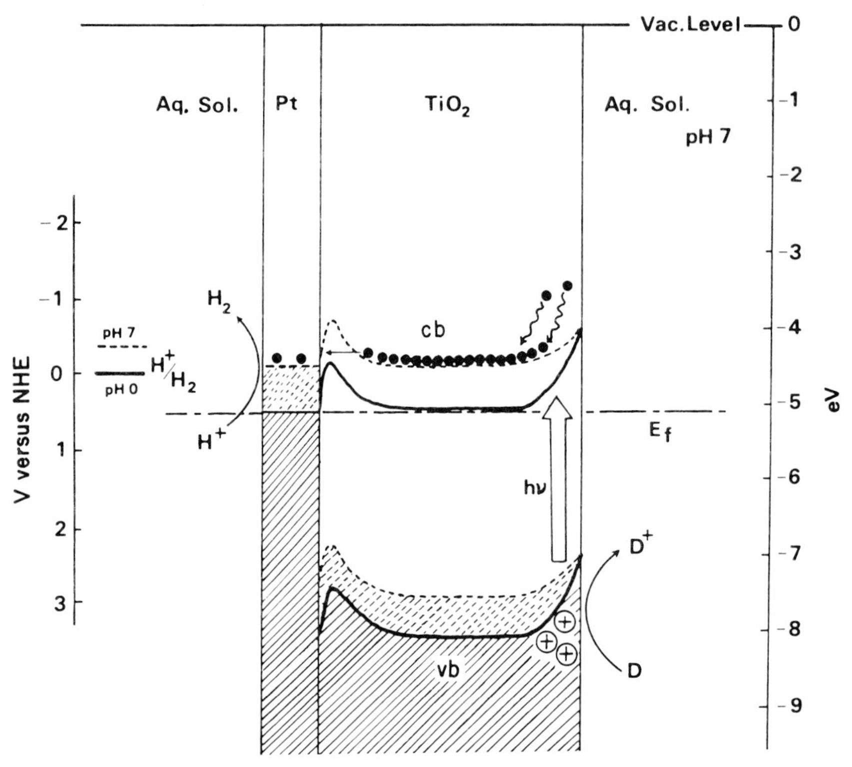

Fig. 8. Energy structure of the photocatalyst TiO_2–Pt in an electrolyte solution (pH 7). The barrier height at the metal–semiconductor interface is modified by the image force of the metal. The solid lines show the energy levels in the dark and the dotted lines those under illumination. From Sakata *et al.* (*31*).

length when monochromatic light was used to excite the semiconductor (Fig. 9). The quantum yield is small at the band-gap wavelength of TiO_2 (410 nm) and grows to a maximum at 350 nm, similar to the wavelength dependence of the photocurrent observed for a single-crystal electrode (*32*).

B. Catalytic Effect of Platinum on the Semiconductor Surface

The hydrogen-evolution rate depends on the kind of metal that is supported on the powdered semiconductor surface. For instance, Pt, Pd, or Rh increase the hydrogen-evolution rate by a factor of 20–100, whereas Fe on TiO_2 suppresses evolution. This remarkable dependence is consid-

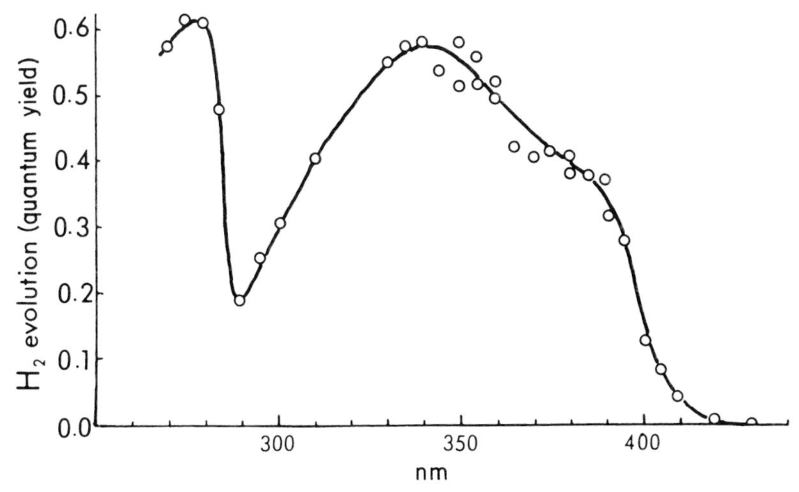

Fig. 9. Dependence of quantum yield of hydrogen evolution on the excitation wavelength. Experimental conditions: TiO_2(rutile)–Pt (300 mg), Pt : TiO_2 = 3.0 mol %, water (15 ml), ethanol (15 ml).

ered to arise partly from the catalytic effect of metals on hydrogen evolution. For instance, Pt, Pd, and Rh are known as good electrode materials, being ~10^4 times more active than the Ti electrode for hydrogen evolution. (*33*). In Fig. 10 the dependence of the hydrogen-evolution rate on the amount of Pt is shown. It can be observed that the rate for the Pt–TiO_2 photocatalyst (5 mol % Pt) is ~100 times larger than that for TiO_2 alone and shows a maximum at ~5 mol % Pt on TiO_2. It can also be observed that the rate is not linear to decreasing amounts of Pt. The activity of the photocatalyst is maintained even at as small a ratio of Pt to TiO_2 as 0.0005, below which it decreases rapidly (Fig. 10b). At the ratio Pt/TiO_2 = 0.0005, Pt is expected to form a 0.001 monolayer on the TiO_2 surface, assuming uniform deposition and a surface area for powdered TiO_2 of 10 m^2/g (the average particle size of TiO_2 is 0.2–0.4 μm). It is remarkable that so small an amount of Pt (10^{-2} monolayer) is sufficient to produce good photocatalytic activity.

C. Size Effect of Photocatalytic Activity

Figure 11 shows the dependence of the rate of hydrogen evolution on the particle size (or surface area) of TiO_2 in an ethanol–water (1 : 1) mixture. This plot shows that the evolution rate is very small, micromoles per hour, for a sandwich-like diode (◇) composed of a TiO_2 single crystal (10

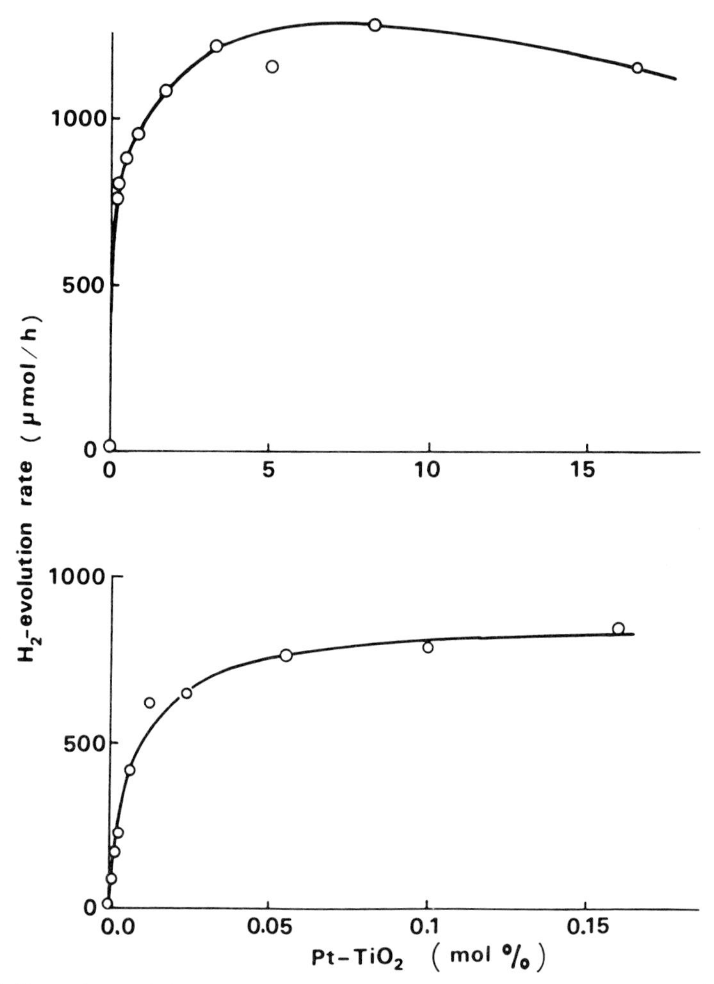

Fig. 10. Dependence of the hydrogen-evolution rate on the amount of Pt supported photochemically on TiO_2. Experimental conditions: water (15 ml), ethanol (15 ml), TiO_2(rutile)–Pt (300 mg), irradiation with a 500-W Xe lamp. From Sakata *et al.* (*31*).

× 10 × 1.5 mm) and a Pt plate (10 × 10 × 0.1 mm) with an ohmic contact, whereas it is more than 800 μmol/h (400 times greater) for photocatalysts with a particle size smaller than 0.4 μm.

At first sight it would appear that this increase is a result of the increased surface area of the photocatalyst, as in the case of thermal catalysts. However, a complete explanation must take other factors into consideration. In the photoelectrode reaction with a TiO_2 single crystal and a light source such as a 500-W Xe lamp, the rate-determining step is known

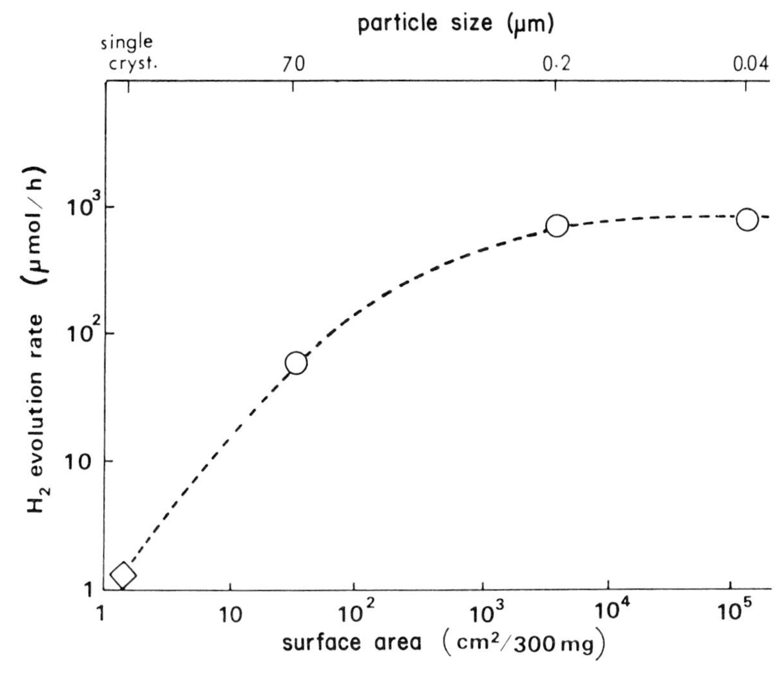

Fig. 11. Dependence of the hydrogen-evolution rate on the particle size and surface area of powdered (○) or single-crystal (◇, 10 × 10 × 0.8 mm) TiO_2. From Sakata *et al.* (*31*).

to be not the photoanodic reaction (surface reaction), but the supply of photogenerated holes to the surface. In this case the anodic photocurrent increases linearly with increasing light intensity. Therefore, the anodic reaction (oxidation) is limited not by the surface area of TiO_2 but by the light intensity. Furthermore, the photocurrent of the electrochemical cell, <TiO_2 |ethanol–water (1 : 1), 1 *M* KCl| Pt>, is small, owing to the small band bending of TiO_2 under short-circuit conditions, and is ~100 μA with a 500-W Xe lamp. This photocurrent corresponds to a hydrogen-evolution rate of 2 μmol/h, in fair agreement with the experimental value (1 μmol/h) for the TiO_2 single-crystal and Pt-plate photochemical diode (Fig. 11).

The following explanations may be advanced for the improved efficiency of charge separation, as shown by the data in Fig. 3, for a small-particle TiO_2–Pt photocatalyst compared with a photochemical diode composed of a TiO_2 single-crystal and Pt plate. First, for a given quantity of photocatalyst, the smaller the size of the semiconductor particle, the larger the total volume of the space-charge layer, where electrons and holes separate efficiently. Second, electrons and holes easily reach the surface within their lifetime when the diffusion length [10^4 Å for TiO_2 (*34*,

35)] is larger than the radius of the particle. Third, if active centers of some sort are assumed to play an important role in the photoanodic reaction, a powdered semiconductor composed of small particles is likely to contain more such centers than a single-crystal semiconductor. Finally, from the viewpoint of the cathodic reaction, the larger the surface area of supported Pt, the more efficiently hydrogen will evolve, because hydrogen evolution is a dark reaction and is proportional to the surface area of the electrode.

D. Correlation between the Efficiency of Hydrogen Evolution and the Irreversibility of the Surface Reaction

The first step in photoelectrochemical energy conversion is charge separation at the space-charge layer of the semiconductor. In this regard the irreversibility of the surface reactions is important when a semiconductor powder or a dye is used as a photocatalyst. Table X shows the rate of hydrogen evolution with TiO_2–Pt as photocatalyst and various electron donors. For H_2O, $S_2O_3^{2-}$, I^-, and Fe^{2+}, the evolution rate is very small even though all of the donors, except H_2O, are easily oxidized. In addition, it is very high for methanol, ethanol, ethylene glycol, and glycerol, which are not readily oxidized. The indication is that the irreversibility of surface reactions plays an important role in attaining an efficient hydrogen evolution with powdered semiconductor photocatalysts.

TABLE X

Correlation between the Hydrogen-Evolution Rate and Reversibility of Reaction

System	Reactant (electron donor)	H_2 evolution rate[a] (μmol/10 h)
Reversible	H_2O	<1
	Fe^{2+}	1
	$S_2O_3^{2-}$	6
	I^-	<1
Irreversible	CH_3OH	9000
	C_2H_5OH	8000
	Glycerol	7500
	Ethylene glycol	5800

[a] 500-W Xe lamp; irradiation for 10 h. Photocatalyst: TiO_2–Pt (300 mg).

VII. Application of Photocatalytic Reaction to Organic Synthesis

Photocatalysis on a semiconductor powder is essentially a redox reaction. Accordingly, not only H_2 production from water, but also the application of photocatalysis to CO_2 and N_2 fixation, has been investigated. Furthermore, a variety of photocatalytic reactions with organic substances has been developed. Table XI shows examples of reactions with organic molecules as starting materials, for example, oxidation of carboxylic acid, alcohol, benzene, lactams, aromatic olifins, and so on.

TABLE XI

Examples of Photocatalytic Reactions of Organic Molecules on a TiO_2 Powder

Reaction	Reference
$CH_3COOH \longrightarrow CH_4 + CO_2 + C_2H_6 + H_2$	*39*
Propionic acid $\longrightarrow$ CO_2 + ethane + $H_2C{=}CH_2 + H_2$	*39*
n-Butyric acid $\longrightarrow$ CO_2 + propane + H_2	*39*
n-Valeric acid $\longrightarrow$ CO_2 + *n*-butane + H_2	*39*
Pivalic acid $\longrightarrow$ CO_2 + isobutane + H_2 + isobutylene	*39*
 $C_6H_5{-}CH_3 + O_2 \longrightarrow C_6H_5{-}CHO + H_2O$	*40*
 $C_6H_6 + \cdot OH \longrightarrow C_6H_5{-}OH$	*41*
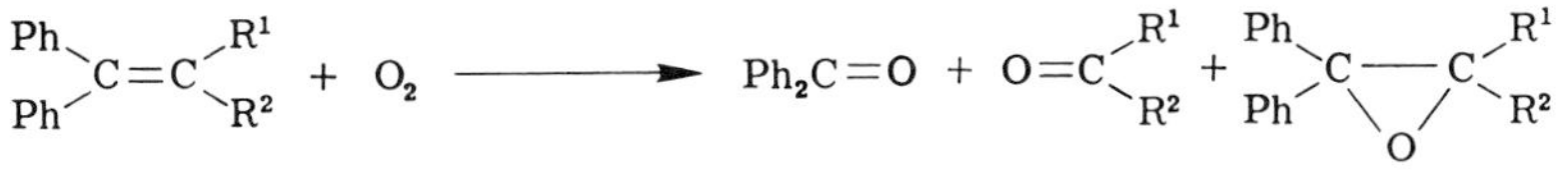 $Ph_2C{=}CR^1R^2 + O_2 \longrightarrow Ph_2C{=}O + O{=}CR^1R^2 + Ph_2C(-O-)CR^1R^2$	*42*
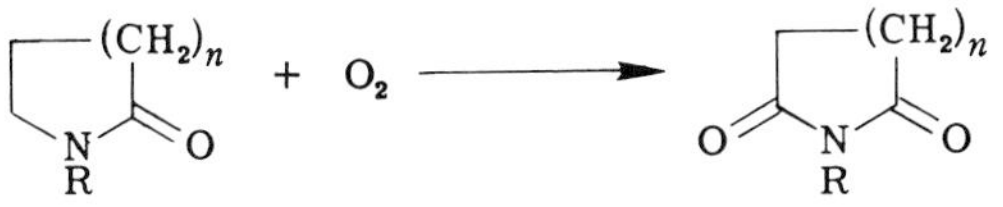 N-R lactam with $(CH_2)_n$ ring $+ O_2 \longrightarrow$ N-R imide with $(CH_2)_n$ ring (O=C–N(R)–C=O)	*43*
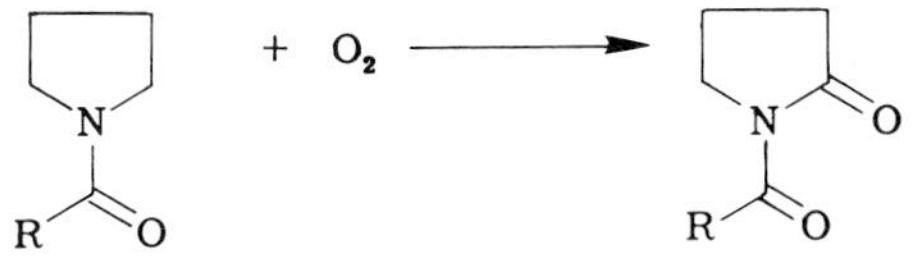 N-acyl pyrrolidine (N–C(=O)R) $+ O_2 \longrightarrow$ N-acyl pyrrolidinone (N–C(=O)R)	*43*
$CH_3OH \longrightarrow HCHO$	*25*
$C_2H_5OH \longrightarrow CH_3CHO$	*26*

The reducing and oxidizing power of the photocatalysts is controlled by the corresponding conduction- and valence-band positions, and because of this effect, a variety of reactions, such as O—C splitting or C≡N hydrogenation, can be realized by choosing an appropriate semiconductor. In an alkaline solution containing a CH_3CN–CdS–Pt mixture, NH_3 is produced, which indicates hydrogenation of the triple bond. Polyethylene glycol produces C_2H_6 and H_2 on CdS–Pt, which indicates the splitting and hydrogenation of the C—O bond.

VIII. Summary

1. Because these reactions occur through the action of electrons and holes produced in a semiconductor, a strong redox reaction is possible if we choose an appropriate semiconductor. The reducing power can be controlled by the conduction-band position and the oxidizing power by the valence-band position.
2. In these reactions, H_2O plays the role of oxidizing agent.
3. Pt, Pd, or RuO_2 are very efficient catalysts for H_2 evolution from water.
4. The quantum yield of H_2 production from water–organic compound mixtures is much higher than that from water alone. For example, it is ~1000 times larger for water–alcohol than for the direct decomposition of water.
5. The free-energy increase in these reactions is not large (it is sometimes negative). From this point of view, they are regarded as a reactions that reform hydrogen fuel from organic resources. This aspect is well illustrated by the decomposition of glucose.
6. Each photocatalyst particle acts as a microphotoelectrode. 300 mg of the powder (diameter 0.1 μm) consists of more than 10^{11} particles.
7. Some reactions mentioned so far are similar to fermentation by bacteria. The photocatalytic reaction, however, occurs even in a strong acidic or alkaline solution. We have found that among the various semiconductor powders so far examined, TiO_2 and CdS have the highest activity in photocatalytic H_2 productions. The requirements for a desirable photocatalyst are (a) utilization of visible light, (b) stability, (c) nontoxicity, and (d) low cost.

If one could succeed in exploiting a new stable semiconductor that functions efficiently with visible light by using sunlight, this method could be applicable to hydrogen production from various biomasses—sewage, human waste, and waste water from various factories such as pulp mills, fermentation plants, and artificial-polymer factories.

References

1. Hall, D. O., *in* "Research in Photobiology" (A. Castellani, ed.), p. 347. Plenum, New York, 1976.
2. Calvin, M., *Photochem. Photobiol.* **23,** 425 (1976).
3. Tributsch, H., *Nature (London)* **281,** 555 (1979).
4. Bard, A. J., *Science (Washington, D.C.)* **207,** 139 (1980).
5. Grätzel, M., *Acc. Chem. Res.* **14,** 376 (1981).
6. Fujishima, A., and Honda, K., *Bull. Chem. Soc. Jpn.* **44,** 1148 (1971).
6a. Fujishima, A., and Honda, K., *Nature (London)* **238,** 38 (1972).
7. Ohnishi, T., Nakato, Y., and Tsubomura, H., *Ber. Bunsenges. Phys. Chem.* **79,** 523 (1975).
8. Schrauzer, G. N., and Guth, T. D., *J. Am. Chem. Soc.* **99,** 7189 (1977).
9. Yoneyama, H., Koizumi, M., and Tamura, H., *Bull. Chem. Soc. Jpn.* **52,** 3449 (1979).
10. Damme, H. V., and Hall, W. K., *J. Am. Chem. Soc.* **79,** 4373 (1979).
11. Sato, S., and White, J. M., *Chem. Phys. Lett.* **72,** 85 (1980).
12. Lehn, J. M., Sauvage, J. P., and Ziessel, R., *Nouv. J. Chim.* **4,** 623 (1980).
13. Domen, K., Naito, S., Soma, M., Onishi, T., and Tamaru, K., *J. Chem. Soc. Chem. Commun.*, p. 543 (1980).
14. Borgarello, E., Kiwi, J., Pelizzetti, E., Visca, M., and Grätzel, M., *Nature (London)* **289,** 158 (1981).
15. Kawai, T., and Sakata, T., *Chem. Phys. Lett.* **72,** 87 (1980).
16. Kawai, T., and Sakata, T., *Proc. Int. Congr. Catal. 7th,* p. 1198 (1981).
17. Duonghong, D., Borgarello, E., and Grätzel, M., *J. Am. Chem. Soc.* **103,** 4685 (1981).
18. Kalyanasundaram, K., Borgarello, E., and Grätzel, M., *Helv. Chim. Acta.* **64,** 362 (1981).
19a. Ellis, A. B., Kaiser, S. W., and Wrighton, M. S., *J. Am. Chem. Soc.* **98,** 1635 (1975).
19b. Wrighton, M. S., Boltz, J. M., Bocarsly, A. B., Palazzotto, M. C., and Walton, E. G., *J. Vac. Sci. Technol.* **15,** 1429 (1978).
20. Minoura, H., Tsuiki, M., and Oki, T., *Ber. Bunsenges. Phys. Chem.* **81,** 588 (1977).
21. Amouyal, E., Keller, P., and Moradpour, A., *J. Chem. Soc. Chem. Comm.,* p. 1019 (1980).
22. Kawai, T., and Sakata, T., *Nature (London)* **282,** 283 (1979).
23. Kawai, T., and Sakata, T., *J. Chem. Soc. Chem. Comm.,* p. 1047 (1979).
24. Hashimoto, K., Kawai, T., and Sakata, T., unpublished results.
25. Kawai, T., and Sakata, T., *J. Chem. Soc. Chem. Comm.,* p. 694 (1980).
26. Sakata, T., and Kawai, T., *Chem. Phys. Lett.* **80,** 341 (1981).
27. Kawai, T., and Sakata, T., *Nature (London)* **286,** 474 (1980).
28. Sakata, T., and Kawai, T., *Nouv. J. Chim.* **5,** 279 (1981).
29. Kawai, T., and Sakata, T., *Chem. Lett.,* p. 81 (1981).
30. Nozik, A. J., *Appl. Phys. Lett.* **30,** 567 (1977).
31. Sakata, T., Kawai, T., and Hashimoto, K., *Chem. Phys. Lett.* **88,** 50 (1982).
32. Yoneyama, H., Tamura, H., *Electrochim. Acta* **20,** 341 (1975).

33. Bockris, J. O'M. and Reddy, A. K. N., "Modern Electrochemistry." Plenum Press, New York, 1974.
34. Wilson, R. H., *J. Appl. Phys.* **48,** 4292 (1977).
35. Maruska, H. P., and Ghosh, A. K., *Sol. Energy* **20,** 443 (1978).
36. Hemminger, J. C., Carr, R., and Somorjai, G. A., *Chem. Phys. Lett.* **57,** 100 (1978).
37. Halmann, M., *Nature* (*London*) **275,** 115 (1978).
38. Inoue, T., Fujishima, A., Konishi, S., and Honda, K., *Nature* (*London*) **277,** 637 (1979).
39. Krautler, B., and Bard, A. J., *J. Am. Chem. Soc.* **100,** 5985 (1978).
40. Fujihira, M., Satoh, Y., and Osa, T., *J. Electroanal. Chem.* **126,** 277 (1981).
41. Fujihara, M., Satoh, Y., and Osa, T., *Nature* (*London*) **293,** 206 (1981).
42. Kanno, T., Oguchi, T., Sakuragi, H., and Tokumaru, K., *Tetrahedron Lett.* **21,** 467 (1980).
43. Pavlik, J. W., and Tantayanon, S., *J. Am. Chem. Soc.* **103,** 6755 (1981).

11 Photoelectrolysis of Water and Sensitization of Semiconductors

Tadashi Watanabe
Akira Fujishima
Kenichi Honda

Department of Synthetic Chemistry
Faculty of Engineering
University of Tokyo
Tokyo, Japan

I. Introduction

The photoassisted splitting of water into hydrogen and oxygen is one of the challenging subjects of science and technology. The interest in it comes from the following aspects:

ENERGY RESOURCES THROUGH
PHOTOCHEMISTRY AND CATALYSIS

ISBN 0-12-295720-2

1. Hydrogen would very probably be an ideal future fuel, and water is an inexhaustible natural resource.
2. It is directly related to solar-energy utilization.
3. It is essentially equivalent, if not identical, with the primary photochemical event in natural photosynthesis.

In the gas phase, water splitting can be carried out only by scission of the H—OH bond, which requires 5.2 eV per molecule or 498 kJ mol^{-1}.

In contrast, electrolysis of water proceeds by the combination of two half-cell reactions (in acidic media)

$$2H^+ + 2e^- \longrightarrow H_2 \tag{1}$$

$$H_2O - 2e^- \longrightarrow \tfrac{1}{2}O_2 + 2H^+ \tag{2}$$

to result in an overall process of

$$H_2O \longrightarrow H_2 + \tfrac{1}{2}O_2 \tag{3}$$

The difference in free energy for Eqs. (1) and (2) amounts to 1.23 eV per electron. Hence the minimum energy required to decompose water to H_2 and O_2 according to Eq. (3) is 2.46 eV per H_2O molecule, or 237 kJ mol^{-1}. Normally this energy is supplied by electrical power from an external circuit.

In photoassisted water splitting, the driving force for electrons is given by light energy, and the electron should be transferred from the donor (H_2O) to the acceptor (H^+) via an appropriate mediating species. The latter is usually called the photosensitizer. Figure 1 schematically shows the flow of electrons in photoassisted water splitting mediated by a sensitizer S. The problem is to suppress effectively the energy-wasting backward (recombination) processes depicted by the broken lines.

A variety of sensitizers can be envisaged for such a process. These include

1. Semiconductor electrodes
2. Semiconductor particles

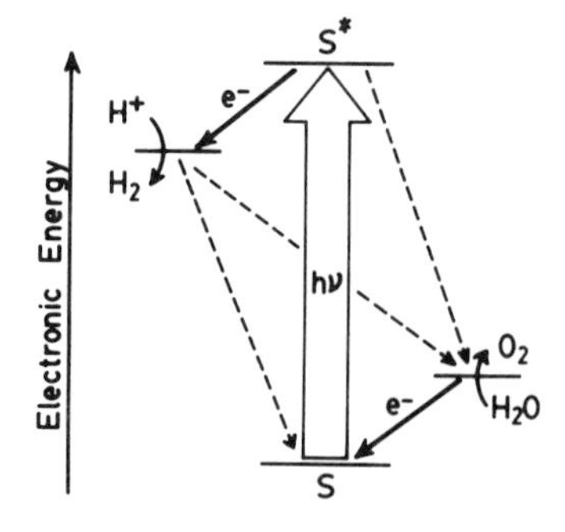

Fig. 1. Schematic illustration of the basic principle of photoassisted water splitting. The solid arrows represent the vectorial electron flow to be promoted, whereas the broken arrows denote the recombination paths to be suppressed.

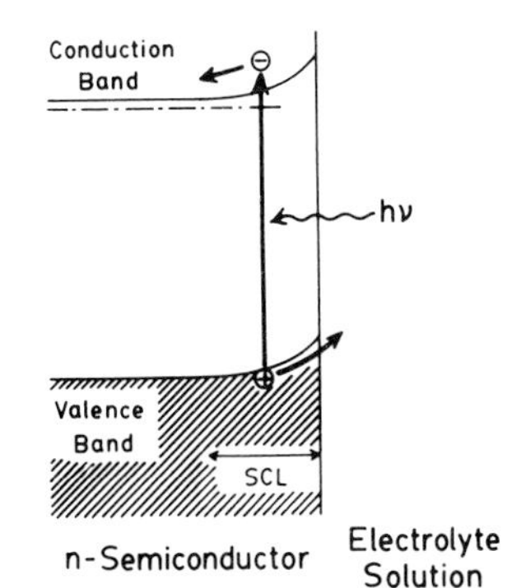

Fig. 2. Photoinduced charge separation in the space charge layer (SCL) of an *n*-type semiconductor in contact with an electrolyte solution.

3. Colored redox species such as dyes and metal complexes in homogeneous and heterogeneous photocatalytic systems
4. Colored redox species adsorbed on semiconductor electrodes

Cases 2 and 3 are described in other chapters of this book. In the present chapter we are primarily concerned with cases 1 and 4.

Section II deals with the fundamental and practical problems associated with case 1, the photoelectrolysis of water. Here the initial charge separation is particularly favored because of the presence of a Schottky barrier (or space-charge layer, SCL) formed at the semiconductor–solution interface. A very simplified picture for this case is illustrated in Fig. 2.

Section III describes our findings concerning case 4, the sensitization of semiconductors. It is, however, very difficult to attain direct water splitting in artificial dye-sensitized processes, owing primarily to the small amount of photon energy absorbed by the sensitizer. Hence emphasis is placed solely on the improvement in the efficiency of the initial photoinduced charge separation, in the hope that such basic research could provide useful information for future implementation of dye-sensitized systems in photoassisted water splitting.

II. Photoelectrolysis of Water with Semiconductors

A. Principles of Photoelectrochemical Cells

When a semiconductor electrode is in contact with an electrolyte solution, thermodynamic equilibration takes place at the interface. This results in the formation of an SCL within a thin semiconductor surface region, where the electronic energy bands are generally bent upward and downward in the cases of *n*- and *p*-type semiconductors, respectively. The thickness of the SCL is usually of the order of $1–10^3$ nm, depending on the carrier density and dielectric constant of the semiconductor. If this electrode receives photons with energy greater than its band gap E_g, electron–hole pairs are generated within the SCL. In the case of an *n*-type semiconductor, the electric field existing within the SCL drives photogenerated holes toward the interfacial region and electrons toward the interior of the electrode. A reverse process takes place in a *p*-type semiconductor electrode. Thus, by combining a semiconductor electrode with an appropriate counterelectrode, a photoelectrochemical cell (PEC) can be constructed.

Figure 3 shows the operational principles of two types of PEC using an *n*-type semiconductor photoelectrode and a metal counterelectrode. If a redox couple (Ox/Red) is present in the electrolyte solution, the holes oxidize Red to Ox on the semiconductor surface. In contrast, the electrons reaching the metal counterelectrode may reduce Ox′ to Red′ of another redox couple (Ox′/Red′).

When the redox potential *E*(Ox/Red) is positive of *E*(Ox′/Red′), the net result of the photoelectrochemical process

$$\mathrm{Red} + \mathrm{Ox}' \longrightarrow \mathrm{Ox} + \mathrm{Red}' \tag{4}$$

is thermodynamically an uphill reaction leading to the storage of chemical energy. Such a system is called a photoelectrolytic cell (Fig. 3a). Among many possibilities, the splitting of water (Red = $2H_2O$, Ox = $O_2 + 4H^+$, Ox′ = $4H^+$, and Red′ = $2H_2$) is probably the most important and promising one.

In contrast, when a common redox couple is reacting at both the photoanode and the metal cathode, no net chemical energy storage occurs, but electrical energy can be withdrawn in the external circuit. This type of PEC is called a regenerative photovoltaic cell (Fig. 3b).

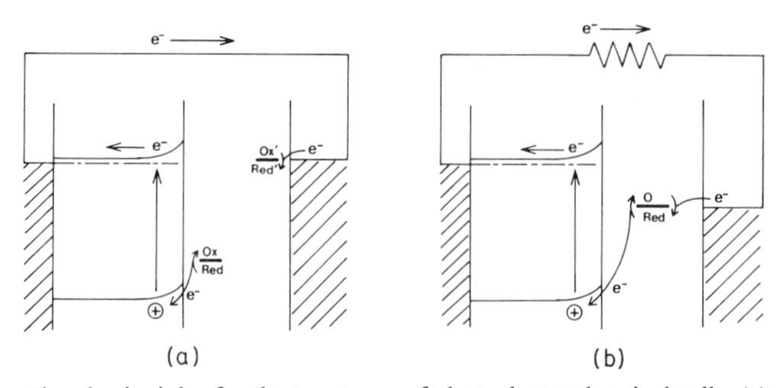

Fig. 3. Operational principles for the two types of photoelectrochemical cells: (a) photoelectrolytic cell and (b) regenerative photovoltaic cell.

The possibility of solar photoelectrolysis of water was demonstrated by us for the first time with a system in which an *n*-type titanium dioxide (TiO_2) electrode was connected to a platinum black electrode, the former exposed to near-uv light (*7, 8*).

Figure 4 shows, as the fundamental characteristics of a PEC, typical current–potential curves observed at the TiO_2 single-crystal electrode under illumination and in darkness. An anodic photocurrent proportional to the incident light intensity is generated at potentials positive of −0.5 V versus SCE in an electrolyte solution of pH 4.7, whereas the cathodic branch is hardly influenced by illumination. From product analyses, the photoanodic reaction has been confirmed to be the oxidation of water leading to oxygen evolution,

$$2H_2O + 4p^+ \longrightarrow 4H^+ + O_2, \tag{5}$$

where p^+ denotes a hole photogenerated in the valence band of TiO_2. Thermodynamically, water oxidation should proceed only at $E > +0.7$ V versus SCE, which is positive of the observed photocurrent onset potential by as much as 1.2 V. Hence this phenomenon has been termed the *photosensitized electrolytic oxidation* of water.

On the basis of this finding we have succeeded in constructing a PEC (Fig. 5). Owing to the chemical stability of TiO_2, continuous water splitting is possible with such a PEC for a very long period (of the order of years). The photocurrent was found to flow at wavelengths shorter than

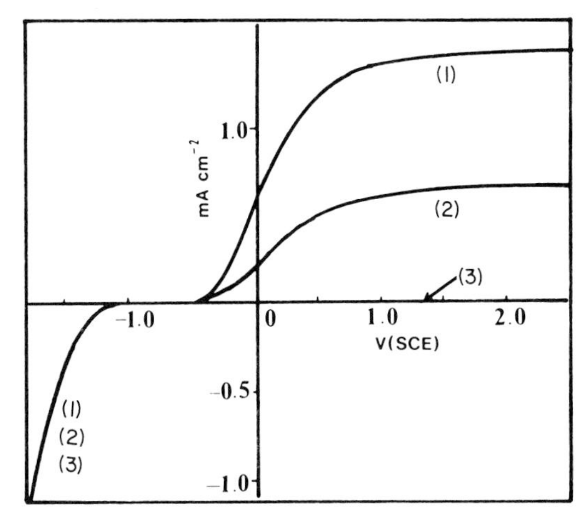

Fig. 4. Current–potential curves at an *n*-type TiO_2 single-crystal (rutile) electrode in contact with an electrolyte solution of pH 4.7: (1) under irradiation (photon flux 100%), (2) under irradiation (photon flux 50%), and (3) in the dark.

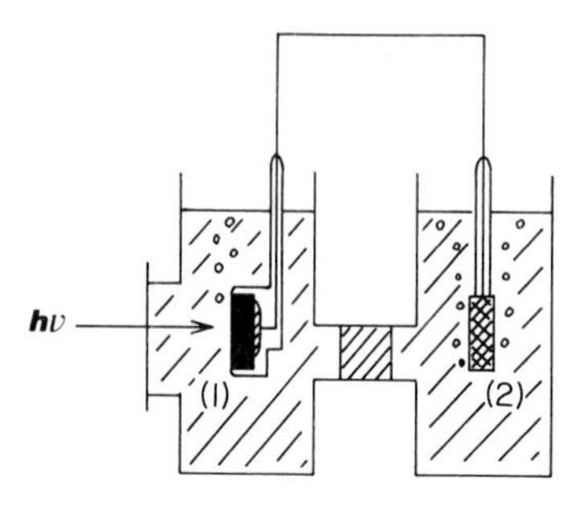

Fig. 5. Photoelectrolytic cell in which a TiO_2 electrode (1) is connected with a platinum electrode (2).

415 nm, corresponding to the intrinsic light absorption of TiO_2 ($E_g \simeq 3.0$ eV).

B. Requirements for Efficient Operation of PECs

From the viewpoint of efficient utilization of solar energy, the "classical" PEC consisting of a TiO_2 photoanode and a Pt counterelectrode has several drawbacks; TiO_2 absorbs only a small fraction of solar radiation, and its onset potential is not negative enough to construct a PEC with a Pt electrode immersed in a common electrolyte. In general, for the construction of an efficient PEC in which water is decomposed to H_2 and O_2, the following conditions are required for *n*-type semiconductor photoanodes: long-term stability during photoelectrolysis, negative flat-band potential, smaller band gap, and high quantum efficiency. A schematic illustration of these conditions is given in Fig. 6. In what follows, these criteria are discussed in more detail.

1. Stability

A number of oxide semiconductors are known to behave as photoanodes on which oxygen evolution occurs. The oxygen gas should come from water electrolysis, as on a TiO_2 electrode, and not from the decomposition of the photoanode itself, as on a ZnO electrode ($ZnO + 2p^+ \rightarrow Zn^{2+} + \frac{1}{2}O_2$ (*12*). The semiconductors should be stable in contact with an electrolyte solution both in darkness and under irradiation. Whether the decomposition of water or dissolution of the semiconductor takes precedence depends on the relative equilibrium potentials for water oxidation and for the dissolution of the semiconductor. A redox reaction having a more negative redox potential will occur by the action of holes in precedence to a redox reaction that has a more positive redox potential. Thermodynamic calculations can provide information concerning this problem

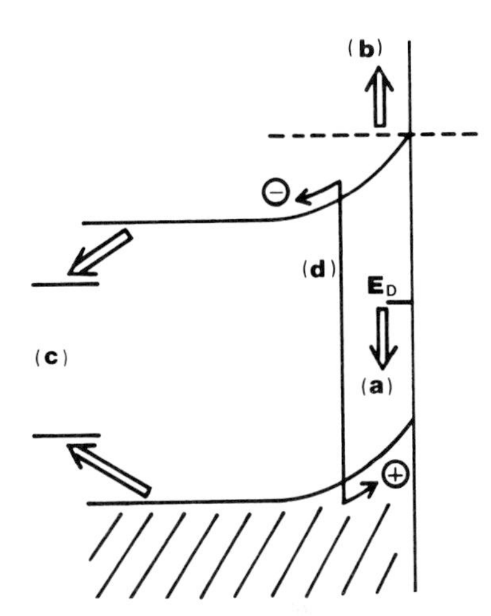

Fig. 6. Favorable characteristics of a semiconductor photoanode for use in a photoelectrochemical cell (PEC): (a) long-term stability, (b) negative flat-band potential, (c) smaller band gap, and (d) high quantum efficiency.

(*2, 14*). In the case of the stable TiO_2 electrode, the potential for water oxidation is negative of that for the decomposition of TiO_2. In contrast, in the case of the ZnO electrode, the potential for water oxidation is positive of the decomposition potential of the semiconductor. Hence the ZnO electrode undergoes dissolution under illumination. Typical stable *n*-type semiconductors thus far reported are TiO_2, $SrTiO_3$, WO_3, SnO_2, $BaTiO_3$, $KTaO_3$, Fe_2O_3, etc.

2. *Flat-Band Potential*

The flat-band potential E_{fb} is one of the most fundamental parameters in semiconductor–electrolyte systems. When a semiconductor is polarized at E_{fb}, there is no space charge or electric field within the semiconductor. The flat-band potentials can be positioned by a variety of methods, including a capacitance measurement as a function of the electrode potential based on the Mott–Schottky equation. In general, the flat-band potential shows a good coincidence with the onset potential of anodic photocurrent and gives information as to how much of the actual electrolysis voltage can be saved for a given electrochemical reaction.

With regard to an *n*-type semiconductor photoanode, the more negative E_{fb} (or onset potential) leads to more efficient output characteristics of a PEC (see arrow b in Fig. 6). When E_{fb} is situated negative of the hydrogen-evolution potential, photogenerated electrons reaching the counterelectrode can reduce protons to result in hydrogen evolution under no applied bias. Among the oxide semiconductors, $SrTiO_3$ and $KTaO_3$ satisfy this condition (*3, 31*). Otherwise, the use of an external bias is indispensable to obtain H_2 evolution, as in the case of a PEC consisting of a WO_3 photoanode and a metal cathode (*17*).

In the case of a TiO_2 photoanode, with which we constructed a PEC for the first time (*9*), the E_{fb} is somewhat positive of the hydrogen-evolution potential. Hence it was necessary to apply an external bias or, equivalently, a chemical bias provided by a pH difference between the anolyte and catholyte (Fig. 5). The two compartments in such a "heterogeneous" PEC are connected via an agar salt bridge. A disadvantage of the heterogeneous PEC is that each electrolyte becomes diluted with passage of time as a result of neutralization.

Another means for overcoming the positive flat-band potential of a TiO_2 photoanode is the use of a *p*-type semiconductor photocathode in place of a metal counterelectrode (*33*).

3. *Smaller Band Gap*

It is desirable that the band gap of the semiconductor be near the band gap needed for optimum utilization of solar energy. Even when photons are completely absorbed, a portion of the photon energy ($E > E_g$) is not utilized in the PEC because vibrational relaxation takes place in the upper excited state prior to charge separation. Thus the fraction $(E - E_g)/E$ of the photon energy is dissipated as heat, and only the fraction E_g/E can be used.

When a semiconductor electrode is excited with AM 1 sunlight, the theoretical maximum energy conversion efficiency is about 30% for an E_g value of 1.4 eV (*32*). However, taking other loss factors (e.g., overvoltages) into account, a minimum value of E_g = 2–2.5 eV would be required for an *n*-type photoanode for water photoelectrolysis.

4. *High Quantum Efficiency*

The quantum efficiency (the number of electrons flowing in the PEC system per one absorbed photon) should be high, ideally 100%, to construct an efficient PEC. It is desirable that the incident photons be totally absorbed within the SCL. Because the thickness of the SCL depends on the carrier density, the latter must be adjusted to an appropriate level (*13*). If the density of recombination centers in the SCL is high, the quantum efficiency will accordingly be lowered. However, it is not so difficult to have photoanodes with reasonably high quantum efficiencies under band-gap illumination. For example, quantum efficiency values of 70–85% have been observed even at polycrystalline oxide semiconductor photoanodes made by thermal treatment of metal sheets (e.g., TiO_2 on Ti) or by sintering semiconductor powders (e.g., ZnO) (*11*). This constitutes, among others, an advantage of the PEC systems over the solid-state photovoltaic cells, for which much care is required in the preparation and crystallinity of the entire material to obtain a high efficiency.

C. Hydrogen Production under Sunlight

For a practical application of PEC, inexpensive methods for the preparation of large-area semiconductor electrodes must be sought. Thus far, the following methods have been proposed:

1. Electrochemical oxidation or codeposition technique, (e.g., for TiO_2, CdS, and CdSe)
2. Chemical vapor deposition (CVD) technique (e.g., for TiO_2 and Fe_2O_3)
3. Thermal formation technique (e.g., for TiO_2, Fe_2O_3, and WO_3)
4. Sintering of semiconductor powders (e.g., for $SrTiO_3$, TiO_2, CdS, and CdSe)
5. Vacuum evaporation technique (e.g., for CdS and TiO_2)

We have constructed and tested a large-scale PEC for outdoor operation by using thermally formed TiO_2 photoanodes (*10*). A Ti plate was heated in a town-gas flame at temperatures between 1100 and 1400°C to obtain an approximately 1-μm-thick TiO_2 surface layer. The crystal structure of the oxide was found by x-ray diffraction to be the rutile type. Anodic photocurrents at these oxide electrodes were nearly as large as those at a single-crystal TiO_2 electrode. Figure 7 shows the geometric arrangement of the PEC with 20 photoanode wafers (85 × 100 mm each) thus prepared. The cell comprises five Pt black cathodes. An alkaline solution in the anodic compartments and an acidic solution in the cathodic compartments were used. The anolyte and catholyte were separated with an agar salt bridge saturated with KCl. In this arrangement, the potential drop through the solution or the salt bridge was negligible. The total anodic surface area was 0.17 m^2. Figure 8 shows the relationship between the operation time and the volume of hydrogen collected under sunlight on a clear summer day in Japan. The intensity of the solar radiation was around 110,000 lux from 11 AM to 2 PM. We were able to collect 1.1 liters of H_2 per day with this PEC (6.6 liters of H_2 per square meter of TiO_2). The solar-energy conversion efficiency has been calculated to be about 0.4%, on the basis of the H_2 collected and the mean solar power incident on this PEC. This efficiency is considered to be relatively high, in view of the limited absorption wavelength range (about 3% of the entire solar spectrum) by TiO_2.

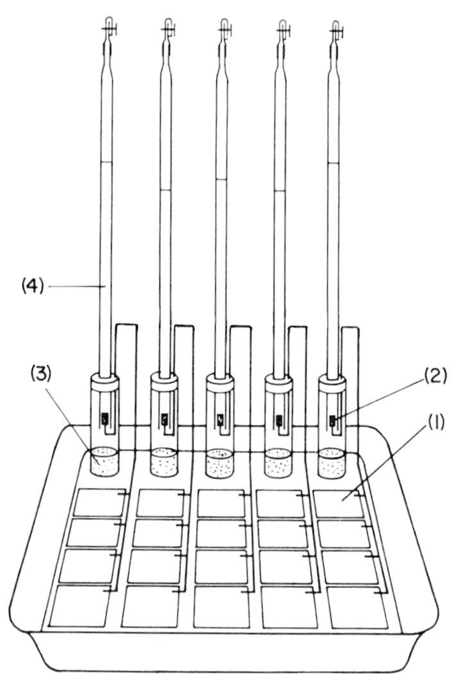

Fig. 7. Assembly of a large-scale PEC for hydrogen production under sunlight: (1) photoanodes (Ti plates with TiO_2 surface layer), (2) Pt black cathodes, (3) agar salt bridges, and (4) gas burettes for collection of evolved H_2.

D. Routes to Improving PEC Characteristics

A large number of simple and mixed oxides have been examined for photoeffect and possible utilization as electrodes for water splitting under solar irradiation. Most of the oxides are *n*-type semiconductors. Only large-band-gap oxides have been found to be stable against photocorrosion with reasonable photoelectrochemical characteristics. Low-band-gap oxides usually have either more positive flat-band potentials or poor SCL properties.

To improve the efficiency of water photolysis in a PEC, it is desirable to use smaller-band-gap semiconductors such as Si and GaAs. However, these nonoxide semiconductors are unstable in electrolyte solutions, easily resulting in surface dissolution or formation of an insulating oxide film. In view of this, coatings of these unstable semiconductors with a film of stable metals or semiconductors have been attempted [for instance, (*23, 25, 26*)]. Figure 9 illustrates some examples of surface treatments of semiconductor electrodes. Such surface treatments have been found effective

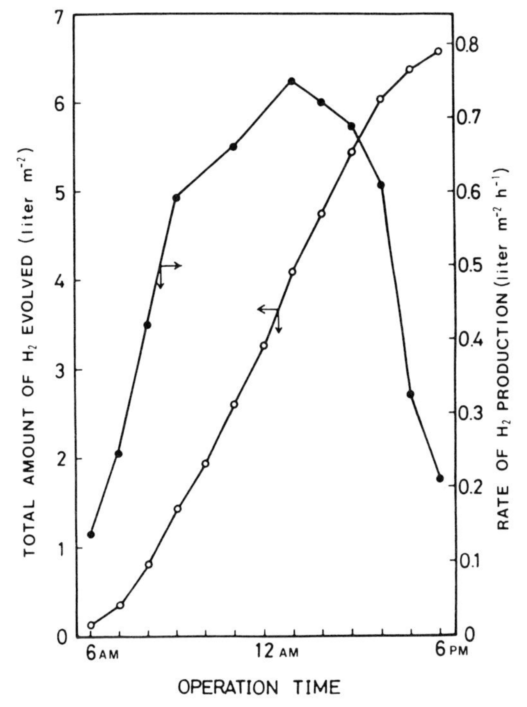

Fig. 8. Hydrogen production under sunlight with the PEC shown in Fig. 7.

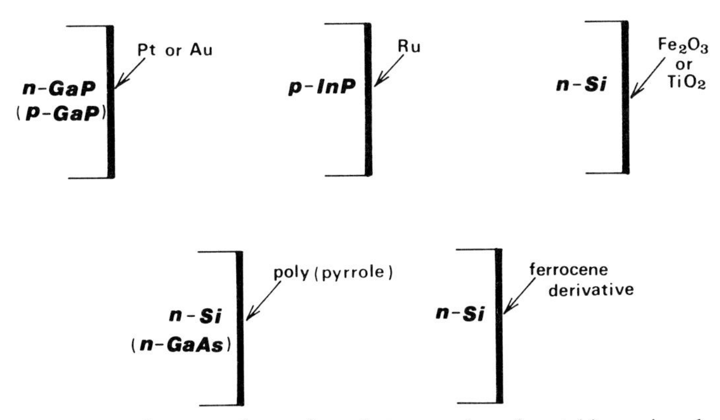

Fig. 9. Various ways for protection against photocorrosion of unstable semiconductor electrodes.

not only in the construction of a PEC for water photoelectrolysis, but also for photoelectrodes used in regenerative photovoltaic cells (Fig. 3b).

p-Type electrodes such as *p*-Si and *p*-GaP can serve as photocathodes, on which photoassisted hydrogen evolution takes place. However, the onset potential for hydrogen evolution on these photocathodes is usually not positive enough to construct a PEC with a metal anode. To improve the kinetics of hydrogen evolution, surface modification of photocathodes has been conducted by coating the surface with a thin layer of catalytic metals such as Pt and Ru (*1, 25, 26*).

Besides these surface treatments, the use of *n*- and *p*-type semiconductors as two photoelectrodes in a single PEC has been proposed by several workers. Here, water decomposition could be made to proceed by the coupling of two different photons without an external bias. However, most *n–p* semiconductor combinations investigated so far suffer from a poor matching of their energetics.

III. Sensitization of Semiconductors: Chlorophyll-Sensitized Semiconductor Electrodes

A. *Primary Processes of Photosynthesis*

Photosynthesis is the largest-scale process taking place on our planet for the collection and storage of solar energy and is really "the power plant and the chemical factory of life" (*29*). It produces a variety of organic matter and molecular oxygen, which are indispensable for all living organisms, from CO_2 and H_2O supplied as raw materials. In the initial stage of green-plant photosynthesis, two photons absorbed in series by two "reaction centers," pump up an electron from an H_2O molecule to $NADP^+$ (the oxidized form of nicotinamide adenine dinucleotide phosphate), leading to the net redox process

$$\tfrac{1}{2}H_2O + \tfrac{1}{2}NADP^+ \xrightarrow{2h\nu} \tfrac{1}{4}O_2 + \tfrac{1}{2}NADPH + \tfrac{1}{2}H^+ \qquad (6)$$

This is a thermodynamically uphill electron transfer by 1.16 eV (from +0.81 V versus NHE for the H_2O/O_2 couple to −0.35 V for the NADPH/$NADP^+$ couple at pH 7). Because the redox potential for hydrogen evolution is −0.42 V versus NHE at pH 7, the energy storage in process (6) is nearly equivalent to that of water splitting into O_2 and H_2. The absorption threshold energy for green-plant photosynthetic apparatus lies at ~1.8 eV (680–700 nm). Hence the monochromatic energy conversion efficiency of process (6) is $100[1.16/(2 \times 1.8)] = 32\%$ at best. Taking the solar spectrum into account, this corresponds to a maximum solar conversion efficiency η_s of about 8%.

One of the most prominent features of photosynthesis, aside from the moderate value of η_s, is the extremely high quantum yield (≃100%) for photoinduced charge separation in the initial stage. No artificial systems with molecular photosensitizers have been capable of achieving this. The following three factors, in cooperation, are presumed to play a crucial role in attaining such a high efficiency:

1. Favorable energetics at the Ox–D*–R interfaces
2. Sufficiently higher rate constant for the initial charge separation ($Ox–D^*–R \rightarrow Ox^-–D^+–R$ or $Ox–D^-–R^+$) than those for other photophysical processes occurring from D*
3. Special intermolecular orientation and arrangement, which could facilitate the unidirectional electron flow while suppressing recombination routes

D*, Ox, and R denote the excited photosensitizer (reaction center), electron acceptor, and donor, respectively. With regard to photosynthetic organisms, there have been several proposals (*5, 27*) that the reaction centers consist of some dimeric or oligomeric forms of chlorophyll *a* (Chl *a*). However, the *in vivo* "molecular machinery" underlying features 1–3 remains far from being unraveled. This constitutes one of the incentives to carry out model experiments with *in vitro* systems.

B. Dye-Sensitized Semiconductor Electrodes

A model system for studying photoinduced charge separation is constructed by using a large-band-gap *n*-type semiconductor as Ox, an adsorbed dye molecule as D, and a reducing agent as R. The latter is dissolved in an electrolyte solution. Such a system is referred to as a dye-sensitized semiconductor electrode and has been under intense investigation (*15, 18*). A general scheme for this system is illustrated in Fig. 10.

An anodic-sensitized photocurrent is generated by electron injection from the excited dye (D*, singlet or triplet) into the conduction band (cb) of the semiconductor. This is possible only when the one-electron oxidation level of D*, $E(D^*/D^+)$, is energetically higher than the level of cb. These two levels can be positioned on a common scale (electronic energy or electrochemical potential) from the knowledge of electrochemical and optical data (*15*). Because of the existence of a rectifying SCL in the surface region of the semiconductor, the quantum yield for photoinduced

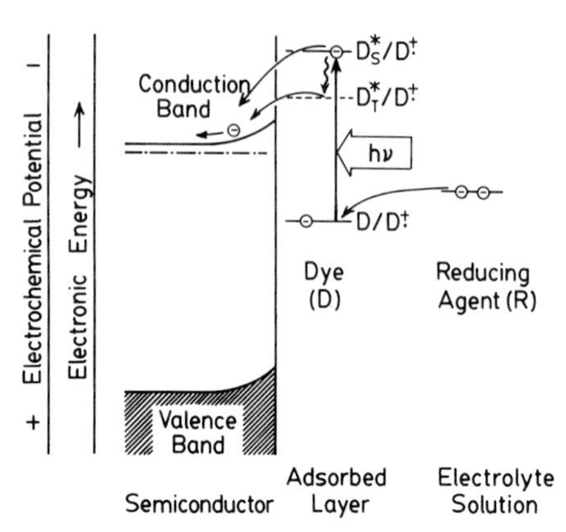

Fig. 10. Simplified scheme for dye sensitization of an *n*-type semiconductor electrode.

charge separation in this type of system is generally higher than those observed in other systems (e.g., dye-based homogeneous photocatalytic systems).

It should be kept in mind that, besides the expected vectorial charge separation, a number of intra- and intermolecular photophysical pathways, radiative or nonradiative, would occur from D*, though they are not depicted in Fig. 10.

C. *An* in Vitro *Approach to Photoinduced Charge Separation at Chlorophyll–Acceptor Interfaces*

1. Outline

In this subsection we present some of our findings from studies on Chl-sensitized semiconductor electrodes. One of the main objectives of these studies has been to answer the following question: What factors determine the efficiency of interfacial charge separation initiated by excitation of dye molecules adsorbed on a semiconductor? Because the present experimental system is not a "molecular machinery" model, we are primarily interested in the factors associated with aspects 1 and 2 given in Section III,B.

For this purpose, we sought to prepare a series of dye molecules, differing widely from each other in redox, photophysical, and intermolecular aggregation properties, and to choose a common substrate (SnO_2) as the semiconductor electrode. To avoid complications arising from different molecular orientations at the interface, the affinity of each dye to the SnO_2 surface should be the same. In the course of our preliminary work we found that metallosubstituted chlorophyll *a*s (M-Chl *a*s)[1] could meet most of these requirements. A drastic change in the physicochemical properties takes place by exchange of one central metal to another, and yet all these M-Chl *a*s have a common molecular structure except for the nature of M. Thus we have undertaken a number of photoelectrochemical measurements on SnO_2–M–Chl *a* interfaces to obtain useful information concerning the question stated in the preceding paragraph.

We further expected that these investigations could also provide a key for understanding a more fundamental question surrounding photosynthesis, namely, the reason for the choice, with no exception, of Mg complexes as photoactive pigments in photosynthetic organisms.

2. *Experimental Methods*

a. Synthesis of M-Chl as. Mg-Chl *a* was extracted from fresh spinach leaves and purified chromatographically (*28*). H_2-Chl *a* (pheophytin *a*) was obtained by treating Mg-Chl *a* with 2 *N* HCl aqueous solution. Metallation of H_2-Chl *a* was achieved by solution phase reaction with metal salts (chloride or acetate) in appropriate solvents (glacial acetic acid, ethanol, or aqueous acetone). The reaction temperature was kept below 30°C. When the reaction was made to proceed at temperatures higher than 50°C for more than 10 h, particularly under air, there occurred, as revealed by high-performance liquid chromatographic (HPLC) analysis, the formation of a substantial amount of unidentified side products hardly distinguishable from each other by simple spectroscopic measurements. Judging from the evolution of visible absorption spectra during the reaction, we were able to metallize H_2-Chl *a* with Mn(III), Co(II), Ni(II), Cu(II), Zn(II), Pd(II), Ag(II), Sn(IV), and Hg(II) ions, but the reactions with Al(III), Ca(II), Ti(IV), Fe(II), Cd(II), Pt(II), Au(III), and Pb(II) have been unsuccessful. Before any measurement, each M-Chl *a* sample was purified by preparative scale HPLC.

b. The Langmuir–Blodgett Technique. A monomolecular layer of M-Chl *a* with controlled surface concentration and orientation was deposited on a 3 × 3 cm SnO_2 optically transparent electrode (OTE) by means of the Langmuir–Blodgett technique. The details for the deposition procedure have been reported elsewhere (*21, 22*). It is supposed that within a M-Chl *a* monolayer the hydrophilic moieties (a keto carbonyl and two ester carbonyls) in the molecule are attached to the hydrophilic SnO_2 surface

[1] Because the naturally occurring Chl *a* is often called Mg-Chl *a*, throughout this section we refer to a metal derivative of pheophytin *a* (M-free Chl *a*) as a M-Chl *a*. The pheophytin *a* itself is referred to as H_2-Chl *a*.

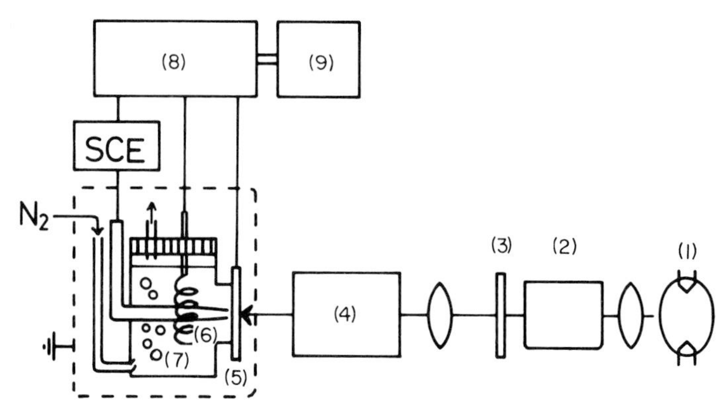

Fig. 11. Experimental setup for photoelectrochemical measurements: (1) light source, (2) water filter for ir removal, (3) shutter, (4) grating monochromator, (5) SnO_2/M-Chl *a* working electrode, (6) luggin capillary, (7) Pt counterelectrode, (8) potentiostat, and (9) recorder.

and that the hydrophobic phytol chain and chlorin ring direct themselves outward from the surface. The surface pressure during deposition was regulated to ~20 dynes cm^{-1}. In experiments with controlled surface concentration of M-Chl *a*, dipalmitoyl L-α-phosphatidylcholine (or dipalmitoyl lecithin, DPL) was used as a two-dimensional diluent.

c. Photoelectrochemical Measurements. Figure 11 schematically shows the experimental setup for photocurrent measurements. The working electrode, a M-Chl *a* monolayer-deposited SnO_2 OTE, is attached as a window of the electrochemical cell. The electrode potential is controlled with a potentiostat against a saturated calomel electrode (SCE). Light from a 500-W xenon arc lamp passes successively through a water filter and a grating monochromator and excites the adsorbed M-Chl *a* monolayer in contact with an electrolyte solution. The latter contains 0.1 *M* Na_2SO_4 as a supporting electrolyte and 0.05 *M* hydroquinone as a reducing agent (Fig. 10) and is flushed with high-purity nitrogen to remove dissolved oxygen. The quantum yield of anodic-sensitized photocurrent is calculated on the basis of the number of photons absorbed by the M-Chl *a* monolayer, which in turn is obtained from the incident photon flux and the absorbance of the monolayer.

D. Characterization of M-Chl *a*s

A brief summary of physicochemical properties of M-Chl *a*s, which are relevant to the discussion of the efficiency of photoinduced charge separation at SnO_2–M-Chl *a* interfaces, is given in the following paragraphs.

1. Spectroscopic Properties

Each M-Chl *a* shows a characteristic absorption spectrum. Several examples are illustrated in Fig. 12. The spectra of divalent Ms are similar in shape to that of Mg-Chl *a*, with a noticeable difference in the λ_{max} value and blue–red absorbance ratio. Mn-Chl *a*, in which the central metal is trivalent under normal conditions, exhibits a somewhat unusual feature in the blue range. The order of λ_{max} and ε_{max} (extinction coefficient) for different Ms parallels that for various metalloporphyrins. This demonstrates that the influence of the central metal ion on the π-electronic structure of the macrocycle is similar in both porphyrin and chlorin rings.

Of the eleven M-Chl *a*s (including Mg- and H_2-Chl *a*), for which we have heretofore succeeded in isolation and purification, five pigments possess fluorescence at room temperature (Fig. 13). In each case the Stokes shift is of the order of 0.02 eV. Spectroscopic parameters for the eleven M-Chl *a*s are given in Table I.

2. Redox Properties

The redox potentials for oxidation and reduction of M-Chl *a*s have been determined by cyclic voltammetry in butyronitrile (BN), acetonitrile (AN), and dimethylformamide (DMF). The potentials were measured against a ferrocene/ferricinium couple and were then positioned against SCE. We noted very little influence of the nature of the solvent on the redox potential values. Figure 14 shows the results obtained in BN.

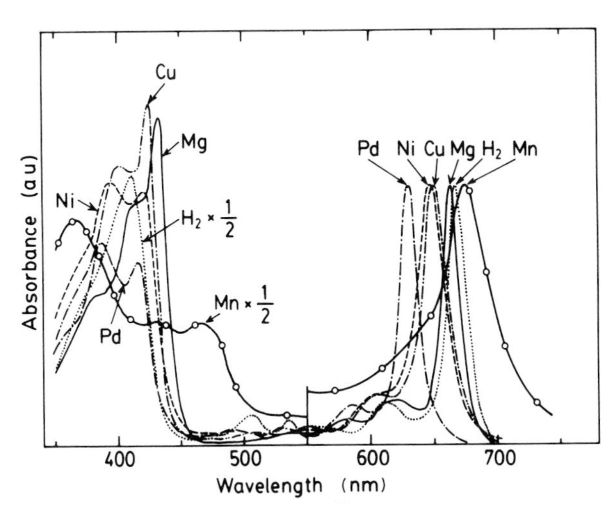

Fig. 12. Visible absorption spectra of several M-Chl *a*s in acetone at room temperature. For convenience, the red bands are arbitrarily scaled to a common height.

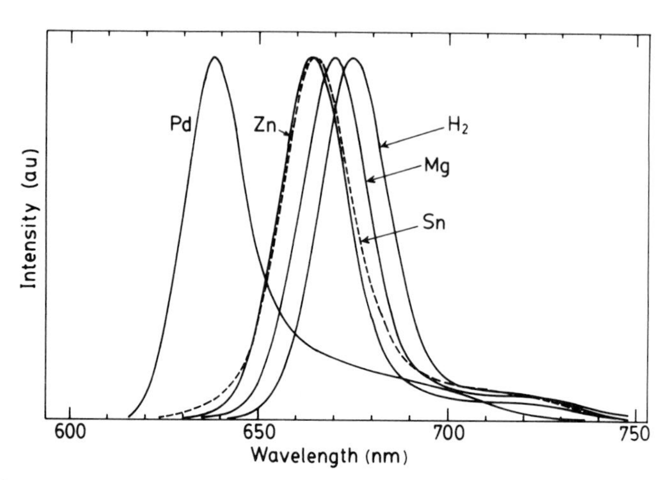

Fig. 13. Fluorescence spectra of M-Chl *a*s in deoxygenated benzene at room temperature. The peaks are adjusted to the same height.

TABLE I

Absorption and Emission Properties of Me-Chl *a*s at Room Temperature

M	Visible Absorption[a]		Fluorescence[b]		
				Quantum yield ϕ_F[c]	
	Red λ_{max}(nm) [$\varepsilon_{max}(M^{-1}\ cm^{-1})$]	Blue λ_{max} (nm)	λ_{max} (nm)	Under air	Under N_2
Pd(II)	631 [95,900]	415	638	0.0025	0.0025
Ag(II)	647 [44,000]	428	—	—	—
Ni(II)	648 [50,700]	419	—	—	—
Cu(II)	651 [60,000]	424	—	—	—
Co(II)	651 [35,900]	423	—	—	—
Sn(IV)	653 [43,400]	422	664	0.096	0.11
Zn(II)	655 [77,600]	422	663	0.21	0.23
Mg(II)	663 [76,600]	432	670	0.28	0.30
H_2(II)	668 [43,000]	411	674	0.22	0.24
Mn(III)	675 [31,400]	465	—	—	—
Hg(II)	678 [32,400]	452	—	—	—

[a] In acetone.
[b] In benzene.
[c] Based on $\phi_F = 0.30$ for Mg-Chl *a* in deoxygenated benzene.

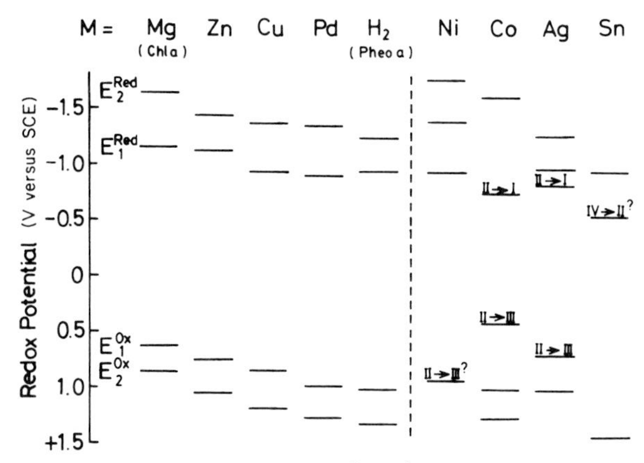

Fig. 14. Oxidation (E_1^{ox}, E_2^{ox}) and reduction (E_1^{red}, E_2^{red}) redox potentials of M-Chl *a*s measured in butyronitrile. See text for the classification by the vertical dashed line.

With regard to the redox behavior, the M-Chl *a*s can be classified into two groups (separated by the dashed line in Fig. 14). For the first group (Mg-, Zn-, Cu-, Pd-, and H_2-Chl *a*), all the redox reactions proceed via electron transfer to or from the chlorin ring. This assessment is based on the following two criteria, generally accepted for metalloporphyrins (*6*): (a) the values of $|E_1^{ox} - E_2^{ox}|$ and $|E_1^{red} - E_2^{red}|$ are around 0.3 V, and (b) a drastic change in absorption spectra occurs by oxidation–reduction processes. Details for the latter aspect are to be published elsewhere. In Fig. 14, the M-Chl *a*s within the first group are arranged in the order of the level for the one-electron oxidation at the singlet excited state, $E_1^{ox} - \Delta E^*/q$ (ΔE^* is the excitation energy and q the elementary charge). This order corresponds relatively well to the order of electronegativity of the central metal. It is seen that Mg-Chl *a* has the strongest capability of electron releasing at the excited state.

In contrast, the M-Chl *a*s gathered in the second group (Ni- to Sn-Chl *a*) appear to involve redox processes on the central metal ion, in analogy to their counterparts in metalloporphyrins. This assignment, however, needs to be further substantiated by spectroelectrochemical investigations to check for the previously mentioned criterion (b). The second and third reduction potentials for Ag-Chl *a* are nearly identical with the first and second reduction potentials for H_2-Chl *a*, respectively. This reflects the demetallation on one-electron reduction Ag(II) → Ag(I), as has been reported for silver(II) porphyrins (*16*).

Because the cb-level edge of SnO_2 is situated at around −0.25 V versus SCE in neutral media (*21*), we could expect, from a simple energetic consideration alone, that most of the M-Chl *a*s are capable of injecting an electron at the excited state into the cb of SnO_2 with considerable quantum yield.

3. Aggregation Properties

In a monomolecular layer, each M-Chl *a* has an adjacent molecule at a distance of ~10 Å. Thus it is probable that dimers and/or higher aggregates are formed and these could act as an energy trap (quenching center). In view of this, we conducted a preliminary study to examine the aggregate-formation capability of each M-Chl *a*, although in environments different from that for photoelectrochemical measurements.

First, the method described by Katz *et al.* (*19*) was employed, in which M-Chl *a* was suspended in *n*-dodecane containing a small amount of water. Only Mg- and Zn-Chl *a* showed an obvious tendency toward aggregate formation, characterized by the appearance of a new absorption peak at ~730 nm. Second, according to Uehara *et al.* (*30*), M-Chl *a* was added to a PVA aqueous solution and the evolution of visible absorption spectra recorded. A rapid growth of 730-nm peaks was again confirmed with Mg- and Zn-Chl *a*. A 700-nm peak, assignable to a dimer or a small oligomer, grew slowly in the cases of Co- and Cu-Chl *a*. For H_2- and Ag-Chl *a*, a new peak appeared at wavelengths 680–690 nm, suggesting the formation of π dimers. Other M-Chl *a*s showed no trend toward dimer or aggregate formation.

These experimental findings demonstrate a drastic change in the physicochemical properties of M-Chl *a* by simple replacement of the central metal ion.

E. Anodic Photocurrents Observed at M-Chl a Monolayer-Sensitized SnO_2 Electrodes

Anodic-sensitized photocurrents i_s were observed at several M-Chl *a* monolayer-coated SnO_2 electrodes. The action spectrum of i_s generally coincided with the absorption spectrum of the M-Chl *a* monolayer. For a given M-Chl *a*, the magnitude of i_s was reproducible to within 20%. Typical action spectra are displayed in Fig. 15. In this figure the level of i_s has been corrected for the spectral distribution of the light source and for the absorbance of each M-Chl *a* monolayer. Hence the height of the peaks reflects the relative quantum yield ϕ_s of anodic photocurrent generation. The absolute values of ϕ_s at the absorption maxima of several M-Chl *a*s are summarized in Table II. In this summary of data we note the following:

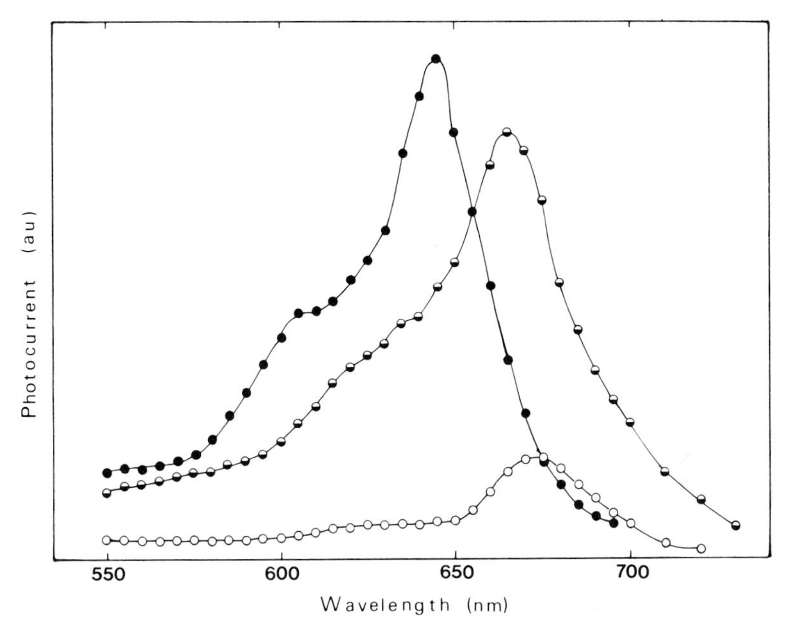

Fig. 15. Typical action spectra for the anodic-sensitized photocurrent at M-Chl *a* monolayer-coated SnO_2 electrodes. Monolayers: ○, Mg-Chl *a*; ●, Pd-Chl *a*; ◒, Cu-Chl *a*.

1. For the M-Chl *a*s classified as the second group in Fig. 14, the value of ϕ_s is practically zero (Ni-, Ag-, and Mn-Chl *a*) or very small (Co-Chl *a*). At the present stage, we have no clear idea of how to rationalize this finding. It is probable, however, that the excited state of the chlorin macrocycle of these M-Chl *a*s is rapidly quenched intramolecularly via *d*–*d* transitions, or it could be a result of the paramagnetic nature of the central metal ion. In this context we note that the excited state of a Ni-

TABLE II

Quantum Yields (ϕ_s) for Anodic Photocurrent Generation at SnO_2–M-Chl *a* Monolayer Interfaces

	Metal								
	Group 1					Group 2			
	Mg	Zn	Cu	Pd	H_2	Ni	Co	Ag	Mn
ϕ_s (%)	6.0	3.6	28.8	33.0	9.0	0.0	1.2	0.0	0.0

porphyrin undergoes intramolecular quenching within 10 ps, with no fluorescence or phosphorescence emission (*20*).

2. Among the M-Chl *a*s undergoing electron transfer exclusively on the chlorin macrocycle (from Mg- to H_2-Chl *a*), we see a substantial difference in the ϕ_s values, though all of these pigments possess electron-releasing levels at an excited state well above the cb level of SnO_2 (Fig. 14). In particular, the ϕ_s values are relatively small for Mg-, Zn-, and H_2-Chl *a*. These three M-Chl *a*s have two characteristics in common: relatively high fluorescence quantum yields (Table I) and substantial tendency toward aggregate formation (see Section III,D,3). Provided that the sensitization efficiency of dimeric or aggregated species is much lower than that for monomeric M-Chl *a*, the present finding could be rationalized by invoking the occurrence of Förster-type energy transfer accompanying quenching (energy trapping) by aggregates within the monomolecular layer. In contrast, remarkably higher ϕ_s values are associated with Cu- and Pd-Chl *a*. The former pigment is characterized by the absence of Förster-type energy transfer ($\phi_F = 0$, see Table I), and the latter by the inability of aggregate formation.

These experimental results indicate that, for the achievement of a highly efficient photoinduced charge separation with closely packed dye molecules on a semiconductor surface, the dye should not possess a high fluorescence quantum yield and, at the same time, a tendency toward aggregate formation.

Concerning the so-called light-harvesting or antenna Chl system in plant photosynthesis, it is generally supposed that a highly efficient singlet–singlet energy transfer takes place to channel the absorbed light energy to reaction centers. Nevertheless, the quantum yield for the initial charge separation *in vivo* is in the vicinity of 100%. This is suggestive of the presence, on or in the thylakoid membrane, of some particular arrangement of antenna Chl *a* molecules such that the intermolecular aggregation is well hindered.

F. Photocurrent Quantum Yield versus Surface Concentration of M-Chl a

When the intermolecular separation of M-Chl *a* is enlarged on the surface of SnO_2, both the efficiency of Förster-type energy migration and the relative abundance of dimer or aggregate species would be lowered. Thus, according to the foregoing qualitative arguments concerning the ϕ_s values, we could expect an increase in ϕ_s by decreasing the surface concentration of Mg-, Zn-, and H_2-Chl *a*. Figure 16 demonstrates, taking Mg-

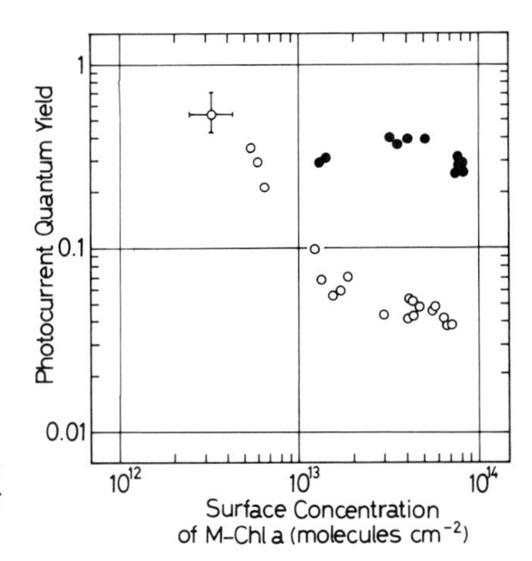

Fig. 16. Dependence of the photocurrent quantum yield on the surface concentration of Mg- (○) and Cu-Chl *a* (●).

Chl *a* as an example, that this is indeed the case. In the course of two-dimensional dilution with DPL, the ϕ_s value shows a monotonic increase from 4 to 6% at a compact monolayer to a level of about 50%. A quantitative analysis of this result is currently under way.

To the contrary, the ϕ_s value for Cu-Chl *a* remains practically constant to a level of 30–40% in a relatively wide range of surface concentration. A preliminary measurement revealed that Pd-Chl *a* behaves in a manner similar to Cu-Chl *a*. These findings support the idea that the simultaneous occurrence of the Förster-type energy transfer and aggregate formation is indeed responsible for the low ϕ_s values for dye molecules with favorable interfacial energetics.

G. *A Kinetic Study with Mixed M-Chl* a *Monolayers*

As shown in Table II, the excited Ni-Chl *a* is completely incapable of injecting an electron to the cb of SnO_2. In contrast, the ϕ_s value for Pd-Chl *a* is relatively high, and there is a sufficient spectral overlap between the fluorescence of Pd-Chl *a* and optical absorption of Ni-Chl *a* (Figs. 12 and 13). Hence, in a mixed monomolecular layer containing both Pd- and Ni-Chl *a*, a Förster-type energy transfer from the former to the latter could take place. This would lead to quenching of Pd-Chl *a*–sensitized anodic photocurrent. In a simplified picture, the photocurrent quantum yield for a single Pd-Chl *a* monolayer $\phi_s{}^O$ and that for a mixed monolayer $\phi_s{}^Q$ can be expressed as

$$\phi_s{}^O = k_s/(k_s + k_d) \tag{7}$$

and

$$\phi_s^Q = k_s/(k_s + k_d + k_{en}), \tag{8}$$

where k_s, k_d, and k_{en} denote the first-order rate constants for electron injection from excited Pd-Chl *a* to SnO_2, recombination at the adsorbed state, and energy transfer from excited Pd-Chl *a* to Ni-Chl *a*, respectively. A typical experimental result verifying such a photocurrent quenching is illustrated in Fig. 17. The open circle represents the photocurrent at a Pd-Chl *a*–coated SnO_2 electrode, with the surface coverage of Pd-Chl *a* controlled to 0.5 by dilution with DPL, and the filled circle represents the photocurrent at a Pd-Chl *a*–Ni-Chl *a* (1 : 1) mixed monolayer. From this result we obtain $\phi_s^O \simeq 0.3$ (Table II) and $\phi_s^Q \simeq 0.06$. Given the value of k_{en}, we could estimate the magnitude of k_s, which is one of the most important kinetic parameters associated with the interfacial charge separation.

However, the Förster-type process in a two-dimensional space does not obey simple exponential kinetics; the rate constant becomes a time-dependent parameter (*24*). Thus we attempted to time average the exact formula and, after a series of calculations using the spectral overlap inte-

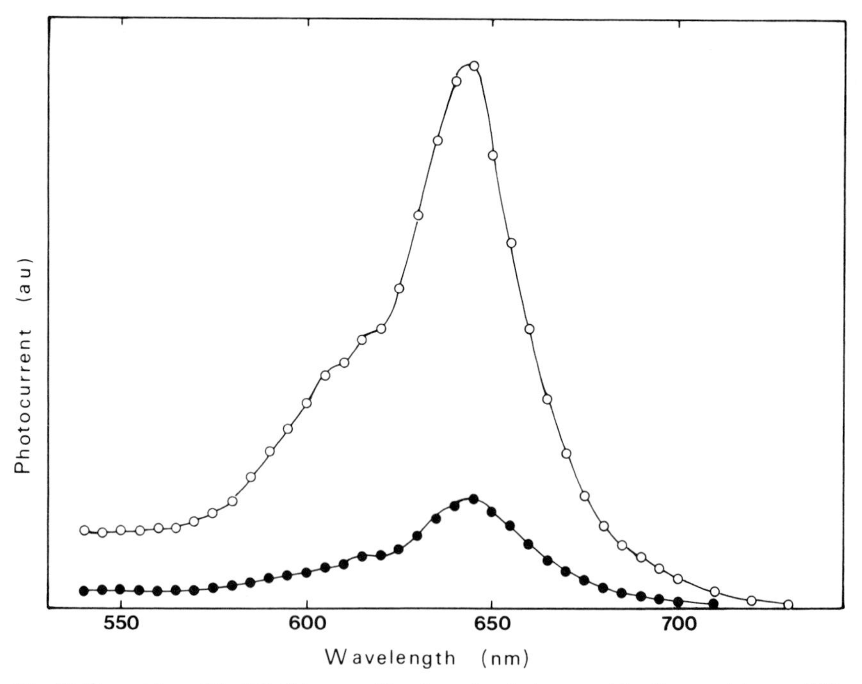

Fig. 17. Quenching of the Pd-Chl *a*–sensitized anodic photocurrent by the coexistence of Ni-Chl *a*. [Pd-Chl *a*] = 5×10^{13} cm^{-2}.

gral and the estimated fluorescence lifetime of Pd-Chl *a*, a value of $k_{en} \simeq 10^{12}\ s^{-1}$ was obtained. This leads, from Eqs. (7) and (8), to a value of $k_s \simeq 10^{11}\ s^{-1}$. This value is in good agreement with the reported rate constant ($1.2 \times 10^{11}\ s^{-1}$) for the primary charge separation in the plant photosystem I (PS-I) (*4*). Improvements of these kinetic treatments are now in progress.

At a SnO_2 electrode coated with a monolayer of Cu-Chl *a*, for which the Förster-type process does not take place owing to the absence of fluorescence (Table I), the ϕ_s value was not, as expected, influenced by the coexistence of Ni-Chl *a*.

Following are the tentative conclusions from the work described in Section III:

1. By changing the central metal ion, we can obtain a series of M-Chl *a*s differing widely from each other in their spectroscopic, redox, and aggregation behaviors.
2. For the purpose of constructing an artificial solar conversion system using dye–semiconductor interfaces, the dye molecule should satisfy the following conditions: (a) a rapid, intramolecular quenching route should not be present and (b) the dye should not possess, simultaneously, a measurable fluorescence quantum yield and aggregation tendency, even when the manner of adsorption and interfacial energetics are favorable for electron transfer. However, condition (b) could be overcome by controlling the intermolecular separation.
3. By means of a mixed monolayer containing a sensitizing ($\phi_s > 0$) M_1-Chl *a* and a nonsensitizing ($\phi_s = 0$) M_2-Chl *a*, we can obtain information on the kinetics of interfacial charge separation through photocurrent measurements, provided that an energy transfer is possible from M_1- to M_2-Chl *a*. As a preliminary result, we estimated the rate constant k_s for the primary charge separation at the Pd-Chl a^*–SnO_2 interface to be of the order of $10^{11}\ s^{-1}$.

References

1. Aharon-Shalom, E., and Heller, A., *J. Electrochem. Soc.* **129,** 2865 (1982).
2. Bard, A. J., and Wrighton, M. S., *J. Electrochem. Soc.* **124,** 1706 (1977).
3. Ellis, A. B., Kaiser, S. W., Steven, W., and Wrighton, M. S., *J. Phys. Chem.* **80,** 1325 (1976).
4. Fenton, J. M., Pellin, M. J., Govindjee, and Kaufmann, K. J., *FEBS Lett.* **100,** 1 (1979).
5. Fong, F. K., Kusunoki, M., Galloway, L., Matthews, T. G., Lytle, F. E., Hoff, A. J., and Brinkman, F. A., *J. Am. Chem. Soc.* **104,** 2759 (1982).

6. Fuhrhop, J.-H., *in* "Porphyrins and Metalloporphyrins" (K. M. Smith, ed.), p. 593. Elsevier, Amsterdam, 1975.
7. Fujishima, A., and Honda, K., *Bull. Chem. Soc. Jpn.* **44,** 1148 (1971).
8. Fujishima, A., and Honda, K., *Nature (London)* **238,** 37 (1972).
9. Fujishima, A., Kohayakawa, K., and Honda, K., *Bull. Chem. Soc. Jpn.* **48,** 1041 (1975).
10. Fujishima, A., Kohayakawa, K., and Honda, K., *J. Electrochem. Soc.* **122,** 1487 (1975).
11. Fujishima, A., Maeda, Y., and Honda, K., *Bull. Chem. Soc. Jpn.* **53,** 2735 (1980).
12. Gerischer, H., *J. Electrochem. Soc.* **113,** 1174 (1966).
13. Gerischer, H., *J. Electroanal. Chem.* **58,** 263 (1975).
14. Gerischer, H., *J. Electroanal. Chem.* **82,** 133 (1977).
15. Gerischer, H., and Willig, F., *Top. Curr. Chem.* **61,** 31 (1976).
16. Giraudeau, A., Louati, A., Callot, H. J., and Gross, M., *Inorg. Chem.* **20,** 769 (1981).
17. Hodes, G., Cahen, D., and Manassen, J., *Nature (London)* **260,** 312 (1976).
18. Honda, K., Fujishima, A., and Watanabe, T., *in* "Surface Electrochemistry" (T. Takamura and A. Kozawa, eds.), p. 141. Japan Sci. Soc. Press, Tokyo, 1978.
19. Katz, J. J., Ballschmeiter, K., Garcia-Morin, M., Strain, H. H., and Uphaus, R. A., *Proc. Natl. Acad. Sci. USA* **60,** 100 (1968).
20. Kobayashi, T., Straub, K. D., and Rentzepis, P. M., *Photochem. Photobiol.* **29,** 925 (1979).
21. Miyasaka, T., Watanabe, T., Fujishima, A., and Honda, K., *J. Am. Chem. Soc.* **100,** 6657 (1978).
22. Miyasaka, T., Watanabe, T., Fujishima, A., and Honda, K., *Nature (London)* **277,** 638 (1979).
23. Morisaki, H., Ono, H., Dohkoshi, H., and Yazawa, K., *Jpn. J. Appl. Phys.* **19,** 148 (1980).
24. Nakashima, N., Yoshihara, K., and Willig, F., *J. Chem. Phys.* **73,** 3553 (1980).
25. Nakato, Y., Abe, K., and Tsubomura, H., *Ber. Bunsenges. Phys. Chem.* **80,** 1003 (1976).
26. Nakato, Y., Tsubomura, H., and Tonomura, S., *Ber. Bunsenges. Phys. Chem.* **80,** 1289 (1976).
27. Norris, J. R., Scheer, H., and Katz, J. J., *Ann. N.Y. Acad. Sci.* **244,** 260 (1975).
28. Omata, T., and Murata, N., *Photochem. Photobiol.* **31,** 183 (1980).
29. Rabinowitch, E., and Govindjee, *in* "Photosynthesis," p. 11. Wiley, New York, 1969.
30. Uehara, K., Nakajima, Y., Yonezawa, M., and Tanaka, M., *Chem. Lett.*, 1643 (1981).
31. Watanabe, T., Fujishima, A., and Honda, K., *Bull. Chem. Soc. Jpn.* **49,** 355 (1976).
32. Wrighton, M. S., *Chem. Eng. News* **57,** 21 (1979).
33. Yoneyama, H., Sakamoto, H., and Tamura, H., *Electrochim. Acta* **20,** 341 (1975).

12 Hydrogen-Generating Solar Cells Based on Platinum-Group Metal Activated Photocathodes

Adam Heller

Bell Laboratories
Murray Hill, New Jersey

ENERGY RESOURCES THROUGH
PHOTOCHEMISTRY AND CATALYSIS

ISBN 0-12-295720-2

I. Scope

The principles of hydrogen generation at photocathodes activated by classical hydrogen-evolution catalysts such as platinum, rhodium, and ruthenium are discussed in this chapter. Photoanode-based cells (*1–6*) are the subject of Watanabe *et al.* (Chapter 11).

To reduce the kinetic barrier or overvoltage for hydrogen evolution at an illuminated photocathode one may use either a metallic catalyst or an ionic redox mediator, usually an organic or metal organic compound. The latter may be incorporated in a surface-bound polymer, bound *in situ* to a photoelectrode, or serve as a solute in a homogeneous solution (*7–9*). This chapter is restricted to photocathodes with metallic catalysts and centers on *p*-InP-based systems, reaching 13.3% solar-to-chemical (hydrogen) conversion Gibbs free-energy efficiency.

II. Requirements for Efficient Solar Hydrogen Generation

For efficient solar-to-hydrogen conversion a photocathode must simultaneously meet eight requirements:

1. The semiconductor employed must have a band gap of 1.0–1.7 eV.
2. The hydrogen-evolving surface must be chemically stable.
3. If stability is attained through the presence of a corrosion-passivating layer of an insulator, this layer must be thin enough (~30 Å or less) to allow free tunneling of electrons photogenerated in the semiconductor to ions adsorbed from the solution.
4. The surface of the semiconductor must not contain species causing excessively rapid radiationless recombination of photogenerated electrons with holes; the rate of carrier recombination must be

slower than any of the steps involved in the photoelectrochemical generation of hydrogen.

5. The overvoltage for hydrogen evolution at the photocathode–electrolyte interface must be low, that is, a catalyst for hydrogen evolution must be present.
6. The catalyst must not absorb or reflect a significant fraction of the solar photons. The thickness of a uniform platinum-group metal layer must not exceed 20 Å.
7. The barrier to the transport of holes to the semiconductor–metal catalyst interface must be high.
8. In photocathodes with porous catalyst films the barrier to the transport of holes from the semiconductor to the solution must also be high.

III. Solar Conversion Efficiency

A. *Definition of the Solar Conversion Efficiency of Hydrogen-Generating Cells*

The preferred definition of efficiency of a hydrogen-generating electrochemical cell is identical with that of the solar-to-electrical conversion efficiency in conventional photovoltaic cells, except that the amount of electrical power conserved in the electrolytic cell, rather than the electrical power produced, is considered as the gain (*10, 11*). The analogy is shown in Figs. 1 and 2. Under a solar irradiance of W_{sun} a conventional cell (Fig. 1) develops an open-circuit voltage V_{oc} and a short-circuit photocurrent density i_{sc}. The maximum power density, represented by the shaded area, is reached at a point (V_{max}; i_{max}), where the solar-to-electrical conversion efficiency is

$$\eta_{max} = V_{max} i_{max}/W_{sun} = V_{oc} i_{sc} \mathrm{ff}/W_{sun}. \tag{1}$$

The fill factor ff is defined by Eq. (1) as

$$\mathrm{ff} = V_{max} i_{max}/V_{oc} i_{sc}. \tag{2}$$

The solar-to-hydrogen conversion efficiency is defined in Fig. 2. Under an irradiance of W_{sun}, the threshold for hydrogen evolution is reduced by ΔV_{oc} with respect to a reversible hydrogen electrode, such as platinum. If the electrolyte in the photoelectrolytic cell is a 1 *M* acid, the reference point is the standard hydrogen electrode (SHE). The short-circuit current density is defined as the photocurrent density at zero bias, that is, when the photocathode is at the potential of a platinum electrode, both elec-

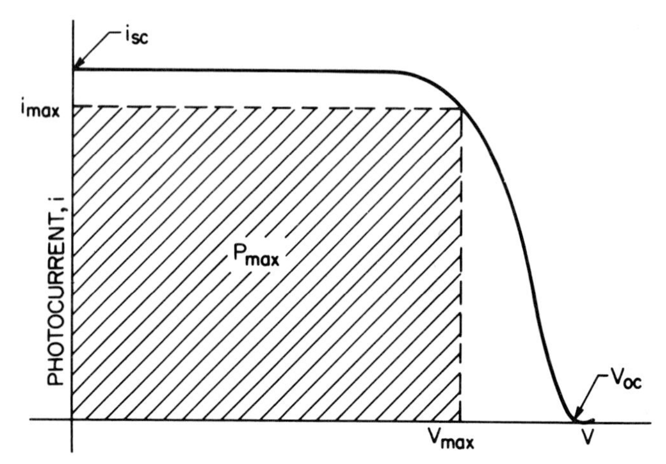

Fig. 1. The efficiency of a conventional photovoltaic cell is the ratio of the electrical power output per unit area $i_{max}V_{max}$ (shaded) and the solar irradiance: $\eta = P_{max}/|W_{sun}|_{\lambda=0}^{\lambda=\infty}$. V_{oc} and V_{max} are, respectively, the open-circuit voltage and the voltage at the maximum power point; i_{sc} and i_{max} are the short-circuit photocurrent and the photocurrent at the maximum power point. The fill factor ff is $(i_{max}V_{max})/(i_{sc}V_{oc})$.

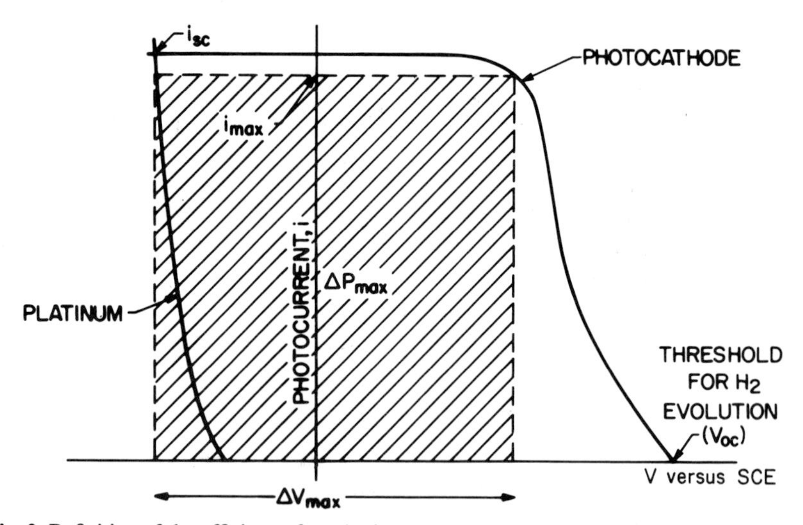

Fig. 2. Definition of the efficiency for a hydrogen-generating photoelectrochemical cell. The efficiency is the ratio of the electrical power per unit area conserved, $i_{max}\,\Delta V_{max}$ (shaded), and the solar irradiance: $\eta = \Delta\, P_{max}/|W_{sun}|_{\lambda=0}^{\lambda=\infty}$. The onset potential for hydrogen evolution is shifted with respect to that of a reversible platinum electrode by ΔV_{oc}. The maximum conversion efficiency is reached when the voltage gain with respect to a platinum electrode operating at the same current density is ΔV_{max} and the photocurrent is i_{max}; i_{sc} is the current density at the intersection of the iV curves of the platinum cathode and the photocathode. The fill factor ff is $(i_{max}\,\Delta V_{max})/(i_{sc}\,\Delta V_{oc})$.

trodes operating at identical current densities in the same electrolyte. The maximum conversion point is (ΔV_{max}; i_{max}), where the bias required to generate hydrogen at a current density i_{max} is reduced by ΔV_{max}. The maximum electrical power conserved per unit area, when a platinum cathode is replaced by a photocathode, is $\Delta V_{max} i_{max}$, and the efficiency of solar conversion is the electrolytic cell is

$$\eta_{max} = \Delta V_{max} i_{max}/W_{sun} = \Delta V_{oc} i_{sc} \mathrm{ff}/W_{sun}, \quad (3)$$

where

$$\mathrm{ff} = \Delta V_{max} i_{max}/\Delta V_{oc} i_{sc}. \quad (4)$$

Note that if ΔV_{max} exceeds the bias required for electrolysis at a photocurrent density i_{max}, photoelectrolysis is spontaneous and electrical power is derived from the cells in addition to chemicals. If ΔV_{max} equals the bias required, photoelectrolysis is still spontaneous, but electrical power is derived simultaneously with the chemicals only by reducing the efficiency. If ΔV_{max} is less than the bias required for electrolysis at a current density i_{max}, an external bias is required, and the process is termed photoassisted electrolysis.

The efficiency definition for a hydrogen-generating cell (Fig. 2) measures the ratio of the electrical power density conserved in the electrolytic process and the solar irradiance. An alternative useful definition measures the Gibbs free-energy efficiency, or the ratio of the electrical power that can be derived from the chemicals generated, on their reaction in an ideal fuel cell, and the solar irradiance (*5, 12*). This definition is similar to that of Fig. 2 and Eq. (3), except that ΔV_{max} is measured with respect to the reversible hydrogen-electrode potential in the electrolyte used. When the electrolyte is a 1 *M* acid, the reference point is the SHE potential.

B. Band Gap and Efficiency

The optimal band gap or absorption edge of a single-stage terrestrial solar converter, such as a photovoltaic cell (*13–18*) or a photosynthesizing system (*19, 20*), is near 1.3 eV (960 μm). Here the theoretical efficiency is 25–30%. The theoretical efficiency exceeds 20% through the 1.0- to 1.7-eV (1250- to 740-nm) range (*13–20*). In multistage converters, based on multiple semiconductors or absorbers, the limiting efficiency is higher (*21*).

A hydrogen-generating photocathode, like any other threshold converter, requires for adequate solar conversion efficiency a semiconductor with a band gap of 1.3–1.7 eV. The maximum gain in threshold potential for hydrogen evolution ΔV_{oc}, the equivalent of the open-circuit voltage in a conventional solar cell, cannot exceed the band gap. Furthermore, the power product ($\Delta V_{max} i_{max}$) in hydrogen-generating cells rarely peaks at a gain in bias exceeding half the band gap. For this reason, efficient single-stage photocathode-based cells photoelectrolyze spontaneously only those electrolytes that require a bias of less than 0.6 V at the ~30 mA/cm^2 photocurrent densities, typical of 1.3-eV band-gap materials in sunlight. Thus, for example, HI can be spontaneously photoelectrolyzed at high efficiency (*11, 12*). Photocathodes can nevertheless be efficient in *photoassisted* electrolysis, irrespective of the bias required: dilute perchloric acid is electrolyzed to hydrogen and oxygen, dilute hydrochloric acid to hydrogen and chlorine, and dilute hydrobromic acid to hydrogen and bromine at 16% efficiency (see Section IX,D) with *p*-InP (platinum-group metal catalyst) photocathodes (*11, 12*).

C. Two-Photoelectrode Cells

ΔV_{oc} and ΔV_{max} can be substantially augmented in photoelectrolytic cells if both a photoanode and photocathode are used (*22–26*). When such cells operate at the current density i_{max}, the photoanode produces a bias gain ΔV_{max}^{anode} and the photocathode produces a bias gain $\Delta V_{max}^{cathode}$.

The sum bias $\Delta V_{max}^{anode} + \Delta V_{max}^{cathode}$ can reach ~1.2 V, even though both the photoanode and the photocathode are made with semiconductors of appropriate band gap for efficient solar conversion (*12*).

In two-photoelectrode cells the anodic and cathodic photocurrents must be equal to maximize the efficiency. [Because these cells are two-stage converters they can be more efficient than single-stage systems (21).] To optimize their efficiency simultaneously with their stability to corrosion, the cells should have a photoanode with a band gap larger than 1.3 eV and a photocathode with a band gap smaller than 1.3 eV. Although no such cell has been made as yet, an efficient two-photoelectrode cell, capable of spontaneously electrolyzing HBr with light of wavelengths up to 950 nm has been reported (Section XI) (*12*).

IV. Chemical Stability of the Photocathode–Solution Interface

A. Cathodic Protection by Photogenerated Electrons

Photocathodes are inherently more stable than photoanodes in electrochemical solar cells. Whereas in photoanodes photogenerated holes arrive at the solution interface and may cause the oxidation of semiconductors (*1–5*), photogenerated electrons arriving at the solution interface protect photocathodes against oxidative corrosion (*26–30*). Because the stabilization of efficient photoanodes is kinetic in nature, the electrodes photocorrode at high levels of irradiance, when the rate of hole transport to the solution can no longer match the rate of their photogeneration (*30*). In contrast, high levels of irradiance cathodically protect, rather than damage, photocathodes (*30*).

B. Formation of Interfacial Oxides by Reaction with Aqueous Solutions

Although intense light does not cause photocorrosion in photocathodes, most semiconductors having appropriate band gaps for efficient solar conversion react with aqueous electrolytes. Typical reactions are those of Si and InP:

$$Si + 2H_2O \longrightarrow SiO_2 + 2H_2$$
$$2InP + 9H_2O \longrightarrow In_2O_3 + 6H_2 + 2H_3PO_3$$

Thus the stability of photocathodes like *p*-Si and *p*-InP depends on whether chemical passivity is or is not attained with an oxide film on the surface. To attain passivity the oxide layer must prevent the transport of both reactants and products to and from the semiconductor surface; the corrosion-passivating oxide film must be a continuous, sealing layer. Such layers form in dilute mineral acids on both *p*-Si and *p*-InP, but in the case of Si a thicker oxide layer is required to avoid mass transport than in the case of InP. Consequently, *p*-Si photocathodes slowly loose efficiency as the internal resistance of the cells increases with the growth of the oxide layer (*20*). In *p*-InP in dilute acids the oxide grows only to a thickness of 6–10 Å (Fig. 3); this allows the free tunneling of photogenerated electrons to the solution (*32*). Furthermore, in acids this layer remains sufficiently thin over a window of potentials both positive and negative with respect to SHE (*33*).

C. Protection of the Semiconductor Surface by Platinum-Group Metal Catalysts

In Section VII,A we shall see that by hydrogen saturation it is possible to reduce the work function of platinum-group metal catalysts to the hydrogen-electrode potential (*34*). For this reason, a hydrogen-saturated platinum alloy, in contrast to pure platinum, does not shunt hydrogen-generating photocathodes; neither the barrier height ψ_B nor the gain in

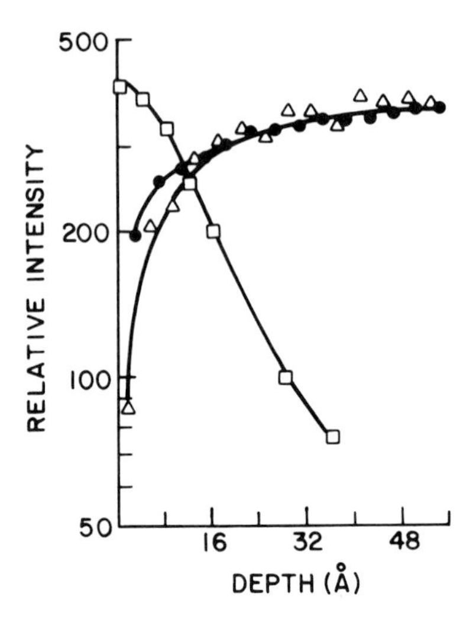

Fig. 3. Depth profile of the composition of a *p*-InP photocathode, determined by low-energy ion-scattering spectroscopy (LEISS). ●, In; △, P; □, O.

bias ΔV_{oc} is decreased when the semiconductor surface is metallized (*34*). Thus metallization with an inert platinum-group metal or platinum-group metal alloy can prevent mass transport to and from the semiconductor surface and prevent corrosion or film growth. Unfortunately, adequately thick ($\gtrsim$100 Å) layers of platinum-group metals absorb or reflect most of the incident solar photons and reduce the solar conversion efficiency. Thin layers ($\leq$50 Å) are porous. Although they slow down the growth of oxides and reduce the rate of corrosion, they do not prevent these reactions (*35*).

V. Radiationless Recombination of Photogenerated Electrons at the Photocathode–Electrolyte Interface

A. Chemical Bonding and Surface Recombination

Progress in the chemical control of radiationless electron–hole recombination processes at surfaces of *p*-type semiconductors is key to new, efficient photocathodes. Among the photocathodes having appropriate band gap for solar conversion, very few (*p*-InP, *p*-Si, and *p*-type layered chalcogenides) have sufficiently low recombination velocities at their aqueous acid interfaces to yield efficient electrochemical solar cells.

Recombination at surfaces and at grain boundaries is associated with the presence of intrinsic or impurity related (extrinsic) "weak" bonds. In these the bonding energy *per valence electron* is less than that in the bulk

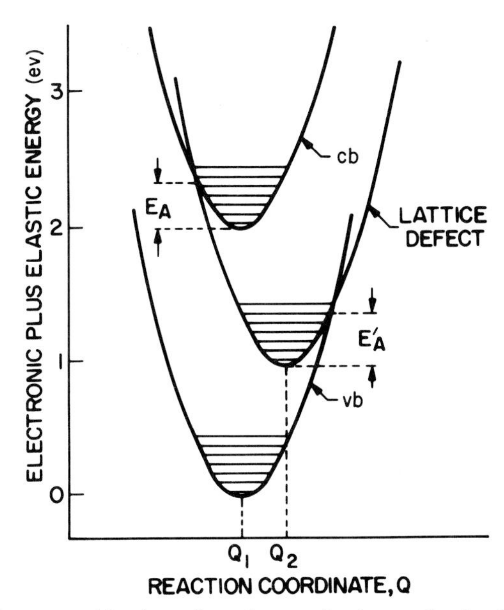

Fig. 4. Radiationless recombination of an electron in the conduction band (cb) involves thermal excitation (E_A) to a defect, surface or grain boundary state (SS), followed by recombination with a hole in a second step that also requires thermal excitation (E'_A). From Heller (*38*).

of the semiconductor lattice. The bonds may be free radicals (''dangling''), unsaturated double or triple bonds, or bonds stressed through their stereochemistry. Such bonds introduce sites with high cross sections for radiationless electron–hold recombination.

The physical basis for the correlation between weak chemical bonding and radiationless recombination is the following. Recombination in defects in semiconductors, of which surfaces represent a special group, involves a multiphonon relaxation process (*36*), shown in Fig. 4, in which the electronic plus elastic (vibrational) energy of the semiconductor lattice is plotted against Q, a reaction coordinate (e.g., interatomic distance). For the radiationless relaxation to occur, a defect or surface state SS must intersect both the conduction and the valence bands. Thermal excitation E_A makes possible the crossing over of an electron in the conduction band cb to SS. Relaxation is completed when, in a second step that also involves thermal excitation E'_A, the electron drops into the valence band vb. A symmetrical picture holds for relaxation by thermal excitation of holes from vb to SS and from SS to cb.

When cb and vb are well separated, that is, when the band gap is large, greater thermal excitation is usually required to reach the intersection

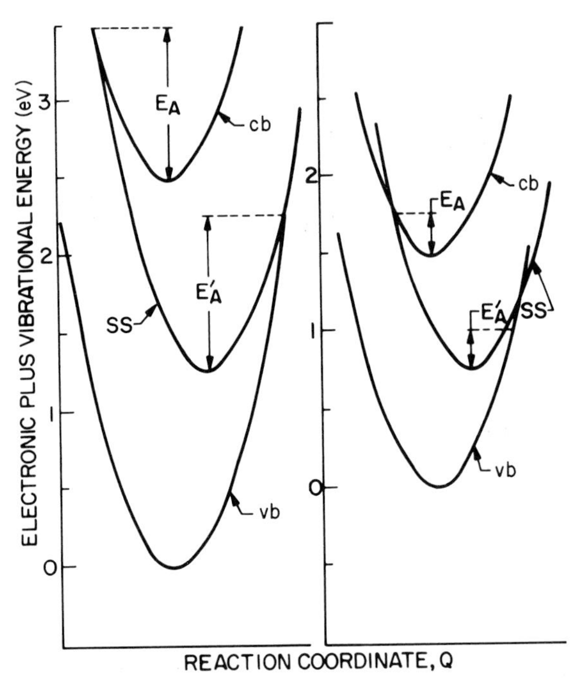

Fig. 5. The activation energies for radiationless electron–hole recombination (E_A and E'_A) depend on the band gap. The larger the band gap the smaller the cross section for radiationless recombination. The figure shows this for defect or surface states (SS) at the center of the gap, undergoing identical shifts in reaction coordinates with respect to those of the bulk lattice. For $E_{BG} = 2.5$ eV (left), $E_A = E'_A = 1.0$ eV. For $E_{BG} = 1.5$ eV (right), $E_A = E'_A = 0.25$ eV.

points (Fig. 5). Consequently, at ambient temperature, surface recombination is much less of a problem in large band-gap semiconductors than in semiconductors having an appropriate band gap for solar conversion. It is for this reason that n-TiO_2 and n-$SrTiO_3$, both with band gaps exceeding 3 eV, were successfully used in photoelectrochemical cells, even as polycrystalline films and as suspended particles, in spite of the high surface-to-volume ratios and of very high densities of surface states (*1–6*).

From Fig. 6 it is evident that the state SS must be "deep," that is, have its minimum near the center of the band gap for the radiationless recombination to have a large cross section. If the state is close to cb, the intersection point with vb is high with respect to both the minimum of SS and vb and well before thermal excitation would lead to radiationless recombination, the electron is thermally reexcited into cb. If the minimum of SS is close to that of vb, the first step of the relaxation process requires excessive thermal excitation and is unlikely at ambient temperature.

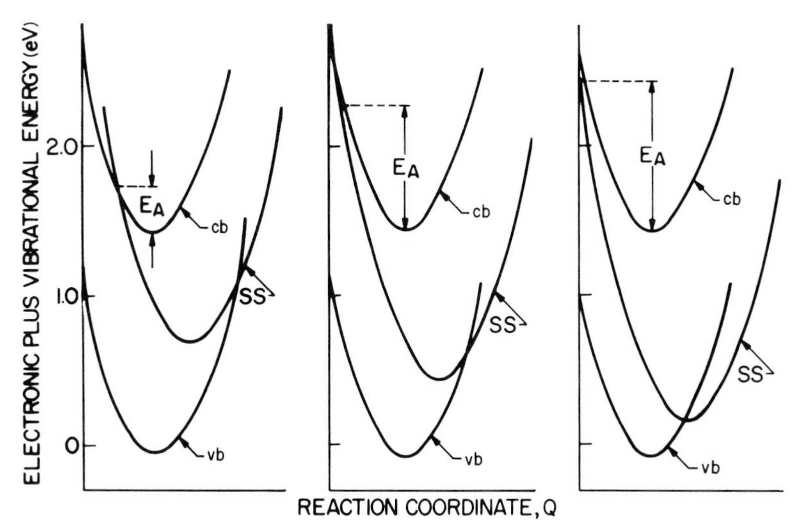

Fig. 6. The activation energy for radiationless electron–hole recombination (E_A) is reduced as the defect or surface state SS approaches the center of the band gap.

For the recombination cross section to be high, the reaction coordinate of the minimum of SS must be either much larger or much smaller than that of vb and cb, that is, the distance between the atoms introducing the surface state must differ from the intrinsic distance in the bulk (Fig. 7). If this condition is not met, intersystem crossing does not take place, except at very high temperature.

Valence and conduction bands derive, respectively, from bonding and antibonding orbitals, and the splitting of these is an approximate measure of the strength of the chemical bonds in the lattice. Dangling, weak, strained, or multiple bonds introduce a lesser splitting than that associated with the normal bulk lattice bonds. The less the binding energy per valence electron, the deeper the states. In homoatomic lattices, such as Si or Ge, a dangling bond introduces a state at the center of the gap. Reduced binding energy per valence electron also causes an increase in Q if the bonds are weak. If a lattice is terminated by strained multiple bonds, Q is reduced. In either case the recombination cross sections increase.

B. Reduction of Recombination at Surfaces of Photocathodes by Chemical Means

Weak bonds at surfaces of semiconductors, whether intrinsic or impurity related, can be chemically reacted by chemisorption. With a species strongly bound to the semiconductor surface, the bonds are

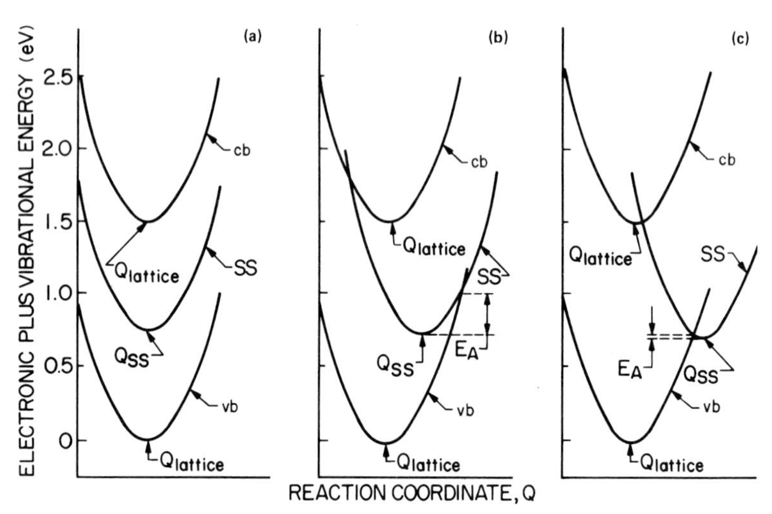

Fig. 7. The activation energy for radiationless electron–hole recombination E_A is larger when the reaction coordinate of the defect or surface state (SS) approaches that of the bulk lattice (cb, vb) and is reduced as the displacement increases. When the reaction coordinates of SS, cb, and vb are identical (left), $E_A \rightarrow \infty$. (a) $Q_{SS} = Q_{lattice}$, $E_A \rightarrow \infty$. (b) $Q_{SS} > Q_{lattice}$, $E_A = 0.025$ eV. (c) $Q_{SS} \gg Q_{lattice}$, $E_A \approx kT$.

strengthened (*30, 37–39*). Figure 8 shows schematically how chemisorption of a reagent R splits a damaging surface state SS, to produce two new states RSS and RSS*, which no longer allow intersystem crossing at moderate temperatures. As discussed in the preceding subsection, the activation energies for recombination E_A and E'_A increase as the states become shallower (i.e., as RSS approaches vb and as RSS* approaches cb) and as the interatomic distance of the surface atoms Q_3 approaches that in the lattice.

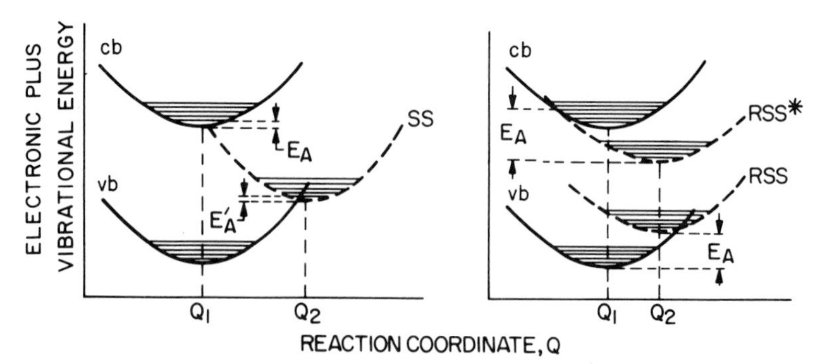

Fig. 8. Reaction of a species introducing a surface state (SS) with a chemisorbed species R splits SS to two new states, RSS and RSS*. For these states the activation energies for radiationless recombination (E_A, E'_A) are higher.

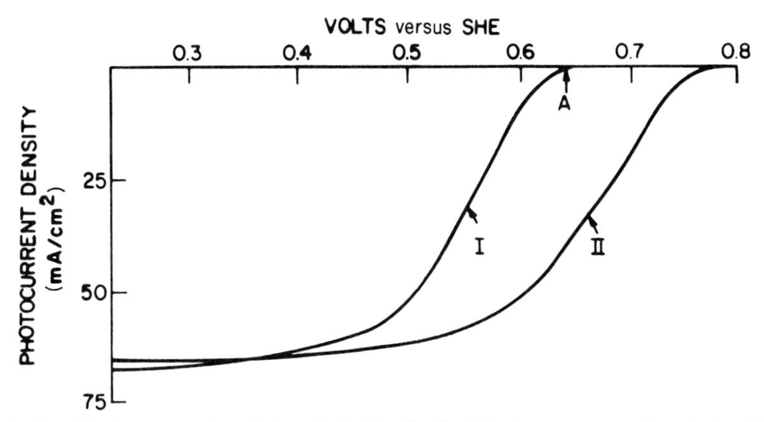

Fig. 9. The *iV* characteristics of the *p*-InP (Rh, H alloy) hydrogen-generating photocathode are improved by oxidation of solution-exposed regions of the *p*-InP surface. On extended cycling of the photocathode in 4 *M* $HClO_4$ between 0.22 V versus SHE and V_{oc}, some of the interfacial oxide is reduced and the characteristics are those of curve I. Following anodization [blocking the light at V_{oc} (point A) followed by reillumination] the photocathode's characteristics are those of curve II.

Examples of surface chemistries representing strengthening of chemical bonds by chemisorbed layers of atoms and consequent reduction in recombination losses in photocathodes are

1. The oxidation of the faces of *p*-InP crystals to form a layer of hydrated indium oxide, (*26, 28, 32, 34*)
2. The chemisorption of submonolayers of silver on faces of *p*-InP (*39*)
3. The oxidation of faces of *p*-Si (*27*)

Figure 9 shows the improvement of a *p*-InP (Rh, H alloy) hydrogen-generating photocathode on mild anodization, a process in which a hydrated indium oxide layer is formed at pores in the catalyst layer (*34*). A similar experiment in which a submonolayer of silver is chemisorbed on *p*-InP, partially covered by islands of hydrogen-saturated rhodium, is shown in Fig. 10 (*39*).

VI. The Relationship between the Fill Factor and the Overvoltage in Hydrogen-Evolving Solar Cells

The process of reduction of protons to hydrogen on a solid electrode involves four steps. First, a proton is adsorbed on the surface; second, an

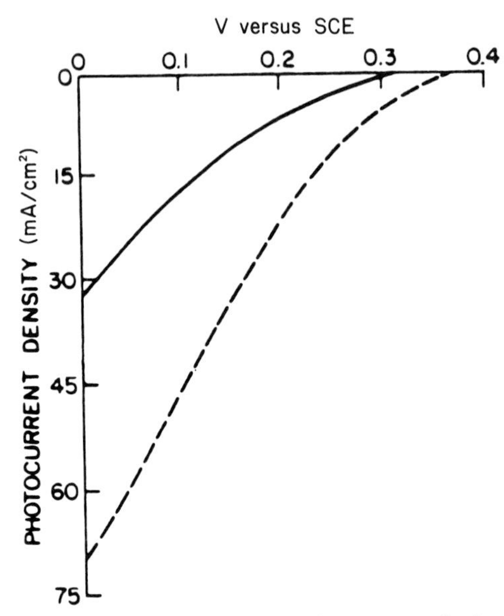

Fig. 10. Chemisorption of a submonolayer of silver ions at pores in the (Rh, H alloy) film on *p*-InP improves the *iV* characteristics of the photocathode. ——, No Ag^+; ---, Ag^+.

electron is transported to and reacts with the proton to produce an adsorbed hydrogen atom; third, two adsorbed hydrogen atoms approach each other by surface or bulk diffusion and react to form a hydrogen molecule; and fourth, the molecule is desorbed. If any of these steps is slow, the bias required to effect electrolysis is increased and electron–hole recombination dominates at the more anodic potentials, where the conversion efficiency would be high in the absence of such recombination.

Platinum-group metals such as Pt, Rh, Ru, and Pd reduce the overvoltage for hydrogen evolution when incorporated in the surface of conventional metallic cathodes. Also, as is evident from a very substantial body of literature on suspended particles of *n*-type semiconductors, incorporation of these metals in their surface facilitates photoelectrolysis. These studies, listed in reference *34*, are reviewed by Grätzel (Chapter 3).

Reduction in overpotential for hydrogen evolution on incorporation of a platinum-group metal in the surface (or in the bulk) of the semiconductor is evident in photocathodes made of a broad variety of *p*-type semiconductors. These include *p*-SiC (*41*), *p*-GaP (*22–26, 40–45*), *p*-CdTe (*24, 41*), *p*-$LuRhO_3$ (*46–47, 47a*), *p*-WSe_2 (*48*), *n*-GaAs (*42*), *p*-InP (*10–12, 29, 30, 34, 42, 49*), and *p*-NiO (*50*).

Figure 11 shows the effect of incorporating ~10 Å of Pt in the surface of

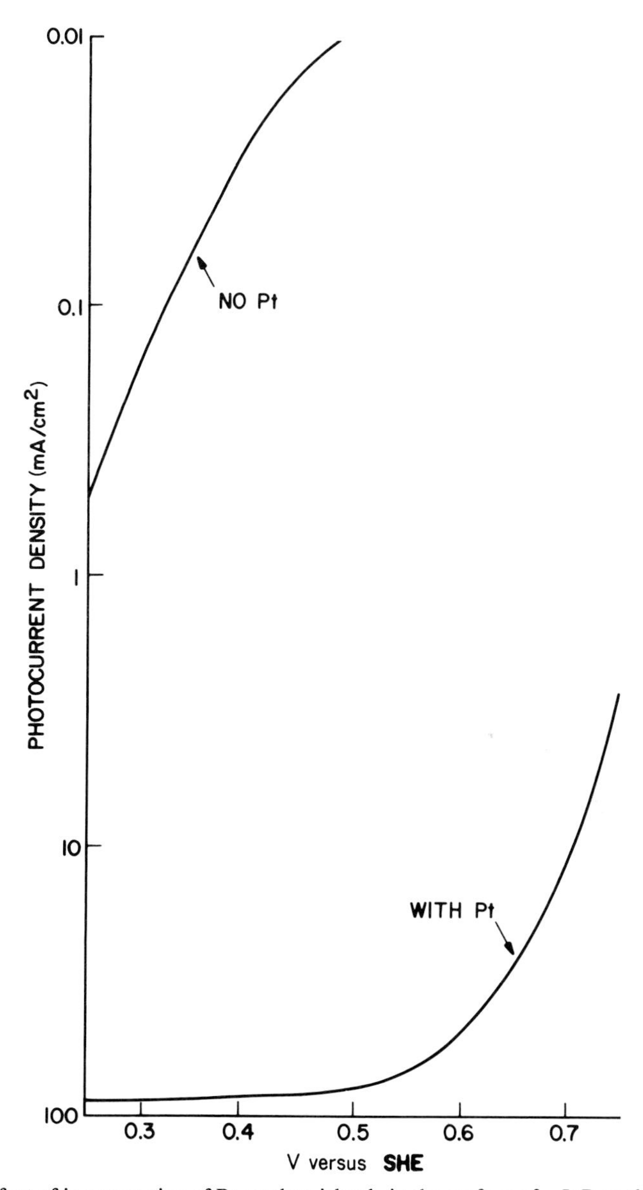

Fig. 11. Effect of incorporation of Pt catalyst islands in the surface of *p*-InP on the current-voltage characteristics of the photocathodes in 3 *M* HCl. Although the threshold potential for hydrogen evolution is not substantially changed, there is a 10^4-fold increase in fill factor, that is, photocurrent, at 0.5–0.6 V versus SHE. From Heller *et al.* (*34*). Copyright 1982 American Chemical Society.

a p-InP photocathode (*34*). Although the threshold potential for hydrogen evolution is not substantially changed, the photocurrent at potentials well positive of SHE is increased by a factor of $\sim 10^4$. From Section III,A and Fig. 2 it is evident that this increase in photocurrent translates to an increase in fill factor by a factor of 10^4. In the absence of a catalyst the fill factor is low; slow kinetics is equivalent to high cell impedance. The catalyst reduces this impedance.

The reason for the lack of substantial change in threshold potential for hydrogen evolution is discussed in Section VII. The limiting photocurrent i_{sc} is controlled by the flux of photons absorbed by the semiconductor and, therefore, by the thickness, uniformity, reflectance, and absorbance of the catalyst layer. Uniform layers of platinum-group metals of 40- to 80-Å thickness absorb about one-half of the incident photons in the visible and near infrared (*51–53*). Thus, to avoid excessive current losses, the catalyst layer must either be thin ($\sim$20 Å) or highly nonuniform in thickness. Macroscopic islands of catalysts spaced within distances comparable to, or less than, the diffusion length of electrons in p-type semiconductors can be formed by etching the catalyst layer so as to remove most of it, yet leaving nucleation sites, at which catalyst islands can be grown photoelectrochemically (*29*).

VII. The Relationship between the Barrier Height and the Gain in Threshold Potential for Hydrogen Evolution

A. Chemistry of the Hydrogen-Generating Junctions and Their Barrier Height

The limit to the realizable gain in threshold potential for hydrogen evolution with respect to a reversible platinum or rhodium electrode, ΔV_{oc}, is the height of the barrier-to-hole transport from the semiconductor to the solution or to the catalyst ψ_B (Fig. 12). ΔV_{oc} approaches ψ_B under high irradiance. For an ideal semiconductor–metal interface,

$$\psi_B = E_f - \phi_m + \Delta_m, \tag{5}$$

where E_f is the Fermi level of the semiconductor, measured relative to the energy of a free electron in vacuum, ϕ_m is the work function of the metal, and Δ_m the potential drop across the interfacial dipole layer (*5, 54*). In the case of a hydrogen-evolving photocathode, without a catalyst, the barrier height is

$$\psi_B = E_f - \phi_H + \Delta_H, \tag{6}$$

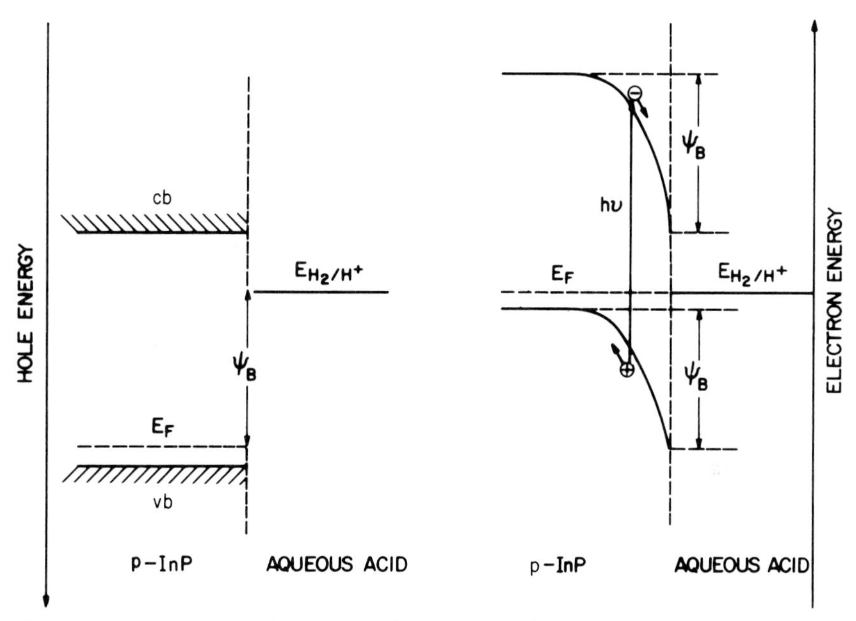

Fig. 12. The theoretical height of the barrier preventing hole transport to the photocathode (p-InP) surface is ψ_B. ΔV_{oc}, the maximum gain in threshold potential for hydrogen evolution, approaches ψ_B under high irradiance. Prior to hypothetical "contacting" of p-InP and of the hydrogen adsorbed on the solid surface (left), the p-InP Fermi level (E_F) and the work function of adsorbed hydrogen (ϕ_H) are separated by ψ_B eV. "Contacting" (right), that is, equalizing E_F and ϕ_H by electron transport between the two, creates a barrier of approximately ψ_B eV.

where ϕ_H is the work function of hydrogen adsorbed on the semiconductor surface and Δ_H the potential drop across the interfacial layer. If the solution redox potential, measured relative to a free electron in vacuum, E_{H_2/H^+} (*55*) equals ϕ_H, one obtains Gerischer's equation (*56*)

$$\psi_B = E_f - E_{H_x/H^+} + \Delta_H, \qquad (7)$$

Δ_H being the potential drop across the Helmholtz layer. Because of the analogy between Eqs. (5) and (6), we shall refer to ϕ_H as the work function of "metallic" hydrogen.

When the catalyst layer at the semiconductor surface is thin enough to avoid light–absorption losses (i.e., when it is <20 Å thick) the layer is porous and there are not only sites with junctions between the semiconductor and the catalyst, but also sites with junctions between the semiconductor and the solution. If the pores are much smaller than the diffusion length of the electrons, the barrier height will be that of a semiconductor–(platinum-group metal–hydrogen alloy) junction [Eq.

(5)]. If this condition is not met, that is, when there are large-diameter areas where the semiconductor and the solution are directly in contact, the barrier height assumes a value intermediate between the values calculated by Eq. (5) and (6) (*34*).

Figure 13 shows the vacuum ϕ_m values of Pt, Rh, and Ru and the value of SHE with respect to the bands and Fermi level of *p*-InP (*34*). Had the catalyst metals retained their vacuum work functions, the barrier height for *p*-InP–platinum would be negligibly small and the barrier height for *p*-InP–rhodium would be smaller than that for ruthenium. One finds, however, that the barrier heights of all three metals are the same and that these barriers are similar to that of a catalyst-free, hydrogen-evolving, *p*-InP photocathode. The cause of these similarities is the dissolution of hydrogen in the platinum-group metals. Such alloying lowers the work

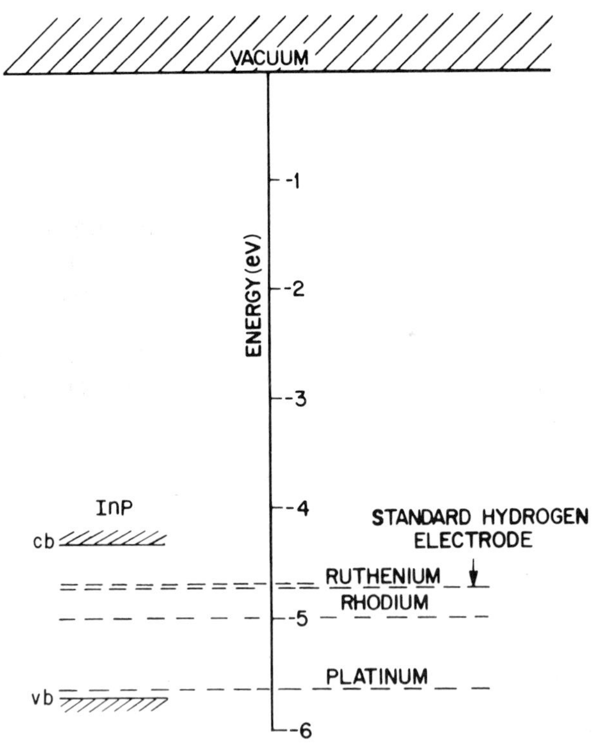

Fig. 13. Work functions of pure Pt, Rh, and Ru; positions of the valence band, conduction band, and Fermi level of *p*-InP; and position of the standard hydrogen electrode, with respect to the energy of a free electron in vacuum. From Heller *et al.* (*34*). Copyright 1982 American Chemical Society.

function of these metals. At hydrogen saturation, the work function reaches that of "metallic" hydrogen.

The reader with a background in electrochemistry may find the case of potential change on hydrogen saturation analogous to the change in electrochemical potential of a lead–sodium alloy. The alloy becomes more reducing as the sodium content is raised. After a series of stoichiometric and nonstoichiometric compositions is passed, sodium saturation is reached. At this point, that is, when the alloy is so rich in sodium that a pure sodium phase separates, the electrode potential equals that of a pure sodium electrode. Similarly, evolution of hydrogen on a photocathode with a very thin catalyst layer represents the point at which the platinum-group metal–hydrogen alloy reaches phase separation. At this point $\phi_M = \phi_H$.

The dependence of work functions and barrier heights on the partial pressure of hydrogen is the basis of hydrogen sensors, usually made of Pd–SiO_2–Si. The dependence has, however, been observed also in a variety of other junctions (*57–73*).

An experiment demonstrating the variation of the photovoltage of a dry Schottky junction cell on the ambient gaseous atmosphere is shown in Fig. 14. From Fig. 13 and Eq. (5) it is evident that in the *p*-InP–(rhodium–hydrogen alloy) junction the barrier height should increase with increasing partial pressures of hydrogen and that the photovoltage of a solar cell

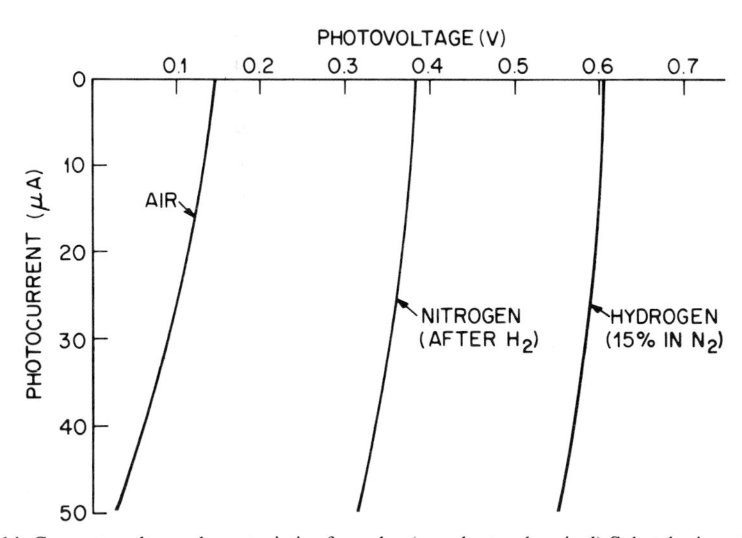

Fig. 14. Current–voltage characteristics for a dry (nonelectrochemical) Schottky junction *p*-InP/Rh photovoltaic cell in different gaseous atmospheres.

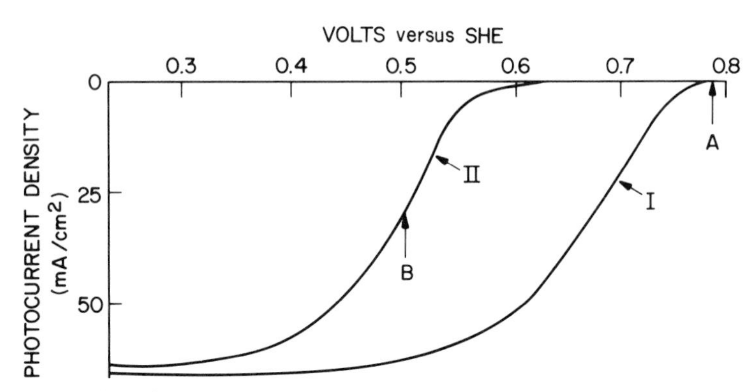

Fig. 15. Oxidation of the hydrogen dissolved in the *p*-InP (Rh, H alloy) photocathode by prolonged anodization at point A (0.8 V versus SHE), where no hydrogen is generated, shifts the current–voltage characteristics from those of curve I to those of curve II. If the anodized electrode is maintained at point B (0.5 V versus SHE), where hydrogen evolution is rapid, the Rh, H alloy is restored and with it the characteristics of curve I (4 *M* $HClO_4$, 3 AMl, tungsten–halogen irradiance).

based on this junction should correspondingly increase. This, indeed, is observed: the cell's photovoltage is 0.14 V in air. It increases to 0.61 V when the rhodium is hydrogen saturated in a 15% H_2–85% N_2 atmosphere and drops to 0.38 V when the partial pressure of hydrogen above the metal is reduced by flushing with nitrogen (*34*).

The key conclusion is that the work function or electrochemical potential of a hydrogen-alloying platinum-group metal catalyst changes drastically when the metal is inserted into a hydrogen-containing environment, that is, when hydrogen evolution starts at the photocathode. Dissolution of hydrogen in the catalyst increases the barrier height and ΔV_{oc} and thereby the solar-to-hydrogen conversion efficiency (*34*). Thus *p*-type semiconductor (platinum-group metal) photocathodes operate best in environments that allow the maintenance of hydrogen saturation of the metal. When the hydrogen is stripped from the photocathode, for example, by anodization or by reaction with oxygen, ΔV_{oc} and the efficiency drop (Fig. 15). The drop is particularly pronounced for metals with large vacuum work functions.

B. pH Invariance of the Barrier Height and of the Gain in Threshold Potential

When the platinum-group metal catalyst covers all or most of the surface of the photocathode, the barrier height ψ_B, according to Eq. (5), is independent of pH. Consequently, the gain in threshold potential for hydrogen evolution of an illuminated photocathode relative to a platinum or rhodium cathode ΔV_{oc} does not depend on pH (*34*). This independence implies that the threshold potential of an illuminated *p*-type semiconductor (platinum catalyst) photocathode V_{oc}, like that of a platinum cathode, follows the Nernst equation. Both vary by $2.303kT/q$ per pH unit, k being the Boltzmann constant, T the absolute temperature, and q the charge of the electron. Figure 16 (*34*) shows that at 298 K the *p*-InP–hydrogen-saturated Rh photocathode has the theoretical 59 mV per pH unit slope, accurately paralleling the slope of a Pt cathode. The two lines of Fig. 16 are separated by a pH-independent gain in potential ΔV_{oc}. ΔV_{oc} depends only on two parameters: the partial pressure of hydrogen (Section VII,A) and the irradiance (Section IX).

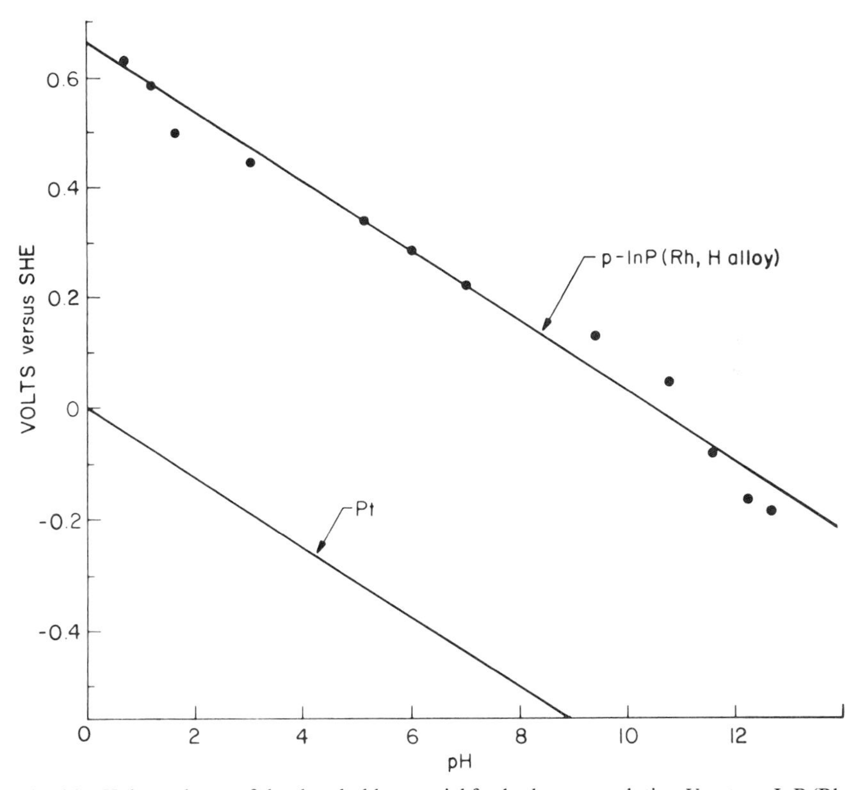

Fig. 16. pH dependence of the threshold potential for hydrogen evolution V_{oc} at a *p*-InP (Rh, H alloy) photocathode. The bottom line is the theoretical curve for a reversible hydrogen electrode at solid Pt or Rh. From Heller *et al.* (*34*). Copyright 1982 American Chemical Society.

VIII. Stability of the Solar Conversion Efficiency

High solar-to-chemical conversion efficiencies can be maintained over periods of months only if

1. The catalyst is not poisoned
2. A steady-state product concentration is maintained in both the anode and in the cathode compartments
3. The platinum-group metal catalyst remains hydrogen saturated
4. The interfacial layer that passivates the recombination of photogenerated electrons and holes is maintained at the operating potential of the photocathode

Poisoning of the catalyst can be avoided by using pure electrolytes and by preferring those catalysts that are less susceptible to impurities.

Because aqueous electrolytes dissolve relatively little hydrogen, a steady-state product concentration is always maintained in the photocathode compartment. A steady-state product concentration is also reached in the anode compartment when the product is oxygen. When the product is a halogen, the halogen or its hydrate can either be precipitated or the halogen can be continuously extracted from the anolyte.

Maintenance of hydrogen saturation of the platinum-group metal catalyst requires that the photocathode, when not evolving hydrogen, be stored under a hydrogen blanket and that it not be brought to oxidizing potentials at which hydrogen is depleted. If exposed to air, or if brought to excessively positive potentials, the photocathode can be readily rejuvenated by operation at sufficiently negative potentials for hydrogen evolution to begin and by gradually increasing the potential as the catalyst approaches hydrogen saturation (*34*).

The chemical nature of the interfacial layer that passivates electron–hole recombination depends on the semiconductor and the electrolyte used. The feasibility of maintaining this layer depends on the specific photocathode chemistry. In *p*-InP photocathodes the passivating layer consists of hydrated In_2O_3 (Section IV,B) (*32*). Potentials negative of +0.42 V versus SHE, at which the layer can be reduced to metallic indium, must be avoided. If the photocathode is accidentally brought to an excessively reducing potential, its activity can readily be restored by brief anodization at +0.8 V versus SHE, at which the oxide layer is reformed.

To maintain both hydrogen saturation of the catalyst, yet avoid reduction of the recombination-preventing oxide, the *p*-InP (Rh, H alloy) photocathode in 4 *M* $HClO_4$ under typical solar irradiance must be operated in the potential range +0.42 to +0.58 V versus SHE. Fortunately the solar-to-hydrogen conversion efficiency of the photocathode is near its maximum in this range (*34*). Furthermore, operation of the *p*-InP (Rh, H alloy) photocathode under high levels of irradiance (Section VIII) allows the generation of hydrogen at more positive potentials, at which the indium oxide interfacial layer is very stable. Thus the stability of the solar-to-hydrogen conversion efficiency, already good under AM1 solar irradiance, is further improved as the irradiance is increased.

IX. Photoelectrolysis at High Levels of Irradiance

A. Variation of ΔV_{oc} with Irradiance

The gain in threshold potential for hydrogen evolution ΔV_{oc} follows the diode equation

$$\Delta V_{oc} = (A_o 2.303kT/q)[\log(I/I_o) + 1] \tag{8}$$

up to very high levels of irradiance, where ΔV_{oc} approaches the barrier height ψ_B, the limit to the realizable potential gain. In Eq. (8), A_o is a constant called the ideality or perfection factor, kT and q have meanings identical with those defined in Section VII,B, I is the irradiance, and I_o is a constant. In the absence of radiationless recombination of photogenerated electrons and holes (Section V), $A_o \cong 1$ and the diode is termed "ideal" or "perfect." That such ideality is achieved in the *p*-InP (Rh, H alloy) diode is seen in Fig. 17, where the slope of ΔV_{oc} versus the irradiance is 2.303kT/q; ΔV_{oc} increases at 298 K by 59 mV per 10-fold increase in irradiance (*34*).

B. Variation of the Quantum Yield with Irradiance

The quantum or current yield of a photocathode is potential dependent (Section IX,C). Figure 18 shows the variation of the quantum efficiency of hydrogen evolution with irradiance for the *p*-InP (Rh, H alloy) photo-

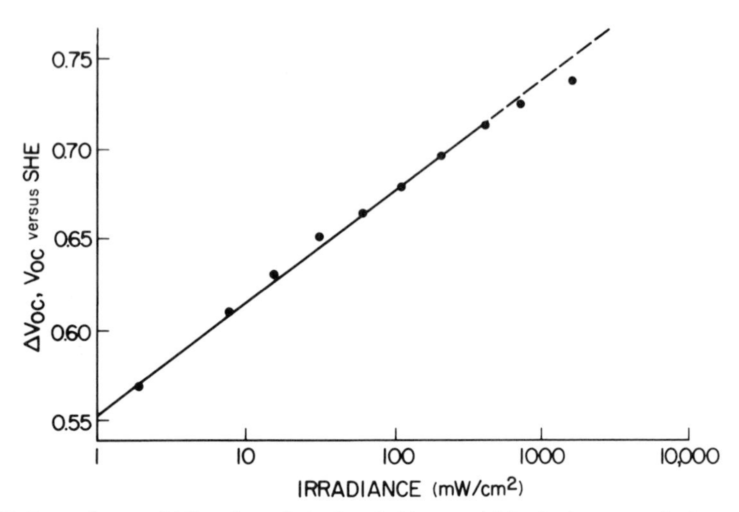

Fig. 17. Dependence of ΔV_{oc}, the gain in threshold potential for hydrogen evolution, on the irradiance [*p*-InP (Rh, H alloy) photocathode; 3 *M* HCl; argon-ion laser source]. From Heller *et al.* (*34*). Copyright 1982 American Chemical Society.

cathode in 3 *M* HCl at +0.24 V versus SHE. The straight line represents the calculated ("theoretical") quantum efficiency. The quantum efficiency remains close to unity up to a current density of ~60 mA/cm^2, corresponding to ~2 AM1 suns. At ~5 AM1 suns the quantum efficiency is 0.65, and at 10 AM1 suns, 0.60 (*34*).

C. Dependence of the Current–Voltage Characteristics on the Irradiance

The current–voltage characteristics at high levels of irradiance are dominated by the transport properties of the electrolyte. The sluggish motion of ions usually causes polarization, which translates to a loss in fill factor [Eq. (4)]. Because of the good mobility of protons, there is less polarization in strong mineral acids, in which photocathodes can be operated at current densities of 100–250 mA/cm^2, similar to those currently employed in industrial, hydrogen-producing electrolytic cells.

Figure 19 shows the current–voltage curves for a *p*-InP (Rh, H alloy) photocathode in 3 *M* HCl (*34*). The curves show that 10-fold concentration of sunlight is warranted, even though there is a loss in quantum yield (Section IX,B). Under 10-fold concentrated sunlight (~1 W/cm^2) the *iV* characteristics approximate those measured with the argon-ion laser at 0.82 W/cm^2. At this point the solar-to-chemical conversion efficiency still exceeds 10% (*49*).

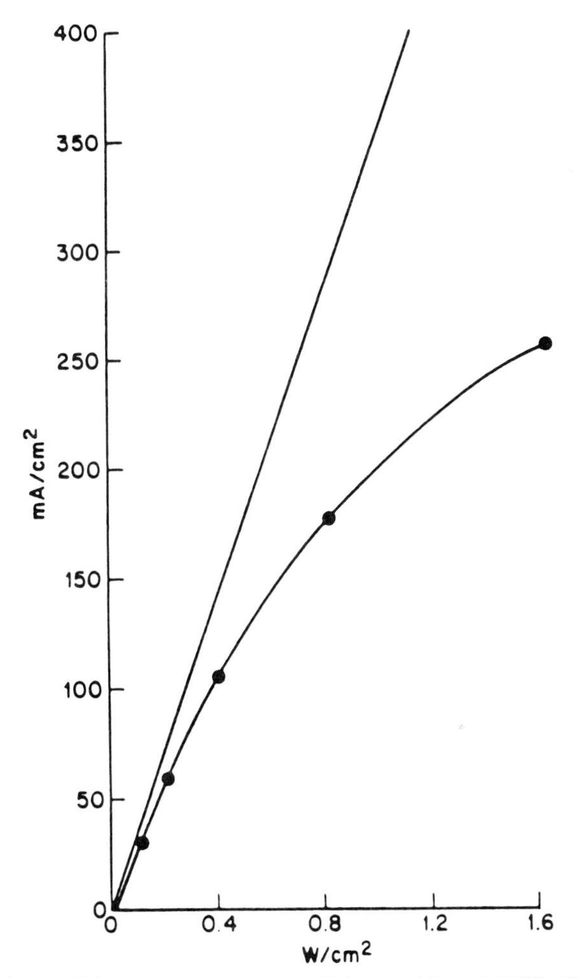

Fig. 18. Dependence of the current or quantum efficiency of the *p*-InP (Rh, H alloy) photocathode (in 3 *M* HCl at +0.24 V versus SHE) on irradiance. The straight line represents the theoretical limit to the photocurrent, unit quantum efficiency. From Heller *et al.* (*34*). Copyright 1982 American Chemical Society.

X. Photoelectrolytic Cells with *p*-InP (Rh, H Alloy) Photocathodes

A. Spontaneous Photoelectrolysis of HI

Hydroiodic acid is spontaneously photoelectrolyzed in the cell *p*-InP (Rh, H alloy)/2 *M* HI, 2 *M* $NaClO_4$/Pt(Rh) (*11*). The cell reaction is

$$2H^+ + 3I^- \longrightarrow H_2 + I_3^-$$

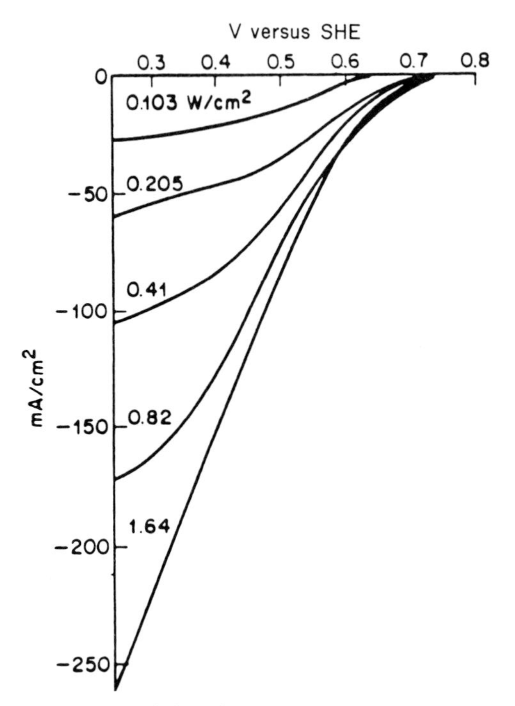

Fig. 19. Current–voltage characteristics of the *p*-InP (Rh, H alloy) photocathode in 3 *M* HCl at varying levels of argon-ion laser irradiance. From Heller *et al.* (*34*). Copyright 1982 American Chemical Society.

As seen in Fig. 20, hydrogen evolution on *p*-InP (Rh, H alloy) starts at potentials 0.25 V positive of the potential required for oxidation of I to I_3^- at moderate ($\sim 10^{-2}$ *M*) I_3^- concentrations. Under tungsten–halogen illumination, producing a light-limited current corresponding to that measured under AM1 sunlight, both the anode and the photocathode reach, at +0.22 V versus SCE (+0.46 V versus SHE), a current density of 25 mA/cm^2, representing a quantum efficiency of 0.8.

B. Photoassisted Electrolysis of HBr

In the cell *p*-InP (Rh, H alloy)/2 *M* HBr/Pt(Rh), under tungsten–halogen illumination corresponding to AM1 sunlight, an external bias of 0.22 V is needed to start electrolysis (Fig. 21). The cell reaction is

$$2H^+ + 3Br^- \longrightarrow H_2 + Br_3^-$$

A quantum efficiency of 0.8 is reached for an external bias of 0.6 V (*11*).

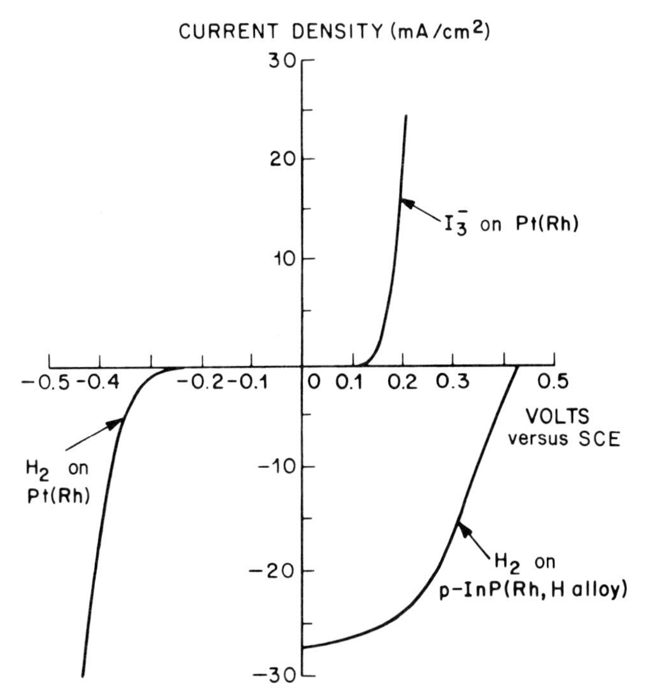

Fig. 20. Current–voltage characteristics of the *p*-InP (Rh, H alloy) photocathode in 2 *M* HI + 2 *M* $NaClO_4$ under tungsten–halogen irradiance corresponding to ~1 AM1 sun. The characteristics of the hydrogen-evolving Pt (Rh) cathode and of the Pt (Rh) anode, where I_3^- is formed, are also shown. From Heller (*11*).

C. *Photoassisted Electrolysis of HCl*

Under tungsten–halogen irradiance equivalent to AM1 sunlight, electrolysis of HCl requires an external bias of 0.6 V and reaches a quantum efficiency of 0.8 at a bias of 0.9 V (Fig. 22) (*11*). The cell reaction is

$$2H^+ + 2Cl^- \longrightarrow H_2 + Cl_2$$

D. *Photoassisted Electrolysis of $HClO_4$*

Unlike the oxidation reactions of the halide ions, which are electrochemically reversible, the oxidation of water to oxygen in acids on Pt is irreversible. Electrolysis at a current density of 25 mA/cm^2 requires, in a cell with a smooth Pt anode and cathode, a 1.9-V bias, and there is little hydrogen or oxygen evolution until the bias exceeds 1.6 V (Fig. 23). With a *p*-InP (Rh, H alloy) photocathode under tungsten–halogen irradiance

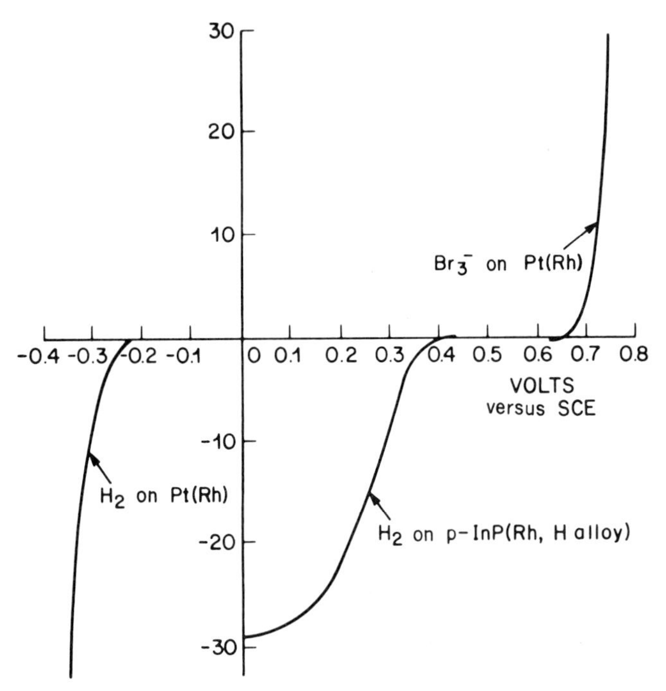

Fig. 21. Current–voltage characteristics of the *p*-InP (Rh, H alloy) photocathode in 2 *M* HBr, under tungsten–halogen irradiance corresponding to ~1 AM1 sun. The characteristics of the hydrogen-evolving Pt (Rh) cathode and of the Pt (Rh) anode, where Br_3^- is formed, are also shown. From Heller (*11*).

equivalent to AM1 sunlight, electrolysis starts at a bias of 0.9 V and reaches a quantum efficiency of 0.8 (current density of 28 mA/cm^2) at a bias of 1.4 V (Fig. 23) (*74*).

E. Solar Conversion Efficiency

Figure 24 shows the *iV* characteristics of a *p*-InP (Rh, H alloy) photocathode illuminated by 81.8 mW/cm^2 sunlight and of a Pt cathode, both in 1 *M* $HClO_4$. At a photocurrent density of 23.5 mA/cm^2 the bias is reduced by 0.565 V. Thus 13.3 mW/cm^2 of electrical power is conserved. The efficiency [Eq. (3)], is 16.2% (*74*), well in excess of the earlier reported 12% for *p*-InP (Ru, H alloy) in 3 *M* HCl (*29*). ΔV_{oc} is 0.65 V, and the quantum efficiency 0.9. The Gibbs free-energy efficiency or the efficiency

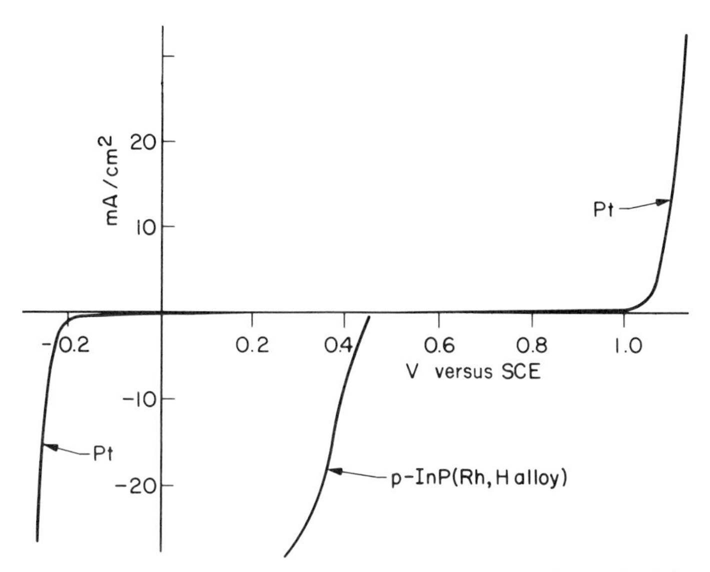

Fig. 22. Current–voltage characteristics of the *p*-InP (Rh, H alloy) photocathode in 2 *M* HCl under tungsten–halogen irradiance corresponding to ~1 AM1 sun. The characteristics of the hydrogen-evolving Pt (Rh) cathode and the Pt (Rh) anode, where Cl_2 and Cl_3^- are formed, are also shown. From Heller (*11*).

based on the electrical power that can be generated, using the photoelectrochemically produced hydrogen in an ideal fuel cell, is 13.3% (*74*).

XI. Spontaneous Two-Photon Photoelectrolysis of HBr

Although prevalent photocorrosion limits the number of photoanodes having appropriate band gap for efficient solar conversion, van der Waals faces of layered compound semiconductors yield stable photoanodes on which halides can be oxidized (*75–79*). These photocathodes are sufficiently stable to be used in the oxidation of chloride ions to chlorine in concentrated lithium chloride solutions (*77*). There is little recombination of photogenerated holes with electrons on the van der Waals planes, though the recombination on planes perpendicular to the van der Waals planes is rapid. Thus microscopic smoothness and solar conversion efficiency are related (*80–82*). Photoanodes producing 10% efficient regenerative solar cells have been made with microscopically smooth faces of *n*-$MoSe_2$ and *n*-WSe_2 (*83, 84*). These photoanodes can be combined with

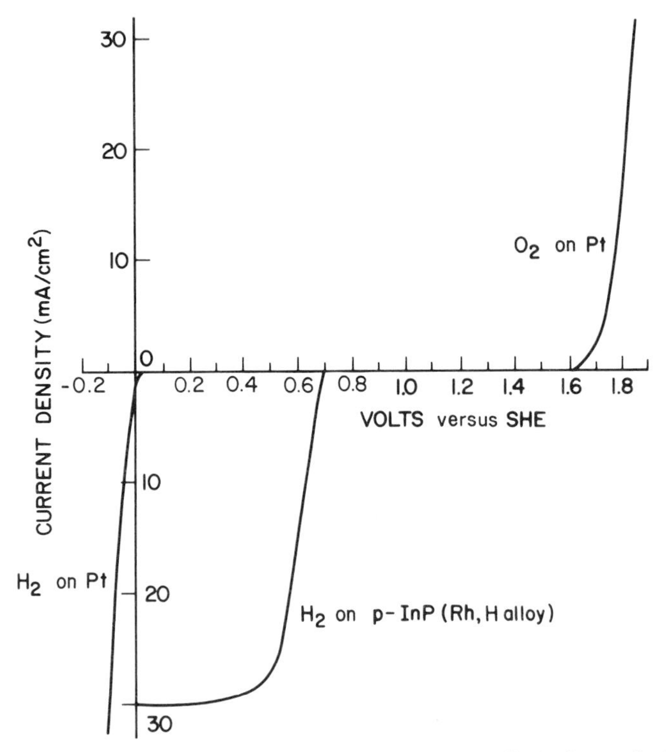

Fig. 23. Current–voltage characteristics of the *p*-InP (Rh, H alloy) photocathode in 2 *M* $HClO_4$ under tungsten–halogen irradiance corresponding to ~1 AM1 sun. The characteristics of the hydrogen-evolving Pt cathode and of the Pt anode, where O_2 is evolved, are also shown. From Heller (*11*).

p-InP (Pt, H alloy) photocathodes (Section III,C) to produce cells with an open-circuit voltage of 1.2 V. Examples of such cells are *p*-InP (Pt, H alloy)/2 *M* HBr/*n*-$MoSe_2$, *p*-InP (Pt, H alloy)/2 *M* HBr/*n*-WSe_2 (*12*). The bias-current behavior of the first cell is shown in Fig. 25. At zero bias, that is, when the two photoelectrodes are shorted, HBr is electrolyzed to hydrogen and tribromide ions at a quantum efficiency of 0.4. At lower quantum efficiencies the bias voltage is negative; the cell produces both chemicals and electrical power.

Like plant photosynthesis, photoelectrolysis with two-photoelectrode cells is a two-photon process. The net solar-to-chemical conversion efficiency is about fivefold higher because a peripheral biological system need not be sustained and because the photoresponse spectrum of the two-electrode photoelectrochemical cell (Fig. 26), unlike the action spec-

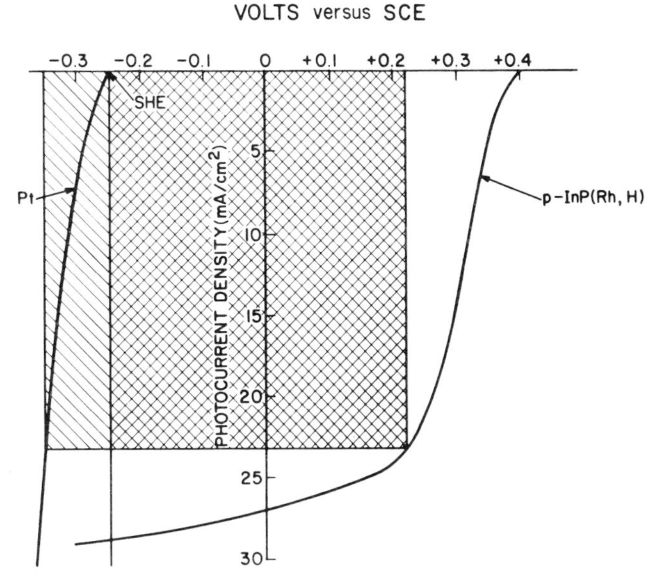

Fig. 24. Current–voltage characteristics of the hydrogen-evolving photocathode *p*-InP (Rh, H alloy) in 1 *M* $HClO_4$ under 78.8 mW/cm^2 sunlight. The characteristics of the hydrogen-evolving Pt cathode are also shown. At the maximum conversion point, $\Delta V_{max} = 0.565$ V and $i_{max} = 23.5$ mA/cm^2, corresponding to 13.3 mW/cm^2 and to a solar conversion efficiency of 16.2%. The doubly shaded area (13.3% efficiency) represents the Gibbs free-energy efficiency or the electrical power that can be generated from the hydrogen produced in an ideal fuel cell. From Aharon-Shalom and Heller (*74*), by permission of the publisher, The Electrochemical Society, Inc.

trum for plant photosynthesis, is flat from 500 to 900 nm, showing that all solar photons in this range are efficiently utilized.

XII. Conclusions

Hydrogen can be generated in semiconductor–liquid junction based solar cells using either photoanodes or photocathodes. The problem of oxidative photocorrosion, prevalent in photoanodes, is avoided in photocathodes.

For efficient solar-to-hydrogen conversion a noble-metal activated photocathode must meet a series of requirements:

1. The semiconductor employed must have an appropriate (1.0–1.7 eV) band gap.

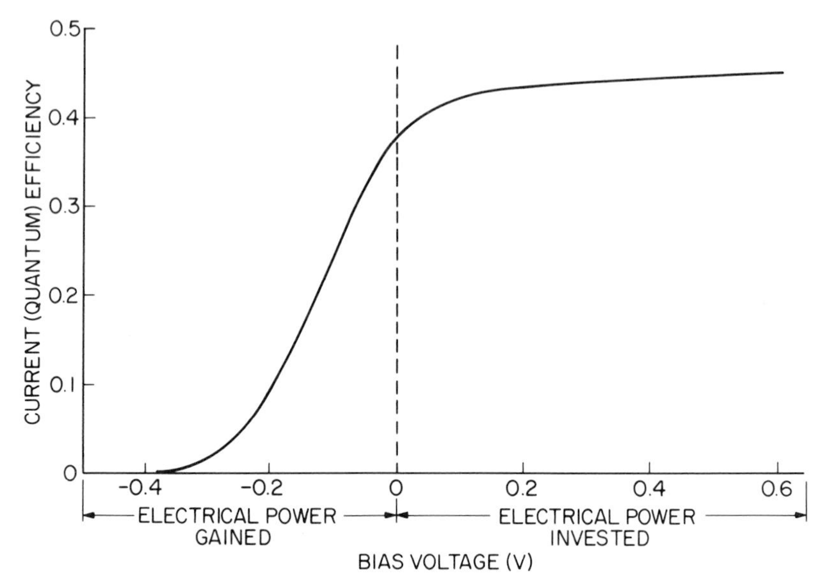

Fig. 25. Dependence of the current (quantum) efficiency of H_2 and Br_3^- generation in the cell *p*-InP (Pt, H alloy)/2 *M* HBr/*n*-$MoSe_2$ on external bias. At zero bias the two photoelectrodes [photoanode *n*-$MoSe_2$; photocathode *p*-Inp (Pt, H alloy)] are shorted. At a negative bias the cell delivers both chemicals and electrical power. From Levy-Clement *et al.* (*12*), by permission of the publisher, The Electrochemical Society, Inc.

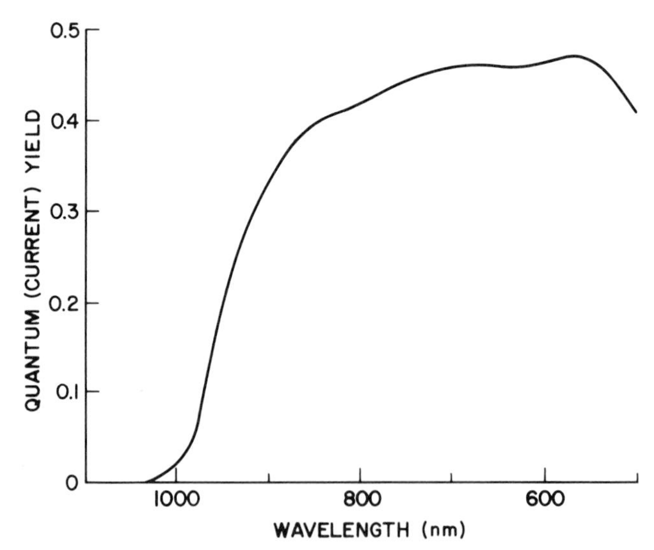

Fig. 26. Photoresponse spectrum of the cell *p*-InP(Pt, H alloy)/2 *M* HBr/*n*-$MoSe_2$ in which HBr is spontaneously photoelectrolyzed. From Levy-Clement *et al.* (*12*), by permission of the publisher, The Electrochemical Society, Inc.

2. The hydrogen-evolving surface must be chemically stable.
3. If insulating, the layer passivating the surface of the semiconductor against corrosion must be sufficiently thin to allow free tunneling of photogenerated electrons from the semiconductor to the solution.
4. The surface of the semiconductor must not contain chemical species causing excessively rapid radiationless recombination of photogenerated electrons with holes. (The electrochemical kinetics of hydrogen evolution must be much faster than the radiationless recombination process.)
5. The overvoltage for hydrogen evolution at the solid–electrolyte interface must be small.
6. The barrier to the transport of holes to the semiconductor–(hydrogen-saturated platinum-group metal catalyst) junction must be high.

The *p*-InP–(hydrogen-saturated platinum-group metal)–aqueous-acid photocathode meets these requirements. InP has a band gap of 1.34 eV. Its interface with dilute aqueous acids consists of a 6- to 10-Å thick layer of hydrated indium oxide, remaining stable at potentials at which hydrogen is rapidly evolved. It allows the free tunneling of photogenerated electrons and passivates their recombination with holes. Incorporation of Pt, Rh, or Ru in the surface lowers the overpotential for hydrogen evolution without lowering the barrier height; the barriers preventing access of holes either to the *p*-InP–H_2/H^+ (solution) interface or to the *p*-InP–(platinum-group metal) interface exceed 0.8 eV, but *only* when the catalyst is hydrogen saturated.

The resultant solar-to-hydrogen conversion efficiency of the *p*-InP (Hydrogen-saturated rhodium) photocathode in 2 *M* $HClO_4$ is 16.2% based on the amount of electrical energy conserved in the electrolytic process and 13.4% based on the amount of electrical power that can be generated in an ideal fuel cell from the hydrogen evolved.

References

1. Harris, L. A., and Wilson, R. H., *Annu. Rev. Mater. Sci.* **8,** 99 (1978).
2. Gerischer, H., *Top. Appl. Physics* **31,** 115 (1979).
3. Nozik, A. J., *Adv. Hydrogen Energy: Hydrogen Energy Systems* **3,** 1217 (1979).
4. Tomkiewicz, M., and Fay, H., *Appl. Phys.* **18,** 1 (1979).
5. Butler, M. A., and Ginley, D. S., *J. Mater. Sci.* **15,** 1 (1980).
6. Pleskov, Yu. V., *Elektrokhimiya* **17,** 3 (1981).
7. Bookbinder, D. C., Bruce, J. A., Dominey, R. N., Lewis, N. S., and Wrighton, M. S., *Proc. Natl. Acad. Sci. USA* **77,** 6280 (1980).

8. Abruna, H. D., and Bard, A. J., *J. Am. Chem. Soc.* **103,** 6898 (1981).
9. Dominey, R. N., Lewis, N. S., Bruce, J. A., Bookbinder, D. C., and Wrighton, M. S., *J. Am. Chem. Soc.* **104,** 467 (1982).
10. Heller, A., *in* "Electrochemical Solar Cells" (K. J. Laidler, ed.), p. 27. Pergamon Press, Oxford, 1982.
11. Heller, A., *Proc. Int. Conf. Photochem. Convers. Storage Sol. Energy, 4th* **A,** 63–75 (1982).
12. Levy-Clement, C., Heller, A., Bonner, W. A., and Parkinson, B. A., *J. Electrochem. Soc.* **129,** 1701 (1982).
13. Fonash, S. J., "Solar Cell Device Physics." Academic Press, New York, 1981.
14. Loferski, J. J., *J. Appl. Phys.* **27,** 777 (1956).
15. Prince, M. B., *J. Appl. Phys.* **26,** 534 (1955).
16. Rittner, E. S., *Phys. Rev.* **96,** 1708 (1954).
17. Rappaport, P., *RCA Rev.* **20,** 373 (1959).
18. Shockley, W., and Queisser, H. J., *J. Appl. Phys.* **32,** 510 (1961).
19. Almgren, M., *Photochem. Photobiol.* **27,** 603 (1978).
20. Bolton, J. R., Haught, A. F., and Ross, R. T., *Proc. Int. Conf. Photochem. Convers. Storage of Solar Energy,* p. 297 (1981).
21. Henry, C. H., *J. Appl. Phys.* **51,** 4494 (1980).
22. Yoneyama, H., Sakamuto, H., and Tamura, H., *Electrochim. Acta* **20,** 341 (1975).
23. Nozik, A. J., *Appl. Phys. Lett.* **28,** 150 (1976).
24. Ohashi, K., McCann, J., and Bockris, J. O'M., *Nature (London)* **266,** 610 (1977).
25. Mattee, H., Otvos, J. W., and Calvin, M., *Sol. Energy Mater.* **4,** 443 (1981).
26. Heller, A., Miller, B., Lewerenz, H. J., and Bachmann, K. J., *J. Am. Chem. Soc.* **102,** 6555 (1980).
27. Heller, A., Lewerenz, H. J., and Miller, B., *J. Am. Chem. Soc.* **103,** 200 (1981).
28. Heller, A., Miller, B., and Thiel, F. A., *Appl. Phys. Lett.* **38,** 282 (1981).
29. Heller, A., and Vadimsky, R. G., *Phys. Rev. Lett.* **46,** 1153 (1981).
30. Heller, A., *Acc. Chem. Res.* **14,** 154 (1981).
31. Miller, B., Menezes, S., and Heller, A., *J. Electrochem. Soc.* **126,** 1483 (1979).
32. Lewerenz, H. J., Aspnes, D. E., Miller, B., Malm, D. L., and Heller, A., *J. Am. Chem. Soc.* **104,** 3325 (1982).
33. Menezes, S., Miller, B., and Bachmann, K. J., *J. Electrochem. Soc.* **130,** (1983).
34. Heller, A., Aharon-Shalom, E., Bonner, W. A., and Miller, B., *J. Am. Chem. Soc.* **104,** 6942 (1982).
35. Menezes, S., Heller, A., and Miller, B., *J. Electrochem. Soc.* **127,** 1263 (1980).
36. Henry, C. H., and Lang, D. V., *Phys. Rev. B: Condens. Matter* **15,** 989 (1977).
37. Heller, A., *ASC Symp. Ser.* No. 146, 57 (1981).
38. Heller, A., *J. Vac. Sci. Technol.* **20,** 2 (1982).
39. Heller, A., Leamy, H. J., Miller, B., and Johnston, W. D., Jr., *J. Phys. Chem.* **20** (1983).
40. Nakato, Y., Tonomura, S., and Tsubomura, H., *Ber. Bunsenges. Phys. Chem.* **80,** 1289 (1976).
41. Bockris, J. O'M., and Uosaki, K., *J. Electrochem. Soc.* **124,** 1348 (1977).
42. Tomkiewicz, M., and Woodall, J. M., *Science (Washington, D.C.)* **196,** 990 (1977).
43. Bourasse, A., and Horowitz, G., *J. Phys. Lett. (Orsay Fr.)* **38,** L-291 (1977).
44. Dare-Edwards, M. P., Hamnett, A., and Goodenough, J. B., *J. Electroanal. Chem.* **119,** 109 (1981).
45. Yoneyama, H., Ohkubo, Y., and Tamura, H., *Bull. Chem. Soc. Jpn.* **54,** 404 (1981).
46. Jarrett, H. S., Sleight, A. W., Kung, H. H., and Gillson, H. H., *J. Met.* **31,** 146 (1979).

47. Jarrett, H. S., Sleight, A. W., Kung, H. H., and Gillson, H. H., *Surf. Sci.* **101,** 205 (1980).
47a. Jarrett, H. S., Sleight, A. W., Kung, H. H., and Gillson, H. A., *J. Appl. Phys.* **51,** 3916 (1980).
48. Kautek, W., Gobrecht, J., and Gerischer, H., *Ber. Bunsenges. Phys. Chem.* **84,** 1034 (1980).
49. Heller, A., Vadimsky, R. G., Johnston, W. D., Jr., Strege, K. E., Leamy, H. J., and Miller, B., *IEEE Photovoltaic Specialists Conf. 15th,* p. 1722 (1981).
50. Koffyberg, F. P., and Benko, F. A., *J. Electrochem. Soc.* **128,** 2476 (1981).
51. Coulter, J. K., Haas, G., and Ramsey, J. B., *J. Opt. Soc. Am.* **63,** 1149 (1973).
52. Kirilova, M. M., Nomerovannaya, L. V., Noskov, M. M., Gorina, N. B., Polyakova, V. P., and Savitskii, E. M., *Sov. Phys. Solid State (Engl. Transl.)* **20,** 994 (1978).
53. Haas, G., and Hadley, L., *in* "Optical Properties of Metals" (D. E. Gray, ed.), p. 6-144-146. McGraw-Hill, New York, 1972.
54. Butler, M. A., and Ginley, D. M., *J. Electrochem. Soc.* **125,** 228 (1978).
55. Gomer, R., and Tryson, G., *J. Chem. Phys.* **66,** 4413 (1977).
56. Gerischer, H., *Top. Appl. Phys.* **31,** 115 (1979).
57. Lundstrom, I., Shivaraman, M. S., and Svensson, C. M., *J. Appl. Phys.* **46,** 3876 (1975).
58. Lundstrom, I., Shivaraman, M. S., Svensson, C. M., and Lundquist, L., *Appl. Phys. Lett.* **26,** 55 (1975).
59. Lundstrom, I., Shivaraman, M. S., Stilbert, L., and Svensson, C., *Rev. Sci. Instrum.* **47,** 738 (1976).
60. Shivaraman, M. S., Lundstrom, I., Svensson, C., and Hammarschen, H., *Electron. Lett.* **12,** 483 (1976).
61. Lundstom, I., Shivaraman, M. S., and Svensson, C., *Vacuum* **27,** 245 (1977).
62. Lundstrom, I., Shivaraman, M. S., and Svensson, C., *Surf. Sci.* **64,** 497 (1977).
63. Lundstrom, I., *Sensors Actuators* **1,** 403 (1981).
64. Armgarth, M., and Nylander, C., *Appl. Phys. Lett.* **39,** 91 (1981).
65. Fonash, S. J., *J. Appl. Phys.* **47,** 3597 (1976).
66. Ruths, P. F., Ashok, S., Fonash, S. J., and Ruths, J. M., *IEEE Trans. Electron. Devices* **ED-28,** 1003 (1981).
67. Steele, M. C., and MacIver, B. A., *Appl. Phys. Lett.* **28,** 678 (1976).
68. Chauvet, F., and Caratge, P., *C. R. Hebd. Seances Acad. Sci. Ser. B* **285,** 153 (1977).
69. Poteat, T. L., and Lalevic, B., *IEEE Trans. Electron. Devices* **ED-29,** 123 (1982).
70. Ito, K., *Surf. Sci.* **86,** 345 (1979).
71. Kawai, T., and Sakata, T., *Chem. Phys. Lett.* **72,** 87 (1980).
72. Yamamoto, N., Tonomura, S., Matsuoka, T., and Tsubomura, H., *Surf. Sci.* **92,** 400 (1980).
73. Yamamoto, N., Tonomura, S., and Tsubomura, H., *J. Appl. Phys.* **52,** 5705 (1981).
74. Aharon-Shalom, E., and Heller, A., *J. Electrochem. Soc.* **129,** 2865 (1982).
75. Tributsch, H., *Z. Naturforsch. A* **322,** 972 (1977).
76. Tributsch, H., *Sol. Energy Mater.* **1,** 257 (1979).
77. Kubiak, C. P., Schneemeyer, L. F., and Wrighton, M. S., *J. Am. Chem. Soc.* **102,** 6898 (1980).
78. Kautek, W., and Gerischer, H., *Ber. Bunsenges. Phys. Chem.* **84,** 645 (1980).
79. Kautek, W., and Gerischer, H., *Electrochim. Acta* **26,** 1771 (1981).
80. Kautek, W., Gerischer, H., and Tributsch, H., *Ber. Bunsenges. Phys. Chem.* **83,** 1000 (1979).
81. Lewerenz, H. J., Heller, A., and DiSalvo, F. J., *J. Am. Chem. Soc.* **102,** 1877 (1980).

82. Lewerenz, H. J., Heller, A., Leamy, H. J., and Ferris, S. D., *ACS Symp. Ser.* No. 146, 17 (1981).
83. Parkinson, B. A., Furtak, T. R., Canfield, D., Kam, K. K., and Kline, G., *Discuss. Faraday Soc.* **70,** 233 (1980).
84. Kline, G., Kam, K. K., Canfield, D., and Parkinson, B. A., *Sol. Energy Mater.* **4,** 301 (1981).

13 Photoelectrochemistry of Cadmium and Other Metal Chalcogenides in Polysulfide Electrolytes

Gary Hodes

Department of Plastics Research
The Weizmann Institute of Science
Rehovot, Israel

and

Ormat Turbines Ltd.
Yavneh, Israel

I. Introduction

The stabilizing effect of a sulfide electrolyte on CdS or CdSe photoanodes was reported independently, and more or less simultaneously, by three different groups working at the Massachusetts Institute of Technology (*16*), Bell Laboratories (*62*), and our group at the Weizmann Institute (*36*, *37*). Prior to this discovery, the emphasis of the research in the field of

ENERGY RESOURCES THROUGH
PHOTOCHEMISTRY AND CATALYSIS

ISBN 0-12-295720-2

photoelectrochemistry was firmly directed toward water splitting into H_2 and O_2, a result of the obvious benefit of producing H_2 by sunlight, as well as the nature of the work by Fujishima and Honda (*22*), which stimulated the emergence of photoelectrochemistry in its present form and which dealt with water splitting using TiO_2. Until 1975, generation of electricity by photoelectrochemical cells (PECs) does not appear to have been given much consideration by most of the few groups working in the field. This was primarily a result of the instability of most semiconductors against photocorrosion when operated as photoanodes in electrolytes used to that time. The fundamental principles of regenerative PECs, as well as some experimental data, have been described by Gerischer (*23*). The cadmium chalcogenide–sulfide system was the first major success in this type of PEC, where the generation of electricity, rather than of chemical products, is the desired goal. Although other potentially practical regenerative PECs have been subsequently described in the literature, it is fair to say that the greatest amount of effort in this branch of photoelectrochemistry has been, and is being, expended toward an understanding of the science involved in the cadmium chalcogenide–sulfide system, as well as toward its development as a practical solar cell.

The purpose of this chapter is to review in detail certain aspects of the work which has been carried out on chalcogenide electrolyte-based photoelectrochemical systems using metal chalcogenide semiconductors. The emphasis is on the general properties of these systems and, in particular, on their stability. This review is limited to a discussion of the photoelectrode–electrolyte subsystem and does not cover other aspects such as counterelectrode, *in situ* storage, and complete cells. In particular, methods of preparation of the photoelectrodes and their efficiencies as solar cells are not considered here, other than where it may be relevant to the subject matter under discussion; to do this in depth would entail a lengthy review in itself.

First, the interaction between the sulfide electrolyte and the semiconductor electrode will be considered. Next, the fundamental principles governing the stability of the CdSe–polysulfide system, which up to now has been the best investigated, serving as a model system for many other chalcogenide semiconductors, will be discussed in detail; CdS will also be included where it assists in understanding the various aspects of stability. Experimental results on stability of the CdSe–S_x^{2-} system will then be described and explained on the basis of the preceding discussion. Other cadmium chalcogenides will be discussed in terms of any differences in their behavior compared with CdSe, with emphasis on the Cd(Se, Te) system. This will be followed by other metal chalcogenide photoelectrodes, covering Zn (and Zn, Cd) chalcogenides, Bi_2S_3, $CdIn_2Se_4$, $CuInS_2$, and $CuInSe_2$. A brief description of work carried out in polyselenide and polytelluride electrolyte will be given. Finally, the effect of a number of surface treatments of the photoelectrodes (etching and ion treatments) will be described.

II. Interaction between CdS and Sulfide Ions in Solution

Although it was not until 1976 that the stabilizing effect of a sulfide electrolyte on CdS and CdSe photoanodes was first reported, hints to this effect had existed for 10 yr. Tygai (*96*) in an investigation of CdS in $Fe(CN)_6^{3-/4-}$ and S_x^{2-}, used capacitance measurements to show that sulfide ions underwent strong specific adsorption on the CdS surface, whereas ferrocyanide did not. In essence, Tygai's results show a cathodic shift of the flat-band potential of CdS when immersed in a polysulfide-containing solution. In fact, it was this mention of specific adsorption of sulfide on CdS that initiated our own investigation into the CdS–S_x^{2-} system (*36, 37*). This was based on the feeling that specific adsorption of the electroactive species may lead to better kinetics for the charge transfer of holes to the electrolyte and, by implication, reduced photocorrosion. Baker (*3*), in a discussion of a paper by Gerischer, asked the question whether photocorrosion of CdS still occurred in a solution containing a common or related ion, such as sulfide. It seems that this suggestion remained no more than that for almost a decade.

The cathodic shift of the flat-band potential of CdS in a sulfide electrolyte, first reported by Tygai, was subsequently examined in more detail, once the CdS–S_x^{2-} system emerged as a forerunner in photoelectrochemical systems at that time [it should be remembered that it was then compared with TiO_2 PECs, so the band gap of CdS (2.4 eV) was considered much superior to that of TiO_2 ($\sim$3.0 eV) for the purpose of utilizing solar radiation, although its band gap is still considerably greater than the broad optimum of $\sim$1.4 eV]. The flat-band potential (fbp) of the (0001) face of CdS had previously been found to be constant at about -0.86 V versus SCE and independent of pH between pH 2 and 13 in an indifferent electrolyte of Na_2SO_4 and either H_2SO_4 or NaOH (*98*). Minoura *et al.* (*65*) reported a shift in the fbp of pressed polycrystalline disks of CdS in the cathodic direction on addition of sulfide to the electrolyte, the degree of the shift being dependent on the sulfide concentration. Minoura *et al.* (*66*) subsequently studied the effect of sulfide and Cd^{2+} ions on the fbp of the (0001) (Cd) and $(000\bar{1})$ (S) faces of single-crystal CdS, using Mott–Schottky plots to determine the fbp in the different electrolytes. They found a fundamental difference in the behavior of the two faces in that for the S face, the fbp was relatively constant at about -1.1 V versus SCE with change in sulfide concentration ($>10^{-5}$ M), whereas for the Cd face, it varied from -1.1 V versus SCE at 10^{-6} M $[S^{2-}]$ to -1.4 V at 10^{-1} M $[S^{2-}]$. It should be noted that the difference between the two faces appears at $[S^{2-}] > 10^{-6}$ M. At 10^{-6} M $[S^{2-}]$, both faces have relatively similar values of fbp (about -1.1 V), which is already shifted considerably negative from the value of about -0.86 V found in an indifferent electrolyte (*98*). For electrolytes containing Cd^{2+} ions, the opposite dependence was noted; for the Cd face, fbp was constant at about -0.85 V and varied with $[Cd^{2+}]$ for the S face, although to a lesser extent than found for S^{2-} on the Cd face. Also, the effect of the Cd^{2+} was to shift the fbp in a positive direction, from about -0.85 V (no Cd^{2+}) to about -0.70 V (10^{-1} M Cd^{2+}), as would be expected from adsorption of a positive ion at the CdS surface. From the lesser shift of fbp by Cd^{2+} compared with sulfide, Minoura *et al.* suggested that Cd^{2+} is less specifically (or less strongly) adsorbed on the S face of CdS than sulfide is on the Cd face. Although not explained by Minoura *et al.*, the different behavior of the two faces in sulfide electrolyte is interesting. For both faces, specific adsorption of sulfide occurs (shift of fbp by about -0.25 V on addition of 10^{-6} M $[S^{2-}]$), but the adsorption varies for $[S^{2-}] > 10^{-6}$ M only for the Cd face. The existence of a cathodic photoeffect at the $(000\bar{1})$ face [but not the (0001) face] of a CdS crystal in a sulfide electrolyte has been attributed to the existence of surface states induced by the interaction of the dangling bond at the S face with sulfide from the electrolyte. The Cd face has no dangling bonds, hence no equivalent surface states and no cathodic photoeffect (*68*). Applying this information to the adsorption of sulfide at the Cd and S faces could lead to the assumption that sulfide chemisorbs at the dangling bonds of the S face, but after saturating these bonds, there is no further specific adsorption, whereas for the Cd face the specific adsorption is dependent on the equilibrium set up between the CdS surface and sulfide in the electrolyte, and the fbp varies with sulfide concentration of pS in the same way that the fbp of oxide semiconductors varies with pH. Of relevance to such an argument is the measurement of the point of zero zeta potential (PZZP) of CdS powder in sulfide solutions (*27*). The PZZP occurs at a unique concentration of sulfide in solution when the potential drop across the Helmholtz layer is zero. Ginley and Butler measured this concentration to be 4×10^{-9} M sulfide. Although this value was found for a powder with many different exposed faces, it indicates that very little sulfide

[$\ll 10^{-6}$ *M*, which was the lowest concentration used by Minoura *et al.* (*66*)] is necessary to cause a major change in the potential drop across the Helmholtz layer, and hence a shift in the fbp. The dependence of the shift of the fbp of CdS with sulfide concentration has been measured also by a number of other groups, and the results are summarized in Fig. 1. Note, in particular, the difference in fbp found by Ellis *et al.* (*19*) when measured by extrapolation of a Mott–Schottky plot or by limiting photopotential under strong illumination; the latter probably gives a more realistic value. Apart from variations in CdS samples (crystal face, polycrystallinity, doping density, etc.), variations may be a result of the potential of a solution containing only sulfide (and no sulfur) being very ill defined because only the reduced form of the redox couple is present. Very small concentrations of S, either as impurity or formed photoanodically, can cause relatively large variations in the equilibrium potential of the solution and also in the fbp.

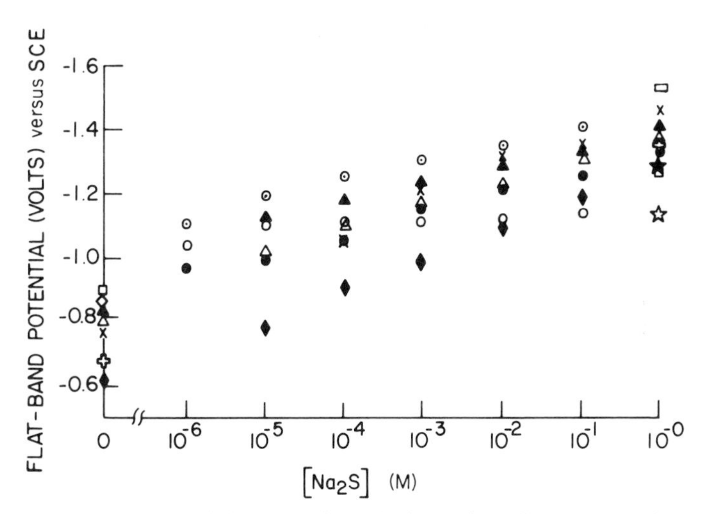

Fig. 1. Flat-band potential of CdS in sulfide solutions of varying concentrations. On the concentration axis, the origin represents an indifferent electrolyte (no sulfide). Unless otherwise stated, the results are for single-crystal CdS, (0001) face (not distinguished between the Cd and S faces) in a sulfur-free solution (i.e., not *poly*sulfide), and flat-band potentials were found from Mott–Schottky (capacitance–potential) plots. □, Polysulfide solution 1 *M* each OH, S^{2-}. S. Ellis *et al.* (*19*). □, As before, but measured by photopotential under intense illumination. ●, Measured by plotting the square of the photocurrent versus potential; the fbp is given by the intercept of the extrapolated $(\text{photocurrent})^2$ on the potential axis. Ginley and Butler (*27*). ✙, (0001) Cd face. Minoura and Tsuiki (*69*). △, Inoue *et al.* (*48*). ◇, Watanabe *et al.* (*98*). ⊙, (0001) Cd face. Minoura *et al.* (*66*). ○, $(000\bar{1})$ S face. Minoura *et al.* (*66*). ◆, Sintered CdS pellet. Minoura *et al.* (*66*). ×, Sintered CdS pellet. Minoura *et al.* (*67*). ▲, Inoue *et al.* (*50*). ☆, S face 1 *M* S^{2-}–S_2^{2-}. Gerischer and Gobrecht (*25*). ★, Cd face 1 *M* S^{2-}–S_2^{2-}. Gerischer and Gobrecht (*25*).

For CdSe in polysulfide solution, Ellis *et al.* (*19*) and Gerischer and Gobrecht (*25*) found values of fbp −1.5 V versus SCE. Values of photocurrent onset for CdSe crystals [(0001) face] in 1 *M* S_2^{2-} solutions as negative as −1.6 V versus SCE have been observed for about AM1 illumination (*87*), the fbp may be somewhat negative of this value. Which faces of the CdSe crystals were used was not always reported in these other investigations, which makes a direct comparison difficult. The work of Ginley and Butler (*27*) did, however, allow them to deduce a great deal concerning the interaction of sulfide species from the electrolyte with the CdS surface. They predicted the fbp of CdS from the atomic electronegativities of Cd and S. They then measured the PZZP for CdS in sulfide solution and also the fbp of CdS as a function of sulfide concentration. The measured fbp, corrected to the PZZP, agreed with the predicted value. They were thus able to separate the specific interaction between sulfide and the CdS surface (or, in other terms, the potential drop across the Helmholtz layer) from bulk properties of CdS. Their measurement of fbp versus sulfide concentration led to a straight-line plot of slope 0.06 V/pS from which they inferred that one of the potential-determining ions must be monovalent (a divalent ion, such as S^{2-}, would give a slope of 0.03 V/pS), and thus must be HS^-. They found a similar slope for fbp versus pH, with a constant $[HS^-]$ in solution, and deduced from this that H^+ must be the other potential-determining ion. [The pH independence reported by Watanabe *et al.* (*98*) was in a sulfide-free solution.] These results are logical in analogy with oxide semiconductors in water, where OH^- and H^+ are the potential-determining ions.

Wilson (*100*) carried out an analysis of the charge transfer in the CdS–sulfide system, based on results reported by Inoue *et al.* (*49*). Wilson replotted their current–voltage curves relative to the fbp and noted that the photocurrent rose at lower voltages on addition of sulfide. Because the voltage at which photocurrent starts is determined by competition between surface recombination of carriers (this includes electrochemical recombination where sulfide species are both oxidized and reduced, i.e., an exchange current) and by charge transfer of holes to the sulfide species, the rise of photocurrent at lower voltages on addition of sulfide could be explained either by an increase in the rate of hole transfer to the electrolyte or by a decrease in the rate of surface recombination. The latter could be owing to adsorption of sulfide at those sites responsible for recombination, which might reduce their efficacy as recombination sites. Wilson also noted that this argument might apply to corrosion sites, which would explain the suppression of corrosion by sulfide and also that the same sites might be responsible for both recombination and corrosion. In addition, Wilson suggested a positive shift of the fbp during photooxidation caused by depletion of sulfide at the surface because of diffusion limitations. This shift, which would lead to a reduced quantum efficiency for the photocurrent owing to reduced band bending and thus less efficient charge separation, could be particularly serious for the polysulfide systems, for the following reason: the composition of the sulfide species at the CdS surface is expected to be sulfur rich, because sulfur is the product of the oxidation. The redox potential of polysulfide solutions becomes rapidly more positive in a non-Nernstian fashion when the [S]:[S^{2-}] ratio exceeds unity (*1, 39*). Although the fbp of CdS (or other cadmium chalcogenides) has not been measured for such sulfur-rich polysulfide electrolytes, the implication here is that a relatively small increase in the sulfur content of the adsorbed species could lead to a considerable positive shift in the fbp and a resulting loss in efficiency. This effect would, of course, be more serious at increased intensities of illumination, where depletion of electroactive species would be greater. A similar effect has been suggested by Minoura and Tsuiki (*69*) for a CdS photoanode in the sulfite (SO_3^{2-}) electrolyte, where photogenerated holes oxidize the electroactive SO_3^{2-} (SO_3^{2-} causes a negative shift in the fbp of CdS), causing the band edges to move in a positive direction toward their position when no SO_3^{2-} is present.

Minoura and Tsuiki (*69*) also described the effect of dipping CdS in either S^{2-}, SO_3^{2-}, or $S_2O_3^{2-}$, followed by rinsing, and then measuring the fbp in a SO_4^{2-} solution. Only S^{2-} was found to have an influence, which points to a much stronger adsorption compared with SO_3^{2-} or $S_2O_3^{2-}$. Also, it was noted that the length of time the CdS was dipped in S^{2-} was important; it took $>$10 min for equilibrium to be set up, indicating that equilibrium adsorption of sulfide on CdS is a slow process.

Gerischer and Gobrecht (*25*) reported that the fbp of CdSe did not shift initially on immersion in a sulfide solution, but required some seconds to stabilize. They suggested that this delay was a result of a surface reaction in the first few seconds that converted the surface to CdS (see Section II,B), and finally established a CdS–S_x^{2-} double layer. Although such a reaction will occur, chemical intuition leads us to expect a shift in fbp of CdSe in sulfide solution, even in the absence of surface exchange. It is probable that this delay is connected, at least in part, with a slow rate of adsorption of sulfide on CdSe, as described previously.

Experiments on the Cd chalcogenide–S_x^{2-} system point to some interesting properties of the interface in this system, although the mechanisms leading to most of these properties are as yet not understood. The technique of relaxation spectrum analysis was used to identify different capacitive elements in this system. Relatively slow-relaxing capacitive elements have been tentatively associated with the CdSe–S_x^{2-} interface because they were absent at Au–CdSe Schottky diodes (*93*). Some interesting variations between single-crystal and polycrystalline (electroplated) CdSe electrodes have been noted. Electrolyte electroreflectance (EER) was used by Silberstein *et al.* (*84*) to detect partial Fermi-level pinning at electroplated CdSe in S_x^{2-}, in contrast to single-crystal material where such pinning was not found to occur. This pinning was found to be a dynamic process that is strongly dependent on the CdSe potential and may be capable of giving information on the nature of surface states present at the polycrystalline CdSe–S_x^{2-} interface. Also found for electroplated CdSe–electrolyte (not only S_x^{2-}) systems, but not for the corresponding single-crystal CdSe systems, was a specific frequency dispersion of electric impedance spectra (*57*). Because this dispersion was not found for Au–CdSe (electroplated) diodes, it was associated with a percolation conduction mechanism at the interface between the very irregular (cauliflower-like) CdSe and electrolyte.

Vainas *et al.* (*97*) studied the CdS–S_x^{2-} system in terms of thermionic emission theory for the transfer of electrons in the dark from the CdS to the electrolyte and found a value for the Richardson's constant A^* about two orders of magnitude lower than that obtained for a Au–CdS solid-state Schottky barrier. Because A^* is a measure of the ease of transfer of an electron from the conduction band edge away from the semiconductor, this implies a blocking step for the transfer of electrons from the CdS into the S_x^{2-}, such as is also found for MIS (metal insulator semiconductor) diodes.

III. Stability of the CdSe–Polysulfide System

A. Fundamental Aspects

The hole reactions occurring at the interface between CdSe (or CdS) and a polysulfide electrolyte can be separated into oxidation of the CdSe or CdS [photocorrosion, Eq. (1) shown for CdSe] or sulfide oxidation [Eq. (2)]. [For simplicity, the electroactive sulfide species is denoted as S_x^{2-},

$$\mathrm{CdSe} + 2\mathrm{h}^+ \longrightarrow \mathrm{Cd}^{2+} + \mathrm{Se} \tag{1}$$

$$\mathrm{S}_x^{2-} + 2\mathrm{h}^+ \longrightarrow \mathrm{S}_x \tag{2}$$

although we do not know its exact nature. Also, follow-up reactions between the products of Eq. (1) and the sulfide electrolyte will be considered later; initially we shall ignore them.] The stability of the system against photocorrosion will be dependent on the competition between reactions (1) and (2). In effect, we can consider the stabilizing effect of S_x^{2-} on CdSe either by a thermodynamic approach or by a kinetic one, as was pointed out by Manassen *et al.* (*60*). The former can be considered as a form of cathodic protection in which the relatively negative redox potential of the polysulfide system (about -0.75 V versus SCE) prevents the corrosion reaction, whereas the latter implies faster kinetics of the sulfide oxidation (2) compared with photocorrosion (1). The possible role of the negative redox potential in protecting CdS and CdSe against photocorrosion had been suggested at the beginning of these studies (*37, 62*). Of interest is the deduction by Ellis *et al.* (*17*) that the apparent fast rate of electron transfer between the electrode and electrolyte suggests a special interaction of sulfide with the CdS surface. This was exactly our reasoning in reverse, as mentioned previously, where the strong interaction between sulfide and CdS reported by Tygai (*96*) might lead to rapid charge transfer between electrode and electrolyte. It should be noted at the outset that the relative importance of the thermodynamic and kinetic factors is still not clear; it does seem that both are important, as will be discussed subsequently.

If we first consider a purely thermodynamic approach to stability, we can use as a starting point the standard potential for anodic dissolution E_D° as described by Bard and Wrighton (*4*) and Gerischer (*24*). According to this description, to be thermodynamically stable against anodic dissolution, E_D° must be more positive than the valence band edge, that is, the photogenerated holes will not have enough energy to oxidize the semiconductor. This is not the case for CdS, nor for any other photoanode material known at present. The next best situation is where E_D° is more positive than the redox potential E_{redox}. In this case, the semiconductor is thermodynamically stable against oxidation by the product of the redox reaction and may be stable if hole transfer to the redox system is fast enough.

To obtain the values of E_D for the Cd chalcogenides in (poly)sulfide solutions, the overall reaction for the corrosion that would occur must be considered, and not just the simple reaction (1). This overall reaction including intermediate steps can be given, for the example of CdSe, as

$$\begin{array}{c} CdSe + 2h^+ \longrightarrow Cd^{2+} + Se \xrightarrow[\text{selenium dissolution}]{S_x^{2-}} S_xSe^{2-} \\ \text{CdS formation} \Big\downarrow S_x^{2-} \\ CdS + S_{x-1} \end{array} \quad (3)$$

The Cd^{2+} ion will be neutralized by a sulfide ion to give CdS. The implications of this reaction are considered in Section II,B. The overall reaction is given by

$$CdSe + 2S_x^{2-} \xrightarrow{2h^+} CdS + S_{(2x-1)}Se^{2-} \quad (4)$$

The standard free energy of this reaction ΔG° can be calculated from the values of ΔG° (in kcal/mol) for the relevant species given in Table I; for $x = 1$, $\Delta G = -29.3$. Other possible reactions can be written. Thus HS^- can be substituted for S^{2-}. This will not change the thermodynamics, because the equilibrium reaction

$$S^{2-} + H_2O \rightleftharpoons HS^- + OH^- \quad (5)$$

has a $\Delta G^\circ \cong 0$, and Eq. (4) can be written as

$$CdSe + 2HS^- + 2OH^- \xrightarrow{2h^+} CdS + SSe^{2-} + 2H_2O \quad (6)$$

where $\Delta G^\circ = -29.3$ as before. For polysulfide solutions, with $x > 1$, the ΔG° of reaction (4) will vary somewhat depending on the value of x. Because the differences will generally be small, we shall ignore them here. In any case, S^{2-} (and HS^-) are normally present even in polysulfide solutions, whereas S_2^{2-} does not exist to any appreciable extent in polysulfide solutions commonly encountered in PEC work; S_3^{2-} and S_4^{2-}, as well as S^{2-} and HS^-, are the most common species in these solutions (*26*).

TABLE I

Standard Free Energies of Formation (at 298 K) ΔG°_{298} of Some Relevant Species.

Species	ΔG°_{298} (kcal/mol)	Species	ΔG°_{298} (kcal/mol)
ZnS (cubic)	−47.5[a,b]	Se^{2-}	+37[c]
ZnS (hexagonal)	−44.2[a]	Se_2^{2-}	+31[c]
ZnSe	−39.0[b]	Se_3^{2-}	+25[c]
ZnTe	−26.4[b]	Se_4^{2-}	+19.7[c]
CdS	−37.0[b]	HSe^-	+23.5[a]
CdSe	−32.4[b]	Te^{2-}	+52.7[a]
CdTe	−23.8[b]	SSe^{2-}	+19.5[e]
S^{2-}	+22.1[a]	SSe_3^{2-}	+17[f]
S_2^{2-}	+21.8[a]	TeS_3^{2-}	+15.1[d]
S_3^{2-}	+21.1[a]	OH^-	−37.6[a]
S_4^{2-}	+19.4[a]	H_2O	−56.7[a]
HS^-	+3.0[a]		

[a] Latimer (1952).
[b] Calculated from ΔH°_{298} values given by Bachmann *et al.* (*2*) and S° values from Latimer (*55*) for the elements and Bachmann *et al.* for the compounds.
[c] Zaitseva and Greiver (*99*).
[d] Greiver and Zaitseva (*28*).
[e] Greiver *et al.* (*29*).
[f] Approximated from other values in this table.

[It should be noted that large variations exist in the reported values of the second dissociation constant of H_2S. If the lowest values ($\sim 10^{-17}$) are used, this would mean that the S^{2-} species is virtually nonexistent in pure sulfide (i.e., no polysulfide) solution, even at pH = 14, and that all of the sulfur is in the form HS^-.]

The value of E°_D (the standard decomposition potential) for CdSe in a sulfide solution can be calculated from the value of $\Delta G^\circ = -29.3$, where $\Delta G^\circ = -nFE^\circ_D$, giving $E^\circ_D = -0.635$ V (all potentials relative to NHE). The values of ΔG° and E°_D (as well as ΔG and E_D under the more realistic condition of low concentration of chalcogen originating from the semiconductor; see Section II,B) are given in Table II for reactions of interest, many of which will be discussed subsequently. It should be noted that where the semiconductor surface is converted to CdS, the thermodynamics of the surface will be determined by that of CdS rather than by the underlying semiconductor.

In all the examples given E°_D (and E_D) are as negative, or more negative, than E_{redox}, which implies instability based on the previous reasoning. The actual situation may be even worse if we consider the argument of Frese *et al.* (*21*). This invokes the concept of microscopic decomposition potential E'_D at a defect where, using a thermodynamic argument, they showed that E'_D is more negative than E_D at a nondefected site. The defects are associated with weakening or breaking of bonds, and it is reasonable to expect such sites to be more prone to decomposition.

On the basis of the values of E_D given in Table II and the thermodynamic instability of these systems that can be inferred from them, we must conclude that other factors play a determining role. The most obvious is that the overall kinetics of the (poly)sulfide oxidation are faster than those of the photocorrosion. This includes not only the inherent electrochemical kinetics of the S_x^{2-} oxidation, but also the possibility that surface states may facilitate hole transfer to the S_x^{2-} in preference to the photocorrosion reaction. This is not so coincidental as it may seem at first sight, because chemisorption of sulfide species on the photoelectrode, insofar as it alters the chemical bonding at the surface, may be looked at as generating surface states at the sulfide potential. The strong interaction between chemisorbed sulfide and the semiconductor suggests that such surface states might be in equilibrium with the semiconductor bulk. Thus such chemisorption-induced surface states could facilitate hole transfer to the sulfide species. We should also consider the possibility that the photocorrosion reaction actually might occur in preference to sulfide oxidation. Such a reaction will convert the surface of the CdSe to CdS. As discussed subsequently, if the exchange of CdSe to CdS does not extend too far into the bulk, degradation of the photoresponse does not occur. If the ex-

TABLE II

Thermodynamic Calculations for Various Possible Photocorrosion Reactions of II–VI Semiconductors in Polychalcogenide Electrolytes[a]

Reaction	ΔG° (kcal/mol)	E°_D (V)	ΔG	E_D
(a) $CdS + 2S^{2-} \underset{-2h^+}{\overset{+2h^+}{\rightleftharpoons}} CdS + S_2^{2-}$	−22.4	−0.486	−29.2	−0.633
(b) $CdSe + 2S^{2-} \underset{-2h^+}{\overset{+2h^+}{\rightleftharpoons}} CdS + SSe^{2-}$	−29.3	−0.635	−36.1	−0.783
(c) $CdTe + S^{2-} + S_3^{2-} \underset{-2h^+}{\overset{+2h^+}{\rightleftharpoons}} CdS + TeS_3^{2-}$	−41.4	−0.898	−48.2	−1.05
(d) $CdTe + S^{2-} \underset{-2h^+}{\overset{+2h^+}{\rightleftharpoons}} CdS + Te$	−35.3	−0.765	No change from ΔG°, solid products	
(e) $ZnS + 2S^{2-} \underset{-2h^+}{\overset{+2h^+}{\rightleftharpoons}} ZnS + S_2^{2-}$	−22.4	−0.486	−29.2	−0.633
(f) $ZnSe + 2S^{2-} \underset{-2h^+}{\overset{+2h^+}{\rightleftharpoons}} ZnS + SSe^{2-}$	−31.7	−0.687	−38.5	−0.835
(g)[b] $ZnTe + S^{2-} + S_3^{2-} \underset{-2h^+}{\overset{+2h^+}{\rightleftharpoons}} ZnS + TeS_3^{2-}$	−47.8	−1.036	−54.6	−1.18
(h)[b] $ZnTe + S^{2-} \underset{-2h^+}{\overset{+2h^+}{\rightleftharpoons}} ZnS + Te$	−41.7	−0.904	As in reaction (d)	
(i) $CdS + 2Se_2^{2-} \underset{-2h^+}{\overset{+2h^+}{\rightleftharpoons}} CdSe + SSe_3^{2-}$	−40.4	−0.876	−47.2	−1.02
Chemical exchange (no net charge transfer)				
(j) $CdSe + S^{2-} \rightleftharpoons CdS + Se^{2-}$	+10.3		+ 3.5	
(k) $CdSe + S^{2-} + S_4^{2-} \rightleftharpoons CdS + SSe^{2-} + S_3^{-2}$	−5.1		−11.9	
(l) $CdTe + S^{2-} \rightleftharpoons CdS + Te^{2-}$	+17.4		+10.6	
(m) $CdTe + S_4^{2-} \rightleftharpoons CdS + TeS_3^{2-}$	−17.8		−24.6	
(n) $CdS + Se_2^{2-} \rightleftharpoons CdSe + SSe^{2-}$	−10.0		−16.8	

[a] The concentration of chalcogen originating from the semiconductor is based very approximately on a 1-cm^2 photoelectrode in 1 ml of electrolyte, with an exchange of some tens of Ångstroms, giving ~10^{-5} *M*. The value of ΔG is derived from the equation $\Delta G = \Delta G^\circ + RT \ln Q$, where Q is the concentration of semiconductor-derived chalcogen; the other reactants are taken to be at unit activity (either solid phase or ~1 *M* concentration). Before any exchange has occurred, ΔG will be very negative but will rapidly move toward the values given in the table as exchange proceeds. The values of ΔG (and E_D) are provided to give an idea of the realistic values under operating conditions; from the preceding consideration, however, they should be considered neither constant nor exact.

[b] Reactions (g) and (h) are unlikely because ZnTe is *p* type.

change is thus limited to the surface layers (which will be CdS eventually), the result will be a continual turnover of the surface to fresh CdS. In this case, the photocorrosion could in theory account for 100% of the photocurrent, and yet the photoelectrode may continue to give a constant output.

An idea of the importance of the value of E_{redox} for the polysulfide system can be obtained from experiments carried out to determine the stabilizing effect of different redox systems as a function of their E_{redox}. Inoue *et al.* (*49*), using rotating-ring-disk methodology, measured the suppression of photodissolution of CdS by various reducing agents and found that the stabilizing effect of the reducing agents correlated with their E_{redox}, with a more negative E_{redox} imparting greater stability to the CdS. In particular, under the conditions of their experiments, SO_3^{2-}, $S_2O_3^{2-}$, and S^{2-} all were capable of essentially stabilizing the CdS. As will become apparent later, stability is relative and the systems we are discussing here are not totally stable (often far from it). This point should be kept in mind from the beginning. Also, these rotating-ring-disk experiments cannot measure a corrosion rate for a system like CdS–S_x^{2-}, where the product of the corrosion (Cd^{2+}) will not reach the ring but will remain in some insoluble form on the semiconductor surface. A step in overcoming this problem was made by Miller and Heller (*63*) who used simultaneous modulation of angular velocity and photocurrent output of a CdS rotating disk to differentiate between mass-transport-controlled processes ($S_x^{2-} \rightarrow S_x$) and corrosion of the CdS (which does not show a modulated component). They found that in a 1 m*M* Na_2S electrolyte, photocorrosion of the CdS begins to become important only at potentials positive of −0.8 V versus SCE. Other investigations of the degree of stabilization of CdS by different reducing agents showed suppression of dissolution of CdS, primarily by S^{2-}, SO_3^{2-}, and $S_2O_3^{2-}$ (*48, 69*), the former by rotating-ring-disk methodology and the latter by measuring the concentration of Cd^{2+} ions in the electrolyte owing to the photocorrosion, using atomic absorption spectroscopy. Minoura and Tsuiki also showed reduced stabilizing ability in the order $S^{2-} > SO_3^{2-} > S_2O_3^{2-}$. Because free Cd^{2+} ions will not exist to any measurable extent in a sulfide solution, however, such experiments would not reveal photocorrosion in a sulfide electrolyte if it does occur. They found that the stabilizing influence of SO_3^{2-} and $S_2O_3^{2-}$ was strongly dependent on potential, being greater at more negative potentials, as might be expected from thermodynamic reasoning. They also measured the fbp of CdS in the various electrolytes and found shifts of fbp with S^{2-}, SO_3^{2-}, and $S_2O_3^{2-}$ (in decreasing order of degree of shift) and no effect on the fbp with I^-, Br^-, or $Fe(CN)_6^{4-}$. This suggests that there is some specific interaction between SO_3^{2-} (and to a lesser extent $S_2O_3^{2-}$) and

CdS. This being the case, the difficulty in separating the effect of E_{redox} from that of charge transfer on the stability remains, because the order of increasing stabilizing influence correlates with both increasing negative value of E_{redox} and with increasing interaction between the reducing species and CdS. Easier to interpret in this context are the results of Inoue (*49*) for redox couples with more positive values of E_{redox} that do not cause a shift in fbp. For these couples, the stabilizing power does correlate with the redox potential, showing that this effect is important apart from any consideration of specific interaction.

B. Sulfur–Selenium Exchange

Returning to the CdSe–S_x^{2-} system and the overall photocorrosion shown in reaction (3) (assuming it occurs at all), we would expect a change of the surface from CdSe to CdS. Such an exchange was suggested by Hodes *et al.* (*37*), where electron-microprobe analysis indicated conversion of some of a polycrystalline CdSe photoelectrode to CdS after having been operated as a photoanode in a polysulfide electrolyte. Manassen *et al.* (*60*) provided further evidence for exchange of Se by S for polycrystalline CdSe electrodes, with the exchange occurring in the top few hundred angstroms of the electrode. Interestingly, some exchange also occurred in the dark, although much less than for a working photoelectrode. This implies a thermodynamic instability of CdSe in polysulfide solution in the absence of photogenerated holes. The thermodynamics of reaction (k) in Table II were considered by Heller *et al.* (*34*). On the basis of the data available to them, it was not possible to decide on the preferred direction of this reaction. Notice the difference in the thermodynamics of the chemical exchange for CdSe in S-free sulfide and polysulfide solutions [reactions (j) and (k) in Table II]. In this case, the presence of polychalcogenide species facilitates the exchange greatly, in contrast to the charge-transfer reactions (under illumination), where the difference is small. From the value of ΔG for the charge-transfer exchange reaction of CdSe in S^{2-} [reaction (b) in Table II], we see that such an exchange is thermodynamically favorable. As noted previously, the substitution of polysulfide species for S^{2-} does not change the results greatly; any change is usually in the direction of more favorable exchange. The chemical exchange (in the dark) in polysulfide solution [reaction (k), Table II] has a negative G and is, therefore, also thermodynamically possible; in S-free S^{2-} [reaction (k)], the exchange needs to be very slight indeed (only 10^{-8} M Se in solution will make G negative) to make this reaction, too, thermodynamically favorable.

Reports on such an exchange were given by Manassen *et al.* (*60*), Cahen *et al.* (*7*), Heller *et al.* (*34*), Gerischer and Gobrecht (*25*), Noufi *et al.* (*75*), and DeSilva and Haneman (*14*). Because much of the information from these experiments is consistent, they will be discussed together, and differences in techniques, results, or interpretations will be treated as they arise.

The methods used to investigate the exchange reaction in most cases were x-ray photoelectron spectroscopy (xps or esca) and Auger spectroscopy. Electron-beam-stimulated luminescence was also used by Heller *et al.* (*34*), and changes in photocurrent spectra and Mott–Schottky plots were used by Gerischer and Gobrecht (*25*). In all cases, the surfaces of CdSe photoelectrodes (both single-crystal and electroplated polycrystalline) were found to undergo exchange with sulfur from the electrolyte. In those investigations where depth profiling was carried out to estimate the depth of exchange into the CdSe bulk, it was found that exchange occurred to 50–150 Å into the bulk, although most of the exchange occurred by ~50 Å. Also, the depth of exchange increased as deactivation of the CdSe progressed (*75*). DeSilva and Haneman (*14*) gave the S : Se ratio at the surface of electroplated CdSe operated as a photoanode in S_x^{2-} as 1.8 : 1 (surface in this case refers to the depth that is sampled by xps, ~15 Å). Exchange was also observed by several groups for CdSe in a S_x^{2-} solution either in the dark or under illumination, but in the absence of external photocurrent flow. In these cases, the depth and amount of exchange was considerably less than for electrodes that had operated as photoanodes; usually just a few atomic layers.

Noufi *et al.* (*75*) also investigated the S–Se exchange for CdS crystals in a polyselenide solution. For the case where no external photocurrent flow occurred (illumination at open circuit), exchange occurred only to a few atomic layers. Reaction (n) in Table II, shown for Se^{2-}, shows such an exchange is thermodynamically possible. The value of ΔG (both for the chemical- and charge-transfer exchange) will depend relatively strongly on the polyselenide species present, owing to the large differences in $\Delta G°$ of Se_x^{2-}, depending on the value of x (Table I). For that reason, an intermediate value has been used here. They found that for CdS that has passed photocurrent, however, extensive exchange occurs to a depth of >800 Å for only 60 C/cm^2 charge passed. From reaction (i) (in Table II), we can see that such an exchange is thermodynamically more favorable than the equivalent CdSe–S^{2-} exchange [reaction (b), Table II]. However, we must be very cautious in using such data, both because of the uncertainties in the values used (primarily because of the uncertainty in the exact nature of the reaction) and because, as shown throughout this work, kinetic factors are usually more important. Experiments carried out on changes in surface morphology of deactivated CdSe and CdS photoelectrodes have shown that CdS in S_x^{2-} undergoes much more severe surface changes than does CdSe (*10*), indicating that CdS is intrinsically less stable than CdSe. This runs counter to the thermodynamic reasoning from a comparison of reactions (a) and (b) in Table II.

Such changes in surface morphology have been inferred by a number of the groups investigating the surface exchange. Cahen *et al.* (*7*) reported a decrease in reflectivity (measured by ellipsometry) of an initially smooth CdSe crystal photoelectrode after operation in a polysulfide solution and interpreted this as being a result of a restructuring of the CdSe surface to a more polycrystalline CdS surface. Gerischer and Gobrecht (*25*) came to a similar conclusion when, using capacitance techniques, they measured a large increase in charge in the space-charge layer after operation of CdSe and CdS as photoelectrodes in S_x^{2-}; they interpreted this result as indicating that the surface-exchanged CdS was in a very disordered state (the disordered CdS layer can trap electric charge at grain boundaries or dislocations).

The change in surface morphology of CdS and CdSe has been treated in detail by Cahen *et al.* (*10*). For CdS in S_x^{2-}, they used thin-film x-ray diffraction to estimate a crystallite size of the exchanged layer on CdS of <500 Å, if not actually amorphous. For CdSe in S_x^{2-}, although the surface crystallinity degraded, this degradation was less than for CdS. Reflection electron diffraction measurements lent further support to these conclusions and also showed the presence of polycrystalline Cd(S, Se) on a deactivated CdSe crystal. Other methods used by Cahen *et al.* (*10*) to probe the morphology and nature of the exchanged layer were scanning electron microscopy (sem), photoacoustic (pa) measurements, and changes in optical spectral response. Scanning electron microscopy measurements, in addition to giving direct verification of a restructuring of the surface [also shown by Noufi *et al.* (*75*) for CdS in Se_x^{2-}], indicated that the surface layer was more highly charged after exchange (charging of the sample in sem is seen as a blurring of the picture). This suggested that the surface layer was relatively insulating compared with the underlying substrate. [Note also the charging reported by Gerischer and Gobrecht (*25*) discussed previously.]

Photoacoustic measurements showed an increase in pa signal after deactivation of CdSe photoelectrodes. Because the pa signal is inversely related to the particle size at the surface, this is further indirect evidence for increase in polycrystallinity at the surface.

Spectral response measurements on CdS and CdSe crystals after deactivation in S^{2-} show a small sub-band-gap response that was absent initially (*10*); this was interpreted as being caused by defect states within the band gap owing to the surface restructuring. Gerischer and Gobrecht (*25*) observed a different type of change in spectral response of CdSe single-crystal photoelectrodes after deactivation. In this case, there was little difference in the response up to the band gap of CdS (~520 nm), but at longer wavelengths the response decreased dramatically. This observation was echoed by parallel measurements of the stability of CdSe in S_x^{2-} at short (supra CdS band gap) and long (sub CdS band gap) wavelengths of illumination. The CdSe showed reasonable stability for the former under conditions where it was much less stable for the latter. These results suggest that a relatively thick CdS layer is formed in these experiments, blocking much of the charges generated in the CdSe itself.

An important development reported by Heller *et al.* (*34*), was that addition of elemental Se to the polysulfide solution (in which it dissolves as $Se_yS_x^{2-}$) resulted in a large increase of stability for CdSe photoanodes. Auger analyses showed that no S–Se exchange occurred in a polysulfide solution containing 75 m*M* Se, in contrast to that occurring under the same conditions but in the absence of added Se. Noufi *et al.* (*75*) reported a considerable reduction in the exchange on addition of 50 m*M* Se to a polysulfide solution, although exchange did still occur. The cause of the stabilizing effect of Se will be discussed in the following text.

At this stage, it is worth summarizing the more important points connected with the exchange:

1. CdSe immersed in a polysulfide solution undergoes exchange with solution sulfur to give a surface layer of CdS down to no more than a few atomic layers.
2. When operated as a photoanode, this exchange becomes greater and extends deeper into the bulk of the CdSe. The longer it is operated as a photoanode (under any specified conditions), the deeper the exchange extends (although the rate of increase in the depth of exchange will decrease with depth, probably reaching a maximum limit; this limit will depend on the conditions).
3. Addition of Se to the electrolyte results in reduction or prevention of the exchange.
4. Surface restructuring occurs on deactivated CdSe and CdS to give a more finely crystalline or amorphous surface compared with the original surface.

The mechanism of processes 2 and 4 can be explained by referring back to reaction scheme (3). If photocorrosion occurs, even to a tiny extent, Cd^{2+} ions will be formed. These will be neutralized by sulfide ions to give CdS, hence the surface of the CdSe will eventually become CdS and exchange will continue between solution sulfur and that from the

CdS. In other words, the surface of a CdSe or CdS photoanode will be in a continuous state of exchange. With increasing time, or by varying any of the experimental parameters so as to decrease the efficiency of the sulfide oxidation step (to be discussed in the following text), the photocorrosion reaction increases relative to the sulfide oxidation and exchange can occur to a greater extent.

In theory, if the Cd^{2+} ion is neutralized immediately as it is formed, then an epitaxial layer of CdS would be expected to form on the CdSe (or CdS). Because this is not the case, as seen from the microcrystallinity of the exchanged layer, it seems that Cd^{2+} can migrate over the surface of the semiconductor before neutralization.

Having shown that exchange does occur as shown previously, the most logical question is, How does this exchange affect the performance of the photoanodes and, in particular, what effect does it have on their stability? This question has been considered by the teams working on the exchange reaction, with the common opinion that a thin layer of CdS on CdSe need not affect it adversely (in fact, it may even improve it), whereas a thick (>20 Å) layer will form a blocking contact. The reason for the blocking action of the CdS layer has been explained by most workers in terms of a barrier for hole flow to the surface at the CdSe–CdS interface (or a barrier for electron flow to the bulk for CdS electrodes in a Se_x^{2-} solution) (*7, 34, 75*). Where the exchange is not extensive (a <20-Å exchanged layer), charges can tunnel through this layer, hence there will be no reduction in efficiency. In fact, Cahen *et al.* (*7*) have suggested that the CdS layer may be relatively insulating owing to neutralization of donors by sulfide ions (e.g., S vacancies could be filled in by solution sulfur). The charging that is sometimes observed during sem studies of deactivated samples, indicative of an insulating layer, as mentioned previously, tends to support this theory (*10*). In this case, the CdSe–CdS–electrolyte system can be looked at in terms of a metal–insulator–semiconductor (MIS) [or more correctly, as an electrolyte–insulator–semiconductor (EIS)] system. In analogy to MIS solar cells, the effect of the insulator (CdS) may be to increase the photovoltage, an increase that is observed in practice. Rajeshwar *et al.* (*77*) invoke this MIS analogy to explain the nonideality of the photovoltage–photocurrent curves they measured for chemically deposited CdSe films in S_x^{2-}. They suggested that the nonideality may be a result of a voltage drop across the relatively insulating CdS. Heller *et al.* (*34*) and DeSilva and Haneman (*14*) have argued the opposite view, that the CdS may be very conducting from consideration of lattice mismatch and defects in the CdSe–CdS interface region. However, whether the layer is insulating or conducting, if it is thick enough the effect will be to reduce hole flow to the electrolyte, both because of the barrier to hole flow at the

CdSe–CdS interface and because of the increased probability of electron–hole recombination either at defects in the region of the CdSe–CdS interface (*34*), or at crystalline boundaries in the highly polycrystalline exchanged layer.

DeSilva and Haneman (*14*) suggested that the CdS layer is leaky, that is, electrolyte can reach CdSe underneath, hence a thin layer of CdS has no adverse effect. As the layer becomes thicker, it causes current obstruction and also light absorption. (The latter should not be serious because a large part of the light that can be absorbed by CdS has already been absorbed by the orange polysulfide electrolyte.)

In fact, CdS photoelectrodes are no more stable than CdSe ones under the same conditions—the reverse may even be true to some extent (*44*); this led to the conclusion that the hole barrier at the CdSe–CdS interface did not account for the decrease in output because CdS, which would not form such a barrier (the interface here is CdS–CdS), was also unstable. Hodes *et al.* (*44*) thus reasoned that the degradation was a result of either recombination at crystallite boundaries in the exchanged layer or a photoelectrochemically inactive and porous CdS layer that could block the mass-transport processes in the electrolyte to or from the active semiconductor surface.

The question may now be asked, If the corrosion reaction occurs because the competing sulfide oxidation is kinetically retarded, then what is the slow step in the overall sulfide oxidation? It could be diffusion of sulfide to the CdSe (it could also be connected with slow specific adsorption of sulfide at the surface), the charge-transfer oxidation of sulfide, or removal of oxidized species from the surface. This problem was addressed by Lando *et al.* (*53*), who used transient measurements in the CdSe–S_x^{2-} system to elucidate rate-limiting processes. In this technique, the time response of the photocurrent on initiation of illumination shows an initial peak photocurrent and reaches a steady-state value in a time that depends on experimental variables, but is usually some seconds. The peak value gives the hole flux to the surface in the absence of any mass-transfer limitations, whereas the steady-state value reflects the rate-limiting process. Lando *et al.* found that the difference between the peak and steady values, which would be zero for no mass-transport limitation, varied with experimental variables. Increased temperature, decreased light intensity (decreased photocurrent density), and increased sulfide concentration all decrease the difference between the peak and steady values, as might be expected for a mass-transport-controlled process. Of particular importance is the effect of $[S]:[S^{2-}]$ ratio on the transient behavior. This is shown in Fig. 2, where the normalized ratio is defined as $(I_{peak} - I_{steady})/I_{peak}$. Thus lower values of normalized ratio indicate less

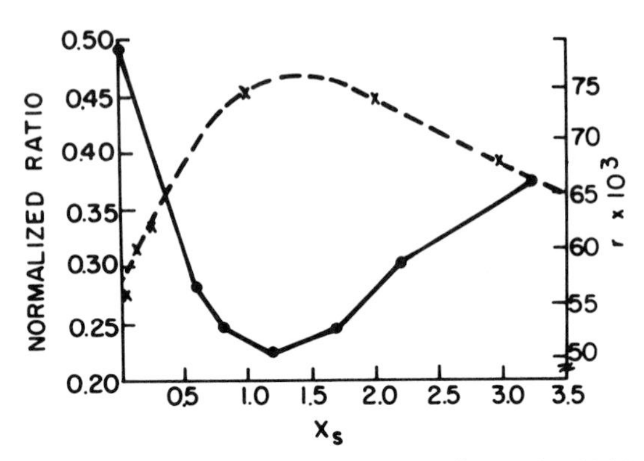

Fig. 2. Comparison between rate of dissolution of S in S_x^{2-} solution (*30*)(---, in min^{-1}; 0.5 *M* $[OH^-]$ and 0.25 *M* $[S^{2-}]$ at 80°C) and normalized ratio (——, 0.2 *M* $[OH^-]$ and 0.25 *M* $[S^{2-}]$ at 34°C) as a function of X_s = added S/S^{2-}. From Lando *et al.* (*53*). Copyright 1979 American Chemical Society.

loss in conversion. Figure 2 shows a minimum loss in conversion at a $[S]:[S^{2-}]$ ratio somewhat greater than unity. This is compared with the rate of dissolution of S in S_x^{2-} that shows a corresponding maximum. This result is strong evidence that the sulfur dissolution is the slowest step in the overall sulfide-oxidation process. Similar results were obtained by Lando *et al.* on metal electrodes, showing that the observed effects were because of solution, rather than semiconductor, processes.

Transient effects were also found by Peter (*76*) for Bi_2S_3 films in 0.1 *M* Na_2S solution. Peter suggested that this effect may be a result of slow surface recombination, a chemical change at the surface, or electrolyte depletion in the pores of the film. It is difficult to compare these results with those of Lando *et al.* (*53*) who used relatively concentrated electrolytes and strong illumination, because the dilute electrolyte used by Peter would lead to a large peak–steady ratio, whereas the low illumination intensity (equivalent to 4.5 $\mu A/cm^2$) would have the opposite effect.

The effect of increasing temperature on the output characteristics of polycrystalline (electroplated) CdSe photoelectrodes in polysulfide electrolyte is to increase both the short-circuit photocurrent (SCC) and fill factor (ff) (the former considerably) while decreasing the open-circuit photovoltage (OCV) (as expected from consideration of either the diode equation or the electrochemical kinetics of the dark forward reaction); for single-crystal CdSe, the effect of temperature is considerably less, although still evident for the ff (*58*). McCann *et al.* (*58*) explained these results by improved kinetics of S dissolution from the CdSe at higher temperatures. For the polycrystalline films that are presumably porous compared with the single crystal, dissolution of S from inside the pores will be more difficult than from a relatively smooth surface, hence the more pronounced effect of temperature on the polycrystalline layer than on the crystal. They also measured the temperature effect on SCC at different light intensities and found it to increase with increasing intensity, as would be expected from a rate-determining S dissolution. They explained the increase in ff as partly resulting from the measured decrease in electrolyte resistivity with increasing temperature. Of experimental interest is the optimum temperature for power output, because of a trade-off between decreasing OCV and increasing (or leveled-off) SCC and ff, of 30–40°C reported by the preceding workers. At the Weizmann Institute of Science (WIS), we have often found an even higher optimum temperature, and it is clear that the value of this temperature can vary depending on other variables—in particular, the form of the CdSe. Mueller *et al.* (*73*) pointed out that the optimum temperature for output is not necessarily the same as that for stability (to be discussed subsequently).

As will be described, the same experimental parameters that reduce the losses in steady-state photocurrents also improve the stability of the CdSe photoanode, as is to be expected if the transient losses reflect an overall reduction in rate of the sulfide oxidation, allowing the photocorrosion step to become more important. Such a connection between transient response and stability was noted by Lando *et al.* (*53*). On this basis, they investigated the effect of added Se on the transient response because Heller *et al.* (*34*) had shown that Se in the electrolyte improved the CdSe stability. Heller *et al.* found that small amounts of Se (in the millimolar range) had a very strong effect on the transient response, minimizing the (photo)current loss (on both CdSe and metal electrodes). Lando (*54*) also found that the time required for a CdSe photoelectrode to relax to a steady-state potential in the dark after it had been illuminated under short-circuit conditions was reduced if Se was present in the electrolyte. This suggested that the role of Se in stabilizing CdSe was to increase the rate of sulfur dissolution and thus the overall sulfide oxidation. They also found that Se increases the rate of bulk S dissolution in sulfide solution, lending further support to this hypothesis.

C. *Experimental Results on CdSe–S_x^{2-} Stability*

Although there are a number of scattered results on the stability of this system in the very early literature (i.e., 1976–1977), these tend to be so widely divergent (for reasons that are better understood now), that we shall not review them here. It should be remembered that when this system was first discovered, a photoelectrode that had not degraded after some minutes was often considered "stable." It became apparent that stability was strongly dependent on a number of factors, in particular, on light intensity (in other words, photocurrent density) and electrolyte composition. We shall cover in this section only those investigations that treat stability in depth. The many other references that exist on stability of this system agree in principle with the results reviewed here. The first reported in-depth study on the stability of the CdSe–S_x^{2-} system was by Heller *et al.* (*34*). They used both single-crystal and hot-pressed polycrystalline CdSe in an electrolyte of 1 *M* each NaOH, Na_2S, and S, with and without added selenium. (For simplicity, polysulfide solutions will be denoted by the nominal molarity of the NaOH (or KOH), Na_2S, and S in that order. Thus the preceding electrolyte can be written as a 1/1/1 solution.) Their results can be summarized as follows:

1. For CdSe single crystals, the (0001) face was much less stable than the $(11\bar{2}0)$ face. The polycrystalline hot-pressed pellets were intermediate in stability between these two faces.
2. At higher photocurrent densities, the stability was less; this dependence was very strong, with a modest increase in current leading to a large decrease in stability.
3. Stirring the electrolyte improved the stability.
4. Addition of Se to the electrolyte improved the stability dramatically (as has been discussed previously).

The reason for result 2 follows naturally from the competition between the photocorrosion and sulfide oxidation reactions. As long as the hole flux to the surface can be accommodated by the sulfide and, importantly, the product sulfur can be dissolved and transported into the electrolyte, photocorrosion will not occur to any destructive degree. (A continuous turnover of CdS confined to the very surface will not lead to degradation.) Once the overall sulfide oxidation is held up, for example, because the sulfur cannot be transported away fast enough, the holes arriving at the surface will be blocked to some degree and will be more likely to cause corrosion. Thus, at low photocurrent densities, a CdSe electrode may be indefinitely stable, whereas at higher currents it will begin to corrode. A small increase in the rate of corrosion is probably sufficient to disturb the balance between the corrosion and sulfide oxidation, hence small increases in photocurrent density beyond the point where the sulfide oxidation can fully cope with the holes will lead to large decreases in stability. The effect of stirring (3) can be explained by an improvement in the mass transport of species to and from the electrode. The effect of crystal orientation (1) cannot be explained so neatly and will be considered subsequently with further results on that aspect.

Most of the WIS group's early work was on electroplated thin layers of CdSe, compared with that of other groups on single crystals or large-grained pressed pellets. Although other factors were different, we realized that our electrodes were considerably more stable than other reported CdSe photoelectrodes. We attributed this to increased surface area of our very rough electrodes compared with relatively smooth single-crystal surfaces, and in parallel to high-surface-area metal electrodes in electrochemistry, this should result in a reduced real photocurrent density at the photoelectrode surface (*7*). It was argued that a reduced photocurrent density should allow the sulfide-oxidation reaction to occur more easily, thus reducing corrosion. The results of Heller *et al.* (*34*) show that a small reduction in photocurrent density would have a large effect on the stability. We studied this effect in more detail by etching the surfaces of CdSe crystals to give controlled morphologies (*45*). Three different crystal faces (all cut from the same CdSe boule) were used [(0001), $(10\bar{1}0)$, and $(11\bar{2}0)$] to be reasonably sure that the effect of etching was not a result of preferential exposure of a more stable plane by the etching. The qualitatively similar behavior of all three faces showed that such an explanation was extremely unlikely. For all three faces, the stability correlated with the degree of surface roughness obtained by using different etchants. This is shown in Fig. 3, which shows the stability of the (0001) and $(10\bar{1}0)$ faces

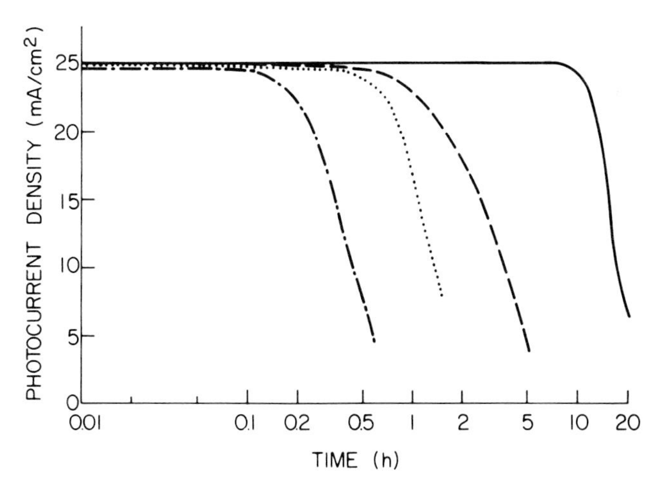

Fig. 3. Output stability of the (0001) or $(10\bar{1}0)$ face of a CdSe single-crystal photoanode in a polysulfide electrolyte (1 *M* each KOH, Na_2S, and S) at 35°C. –.–.–, Polished and Br_2–MeOH [for (0001) face] or polished and $CrO_3 : HCl : H_2O$ in 6 : 10 : 4 w/w ratio ("CrO_3") [for $(10\bar{1}0)$ face] etched; ·····, aqua regia (AR) etch followed by "CrO_3" etch; -----, AR etched; ——, AR etch followed by photoetch. (Both Br_2–MeOH and "CrO_3" act as polish etchants). From Hodes *et al.* (*45*), by permission of the publisher, The Electrochemical Society, Inc.

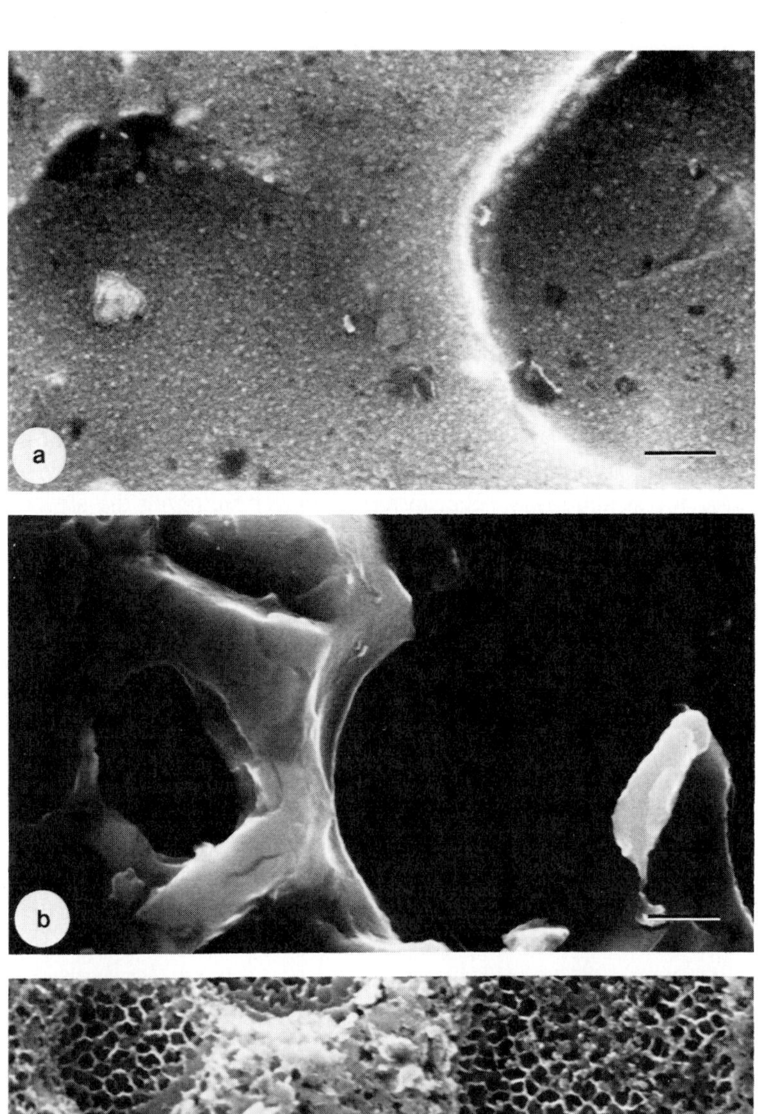
a
b

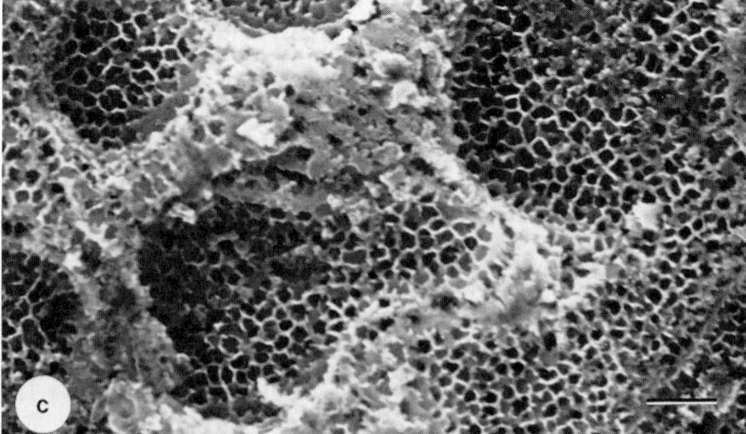
c

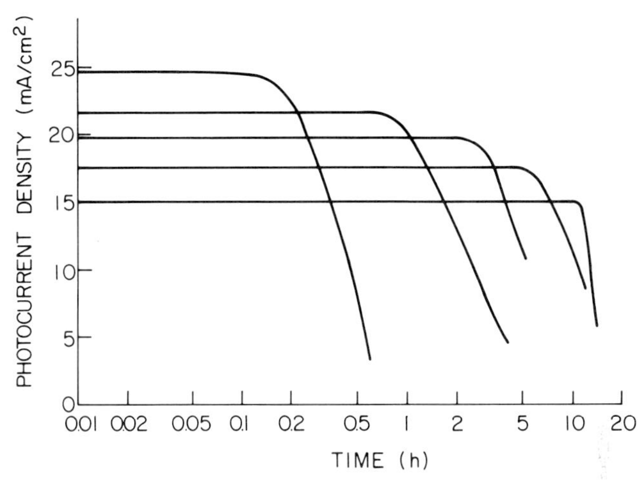

Fig. 5. Effect of photocurrent density variation (by varying illumination intensity) on the stability of a polished and Br_2–MeOH-etched (0001) face of a CdSe single-crystal photoanode. Electrolyte 1 *M* each KOH, Na_2S, and S at 35°C. From Hodes *et al.* (*45*), by permission of the publisher, The Electrochemical Society, Inc.

(which behave similarly) with different etchants, whereas Fig. 4a,b,c shows sem photographs of the surface of the (0001) face after the different etchings (the polished and Br_2–MeOH etched sem picture is not shown—it was completely featureless on the same scale). Figure 5 shows the dependence of stability of the (0001) face (polished followed by Br_2–MeOH etch) on photocurrent density. From this, it can be seen that reducing the real photocurrent density from 25 to 15 mA/cm^2 increases the lifetime by a factor of ~50. Thus the increases in surface area of the variously etched electrodes are more than sufficient to explain the increase in stability. It should be pointed out that the most stable (and most rough) form of the crystals is obtained by photoetching (to be described under surface treatments). Photoetching removes surface recombination centers, and, according to the arguments of Wilson (*100*) mentioned previously, such recombination centers may also be corrosion sites. Taken by itself, the increased stability of photoetched samples could also be explained on this basis. That it fits neatly into the other results on surface etching suggests that the increased surface area is the determining factor,

Fig. 4. Scanning electron micrographs (25 kV; magnification 9400×) of the (0001) face of a CdSe crystal after (a) aqua regia (AR) etch followed by Br_2–MeOH etch, (b) AR etch, and (c) AR etch followed by photoetch. From Hodes *et al.* (*45*), by permission of the publisher, The Electrochemical Society, Inc.

although the possibility of removal of corrosion sites by the photoetching cannot be rejected.

Reichman and Russak (*78*) measured the charge-transfer velocity v_t for hole transfer into the polysulfide electrolyte from vacuum evaporated CdSe films by analysis of *IV* curves. They found values for v_t an order of magnitude higher than for single-crystal CdSe photoelectrodes. This was interpreted as possibly a result of the higher surface area of the evaporated films and was correlated with the greater stability of polycrystalline CdSe photoelectrodes compared with single-crystal ones.

Hodes *et al.* (*45*) also compared a slurry-painted polycrystalline CdSe electrode (photoetched) with a photoetched $(11\bar{2}0)$ face crystal (the most stable known form of single-crystal CdSe). Although the polycrystalline electrode was more stable, the difference was less extreme than previously noted when comparing polycrystalline and single-crystal CdSe, neither of which had been photoetched (*10, 44*). A possible reason for the greater stability of the $(11\bar{2}0)$ face compared with the (0001) face, first reported by Heller *et al.* (*34*), and corroborated, with the addition of the $(10\bar{1}0)$ face, by Hodes *et al.* (*45*), has been suggested by the latter authors [see also Cahen *et al.* (*9*)]. According to them, the differences in stability of different faces might be connected with the number of bonds which that to be broken to free a Cd ion, with different interaction between adsorbed sulfide and the various crystal faces or with varying rates of diffusion of Cd^{2+} in the bulk in different crystal directions [the latter if photocorrosion is a result of diffusion of Cd^{2+} through the bulk to the surface, which has been reported to occur for CdTe in a $Sn^{2+/4+}$ electrolyte (*101*), but has not been directly observed for CdSe]. In short, the variation in stability with crystal face is still open for study and interpretation. It would seem worthwhile to pursue this line because an understanding of the effect may allow further control over the stability of photoelectrochemical systems.

The effect of polysulfide composition on the stability has been reported in a number of studies. More concentrated (both in Na_2S and in S) solutions show much better stabilizing power. The effect of increased S^{2-} concentration has not been studied in detail but is obvious from the various studies in which it has been mentioned (*7, 53, 64, 67, 80*). The effect of S concentration (or more correctly, [S] : [S^{2-}] ratio) has been described by Hodes *et al.* (*44*). A maximum stability of CdSe has been found in an electrolyte with a [S] : [S^{2-}] ratio of ~2. At both higher and lower ratios, the stability decreases.

The stability of the CdSe–S_x^{2-} system has been found to increase with increasing temperature (*44, 73*). This is contrary to what is observed with solid-state cells but is consistent with rate-determining dissolution of S (or, for that matter, any electrochemical or solution reaction) that increases in rate with increasing temperature.

Tenne (*90*) also described an increase in the stability of this system (particularly in dilute polysulfide electrolytes) on addition of high concentrations of NaCl as an "indifferent" electrolyte. Tenne suggested that this effect arose from a decrease in the Debye radius of the screening atmosphere of the sulfide ions because of the added NaCl, thus reducing the interaction between sulfide ions and increasing their mobility, and presumably their concentration at the CdSe surface.

To further complicate the wide variation of stability with different experimental parameters, it has been found that the stability of the CdSe–S_x^{2-} system is minimal at its maximum power point (*91*). This is contrary to what would be expected according to thermodynamic reasoning because the CdSe should be cathodically more protected at maximum power than at short circuit (at which most stability experiments have been run). This effect has been explained by Tenne *et al.* (*91*) as being a result of reduced adsorption of negative sulfide ions on the CdSe surface at more negative potentials, resulting in a lower concentration of electroactive species, and thus lower stability. Transient measurements carried out as a function of potential of the CdSe show a maximum peak-to-steady ratio (thus maximum losses) at the maximum power point, which suggests that the sulfur dissolution is adversely affected by the reduced sulfide concentration at the surface.

To conclude this section, it is of interest to note the long-term stability of CdSe that we have investigated using photoetched slurry-painted CdSe layers in a 1/1/1 electrolyte (J. Rotman, J. Manassen, G. Hodes, and D. Cahen, unpublished). A battery of 64 of these cells, each one of 3-cm^2 active (based on photoelectrode) area and efficiencies of approximately 2.5%, operated under outdoor conditions for 5 months under approximately maximum power conditions. Subsequent degradation was because of leakage of some of the cells. If we assume 5 months as the minimum lifetime under these conditions (it may be more), and by improving the electrolyte composition (greater S^{2-} and S content, and added Se), then it is not unreasonable to expect a lifetime of at least several years, if not more, for CdSe–S_x^{2-} PECs operating at about 2.5% efficiency. As to how much higher efficiencies (and thus higher photocurrent densities; therefore, less stability) affects these results remains to be seen, although some results with higher efficiency Cd(Se, Te) cells will be discussed in the following text.

Previously published results for slurry-painted CdSe operating at almost 4% efficiency showed complete stability for 2 months before the onset of gradual degradation (*44*). The photoelectrode in this study was not photoetched, and it is probable that this was the major factor in the difference between the two studies. Although the operating efficiency of the latter was higher (therefore less stable because of higher photocurrent), the electrolyte was considerably more favorable for stability than in the case of the former study.

IV. Other Cadmium Chalcogenide–Polysulfide Systems

Although CdSe has been the most thoroughly studied of the Cd chalcogenides, other chalcogenides and mixed chalcogenides have also been investigated, in particular, the Cd(Se, Te) solid solutions. Looking first at the binary chalcogenides CdS and CdTe, the former has already been described to some extent and does not differ radically in principle from CdSe (apart from differences arising from its higher band gap). CdTe, however, is very different in being relatively unstable in S_x^{2-} electrolytes. Polycrystalline CdTe, both *n*- and *p*-type, was reported to be stable in S_x^{2-} by Hodes *et al.* (*36, 37*), but single-crystal *n*-CdTe was reported to be unstable by Ellis *et al.* (*18*). This discrepancy can probably be attributed to the polycrystalline specimens employed by Hodes *et al.* compared with the single-crystal ones used by Ellis *et al.* (in the light of comparable results described previously for single-crystal and polycrystalline CdSe stability). Also, the current densities measured by Hodes *et al.* for their early samples of CdTe were <10 mA/cm^2 [e.g., see Manassen *et al.* (*61*)], and at those current densities, polycrystalline CdTe photoanodes can be stable for at least tens of minutes, if not hours, in S_x^{2-}. Such stability, although poor by present standards, was considered very good in those early days. Danaher and Lyons (*12*) reported results of electroplated CdTe photoanodes in S_x^{2-} electrolyte but did not give an indication of their stability.

Although it has not been unequivocably reported, S–Te exchange presumably occurs at the surface of a CdTe photoanode in S_x^{2-}. Under conditions where CdSe photoanodes remain dark, CdTe photoanodes become covered with a very visible yellow layer of CdS that is soluble in HCl but insoluble in S_x^{2-}. It is, therefore, not sulfur (R. Tenne, private communication). The decomposition potential E_D° of CdTe is the most negative of all three Cd chalcogenides [Section II,A; Table II, reactions (c) and (d)], making CdTe thermodynamically the least stable of the series. These reactions are based on the dissolution of Te in S_x^{2-}, where $x >$

1, to give TeS_3^{2-} as well as the insolubility of Te in S^{2-} (*28*) and also on the reported absence of S_2^{2-} under most conditions (*26*). In view of the lack of success of the thermodynamic data in explaining that CdS is no more stable (in S_x^{2-}) than CdSe (maybe less so), we should consider other factors, especially kinetic ones, to explain the relative instability of CdTe.

The electrochemical reactions occurring at the Cd chalcogenide–S_x^{2-} interface occur in reality at a CdS–S_x^{2-} interface in all cases because all the cadmium chalcogenides will become converted to CdS at the surface. From such a consideration alone, there should be no difference in the kinetics of sulfide oxidation at any of the Cd chalcogenides. Any differences that do exist would probably reflect the hole flux reaching the electrolyte, which could depend on the thickness of the exchanged layer, as well as properties of the CdS–Cd chalcogenide heterojunction (barriers to charge flow and interface defect states). Another consideration is the much slower rate of solubility of Te in S_x^{2-} that we find, compared with S or Se. Te can block the surface, leading to increased photocorrosion. This last point may well be the most important one in determining the stability of the CdTe/S_x^{2-} system.

CdTe can be both *n*- and *p*-type (the only Cd chalcogenide that can exist as *p*-type under normal conditions). Although we have only considered the stability of *n*-type semiconductors up to now and do not plan to discuss the general aspects of stability of *p*-type materials, because the vast bulk of work in this system is on *n*-type materials, it is reasonable to expect *p*-type photocathodes to be more stable to reduction by electrons (the minority carriers) than *n*-type photoanodes to be stable against oxidation. This appears to be the case for *p*-CdTe, which has been described in a few cases in S_x^{2-} electrolytes and appears to be stable (*6, 37, 83*). The open-circuit voltage (OCV) of *p*-CdTe is very low—about 150 mV—and the short-circuit quantum yields are about 0.2% (*6*), which compares very unfavorably with the almost 100% quantum yield obtainable from *n*-type Cd chalcogenide photoanodes. It appears that the band bending in *p*-CdTe–S_x^{2-} is much lower than in the case of *n*-CdTe–S_x^{2-}. This is not unexpected, because if we take the band gap of ~1.4 eV, subtract ~0.3 V from this [the sum of the energy difference between the Fermi levels and respective bands for *n*- and *p*-type will be of this order—even greater for the low-doped *p*-CdTe used by Bolts *et al.* (*6*)], and a further ~0.8 V for the band bending of *n*-CdTe in S_x^{2-} [from the fbp of *n*-CdTe in S_x^{2-} given by Ellis *et al.* (*19*)], this leaves ~0.3 V for the band bending in *p*-CdTe (assuming there is no difference in the interaction of sulfide with *n*- and *p*-CdTe). This would explain the low values of OCV observed.

All the Cd chalcogenides are miscible among themselves to give solid solutions. CdSe and CdS, both of which are normally hexagonal, show a monotonic variation of band gap over the complete composition range from CdS (E_g, 2.4 eV) to CdSe (E_g, 1.75 eV). The only reported use of Cd(S, Se) solid solutions (or, as we shall call them here, alloys) was by Noufi *et al.* (*74*), who prepared Cd(S, Se) photoanodes by evaporation and by sintering pressed pellets. Their most important observation was that the fbp (and OCV) of the photoelectrodes went through a maximum at a composition of $CdS_{0.9}Se_{0.1}$, with a difference of ~0.23 V between the fbp of this material and pure CdS. Because they found this maximum also in sulfide-free alkaline electrolytes, they inferred that the effect was not a result of differences in the specific adsorption of sulfide on the different compositions and may be a result of either a change in the electron affinity of the Cd(S, Se) with varying composition or a junction potential between the surface (exchanged) layer and the underlying bulk.

CdSe and CdTe, unlike the Cd(S, Se) alloys, show a broad minimum in the band gap at a composition of about $CdSe_{0.5}Te_{0.5}$ and a change of normal crystal structure in the region of that composition from hexagonal for Se-rich alloys to cubic (the normal form of CdTe) for Te-rich ones (*86*). Because CdSe is stable in S_x^{2-}, whereas CdTe is relatively unstable, it was reasonable to expect that Cd(Se, Te) alloys would be stable for some compositions, but with a lower E_g than CdSe, thus higher photocurrents. This was shown to be the case by Hodes (*40*) and Hodes *et al.* (*46*), who prepared electrodes of these alloys by a slurry-painting method, as well as by electroplating (*47*). A composition of $CdSe_{0.65}Te_{0.35}$ was found to be both relatively stable and more efficient than CdSe because of an increase in photocurrent. Although the E_g of this alloy is ~0.2 eV less than that of CdSe, the photovoltages are similar. In fact, it is interesting to note that, both from the literature and from our own experience, there is no consistent and appreciable difference in OCV between CdS (E_g = 2.4), CdSe (E_g = 1.75), and Cd(Se, Te) (E_g = 1.45 for alloys normally used by us). This may be connected with the formation of a CdS layer at the surface of all of these materials in S_x^{2-}; thus if the band bending is deter-

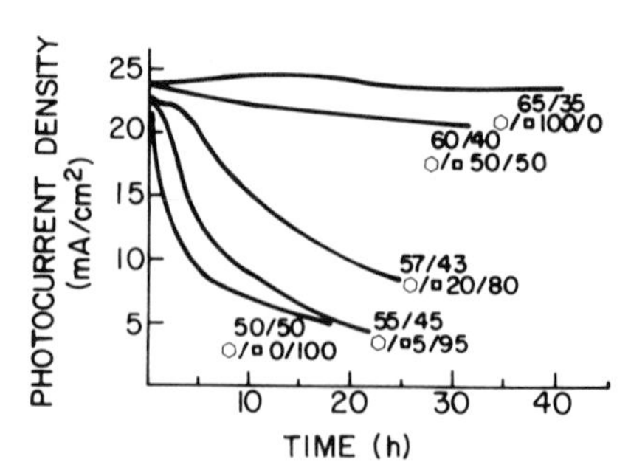

Fig. 6. Output stability of Cd(Se, Te) slurry-painted polycrystalline photoanodes in an electrolyte 2 *M* each in KOH, Na_2S, and S at 35°C, as a function of composition (Se : Te ratio) and crystal structure (hexagonal–cubic ratio) measured by x-ray diffraction. From Hodes *et al.* (*41*). Copyright 1980 American Chemical Society.

mined primarily by the interaction between sulfide and the surface, there should be little difference with differing substrates [although note the change in fbp found by Noufi *et al.* (*74*) for different Cd(S, Se) alloys, which cannot be explained by this reasoning].

Because we are concerned primarily with aspects pertaining to stability at this stage, it is interesting to note how the stability of these alloys varies with Se and Te content. Such a study (for slurry-painted electrodes) was carried out by Hodes *et al.* (*41*). The main results of this study are shown in Fig. 6, in which stability of the photoelectrodes is shown as a function of Se : Te ratio and also of the ratio of hexagonal and cubic phases, both of which can coexist within a certain composition range (Fig. 7). From Fig. 6, it can clearly be seen that the determining factor for stability seems to be crystal structure rather than composition. This was confirmed by varying the annealing conditions of a fixed composition, so as to vary the hexagonal–cubic ratio. The stability of the different electrodes was less as the cubic content increased at the expense of the hexagonal. Although no firm explanation was given for this effect, it was suggested that it might be connected with a difference in bond strengths between the hexagonal and cubic modifications [the lower band gaps of the cubic, compared with the hexagonal, structure (Fig. 5a) suggest a lower value for the bond strength of the former]. Results by Hodes *et al.* (*47*) show no difference between the stability of hexagonal and cubic structures of electroplated CdSe films in S_x^{2-}—both are relatively stable. This result indicates that consideration only of the crystal structure will not explain the difference in stabilities of the type shown in Fig. 6b.

Although we have less data on the long-term stability of Cd(Se, Te)–S_x^{2-} PECs compared with CdSe ones, we have found complete stability for cells (using slurry-painted photoelectrodes) of composition about Cd-$Se_{0.75}Te_{0.25}$ with efficiencies between 4 and 5%, operating in the region of maximum power outdoors for periods of several months, with subsequent degradation after this time that, although slow, varied considerably

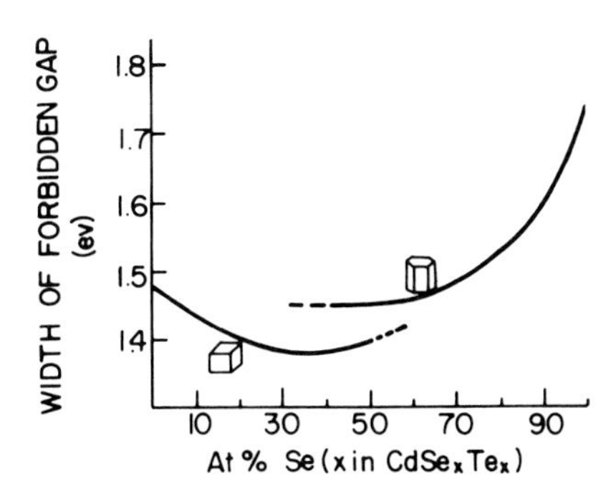

Fig. 7. Variation of room temperature optical band gap of Cd(Se, Te) as a function of Se content and crystal structure for well-annealed samples. The two phases can coexist in the intermediate composition range, and for certain methods of preparation both crystal structures can exist over the entire composition range. From Hodes *et al.* (*41*). Copyright 1980 American Chemical Society.

among different samples. $CdSe_{0.65}Te_{0.35}$ was found to be less than completely stable in such long-term tests.

Cd(Se, Te) alloy photoelectrodes in S_x^{2-} have also been investigated by Inoue *et al.* (*51*) using sintered pressed disks of CdSe–CdTe mixtures. No reference to the stability of the photoelectrodes was made.

Finally, although there have been no reports on Cd(S, Te) alloy photoelectrodes in the literature, it is worth briefly mentioning our unpublished attempts at making photoelectrodes of this alloy. As with Cd(Se, Te), alloys of CdS and CdTe also show a minimum in E_g at a composition of about $CdTe_{0.75}S_{0.25}$ (*86*), with a minimum value of ~1.4 eV [the same as for Cd(Se, Te), the value depending on the crystal structure]. Using our slurry-painting technique (*42*), photoelectrodes of varying composition of CdS_xTe_{1-x} were prepared. In all cases, very poor photoelectrodes were obtained—much inferior to Cd(Se, Te) or even CdSe, although the band gaps were low.

V. Zinc (and Zinc–Cadmium) Chalcogenides

Because of the high band gaps of the Zn chalcogenides (3.7, 2.7, and 2.2 eV for ZnS, ZnSe, and ZnTe, respectively), little work has been carried out on these semiconductors in (poly)sulfide solution. ZnSe has been reported to be stable in sulfide electrolyte (*36, 37*), as has ZnS (*38*). Of interest in the context of surface exchange is the stability reported by Hodes (*38*) for ZnO in sulfide solution, which was explained by sulfur–oxygen exchange at the ZnO surface, giving a surface layer of ZnS. Such an exchange was later confirmed by Russak and Reichman (*81*). That this was not a bulk effect was shown by the photocurrent spectrum of ZnO in the sulfide solution, which corresponded to the band gap of ZnO (3.2 eV) and was very different from the photocurrent spectrum of ZnS (E_g = 3.7 eV). In both cases, the spectra were measured for thin layers prepared from thin layers of Zn by heating in air (ZnO) or anodizing in sulfide solution (ZnS). Although the stability tests were not very stringent by present standards, enough photocurrent was passed to entirely dissolve the electrodes several times over if the current had been due only to photocorrosion.

Mixed Cd–Zn chalcogenides offer an option for modification of Cd chalcogenides because band gaps can be tailored over the range of 3.7 eV (ZnS) to >1.4 eV [Cd(Se, Te) or Cd(S, Te) alloys]. Also, crystal structures can be chosen over a large part of this range because the Zn chalcogenides and CdTe are normally cubic, whereas CdS and CdSe are normally hexagonal (although most of these materials do exist in either form, sometimes only one can be prepared in a relatively simple manner). This tailoring of crystal structure and composition has been used by Hodes *et al.* (*41*) to show that, in common with the stability of both cubic and hexagonal CdSe, cubic (Cd, Zn)(Se, Te) alloys show stability comparable with the hexagonal Cd(Se, Te) alloys. In fact, Hodes (*43*) showed that the long-term stability of $CdSe_{0.65}Te_{0.35}$ photoelectrodes can be improved considerably by substitution of 5% Zn for Cd. The thermodynamics of the exchange reactions of ZnX with S^{2-} (Table II) lead to the expectation that the Zn chalcogenides will be somewhat less stable than the corresponding Cd ones. How this will affect the Cd(Zn)X alloys, however, is not clear. As for the difference in stability between cubic and hexagonal Cd(Se, Te) discussed previously, this could possibly be connected with stronger bonding in the zinc-containing materials, thus making them less prone to decomposition. In contrast to the thermodynamic picture, this can be interpreted in terms of kinetics; stronger bonding will make the first step in the decomposition more difficult. We have found, however, that substitution of Zn for Cd in the mixed selenide–tellurides does lead to a drop in photoactivity, even more than would be expected from the increase in band gap (G. Hodes, unpublished). Whether this is an intrinsic property of these compounds or a result of nonoptimization of our preparative technique for these materials (we used slurry painting), we cannot say.

Tsuiki *et al.* (*95*) prepared thin films of (Cd, Zn)(S, Se) alloys by vacuum evaporation for use as photoelectrodes in S_x^{2-} solution. They found that the band gap of an equimolar solid solution of CdS and ZnSe ($Cd_{0.5}Zn_{0.5}S_{0.5}Se_{0.5}$) was lower than that of either of the individual starting materials and that photoelectrodes of this composition showed photocurrent response out to ~650 nm. Although no information on the stability of these electrodes was given, it may be expected from the preceding discussion that they will be very stable.

Chen *et al.* (*11*) disclosed that a sputtered thin film CdSe + 1% ZnSe photoelectrode gave a higher open-circuit photovoltage than CdSe itself (see also Section VII,B).

VI. Other Metal Chalcogenides

A. Bi_2S_3

Along with CdS and CdSe, Bi_2S_3 was recognized to be stable as a photoanode in S_x^{2-} solutions from the start (*36, 37, 61, 62*). Although the band gap of Bi_2S_3 is low (~1.3 eV), the photoactivity of all samples reported were low, both because of low photovoltages and because of, even more obviously, low photocurrents (fractions of a mA/cm^2 for approximate solar irradiances). This may be due in part to the method used by the preceding workers (anodization of Bi metal in sulfide solution). However, using the same basic method of anodizing Bi metal in sulfide solution, Peter (*76*) reported Bi_2S_3 photoanodes with high quantum efficiencies, close to unity in some cases. Both the photovoltage and fill factor were poor for these electrodes. Peter measured onset of photocurrent at -0.8 V versus NHE, which gave an indication of the fbp that should be at this value or slightly negative of it, thus explaining the low open-circuit photovoltages observed (~300 mV). A study of thin-film Bi_2S_3 photoanodes prepared by chemical deposition (*5*) also gave conversion efficiencies in polysulfide electrolyte of approximately 0.1%, primarily because of lower photocurrents. In this study, the Bi_2S_3 films were characterized by various physical methods, and among other properties, they measured an E_g of 1.47 eV, specific resistivity of 20–40 MΩ cm, and mobility (presumably of electrons, though this was not stated) of 0.07 cm^2/V s. The high resistivity and low mobility are likely to be responsible for the low efficiency. No work on single-crystal Bi_2S_3 photoanodes has been reported; because single crystals of higher charge mobility and much lower resistivity can be made, this would be an important study to know whether Bi_2S_3 is an intrinsically poor photoanode material or if major improvements are possible. These same authors also noted that the photocurrent obtained from the chemically deposited Bi_2S_3 films (~0.34 mA/cm^2) was stable for 9 h. Although this is a mild test for stability, it seemed to be the only quantitative one reported for this material.

B. $CdIn_2Se_4$

This material has been reported as a photoanode in S_x^{2-} electrolyte (*92*). It will be mentioned further in the section on surface treatments. Of relevance to this section is its structure, which is made up of an incomplete lattice of electropositive ions. Tenne *et al.* suggested that such a structure might cause this material to be prone to photoanodic decomposition compared with CdSe, if the decomposition mechanism was connected with ionic movement through the bulk, because the incomplete lattice might allow freer ionic movement in the bulk. Their results on the stability of photoetched single crystals of $CdIn_2Se_4$ (no degradation of output at ~30-mA/cm^2 short-circuit current in a 2/2/2 sulfide electrolyte for more than 15 h) indicate that this material is no less stable than CdSe.

x-Ray photoelectron spectroscopy analysis of $CdIn_2Se_4$, which operated as a photoanode in S_x^{2-} solution, showed S–Se exchange at the surface comparable with that occurring in CdSe under the same conditions (Y. Mirovsky, R. Tenne, D. Cahen, G. Sawatzky, and M. Pollak, unpublished).

C. I–III–VI_2 Compounds

$CuInS_2$ was first mentioned in the context of S_x^{2-}-based PECs by Manassen *et al.* (*61*), who found a photoresponse for *n*-$CuInS_2$ in S_x^{2-}. That the $CuInS_2$–S_x^{2-} system was a particularly stable one was shown by Robbins *et al.* (*79*), who found no sign of photocorrosion, and constant output, after 20,000 C of operation (both at short-circuit and under resistive load) for single-crystal or hot-pressed polycrystalline samples of $CuInS_2$. They attributed this stability to the common anion (sulfur) in both semiconductor and electrolyte, and hence lack of ion exchange. That CdS is not completely stable in its common anion S_x^{2-} electrolyte (*10, 44*) as noted previously suggests that there is some other factor involved here. Mirovsky *et al.* (*70*), in a subsequent study of the $CuInS_2$(both *p*- and *n*-type)–S_x^{2-} system, also showed that *n*-$CuInS_2$ was particularly stable in S_x^{2-}. They also found the fbp of *n*-$CuInS_2$ in S_x^{2-} to be comparable with that of Cd chalcogenides in S_x^{2-}. The very low photovoltages (~100 mV) found for *p*-$CuInS_2$ in this study may be explained in the same way as was done for *p*-CdTe previously because the band gap of $CuInS_2$ (1.5 eV) is much the same as that for CdTe.

There are a number of differences between the $CuInS_2$ and Cd chalcogenide systems that point to a basic difference in the electrochemistry involved. The $CuInS_2$ system shows a strong temperature dependence, considerably different in degree and manner from that obtained with CdSe–S_x^{2-} systems (*58*). Both the ff and SCC increase with temperature (up to at least 70°C), giving a greater than 50% increase in efficiency in both studies on $CuInS_2$. Robbins *et al.* found that the ff is the main factor that increases, whereas Mirovsky *et al.* found more of the increase due to SCC. On the basis of double beam experiments, Mirovsky *et al.* suggested that the limiting step in the overall oxidation was either faster than 25 μs or slower than 2 ms; large transient photocurrents, which decrease (reversibly) over seconds–minutes, which we usually find with $CuInS_2$ photoanodes and which do not seem connected with the transient photocurrents described previously for CdSe, suggest that a slow readjustment takes place, possibly connected with some chemical restructuring of the surface or other surface change as suggested by Mirovsky *et al.* (*70*). The well-known high mobility of the Cu^+ ion is one factor that may be suspected to play a role.

Extension of these studies to $CuInSe_2$ ($E_g \simeq 1.0$ eV) has given further information on these systems. Mirovsky and Cahen (*71*) have found that *n*-$CuInSe_2$ is similar in many ways to $CuInS_2$ in S_x^{2-}, but with higher SCC and lower OCV expected from its lower E_g. The temperature behavior parallels that of $CuInS_2$–S_x^{2-}, with increases in both ff and SCC. Stability has been found in this system with single-crystal *n*-$CuInSe_2$ in S_x^{2-} at 50°C under SCC of 40 mA/cm^2. No degradation of output was found after >20,000 C/cm^2 photocharge passage under short-circuit conditions. In this case, the semiconductor is not in its common anion electrolyte. x-Ray photoelectron spectroscopy studies of $CuInSe_2$, which has operated as a photoanode in S_x^{2-}, show no sign of S–Se exchange, in sharp contrast to CdSe or $CdIn_2Se_4$—surprisingly, not even a very localized surface exchange (*71*). This result emphasizes the connection between stability and exchange. The reason for the greater stability (and lack of surface exchange) of $CuInS(Se)_2$ compared with CdS(Se) is an open and intriguing question with obvious importance. One possible reason may be found in the suggestion by Manassen *et al.* (*59*) and Cahen *et al.* (*8*) that the most stable photoanodes are those that have a component element capable of a change of valence state. Both Cu and In exist in two different valency states that could accommodate holes by a change in valence state. [It has been claimed that in the copper sulfides, only Cu(I) exists (*20*).] Cadmium, however, is only known to exist as Cd(II), and a parallel mechanism will not apply. Also, the valence band of $CuInS_2$ may have a high degree of *d* character (*82*). In this case, the argument given for the stability of MoS_2-type photoelectrodes, namely, that no bonds are broken because *d–d* nonbonding transitions are involved (*94*), may be relevant to some extent here. In fact, there is much in common between changes of valence state of transition metals and the *d–d* concept proposed by Tributsch (*94*) in their application to photoelectrodes. x-Ray photoelectron spectroscopy of both $CuInS_2$ and $CuInSe_2$ photoanodes, which have operated in polysulfide electrolytes, shows that the surface of these photoanodes is composed largely of In_2O_3 (*102*). The stability of this In_2O_3 layer in the presence of sulfide species may have much relevance to the stability of the photoanodes and can explain why no S–Se exchange occurs in the case of $CuInSe_2$.

Deng and Li (*13*) have reported the use of solid solutions of $CuInS_2$ and $CuInSe_2$ as photoanodes in S_x^{2-}. They give no information on the stability of their polycrystalline samples.

VII. Polyselenide and Polytelluride Electrolytes

The only comprehensive study on Cd chalcogenides in polychalcogenide solutions other than S_x^{2-} is by Ellis *et al.* (*18*) on CdTe in Te_x^{2-} and a relatively detailed comparison by Ellis *et al.* (*19*) of all three Cd chalcogenides (single crystals) in the three different polychalcogenide solutions (S_x^{2-}, Se_x^{2-}, and Te_x^{2-}). Apart from CdTe–S_x^{2-} (as already mentioned), all other combinations were found to be stable under the conditions of the experiments (typically some mA/cm^2 for 1–2 days in electrolytes with concentrations down to tens of millimoles). They measured the positions of the energy bands of all the combinations relative to the redox levels [by capacitance, onset photocurrent, and photovoltage measurements (the last under intense illumination—up to ~40 equivalent suns)]. Photovoltages greater than 700 mV were found for all CdTe-based systems as well as all polysulfide-based systems, whereas the other combinations (CdS and CdSe in Se_x^{2-} and Te_x^{2-}) gave much lower photovoltages (400–500 mV). (Because we now know that any particular semiconductor–electrolyte combination can give very widely differing photovoltages, depending on preparation of semiconductor, etching technique, etc., relatively small differences in reported values of photovoltage are probably not of general significance.) The reason for the considerably lower amount of band bending in the four combinations that give <500 mV maximum photovoltages is not known. In three of these combinations (CdS in Se_x^{2-}; CdS and CdSe in Te_x^{2-}) surface exchange would lead to a narrower band-gap semiconductor on top of a wider band-gap substrate. For all combinations that give high photovoltages, the exchanged surface has an equal or higher band gap than the substrate. It is tempting to connect this with the degree of band bending, although the CdSe–Se_x^{2-} system is an exception to this reasoning.

Of practical interest are the absorption characteristics of the various polychalcogenide electrolytes in the visible region reported by Ellis *et al.* (*19*). Polysulfide solutions typically transmit radiation with $\lambda > 500$ nm (solutions with very low concentrations of added S transmit considerably shorter wavelengths, but are not very stabilizing, as discussed previously). Polyselenide solutions absorb from ~425 nm, and although the long-wavelength cutoff was not given in this study, its deep red color suggests that it is >600 nm. Polytelluride shows an absorption band with a peak at 512 nm and extending from ~450 to ~650 nm (it is purple). Ellis *et al.* point out that all of these solutions transmit an appreciable fraction of the entire visible spectrum for short path lengths. For more concentrated solutions than generally used in their study (>0.1 *M* equivalent added chalcogen), this will be increasingly less the case. It is important to point out that the colors of the polychalcogenide solutions are a result of the polychalcogen species; the single-atom ions (S^{2-} as opposed to S_2^{2-}, for example) are all colorless. The high absorption of Se_x^{2-} and Te_x^{2-} solutions probably preclude their use in any future front-wall cell (apart from considerations of availability, cost, toxicity, and sensitivity to oxidation by air), although their use in back-wall cells would overcome this particular objection.

VIII. Surface Treatment of CdX Photoelectrodes

A. *Surface Etching*

Etching of semiconductor surfaces is a common step in many processes involving semiconductors, and the CdX photoelectrodes are no exception in this respect. The purpose of etching is usually to remove damaged surface layers. Such damage is caused, for example, by cutting and polishing single crystals. Ellis *et al.* (*16*) noted that etching their CdS and CdSe crystals led to a very large improvement in the output parameters of these materials in S_x^{2-}. Heller *et al.* (*33*) found as much as three orders of magnitude increase in output parameters after chemical etching of single-crystal CdS and CdSe photoanodes, the difference being attributed to removal of charge-trapping surface states arising from the mechanical processing steps; the trapping of carriers in these surface states can be followed by recombination, hence the heavy losses in the unetched crystals. They used a two-beam method to expose the effects of these surface states, whereby a weak monochromatic light source of varying wavelength was used to probe the spectral response in both the presence and absence of intense laser illumination. The latter populated surface states, if present, with electrons, thus increasing the recombination probability for the photogenerated holes and leading eventually to photocurrent saturation. Short-wavelength radiation absorbed close to the surface, where there is a high trap density, will thus lead to less photocurrent than long-wavelength radiation absorbed more in the bulk, and further away from the surface traps. These damage-induced surface states also showed up as a sub-band-gap response that was absent in the etched crystals. Similar effects were found using this two-beam method for the *n*-CdTe–Se_x^{2-} system (*35*). Pulsed-light techniques were applied by Harzion *et al.* (*31*) to the single-crystal CdSe–S_x^{2-} system and showed that the decay time for the short-circuit photocurrent on interruption of illumination decreased by up to 10-fold (from $\sim 10^{-6}$ to $\sim 10^{-7}$ s) after etching. This change was explained by the removal of surface traps by etching, thereby allowing a faster charge transfer through the semiconductor surface. They also described the damaged surface layer resulting from polishing the crystal in terms of an additional ohmic component to the equivalent circuit of the cell, which caused the increase in decay times and, to an extent, decrease in photocurrent (*32*).

DeSilva and Haneman (*15*) carried out surface analyses (xps) of non-etched and etched (by dilute aqua regia) CdSe and found a considerable increase in the Se : Cd ratio after etching. It is possible that this increase is a result of free Se on the surface, which is one of the products of the reaction of aqua regia with CdSe.

A technique of photoelectrochemical etching (now commonly called photoetching), was developed specifically for CdX photoelectrodes, although it is likely to find applications outside of photoelectrochemistry. This technique, whereby the CdX photoanode is deliberately operated in an electrolyte in which it is unstable under illumination, was used by Hodes (*40*) for slurry-painted $CdSe_{0.65}Te_{0.35}$ layers to obtain solar conversion efficiencies up to 8%. Photoetching of CdSe photoanodes was subsequently described in more detail by Tenne and Hodes (*87*). In the process of photoetching, Cd^{2+} ions go into solution, leaving a thin Se layer on the CdSe. Subsequent immersion of the CdSe in S_x^{2-} dissolves off this Se. The resulting CdSe surfaces were found, on sem investigation, to have undergone pitting, with the diameter of the pits, as well as their distance from each other, to be of the order of 10^3 Å. (This showed up as a very distinct blackening of the surface because of reduction in reflectivity.) This nonhomogeneous etching means either that the photogenerated holes preferentially migrate to certain sites (or areas) at the surface or else that they undergo surface diffusion or migration to such sites. The effect of photoetching on CdSe and $CdSe_{0.65}Te_{0.35}$ was shown by Tenne and Hodes (*87*) and Hodes (*40*), respectively, to result in an increase in short-circuit photocurrent and fill factor of the CdX–S_x^{2-} PECs (the effect was greater for thin-film polycrystalline CdSe than for single-crystal material), and this was presumed to be a result of removal of surface (or near surface) defects acting as recombination centers. Electron-beam-induced current (ebic) measurements on Au–CdS Schottky junctions have shown this to be the case (*103*). The nature of these recombination centers is not established at present. On the basis of the dense and even distribution of pits on the surface, Hodes *et al.* (*103*) suggested that these centers may be donor sites (the density of the pits corresponds to doping levels of $\sim 10^{14}$ to $\sim 10^{15}$ per cubic centimeter, assuming every pit represents one donor atom). The holes could attack such sites preferentially because they are probably more chemically reactive than surrounding stoichiometric regions. It is also possible that local electric fields in the vicinity of these sites are such that they attract holes moving toward the surface toward them.

Photoetching has been shown to have a dramatic effect on the photoluminescence (PL) of CdSe single crystals (*85*). The PL intensity was increased by an order of magnitude or more after photoetching. Possible reasons suggested for this enhancement were production of a more emissive surface morphology by the etch or formation of a surface layer that insulates the near-surface region from quenching of emission by electrolyte species or by surface states. In view of the effect of photoetching in removing surface recombination centers discussed previously, it seems probable that these same recombination centers are responsible for a loss in PL efficiency.

It is interesting to note that photocurrent and PL are normally competitive, because the former is a measure of e^{-}–h^{+} separation, whereas PL is a probe of their recombination (*52, 85*). The quenching of PL has been correlated with the photocurrent quantum efficiency by these workers. In the preceding case, photoetching leads to an increase both in PL and photocurrent efficiency.

Both CdS (*88*) and *n*-CdTe (*72*) have shown improvement in their performance as photoanodes in S_x^{2-} electrolytes (as well as in other electrolytes), in a manner similar to CdSe and Cd(Se, Te), after photoetching. It is of interest to note that pitting was not observed on CdS. The photoetching solution used for CdTe was a dilute $K_2Cr_2O_7$–HNO_3 mixture, which was found to give larger photocurrents during etching compared with the dilute HCl or aqua regia often used for other Cd chalcogenides. The redox potential of the dichromate mixture was more positive than that of the HCl or aqua regia and probably formed a larger Schottky barrier with the CdTe, thus producing greater band bending. If the CdTe was given an additional bias of up to +1 V during photoetching in dilute HCl or aqua regia, the photoetching currents were improved, as would be expected.

Photoetching of CdSe films prepared by spray pyrolysis, using neutral or weakly acidic NaCl solutions, has been described by Liu *et al.* (*56*). As before, the photocurrent and ff of CdSe photoelectrodes in S_x^{2-} improved after this treatment, and the surface of the CdSe became darker and less reflective. They found that the pH of the photoetching solution had to be <12, otherwise little photoetching occurred, probably because of the precipitation of Cd hydroxide or oxide species on the surface, which blocked further photocurrent flow.

Within the limitations of low enough pH to prevent the preceding reaction, but not so acidic that the CdSe is rapidly etched in the dark, the composition of the photoetching solution does not seem to be of critical importance. Likewise, the choice of counterelectrode for the photoetching is also not critical, as long as it can complete the flow of current. In fact, the substrate on which the semiconductor lies may itself act as a counterelectrode. In this case, simple immersion of the CdX electrode in the photoetching solution under illumination is sometimes sufficient to photoetch the electrode (R. Tenne, private communication).

Photoetching has also been used successfully with $CdIn_2Se_4$ photoanodes (*91*). This step is essential for this material to achieve a reasonable photoresponse. In this case, anodic bias was necessary for successful photoetching; this is probably for the same reason that it was required during photoetching of CdTe in HCl or aqua regia mentioned previously, namely, that the band bending at the semiconductor–photoetch electrolyte was insufficient without applied bias.

Photoetching has also been used for electroplated $CdSe_{0.65}Te_{0.35}$ electrodes (*47*). In this case, only photoetching was used [as was also the case for the thin films of Liu *et al.* (*56*).] This was found to give somewhat better results than chemical etching followed by photoetching. Photoetching is easier to control than chemical etching and in general removes less of the surface, which is preferable for very thin films.

B. Specific Ion and Associated Surface Treatments

Several ion treatments (or "dips") have been found to improve the photovoltages of CdX photoelectrodes in S_x^{2-}. In no case is the mechanism of such treatments clear, although suggestions have been forthcoming. Hodes *et al.* (*42*) employed a $ZnCl_2$ dip to increase the photovoltage of slurry-painted CdSe photoelectrodes. Tenne (*89*) showed that the effect of this Zn^{2+} ion treatment is to decrease the dark forward (cathodic) current, hence the increase in photovoltage, and explained this effect in an analogous way to metal–insulator–semiconductor (MIS) cells, in which the insulator in this case was the high-band-gap ZnS formed on immersion of the Zn^{2+} on the treated CdSe surface in the S_x^{2-} electrolyte. Russak and Reichman (*81*) also explained the improved photovoltage of a ZnSe-covered CdSe photoanode by analogy with an MIS device. This heterojunction was made by vacuum evaporation of ~100-Å ZnSe onto evaporated CdSe, followed by annealing in air, which changed the surface ZnSe into ZnO, as deduced from Auger surface analysis. After cycling in polysulfide solution, the ZnO was converted to ZnS. Thus the structure of the final photoelectrode was ZnS(~20 Å)–$Zn_xCd_{1-x}S_ySe_{1-x}$ (~80 Å, with x and y decreasing to zero in this range)–CdSe (bulk). Dark-current measurements of this photoelectrode in polysulfide showed a considerably reduced dark forward current and reverse saturation current compared with the same structure, after etching (which removed most of the Zn). The similarities of the photo *IV* characteristics of etched and nonetched devices after subtraction of the dark currents in the potential region near the fbp led Russak and Reichman to the conclusion that the increased open-circuit voltage of this structure was not a result of a shift in fbp; such a shift, they noted, does occur for a $ZnCl_2$-dipped CdSe photoelectrode.

A similar increase in photovoltages of Cd(Se, Te) alloy photoelectrodes has been obtained with a chromate dip (*40*). It is interesting to note that the zinc dip has no effect on Cd(Se, Te) photoelectrodes, when Se : Te $<$ 7 : 3. For higher Se : Te ratios, the zinc dip has an increasingly strong effect.

Tomkiewicz *et al.* (*104*) have found that a Ga^{3+}-ion dip has an effect on CdSe photoelectrodes similar to Zn^{2+}, both on photovoltage and on dark forward current. The rationale behind using Ga was that this metal has a high overpotential for H_2 evolution, and if the dark forward current was a result in part of this reaction, then Ga on the surface could reduce it.

Finally, Tenne (*89*) described the treatment (although not an ion treatment) of polycrystalline CdSe photoelectrodes with phenyl glycidyl ether and subsequent alkaline polymerization of the ether in the pores and pinholes of the CdSe layer (following a suggestion by A. Heller, Bell Laboratories). This treatment reduced the dark forward current at low forward bias, increasing the photocurrent in this region. At larger biases, the treatment had no effect, and thus the open-circuit photovoltage remained unchanged.

Acknowledgments

The numerous comments and suggestions for improvement of this manuscript by David Cahen are gratefully appreciated. Thanks are also accorded to Yehudit Mirovsky and Barry Miller for discussions and assistance with the thermodynamic arguments. Much of the work cited was supported by the United States–Israel Binational Science Foundation, Jerusalem.

References

1. Allen P. L., and Hickling, A., *Chem. Ind. (London)* **51,** 1558 (1954).
2. Bachmann, K. J., Hsu, F. S. L., Thiel, F. A., and Kasper, H. M., *J. Electron. Mater.* **6,** 431 (1977).
3. Baker, G. C., *J. Electrochem. Soc.* **113,** 1174 (1966).
4. Bard, A. J., and Wrighton, M. S., *J. Electrochem. Soc.* **124,** 1706 (1977).

5. Bhattacharya, R. N., and Pramanik, P., *J. Electrochem. Soc.* **129,** 332 (1982).
6. Bolts, J. M., Ellis, A. B., Legg, K. D., and Wrighton, M. S., *J. Am. Chem. Soc.* **99,** 4826 (1977).
7. Cahen, D., Hodes, G., and Manassen, J., *J. Electrochem. Soc.* **125,** 1623 (1978).
8. Cahen, D., Manassen, J., and Hodes, G., *Sol. Energy Mater.* **1,** 343 (1979).
9. Cahen, D., Hodes, G., Manassen, J., and Tenne, R., *ACS Symp. Ser. No.* **146,** 369 (1980).
10. Cahen, D., Vainas, B., and Vandenberg, J. M., *J. Electrochem. Soc.* **128,** 1484 (1981).
11. Chen, S., Russak, M. A., Witzke, H., Reichman, J., and Deb. S. K., U. S. Patent No. 4,172,925 (1979).
12. Danaher, W. J., and Lyons, L. E., *Nature (London)* **271,** 139 (1978).
13. Deng, E., and Li, P., *Acta Energiae Solaris Sinica* **2,** 182 (1981).
14. DeSilva, K. T. L., and Haneman, D., *J. Electrochem. Soc.* **127,** 1554 (1980).
15. DeSilva, K. T. L., and Haneman, D., *Thin Solid Films* **79,** L69 (1981).
16. Ellis, A. B., Kaiser, S. W., and Wrighton, M. S., *J. Am. Chem. Soc.* **98,** 1635 (1976).
17. Ellis, A. B., Kaiser, S. W., and Wrighton, M. S., *J. Am. Chem. Soc.* **98,** 6855 (1976).
18. Ellis, A. B., Kaiser, S. W., and Wrighton, M. S., *J. Am. Chem. Soc.* **98,** 6418 (1976).
19. Ellis, A. B., Kaiser, S. W., Bolts, J. M., and Wrighton, M. S., *J. Am. Chem. Soc.* **99,** 2839 (1977).
20. Folmer, J. C. W., Ph.D. Thesis, Univ. of Groningen, The Netherlands, 1981.
21. Frese, K. W., Jr., Madou, M. J., and Morrison, S. R., *J. Phys. Chem.* **84,** 3172 (1980).
22. Fujishima, A., and Honda, K., *Nature (London)* **238,** 37 (1972).
23. Gerischer, H., *J. Electroanal. Chem. Interfacial Electrochem.* **58,** 263 (1975).
24. Gerischer, H., *J. Electroanal. Chem. Interfacial Electrochem.* **82,** 133 (1977).
25. Gerischer, H., and Gobrecht, J., *Ber. Bunsenges. Phys. Chem.* **82,** 520 (1978).
26. Giggenbach, W., *Inorg. Chem.* **11,** 1201 (1972).
27. Ginley, D. S., and Butler, M. A., *J. Electrochem. Soc.* **125,** 1968 (1978).
28. Greiver, T. N., and Zaitseva, E. G., *Zh. Prikl. Khim. (Leningrad)* **40,** 1920 (1967).
29. Greiver, T. N., Zaitseva, E. G., and Sal'dau, E. P., *Zh. Prikl. Khim. (Leningrad)* **44,** 1689 (1971).
30. Hartler, N., Libert, J., and Teder, A., *Ind. Eng. Chem. Prod. Res. Dev.* **6,** 398 (1967).
31. Harzion, Z., Croitoru, N., and Gottesfeld, S., *Bull. Isr. Phys. Soc.* **26,** 61 (1980).
32. Harzion, Z., Croitoru, N., and Gottesfeld, S., *J. Electrochem. Soc.* **128,** 551 (1981).
33. Heller, A., Chang, K. C., and Miller, B., *J. Electrochem. Soc.* **124,** 697 (1977).
34. Heller, A., Schwartz, G. P., Vadimsky, R. G., Menezes, S., and Miller, B., *J. Electrochem. Soc.* **125,** 1156 (1978).
35. Heller, A., Chang, K. C., and Miller, B., *J. Am. Chem. Soc.* **100,** 684 (1978).
36. Hodes, G., Manassen, J., and Cahen, D., *Bull. Isr. Phys. Soc.* **22,** 100 (1976).
37. Hodes, G., Manassen, J., and Cahen, D., *Nature (London)* **261,** 403 (1976).
38. Hodes, G., *Abstr. Belgian-Israel Symp. Photochem. Photobiol. Energy Convers. Storage, Rehovot, Israel,* No. 15 (1978).
39. Hodes, G., Manassen, J., and Cahen, D., *J. Electrochem. Soc.* **127,** 544 (1980).
40. Hodes, G., *Nature (London)* **285,** 29 (1980).
41. Hodes, G., Manassen, J., and Cahen, D., *J. Am. Chem. Soc.* **102,** 5962 (1980).
42. Hodes, G., Cahen, D., Manassen, J., and David, M., *J. Electrochem. Soc.* **127,** 2252 (1980).
43. Hodes, G., "Final Report to the U.S.-Israel Binational Science Foundation, Jerusalem, Israel," Project No. 1314, 1980.
44. Hodes, G., Manassen, J., and Cahen, D., *Sol. Energy Mater.* **4,** 373 (1981).
45. Hodes, G., Manassen, J., and Cahen, D., *J. Electrochem. Soc.* **128,** 2325 (1981).

46. Hodes, G., Cahen, D., and Manassen, J., U.S. Patent No. 4,296,188 (1981).
47. Hodes, G., Manassen, J., Neagu, S., Cahen, D., and Mirovsky, J., *Thin Solid Films* **90,** 433 (1982).
48. Inoue, T., Watanabe, T., Fujishima, A., and Honda, K., *Proc. Electrochem. Soc.* **77-3,** 210 (1977).
49. Inoue, T., Watanabe, T., Fujishima, A., Honda, K., and Kohayakawa, K., *J. Electrochem. Soc.* **124,** 719 (1977).
50. Inoue, T., Watanabe, T., Fujishima, A., and Honda, K., *Bull. Chem. Soc. Jpn.* **52,** 1243 (1979).
51. Inoue, T., Kaneko, R., Fujishima, A., and Honda, K., Personal communication (1980).
52. Karas, B. R., and Ellis, A. B., *J. Am. Chem. Soc.* **102,** 968 (1980).
53. Lando, D., Manassen, J., Hodes, G., and Cahen, D., *J. Am. Chem. Soc.* **101,** 3969 (1979).
54. Lando, D., Ph.D. Dissertation, Feinberg Graduate School, Weizmann Inst. Science, Rehovot, Israel, 1982.
55. Latimer, W. M., "Oxidation Potentials," 2nd ed. Prentice-Hall, New York, 1952.
56. Liu, C. J., Olsen, J., Saunders, D. R., and Wang, J. H., *J. Electrochem. Soc.* **128,** 1224 (1981).
57. Lyden, J. K., Cohen, M. H., and Tomkiewicz, M., *Phys. Rev. Lett.* **47,** 961 (1981).
58. McCann, J. F., Skyllas Kazacos, M., and Haneman, D., *Nature (London)* **289,** 780 (1981).
59. Manassen, J., Cahen, D., Hodes, G., and Sofer, A., *Nature (London)* **263,** 97 (1976).
60. Manassen, J., Hodes, G., and Cahen, D., *Proc. Electrochem. Soc.* **77-6,** 110 (1977).
61. Manassen, J., Hodes, G., and Cahen, D., U.S. Patent No. 4,064,326 (1977).
62. Miller, B., and Heller, A., *Nature (London)* **262,** 680 (1976).
63. Miller, B., and Heller, A., *Proc. Electrochem. Soc.* **77-6,** 91 (1977).
64. Miller, B., Heller, A., Robbins, M., Menezes, S., Chang, K. C., and Thomson, J., Jr., *J. Electrochem. Soc.* **124,** 1019 (1977).
65. Minoura, H., Oki, T., and Tsuiki, M., *Chem. Lett.* p. 1279 (1976).
66. Minoura, H., Watanabe, T., Oki, T., and Tsuiki, M., *Jpn. J. Appl. Phys.* **16,** 865 (1977).
67. Minoura, H., Tsuiki, M., and Oki, T., *Ber. Bunsenges. Phys. Chem.* **81,** 588 (1977).
68. Minoura, H., and Tsuiki, M., *Chem. Lett.* p. 205 (1978).
69. Minoura, H., and Tsuiki, M., *Electrochim. Acta* **23,** 1377 (1978).
70. Mirovsky, Y., Cahen, D., Hodes, G., Tenne, R., and Giriat, W., *Sol. Energy Mater.* **4,** 169 (1981).
71. Mirovsky, Y., and Cahen, D., *Appl. Phys. Lett.* **40,** 727 (1982).
72. Mueller, N., and Tenne, R., *Appl. Phys. Lett.* **39,** 283 (1981).
73. Mueller, N., Tenne, R., and Cahen, D., *J. Electroanal. Chem. Interfacial Electrochem.* **130,** 373 (1981).
74. Noufi, R. N., Kohl, P. A., and Bard, A. J., *J. Electrochem. Soc.* **125,** 375 (1978).
75. Noufi, R. N., Kohl, P. A., Rogers, J. W., Jr., White, J. M., and Bard, A. J., *J. Electrochem. Soc.* **126,** 949 (1979).
76. Peter, L. M., *J. Electroanal. Chem. Interfacial Electrochem.* **98,** 49 (1979).
77. Rajeshwar, K., Thompson, L., Singh, P., Kainthla, R. C., and Chopra, K. L., *J. Electrochem. Soc.* **128,** 1744 (1981).
78. Reichman, J., and Russak, M. A., *J. Electrochem. Soc.* **128,** 2025 (1981).
79. Robbins, M., Bachmann, R. J., Lambrecht, V. G., Thiel, F. A., Thomson, J., Jr., Vadimsky, R. G., Menezes, S., Heller, A., and Miller, B., *J. Electrochem. Soc.* **125,** 831 (1978).

80. Russak, M. A., Reichman, J., Witzke, H., Deb, S. K., and Chen, S. N., *J. Electrochem. Soc.* **127,** 725 (1980).
81. Russak, M. A., and Reichman, J., *J. Electrochem. Soc.* **129,** 542 (1982).
82. Shay, J. L., and Wernick, J. H., "Ternary Chalcopyrite Semiconductors, Growth, Electronic Properties and Applications." Pergamon, Oxford, 1975.
83. Skotheim, T., *Appl. Phys. Lett.* **38,** 712 (1981).
84. Silberstein, R. P., Pollak, F. H., Lyden, J. K., and Tomkiewicz, M., *Phys. Rev. B: Solid State* **24,** 7397 (1981).
85. Streckert, H. H., Tong, J., and Ellis, A. B., *J. Am. Chem. Soc.* **104,** 581 (1982).
86. Tai, H., Nakashima, S., and Hori, S., *Phys. Status Solidi A* **30,** K115 (1975).
87. Tenne, R., and Hodes, G., *Appl. Phys. Lett.* **37,** 428 (1980).
88. Tenne, R., *Appl. Phys.* **25,** 13 (1981).
89. Tenne, R., *Ber. Bunsenges. Phys. Chem.* **85,** 413 (1981).
90. Tenne, R., *J. Electrochem. Soc.* **129,** 143 (1982).
91. Tenne, R., Lando, D., Mirovsky, Y., Mueller, N., Manassen, J., Cahen, D., and Hodes, G., *J. Electroanal. Chem. Interfacial Electrochem.* **143,** 103 (1983).
92. Tenne, R., Mirovsky, Y., Greenstein, Y., and Cahen, D., *J. Electrochem. Soc.* **129,** 1506 (1982).
93. Tomkiewicz, M., Lyden, J. K., Silberstein, R. P., and Pollak, F. H., *ACS Symp. Ser.* No. 146, 267 (1981).
94. Tributsch, H., *Ber. Bunsenges. Phys. Chem.* **81,** 361 (1977).
95. Tsuiki, M., Ueno, Y., Nakamura, T., and Minoura, H., *Chem. Lett.* p. 289 (1978).
96. Tygai, V. A., *Elektrokhimya* **1,** 387 (1965).
97. Vainas, B., Hodes, G., Manassen, J., and Cahen, D., *Appl. Phys. Lett.* **38,** 458 (1981).
98. Watanabe, T., Fujishima, A., and Honda, K., *Chem. Lett.,* p. 897 (1974).
99. Zaitseva, E. G., and Greiver, T. N., *Zh. Prikl. Khim. (Leningrad)* **40,** 1923 (1967).
100. Wilson, R. H., *J. Electrochem. Soc.* **126,** 1187 (1979).
101. Vázquez-López, Sánchez-Sinencio, F., Helman, J. S., Peña, J. L., Lastras-Martínez, A., Raccah, P. M., and Triboulet, R., *J. Appl. Phys.* **50,** 5391 (1979).
102. Mirovsky, Y., Ph.D. Dissertation, Weizmann Inst. of Science, Rehovot, 1983.
103. Hodes, G., Cahen, D., and Leamy, H. J., *J. Appl. Phys.,* in press.
104. Tomkiewicz, M., Ling, I., and Parsons, W. S., *J. Electrochem. Soc.* **129,** 2016 (1982).

14 Electrically Conductive Polymer Layers on Semiconductor Electrodes

Arthur J. Frank

Solar Energy Research Institute
Golden, Colorado

I. Introduction

Utilization of solar energy based on photoelectrochemical devices is an important but long-term option for meeting the future energy needs of mankind. Such systems could generate electrical power and synthesize fuels and desired chemicals from abundant, renewable resources such as water, nitrogen, and carbon dioxide. An economically viable photoelectrochemical solar cell requires solar energy conversion efficiencies be-

ISBN 0-12-295720-2

yond the 10% break-even point (*20*) as well as long-term stability. A major scientific hurdle associated with these goals is that high conversion efficiencies require narrow-band-gap semiconductors, and these materials are generally susceptible to photodegradation. This photoinstability is a fundamental materials problem that is particularly acute for *n*-type semiconductor electrodes (*1*, *32*).

A promising approach to alleviate the photodegradation problem at the semiconductor–electrolyte interface has emerged. The key component of this strategy is an electrically conductive polymer film that works synergistically with the redox electrolyte and/or a catalyst to stabilize the semiconductor surface from photodegradation and to promote desired reactions at the electrode surface. This chapter presents an examination of the viability of this approach and its application to the generation of electrical power and the photoelectrolysis of water. The emphasis in this article is on *n*-type semiconductors and electrically conductive films of polypyrrole, because this electrode–polymer combination has been the most thoroughly studied. The discussion covers how the physical properties and the chemical environment of the polymer together with the polymer–semiconductor interactions determine the electrode stability and the light-quantum conversion efficiency. The possible application of electrically conductive polymers to control surface states is also discussed. Although the focus is on polypyrrole and *n*-type semiconductors, the potential for applying electrically conductive polymers of various kinds to both *n*- and *p*-type semiconductors becomes evident. The application of electrically conductive polymers as semiconductors (*12*, *24*) is not discussed.

The basic principles of operation of photoelectrochemical devices are reviewed in Section II. In Section III, an analysis of the photoinstability problem and strategies to control it is presented, followed by further discussion of conductive polymers.

II. Photoelectrical Devices: Principles and Definitions

A. *Electrochemical Photovoltaic Cells*

Figure 1a shows a typical configuration of an electrochemical photovoltaic cell, the net product of which is electrical power. The cell consists of a reversible redox couple C^+/C, a counterelectrode, and an *n*-type photoanode. Cell configurations with a metal anode and *p*-type photocathode, *n*- and *p*-type electrodes in combination, or two photoelectrodes in series with a counterelectrode are also possible but will not be considered further

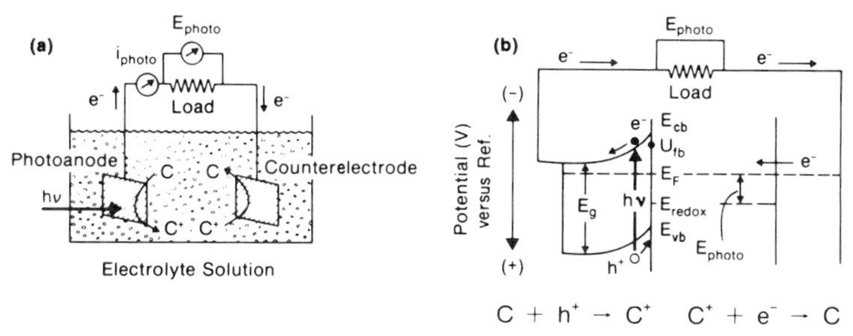

Fig. 1. An electrochemical photovoltaic cell (a) in operation to convert light to electricity and (b) the associated interface energetics.

here. Immersion of the *n*-type semiconductor into the redox electrolyte results in the interfacial energetics illustrated in Fig. 1b. The diagram shows the chemical potential of electrons in the semiconductor (the Fermi level) E_f in equilibrium with the chemical potential of electrons in the electrolyte (the redox potential) E_{redox}. At equilibrium, the conduction band (cb) and the valence band (vb) are bent by up to the difference between the flat-band potential U_{fb} and the redox potential of the electrolyte. The Fermi energy E_f equals U_{fb} when the semiconductor bands are flat (zero space charge in the semiconductor). E_g is the optical band gap of the semiconductor. On illumination of the semiconductor with band-gap radiation, electrons are promoted from the valence band to the conduction band, creating electron–hole pairs at or near the interface. Under the influence of the electric field in the space-charge region, the holes migrate to the semiconductor–electrolyte junction, where they accumulate. The holes will oxidize the semiconductor unless prohibited thermodynamically or unless scavenged by the reduced half of the redox couple C at the interface. Simultaneously, the electrons are driven into the bulk of the semiconductor, exit at the rear ohmic contact, flow through an external load, and then are injected at the counterelectrode–electrolyte interface to the adsorbed oxidized half of the redox couple C^+. The complementary reactions at the photoanode and the counterelectrode produce no net chemical change in the liquid electrolyte (i.e., $\Delta G = 0$), and electrical power is produced in an external load as a result of the optical excitation of the semiconductor.

The open-circuit photovoltage V_{oc} produced between the illuminated semiconductor and the metal electrode is equal to the difference between the Fermi level of the semiconductor and the redox potential of C^+/C. The photocurrent depends on the absorption coefficient of the semiconductor, width of the space charge region, hole-diffusion length, area of the electrode illuminated, photon energy, and radiation intensity. Under short-circuit conditions, the Fermi level of the semiconductor and the potential of the redox couple of the solution are equalized, and a net charge flows during illumination.

The maximum power output of the cell depends on the open-circuit photovoltage V_{oc}, the short-circuit photocurrent I_{sc}, and the fill factor ff defined as $I_{max} V_{max}/I_{sc} V_{oc}$, where I_{max} and V_{max} are the photocurrent and the photovoltage, respectively, at the maximum power point. The fill factor relates to how well the current–voltage curve approximates a rectangle. Efficient conversion of light energy to electrical power requires the optimization of the product of the external photovoltage and photocurrent, a criterion that is satisfied by materials with band gaps between 1.0 and 1.6 eV, that is, light absorption beginning between 1250 and 750 nm. Unfortunately, all known *n*-type semiconductors with band-gap energies in this range are susceptible to photodecomposition.

B. Photoelectrolysis Cells

In photoelectrolysis cells, at least two different redox couples are present in the electrolyte, and optical energy induces a current flow that produces a net chemical change in the electrolyte ($\Delta G \neq 0$). If ΔG of the net electrolyte reaction is negative, the process is exergonic, and light energy provides only the activation energy for the thermodynamically downhill reaction. If ΔG of the net reaction is positive, the process is endergonic, and light energy is converted into chemical energy.

In this subsection, emphasis is placed on the photoelectrolysis of water, which is normally an endergonic process. Several reviews (*2, 8, 33, 39, 60, 83*) have discussed the important energy parameters in the photoelectrolysis of water. In the absence of significant hot-carrier injection (*6, 62, 78*), the flat-band potential U_{fb} must lie negative of the redox potential of the H^+/H_2 or H_2O/H_2 couples plus any overpotential at the cathode (η_c) to generate H_2:

$$U_{fb} \leq -0.059 \text{ pH} - \eta_c \quad \text{versus NHE.} \tag{1}$$

Similarly, the valence-band edge must lie positive of the redox potentials of the O_2/H_2O or O_2/OH^- couples plus any overpotential at the semiconductor anode (η_a) to drive the O_2-evolution reaction:

$$E_{vb} \geq 1.23 - 0.059 \text{ pH} - \eta_a \quad \text{versus NHE.} \tag{2}$$

Figure 2 illustrates the interface energetics for two different photoelectrolysis cells consisting of a photoanode and a metal cathode. The photoelectrochemical cell depicted in Fig. 2a satisfies Eqs. (1) and (2). The

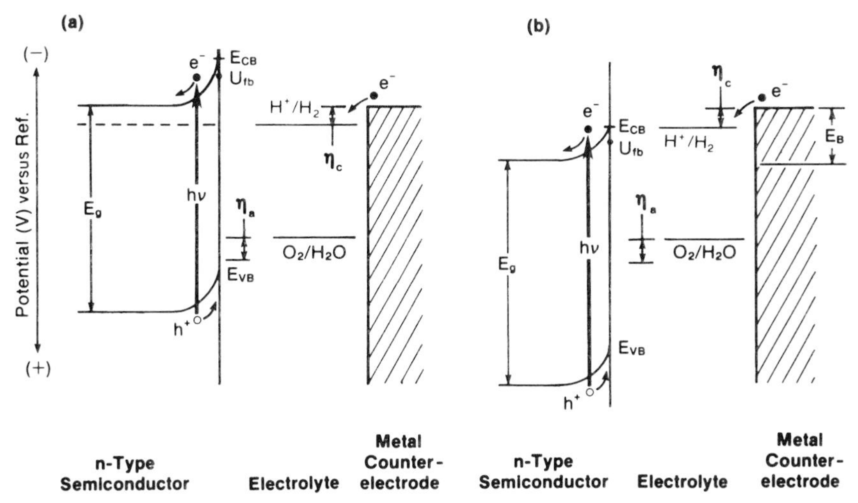

Fig. 2. Energy-level diagrams for semiconductor–metal water-photoelectrolysis cells. (a) The flat-band potential and valence-band edge bracket the redox potentials of the water half-reactions. (b) An external bias E_B is applied to raise the Fermi level in the metal counter electrode above the H^+/H_2 potential.

holes at E_{vb} are capable of oxidizing water and the electrons at the metal cathode are capable of reducing water with no external bias. Several relatively stable semiconductors such as $SrTiO_3$, $KTaO_3$, and Nb_2O_5 require no external bias to generate H_2 and O_2. However, these semiconducting oxides have large band gaps (3.4–3.5 eV; 365–354 nm) and absorb very little of the terrestrial solar spectrum. In contrast, several nonoxide *n*-type semiconductors with suitable band gaps for high solar-light absorption are also capable of generating H_2 and O_2 simultaneously with zero or near-zero applied bias. However, these materials undergo photodecomposition, as discussed further in Section III.

Figure 2b illustrates a photoelectrolysis cell for which Eq. (2) is satisfied but not Eq. (1). A positive bias E_B is required to raise the Fermi level in the metal cathode negative of $E(H^+/H_2)$ and to provide the overpotential (η_c) to sustain the water reduction at a reasonable rate. The situation portrayed in Fig. 2b typifies the stable semiconducting oxides. Their stability in aqueous solution is because of the very positive redox energy of their valence band owing to the high electron affinity of the oxygen atom. In general, holes emerging at the valence-band edge of semiconducting oxides have substantially more energy than that required to sustain O_2 evolution from water. The excess energy is thus wasted and diminishes the efficiency of the device. Moreover, because of the positive valence-band edge, relatively large biases are required to reduce water. Thus iron oxide with a band gap of 2.2 eV requires a relatively large external bias (~700 mV) to photoelectrolyze water (*37*).

The instantaneous power efficiency η of a water photoelectrolysis cell can be defined (*48*) as

$$\eta = (E - E_B)i/P_{h\nu}, \tag{3}$$

where E is the potential for the decomposition of water, E_B the bias voltage, i the current density at the bias voltage, and $P_{h\nu}$ the power density of the incident light. The decomposition potential of water E can be identified as the effective redox level of the electrolyte and can be calculated from the Nernst equation

$$E = E° - (RT/F) \ln(a_{H_2}a_{O_2}), \tag{4}$$

where $E°$ is the standard potential of the cell and the "a" terms are the activities of H_2 and O_2. Equation (4) determines the electrolyte redox level for the reaction

$$2H_2O \longrightarrow 2H_2 + O_2 \tag{5}$$

and will vary with the product of the H_2 and O_2 activities. In the dark at equilibrium, E determines the Fermi level of the metal cathode and, hence, that of the semiconductor anode. Variation of the relative concentrations of H_2 and O_2 at the metal cathode will thus affect the amount of band bending and thus the photovoltage E_{photo}. In general, the photovoltage, which depends on the difference between the conduction-band edge and the energy level of the redox couple, is not the potential available for the water-oxidation reaction. The thermodynamic driving force for the oxidation of water is determined by the difference in the potential between the valence-band edge and the O_2/H_2O couple. What is important in photoelectrolysis is maximizing the short-circuit photocurrent.

Thermodynamically, the electrolysis of water at standard conditions requires 1.23 eV per mole, and depending on the current density of the cell, a minimum of about 0.3–0.4 eV additional energy is required to sustain the reaction. Thus the minimum band gap of a semiconductor is at least 1.5 eV. The maximum band gap to achieve solar conversion efficiencies beyond the 10% break-even point of an economically viable cell is less than 2.4 eV for a single photoanode with a metal cathode (*33*). In principle, higher conversion efficiencies are possible when a photoanode is combined with a suitably matched photocathode (*59, 85*). The conduction-band edge of the photocathode must lie negative of $E(H^+/H_2)$ plus any overpotential, and the sum of the band gaps must probably be larger than 1.5 eV. In addition, the valence-band edge of the photocathode must be positive of the conduction-band potential of the photoanode. Alternatively, two-photo-electrodes with different band gaps can be operated in series (*79*).

III. Instability of *n*-Type Semiconductor Electrodes

A. *Models*

Gerischer (*32*) and Bard and Wrighton (*1*) have developed a thermodynamic model for predicting the stability of a photoelectrode. The anodic decomposition potential of the semiconductor is calculated and compared with the potential of the valence-band edge. Total thermodynamic stability of the electrode is expected when the oxidative decomposition potential lies positive of the valence-band edge. To date no known semiconductor obeys this criterion. Instead, all semiconductors are predicted to exhibit instability toward anodic photodecomposition. This expectation is based on the thermodynamic analysis in which the redox potential of the decomposition reaction always seems to lie negative of the electrochemical potential of the valence-band edge. Whether or not an electrode is photostable then depends on the competitive reaction rates of the thermodynamically possible reactions: the semiconductor decomposition reaction and the electrolyte reactions.

The mechanism for the photoanodic decomposition reaction can be very complex, as suggested by the examples compiled in Table I. Photoanodic instability of the semiconductor entails ionic dissolution, gas evolution, or formation of a new phase of the electrode. In each case, solvation effects and multiple-hole reactions are involved. For illustration, let us consider the photoanodic destruction of CdS. The valance band of CdS is composed of overlapping S^{2-} bonding *p*-type orbitals and the conduction band is made up of Cd^{2+} antibonding *s*-type orbitals. Pho-

TABLE I

Examples of Photoanodic Decomposition Reactions of Various Semiconductor Electrodes

Semiconductor	Decomposition photoanodic process	References
Si	$Si + 4h^+ + 2H_2O \longrightarrow SiO_2 + 4H^+$	*50*
GaAs	$GaAs + 6h^+ + 5H_2O \longrightarrow Ga(OH)_3 + HAsO_2 + 6H^+$	*51*
GaP	$GaP + 6h^+ + 6H_2O \longrightarrow Ga(OH)_3 + H_3PO_3 + 6H^+$	*51*
CdS	$CdS + 2h^+ \longrightarrow Cd^{2+} + S$	*81*
CdSe	$CdSe + 2h^+ \longrightarrow Cd^{2+} + Se$	*81*
MoS_2	$MoS_2 + 18h^+ + 12H_2O \longrightarrow MoO_4^{2-} + 2SO_4^{2-} + 24H^+$	*63, 76*
WO_3	$WO_3 + 2h^+ + 2H_2O \longrightarrow WO_4^{2-} + \frac{1}{2}O_2 + 4H^+$	*1*
TiO_2	$TiO_2 + 2h^+ \longrightarrow TiO^{2+} + \frac{1}{2}O_2$	*1*

toexcitation of CdS promotes electrons from the bonding valence band to the antibonding conduction band. The resulting positive lattice vacancies or holes formed in the space charge region are driven to the semiconductor interface, leaving surface Cd atoms with a net positive charge that can partially coordinate with solvent dipoles. The energy of the system may thus be lowered by complete solvation of the Cd^{2+} ions with concomitant restructuring of the electrode surface.

The rate of photodecomposition depends on the steady-state concentration of holes at the interface, which in turn is determined by the difference in the rates of charge production and consumption. Competition between the redox reactions, the surface recombination of holes and electrons, and the corrosion of the electrode principally account for the decline in the hole population. The rate of interfacial charge transfer will depend on the overlap between the distribution of energy levels of the holes and the distribution of available energy levels of the redox electrolyte (e.g., O_2/H_2O). Holes may undergo internal transitions to surface states that can participate in the solvent-assisted lattice decomposition reaction or in the interfacial charge transfer to the redox electrolyte. Hole removal through surface recombination with the majority charge carriers will also reduce the steady-state hole concentration at the interface. The specific decomposition rate will depend on the semiconductor and the nature and the composition of the electrolyte (solvent, counterions, or complexing ligands). Such kinetic factors, in addition to thermodynamic considerations, determine the relative stability of the semiconductor.

B. Approaches to Prevent Photodegradation

To protect the semiconductor surface from photodecomposition, strategies have been used, alone or in combination, based on limiting the contact between the semiconductor and the aqueous solvent, that promote rapid interfacial charge transfer to redox species, or that alter the interfacial energetics to disfavor electrode decomposition with respect to the desired redox reaction. The effectiveness of the approaches depends on manipulation of both thermodynamic and kinetic factors.

1. Electrochemical Photovoltaic Cells

Figure 3 illustrates unstable and stable energy situations for semiconductors in electrochemical photovoltaic cells. In Figure 3a, the redox potential of the electrode decomposition reaction lies negative of the electrochemical potential of the redox reaction in the electrolyte. This means that the reaction of the photogenerated holes at the valence-band edge

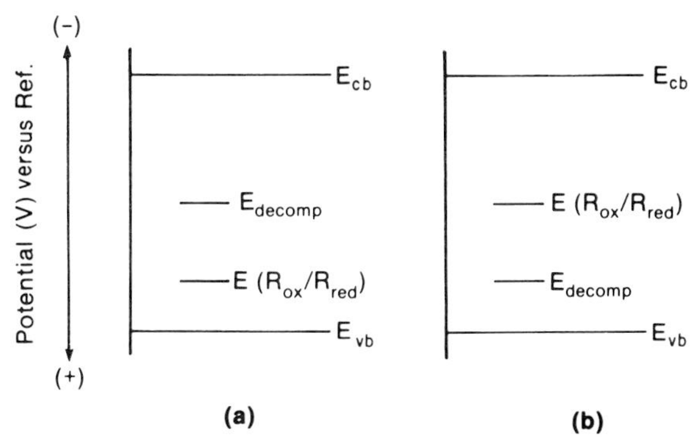

Fig. 3. Thermodynamics of oxidative photodecomposition of *n*-type semiconductor electrodes in electrochemical photovoltaic cells for relatively (a) unstable and (b) stable situations. E_{decomp} is the decomposition potential for the oxidation of the semiconductor and $E(R_{ox}/R_{red})$ the electrochemical potential of the R_{ox}/R_{red} couple.

with the semiconductor lattice is thermodynamically more favorable than the reaction of the holes with reductant R_{red} in the electrolyte. Barring kinetic inhibition for the decomposition reaction, the electrode is expected to exhibit instability. Figure 3b illustrates the converse situation in which hole transfer to the redox couple is thermodynamically more favorable than self-oxidation of the semiconductor. The important implication deduced from this diagram is that an unstable electrode in a given electrolyte may be stabilized with the addition of a reversible redox couple having more favorable energetics than the decomposition reaction. This thermodynamic expectation is kinetically fulfilled by many semiconductors, of which CdS is representative. Hole transfer to the reduced form of the polychalcogenide S_n^{2-} is thermodynamically more favorable than the decomposition reaction of CdS (*1, 32*). The mechanism of stabilization is, however, more complicated than a simple consideration of energetics would suggest. Among the factors implicated are preferential adsorption of S_n^{2-} on CdS and the concomitant shielding of the surface atoms from the solvent; the common-ion effect; the favorable kinetic behavior of the chalcogenide redox couple, which facilitates hole abstraction; and the surface morphology (*9, 10, 21, 34*). The addition of one or more polychalcogenide ions (S_n^{2-}, Se_n^{2-}, or Te_n^{2-}) has been used to stabilize CdS, CdSe, CdTe, GaAs, and InP. Various other reducing agents [I_3^-, $Fe(CN)_6^{4-}$, Fe^{2+}, Ce^{3+}, etc.] have also been employed to scavenge the photogenerated holes at rates that suppress anodic decomposition. In optimizing the cell stability, consideration must be given to a number of factors (*8*), such as the open-circuit photovoltage, which depends on the separation between the conduction-band edge and the redox potential; the presence of oxygen, which can lead to oxide formation on the electrode surface; and the oxidative instability of the redox couple.

Other methods for suppression of photodecomposition have used nonaqueous solvents, molten salts, and high concentrations of electrolytes. In part, like preferential adsorption of redox reagents to the electrode surface, these methods are intended to reduce the solvation effects of water (see Table I) and thus shift the photodecomposition potential to positive values. Other approaches to stabilize the semiconductor have relied on a high concentration of a redox couple in the electrolyte, specific adsorption of a species acting as a charge relay, or covalent attachment of a charge mediator to the electrode surface. These strategies are designed to facilitate charge removal and thus reduce the steady-state population of photogenerated holes at the interface.

2. *Photoelectrolysis Cells*

Figure 4 compares unstable and stable energy relationships for semiconductors in photoelectrolysis cells. Figure 4a represents an unstable condition in which the decomposition potential of the semiconductor lies negative of the electrochemical potential of the water-oxidation reaction. Thus the thermodynamic driving force for the reaction of the photogenerated holes at the valence-band edge with the semiconductor lattice is greater than that with water. This relatively unstable situation typifies the nonoxide semiconductors such as the II–VI and III–V group compounds and several oxides such as ZnO. Figure 4b illustrates the stable condition where thermodynamics favor water oxidation relative to the decomposition reaction of the semiconductor. Several semiconducting oxides such as TiO_2 are representative of Fig. 4b.

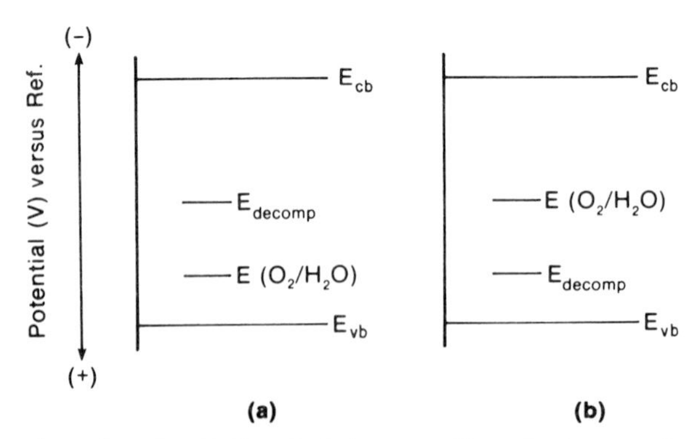

Fig. 4. Thermodynamics of oxidative photodecomposition of *n*-type semiconductor electrodes in water-photoelectrolysis cells for relatively (a) unstable and (b) stable situations. E_{decomp} is the decomposition potential for the oxidation of the semiconductor and $E(O_2/H_2O)$ the electrochemical potential of the O_2/H_2O couple.

The range of possible approaches for suppressing photocorrosion in electrochemical photovoltaic cells is much broader than that for photoelectrolysis cells where the solvent must include water, and the redox couples (H^+/H_2 and O_2/H_2O or H_2O/H_2 and O_2/OH^-) are unalterable. In this subsection the focus is on strategies to control the photoinstability of the semiconductor–electrolyte interface using surface coating techniques.

To stabilize the semiconductor surface from photodecomposition, noncorroding layers of metals or relatively stable semiconductor films have been deposited onto the electrode surface. Nakato *et al.* (*54*) and Harris *et al.* (*38*) reported that continuous metal films that block solvent penetration can protect *n*-type GaP electrodes from photocorrosion. However, if the films are too thick for the photogenerated holes to penetrate without being scattered, they assume the Fermi energy of the metal. Then the system is equivalent to a metal electrolysis electrode in series with a metal–semiconductor Schottky barrier. In such a system, the processes at the metal–semiconductor junction control the photovoltage and not the electrolytic reactions. In general, a bias is required to drive the water oxidation reaction because of insufficient photovoltage developed at the metal–semiconductor junction. In contrast, Harris and Hugo (*40*) observed that an ohmic contact is formed between a continuous Pt film and *n*-type Si that leads to loss of photoactivity.

Discontinuous metal coatings affect the semiconductor differently than continuous metal films in that the electrolyte contacts the semiconductor, and electronic equilibrium directly between the electrode and the redox electrolyte is possible. Discontinuous gold films do not seem to protect *n*-type GaP from photocorrosion (*38*), contrary to an earlier report (*54*). However, if the thin metal layer is also highly catalytic, rapid removal and trapping of holes via chemical processes can impart protection to the semiconductor. Thus Frank and Honda (*27, 28*) showed that RuO_2 powder dispersed on CdS electrodes provides partial protection against photocorrosion while providing for the catalysis of water. In contrast, Gissler and McEvoy (*35*) reported that a thick radiofrequency- (rf)-sputtered RuO_2 film forms a Schottky barrier with the CdS electrode, and the only role of the electrolyte is to make contact with the metal; no water photoelectrolysis occurs. In the case of a RuO_2 film, the thickness of the metal oxide layer, the oxidation state of ruthenium, and the extent of dispersion and hydration of the metal oxide determine not only its ability to protect the semiconductor surface, but also its catalytic activity.

Corrosion-resistant wide-band-gap oxide semiconductor (TiO_2 and titanates mostly) coatings over narrow-band-gap *n*-type semiconductors such as GaAs, GaAlAs, Si, CdS, GaP, and InP have been shown to impart protection from photodecomposition (*46, 75*). One of two problems is associated with the use of optically transparent wide-band-gap semiconducting oxide coatings: either a thick film blocks charge transmission or a thin film still allows photocorrosion. The charge transmission through a thick film is impeded presumably because the valence-band electronic energy of the stable semiconductor oxide lies at an appreciably more positive potential than that of the nonoxide semiconductor substrate. Thus a large energy barrier between the valence bands effectively blocks hole conduction through the film. One solution to this problem is to create a heterojunction between the inner and outer semiconductor (*79*) that also serves as a protective barrier against photodecomposition. For optimal efficiency, both semiconductors absorb the same number of photons. The photovoltage produced at the heterojunction adds in series with the photovoltage generated at the semiconductor–electrolyte junction. The additional photovoltage created at the heterojunction lowers any bias requirement with respect to the single semiconductor cell. The principle of the heterojunction electrode for photoelectrolysis has been demonstrated for electrodes consisting of n-TiO_2 on n-GaAs, n-SnO_2 on n-Si, n-Fe_2O_3 on n-Si, and n-TiO_2 on n-Fe_2O_3 (*49, 53, 61*).

Wrighton *et al.* (*82, 84*) have shown that chemical bonding of electroactive polymers to the semiconductor surface affects the interfacial charge-transfer kinetics such that the less thermodynamically favored redox reaction in the electrolyte predominates over the thermodynamically favored semiconductor decomposition reaction. For photoelectrolysis cells, covalently derivatized electrodes offer the possibility of preventing self-oxidation of n-type semiconductors and affording water oxidation. To date, however, emphasis has been placed on improving the catalytic properties of p-type electrodes, where photocorrosion by reductive processes is not a major problem. The overvoltage for the evolution of hydrogen from p-type electrode surfaces is normally very large. It has been demonstrated, however, that the catalytic property of a p-type Si photocathode is markedly enhanced for hydrogen evolution when a viologen derivative is chemically bonded to the electrode surface and Pt particles are dispersed within the polymer matrix (*5, 19*). The viologen mediates the transfer of the photogenerated electron to H^+ by the platinum to form H_2. A thin platinum coating directly on the p-type Si surface also improves dramatically the catalytic performance of the electrode (*19, 55*).

Modification of the semiconductor surface with thin films of electrically conductive polymers is an alternative to electroactive polymers, that is, polymers with electroactive pendant groups. The conceptual distinction is the emphasis on charge conduction, which is generally much higher in electrically conductive polymers than in typical electroactive polymers. Other differences between conductive and electroactive polymers are pointed out in Section V,B. The point here is that electrically conductive polymers provide a conducting pathway for the rapid channeling of the photogenerated charge in the semiconductor to the redox electrolyte before photodecomposition occurs. In the subsequent sections, we consider the synthesis and applications of electrically conductive polymer films of polypyrrole to photoelectrochemical solar energy conversion.

IV. Experimental Considerations

A. *Preparation of Polypyrrole*

The synthesis of polypyrrole on semiconductor electrodes is, with some variations, analogous to the electrochemical procedure developed for coating platinum electrodes (*44*). Typically in our work (*57*), three electrodes are positioned in a three-compartment cell containing 1.0 *M* pyrrole and 0.3 *M* electrolyte in an inert-gas-purged acetonitrile solution of 0.1–1.0% water. Thin films of the polymer are deposited on the surface of an *n*-type semiconductor electrode at constant potential between 0.6 and 0.9 V versus SCE under illumination from a tungsten–halogen lamp. Initially, the photocurrent is low, but it gradually increases with growth and oxidation of the electrode-bound polymer. The maximum photocurrent density is generally maintained between 100 and 200 μA cm^{-2} by adjusting the electrode potential or the light intensity.

The mechanism for the electrochemical polymerization of pyrrole on platinum electrodes in acetonitrile solution in the presence of BF_4^- has been discussed elsewhere (*17*). Briefly, the polymerization is believed to involve the production of a labile pyrrole π radical-cation, which then reacts with neighboring pyrrole species. Tetrafluoroborate may also play a role in the initiation of the polymerization (*64*). A series of oxidation and deprotonation steps occur at a rate exceeding the diffusion of reactive intermediates away from the electrode surface. The net effect is that the polymer is formed directly on the electrode surface rather than in the bulk of the electrolyte. Furthermore, because the oxidation of polypyrrole ($E° \approx -0.2$ V versus SCE) occurs more readily than that of the pyrrole monomer ($E_p \approx 1.2$ V), the polymer is formed in the oxidized (conducting) state **II,** whereas the neutral form is in the nonconducting state **I**.

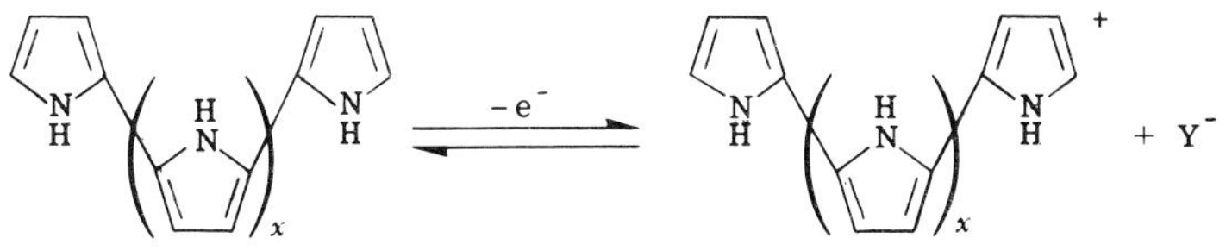

B. Growth-Condition Effects

1. Stoichiometry

The actual stoichiometry of "polypyrrole" depends on the growth conditions, but no comparative study of reaction conditions has yet been reported. When the electrolyte is tetraethylammonium fluoroborate, elementary and electrochemical analyses (*14, 44*) suggest one positive charge for every four pyrroles; that is, x of structure **II** is two. When silver perchlorate is used as the electrolyte in dry acetonitrile, the stoichiometry is consistent with one positive charge for every three pyrroles and $x = 1$ (*73*). When the film is prepared in aqueous sulfuric acid solution, 3–4 pyrrole units for every positive charge are implicated (*18*). Infrared studies (*73*) indicating the existence of C—H bonds in the polymer film suggest that the ring structure is not completely aromatic. Moreover, in aqueous or oxygen-containing electrolytes, it is reasonable to expect some oxygenation of the pyrrole units, principally at the β position or at the terminal α position because evidence (*31*) indicates that pyrrole is linked primarily in the α position in the polymer.

2. Effects of the Electrode Surface on the Polymer

The substrate surface affects the growth pattern of the polymer (*25*). The polymer morphology is important in determining the extent of protection imparted to the semiconductor substrate; a uniform, compact film coating is desirable for electrode stability. In Figure 5, a scanning electron micrograph reveals that the topography of the growing surface of a 4000-Å-thick film on platinum is more compact and featureless than it is on In–SnO_2. On In–SnO_2 the film has an uneven surface of hillocks resulting from the preferred growth. The difference in the morphologies of the polymer grown on the two substrates may be associated with variations in the uniformity of the charge density across the electrode surfaces. The surface of In–SnO_2 is irregular; these irregularities produce local regions of different resistivities that exist even under constant-current conditions. Regions of high-current densities produce higher rates of polymerization than those of low-current densities. The thickness of the polymer film thus varies nonuniformly over the electrode surface. Such morphological change as a function of local variations in the charge density is consistent with the view that the polymerization of pyrrole does not entail ion or free-radical chain processes.

The polarity of the electrode surface affects the adhesion of the polymer (*25*). For example, the binding strength of polypyrrole to antimony-

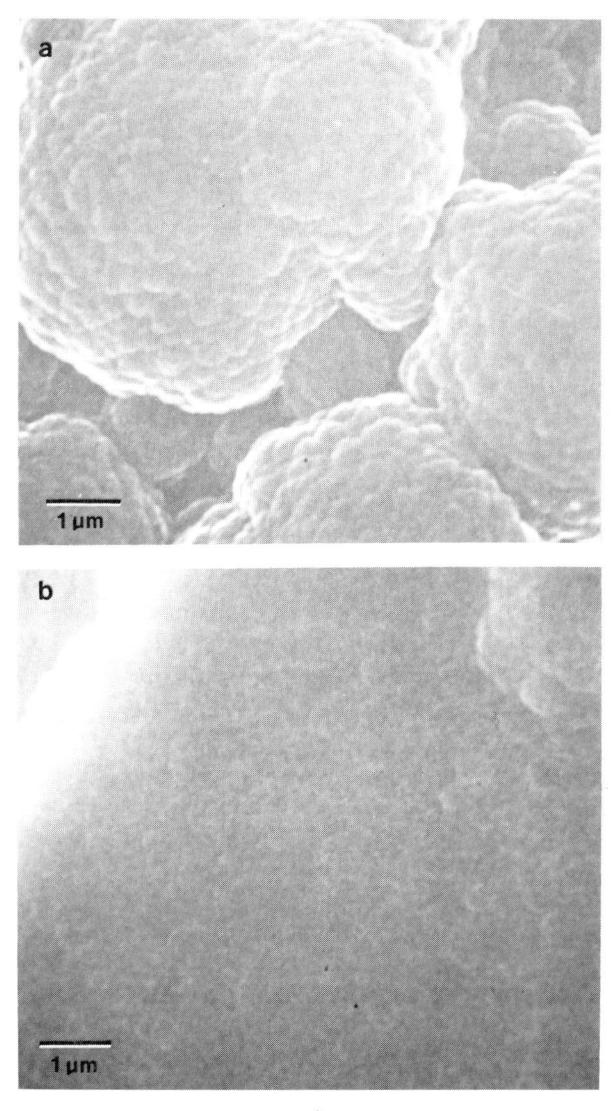

Fig. 5. Scanning electron micrograph of 4000-Å polypyrrole films grown on (a) platinum and (b) indium tin oxide. From Frank (*25*).

doped tin oxide is increased after an acid treatment that produces a clean and relatively hydrophilic electrode surface. Without the acid treatment, free-standing films are readily produced from the relatively hydrophobic electrode surface.

3. *Effects of Water on the Polymer and Semiconductor*

The amount of water present in the acetonitrile affects the uniformity and adherence of the polymer to the electrode surface. Increasing the water concentration improves the uniformity and the adherence of polypyrrole films to platinum electrodes (*44*) and to single-crystal *n*-type Si electrodes overlaid with a thin coating of platinum (*70*). Water also, however, increases the rate of photodecomposition of the semiconductor, and thus a certain amount of compromise is required between optimizing the growth conditions of the polymer and minimizing the corrosion reaction. With some corrosion of the electrode, for instance, we have deposited uniform polypyrrole films on CdS in 0.1 *M* H_2SO_4. In correlating the charge density passed with the film thickness, the contribution from corrosion of the semiconductor to the photocurrent must be taken into account; this is not the case for the synthesis of polypyrrole on noncorroding metal electrodes (*14*).

Normally, the polymer film is unstable in aqueous electrolytes under oxidative conditions, such as at electrode potentials greater than 0.6 V versus SCE (*7, 24*), or in the presence of halogens (*7*). However, as pointed out in Section VIII, we have shown that it is possible to improve the stability of the polymer under oxidative conditions (*13, 27, 28*).

V. Transport Properties

A. *Solid State*

The specific conductivity σ (in $\Omega^{-1}\ cm^{-1}$) of a material is given by the product of the concentration of charge carriers n (in cm^{-3}), their electronic charge q, and their mobility μ (in $cm^2\ s^{-1}\ V^{-1}$), which is a measure of the average length of time they can move before being scattered:

$$\sigma = q\mu n. \tag{6}$$

The electrical conductivities of polypyrrole and various common materials at room temperature are displayed in Fig. 6. Although conductivity alone is insufficient to characterize a material as a metal, a semiconductor, or an insulator, metals are normally associated with high conductivity, $\sigma > 10^2\ \Omega^{-1}\ cm^{-1}$, and insulators with low conductivity, $\sigma < 10^{-7}\ \Omega^{-1}\ cm^{-1}$; the conductivity of semiconductors typically falls somewhere between 10^{-7} and $10^2\ \Omega^{-1}\ cm^{-1}$. The conductivity of polypyrrole can be adjusted to span a range of $\sim 10^{13}$ from that of a metal to an insulator. At

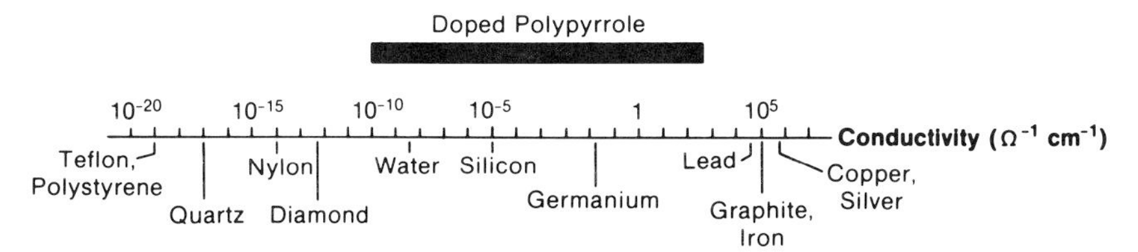

Fig. 6. Room-temperature conductivities of polypyrrole and common materials.

the low end, the conductivity of neutral polypyrrole is $\sim 10^{-10}\ \Omega^{-1}\ \text{cm}^{-1}$, whereas the upper limit is of the order of $10^3\ \Omega^{-1}\ \text{cm}^{-1}$ (*73*). To obtain good conduction, both high mobility and large carrier concentrations are required. In general, the concentrations of carriers in organic and inorganic semiconductors are comparable, between 10^{12} and $10^{19}\ \text{cm}^{-3}$, which is markedly below that of metals, 10^{20}–$10^{23}\ \text{cm}^{-3}$. The limiting factor is generally the mobility, which is much lower in polymers than in inorganic materials.

Little is known about the electronic conduction mechanism of polypyrrole, although theoretical calculations (*23, 36*) suggest that charge transport proceeds through the extended π system on the carbon backbone and not through the nitrogen. Intuition suggests that long or interconnecting π segments should favor efficient charge transport, and conversely, short nonconnecting π segments should disfavor it. Charge transfer between chains or highly conductive regions may involve either a hopping or a tunneling mechanism, although only a hopping model has been considered for polypyrrole (*74*). Because chemical oxidation of the pyrrole rings will interrupt the delocalization of the π-electron system, it is expected to affect adversely the electrical conductivity of the polymer. The good chemical stability and the delocalized, extended states in polypyrrole have been attributed to the very small excess charge on the α carbons (*23*).

Several studies (*15, 17, 66*) have shown that the nature of the counterion significantly influences the electrical conductivity of the dry polymer as illustrated in Table II. The data in the table (*17, 66*) reveal an increase in conductivity of nearly five orders of magnitude when ClO_4^- is substituted for $HC_2O_4^-$. Both the polarizability and the rotational–vibrational dynamics of the anion are considered to be important in determining the electrical conductivity of the oxidized polymer (*66*). Although no mechanism has been reported, the anions may affect the electrical conductivity by stabilizing the delocalized positive charge of the oxidized polymer and by forming a conducting bridge for intersegmental charge transport. Such factors may also play a role in the charge-transfer kinetics of the film in

TABLE II

Effect of Counterion on Conductivity σ of Polypyrrole at Room Temperature[a]

Counterion	σ ($\Omega^{-1}cm^{-1}$)
ClO_4^-	60–200
BF_4^-, PF_6^-, AsF_6^-	30–100
$CH_3C_6H_4SO_3^-$, $BrC_6H_4SO_3^-$	20–100
CF_3COO^-	12
HSO_4^-, $CF_3SO_3^-$	0.3–1
FSO_3^-	10^{-2}
$HC_2O_4^-$	10^{-3}–10^{-2}

[a] Data from Diaz and Kanazawa (*17*) and Salmon *et al.* (*66*).

solution. Neutralization of the polymer results in the complete removal of the anion (*17*).

B. Photoelectrochemical and Electrochemical Considerations

Although the conductivity of polypyrrole is metallic-like, measurements (*24, 30*) suggest that the polymer displays properties more characteristic of a degenerate semiconductor than that of a metal. This conclusion is based on optical and differential capacitance measurements.

The optical spectrum of the oxidized form of polypyrrole tetrafluoroborate (*24*) is displayed in Fig. 7. The positions of the spectral bands and the associated values of the absorption coefficients are consistent with the results of Kanazawa *et al.* (*44*), although other studies indicate that the

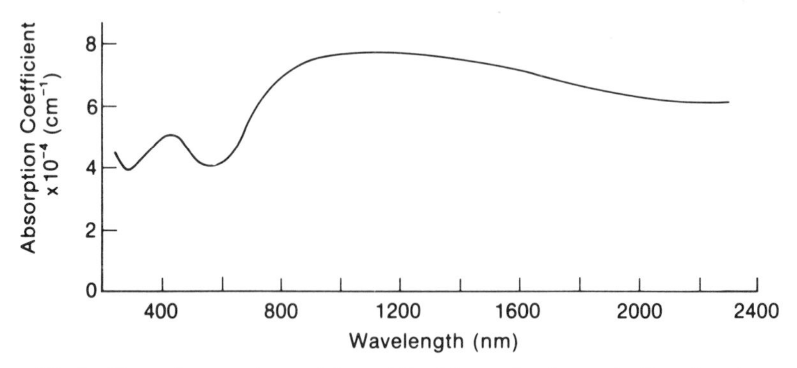

Fig. 7. Absorption spectrum of the oxidized form of polypyrrole fluoroborate.

spectral features are influenced by various factors such as the degree of oxidation of the film, its exposure to oxygen, and the conditions of synthesis (*45, 65, 73*). The band at ~420 nm has been assigned to an interband π–π^* transition (*23, 44, 73*). The presence of the near-ir band at ~1000 nm depends on the doping density; in the neutral film this band is absent (*17, 73*). The presence of such absorption bands is more characteristic of a semiconductor than a metal.

Further evidence of a semiconductor-like nature of the polymer is found in Mott–Schottky measurements of the differential capacitance of polypyrrole on platinum electrodes (*24, 30*). These measurements show a linear relationship between the reciprocal of the square of the capacitance, $1/C^2$, and the voltage between −0.5 and 0.4 V at 50 Hz (Fig. 8). Such behavior is consistent with polypyrrole having a depletion region and exhibiting *p*-type conduction. Thermopower measurements also indicate *p*-type conduction (*44*). Although the *p*-type conduction is consistent with the polymer's being partially oxidized, the existence of a depletion region is a salient feature of semiconductors and not of metals. To exhibit a depletion or space-charge region, the dopant ions in the polymer must be relatively stationary under the conditions of measurement (*30*). With time the relationship between the differential capacitance and voltage changes and becomes more complex. Bull *et al.* (*7*) have studied polypyrrole on tantalum electrodes and have offered an explanation of their complex capacitance–voltage and impedance data based on circuit theory.

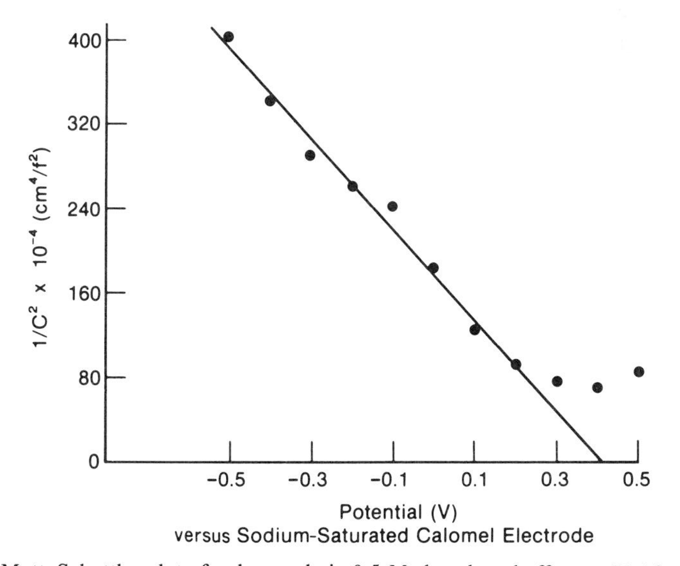

Fig. 8. Mott–Schottky plot of polypyrrole in 0.5 *M* phosphate buffer at pH 6 for 50 Hz.

Electrochemical studies (*16, 66*) have found that the nature of the anion influences the charge-transfer kinetics of the film, although the basis for the observed effects has not been discussed. Ion diffusion is the limiting factor in the redox reactions of the polymer film (*17, 66*). The ionic diffusion coefficient and the free-volume content of the polymer determine the ion velocity in electroactive polymers (*67*) and, probably, in polypyrrole. The greater the free-volume content or intersegmental spacing, the more rapid is the ion motion. In electroactive polymers, a large free-volume content is generally associated with slow charge transport because of the effect of distance on electron hopping between pendant groups. Rapid ion motion and fast charge transport are, to some extent, mutually exclusive in many electroactive polymers (*67*). In the extended π-chain polymers with large open structures, however, it may be possible to allow both rapid ion motion and charge transport.

The free-volume content of polypyrrole is affected by the charge on the polymer. Polypyrrole is more swollen in the oxidized state than in the neutral form. During oxidation, the segmental components of the polymer structure unravel and elongate (*17*) and thus enlarge the free-volume content. The anions studied to date appear to have no effect on the intersegmental spacing because the density of the polymer does not vary with the nature of the counterion (*66*). Presumably, the negligible effect of the anion is related to the amount of cross linking of polypyrrole.

Enhanced cross linking, it seems, should be deleterious to rapid ionic, and thus charge, transport. Diffusion of ions and solvent in the polymer matrix is an activated process. Consequently, a low photocurrent density may result from a large activation energy associated with ionic diffusion through a highly cross-linked polymer. Furthermore, in a highly cross-linked polymer, gaseous bubble formation, such as during the electrolysis of water (*7, 24*), is disruptive to the polymer structure and can reduce the protection of the semiconductor against corrosion. This may not be a serious problem, however, as we have shown it is possible to restrict gas evolution to the surface of the polymer (*13, 27, 28*).

Two conclusions can be drawn from this discussion. The first is that the morphology of the polymer influences ion and charge transport, which are necessary for electrochemical activity and, hence, the extent of protection of the semiconductor against corrosion. Polymer morphology also affects the suitability of the polymer for the specific photoelectrochemical process (e.g., current generation versus gaseous-fuel production).

In the subsequent sections, we examine how it is possible to reduce the oxidative instability and morphological limitations of the polymer and how the polymer affects surface states and interface energetics.

VI. Electrochemical Photovoltaic Cells

Electrically conductive polypyrrole films have been demonstrated to enhance the stability of *n*-type Si, GaAs, CdS, CdSe, and CdTe electrodes operating in contact with aqueous electrolytes in electrochemical photovoltaic cells (*22, 25, 26, 42, 56–58, 68, 69, 70–72*). Applications of polypyrrole to *n*-type Si photoanodes have been the most extensively studied and are emphasized here. The photodegradation of Si electrodes in aqueous electrolytes involves passivation owing to the formation of an insulating oxide layer and not ionic dissolution, as in the case of the cadmium chalcogenides.

In the absence of stability, the photogenerated holes reaching the surface contribute to the oxidation of Si. The growth of the insulating oxide layer is observable as a decline in the photocurrent with time. A reduced photocurrent is associated with a lower tunneling probability through the oxide. If, however, the polymer is effective in stabilizing the *n*-type Si electrode, the photocurrent will remain at a constant level.

Figure 9 presents the photocurrent-time behavior of the bare and the polypyrrole-coated polycrystalline *n*-type Si electrode in aqueous iron sulfate at pH 1 in the presence of air (*57*). The unprotected electrode stops producing photocurrent within one minute, whereas the electrode coated with polypyrrole increases current production about 6% for the first 5 h and shows a 30% decline in current output over 122 h of irradiation. During this period of measurement, ~3100 C cm^{-2} were transmitted at a

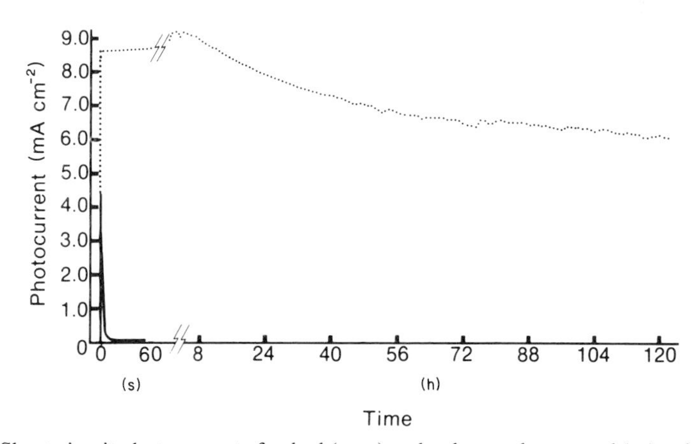

Fig. 9. Short-circuit photocurrent of naked (——) and polypyrrole-covered (···) polycrystalline *n*-type Si electrodes in aqueous 0.15 *M* $FeSO_4$/0.15 *M* $FeNH_4(SO_4)_2 \cdot 12H_2O$/0.1 *M* Na_2SO_4 at pH 1 under tungsten–halogen illumination of 143 mW cm^{-2}. From Noufi *et al.* (*57*). Copyright 1981 American Chemical Society.

current density between 6 and 9.2 mA cm^{-2}; the charge passed was nearly 10^5 times greater than that involved in the photoelectrochemical polymerization of the film. Moreover, the polymer film was absorbing at least 50% of the incident light. Compared with the naked electrode, the polymer-coated electrode exhibits remarkable stability.

Figure 10 records the power characteristics of a polycrystalline *n*-type Si electrode coated with a 1200-Å-thick polypyrrole film (*25*). In an unoptimized cell, the power conversion efficiency was 3%, corresponding to a short-circuit current of 3.2 mA cm^{-2}, an open-circuit voltage of 0.39 V, and a fill factor of 0.6 under tungsten–halogen illumination at 24.5 mW cm^{-2}. Presumably thinner films and redox couples of more positive potential than that of the Fe^{3+}/Fe^{2+} couple would have produced higher photocurrents and photovoltages, respectively. Unfortunately, the photocurrent–voltage curve under intense illumination shows a loss of fill factor and thus efficiency.

The onset potential of the photoanodic current of the coated electrode corresponds to the flat-band potential of the naked *n*-type Si electrode

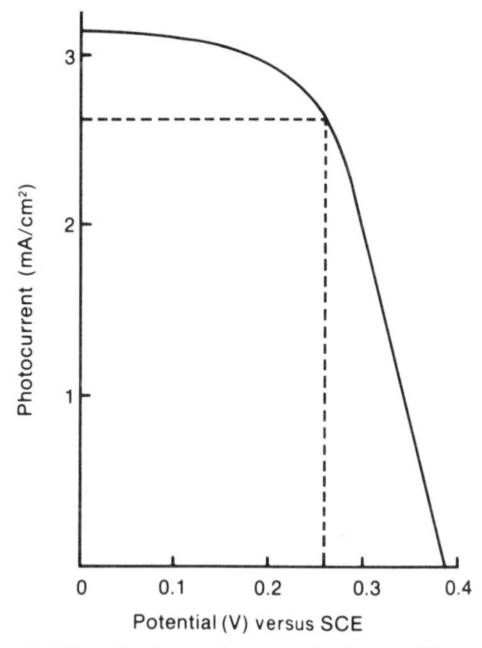

Fig. 10. Power characteristics of polypyrrole-covered polycrystalline *n*-type Si electrode in aqueous 0.15 *M* $FeSO_4$/0.15 *M* $FeNH_4(SO_4)_2 \cdot 12H_2O$/0.1 *M* Na_2SO_4 at pH 1 under tungsten–halogen illumination of 24.5 mW cm^{-2}. The radiant power was corrected for absorption owing to the electrolyte solution and absorption (primarily $\lambda \leq 1200$ nm) of a 4-cm-long water filter. From Frank (*25*).

(~0.1 V), which agrees with capacitance measurements (*57*). These results and others (*24*) indicate that the solvent and ions penetrate polypyrrole films and that the Fermi level of the semiconductor equilibrates with the redox electrolyte. This conclusion that the polymer is permeable to the electrolyte has been verified in other studies (*7*).

The picture that emerges from these studies is that the polymer covers potential oxide-forming sites on the Si surface. Under illumination, the polymer transmits the photogenerated holes in the semiconductor to ferrous ions at a rate much higher than the thermodynamically favored rate of growth of the insulating oxide layer. The high conductivity of the polymer facilitates removal of the photogenerated holes from the semiconductor–electrolyte interface.

In addition to high conductivity, another component of the mechanism for rapid hole removal may be understood from current–voltage measurements of polypyrrole-coated platinum electrodes (*26*). The current–voltage curve of Fig. 11 shows that during voltage excursions from −0.5 to 0.7 V, the current increases sharply and then remains nearly independent of voltage except at the positive extreme. When the sign of the voltage scan is reversed, the current declines sharply and then remains relatively constant. Such behavior is reminiscent of the charging and discharging of a capacitor. The capacitance of polypyrrole is one to two orders of magnitude higher than that of a bare electrode, which is typically tens of micro-

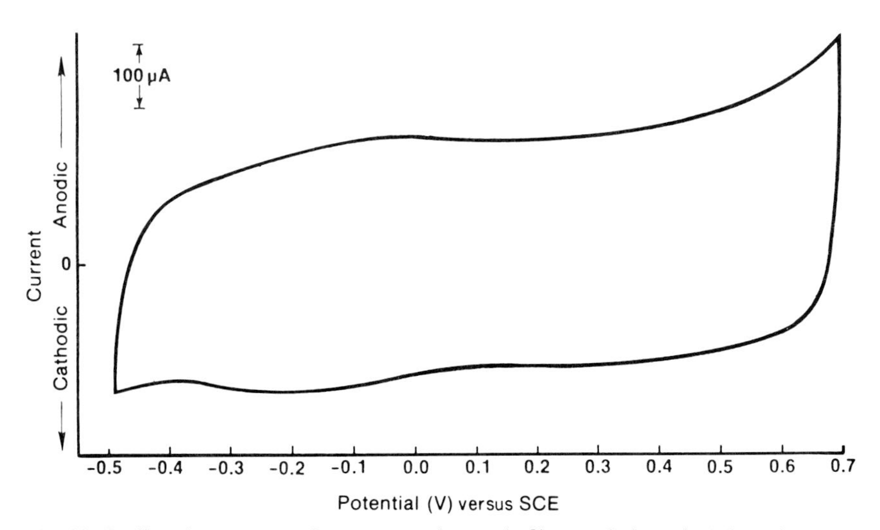

Fig. 11. Cyclic voltammogram for 0.1-μm polypyrrole film on platinum in 0.5 *M* phosphate buffer at pH 1.2 at a scan rate of 200 mV s^{-1}. From Frank (*26*). This figure was originally presented at the Fall 1981 meeting of The Electrochemical Society, Inc., held in Denver, Colorado.

farads per square centimeter. The high capacitance is related to the large surface area of the polymer. In effect, the large capacitance means that the polymer can behave as a sink for the photogenerated holes.

Figure 12 shows that polypyrrole stabilizes single-crystal *n*-type Si with respect to the bare electrode (*25*). However, the polypyrrole-coated single-crystal Si produces a less stable photocurrent than that of the polycrystalline-covered material. The photocurrent of the polymer-protected single-crystal Si declines continuously, whereas over the same period the photocurrent output of the polycrystalline Si is constant. In this case, the difference in photocurrent stability is attributed to the strength of the adhesion of the polymer to the electrode surface. The roughness of the Si surface is one factor affecting polymer adhesion. Polishing the single-crystal Si surface with fine abrasives before film deposition made mechanical removal of the film more difficult. The improved adhesion of the film is owing, in part, to more intimate contact between the surface atoms of the Si substrate and the chemical species involved in the polymerization. Nevertheless, the adhesive strength of the polymer film to the single-crystal Si surface was always less than that of the polymer film to the polycrystalline Si surface.

A different aspect of surface morphology may also have affected adhesion (*25*). Unlike single-crystal Si, which is characterized by a single plane, polycrystalline Si consists of various silicon planes with different spacings between the Si atoms along with dangling bonds at the grain boundaries. The surfaces of both forms of Si are expected to be covered

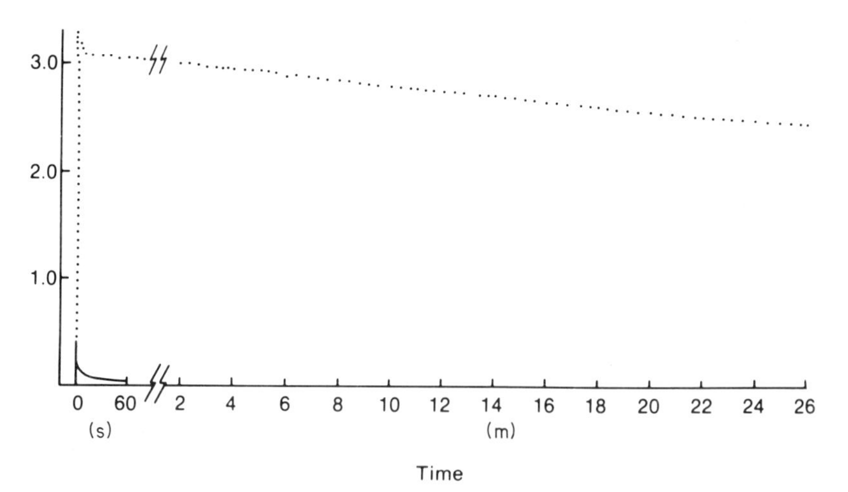

Fig. 12. Short-circuit photocurrent of naked (——) and polypyrrole-covered (···) single-crystal *n*-type Si electrodes in aqueous 0.15 *M* $FeSO_4$/0.15 *M* $FeNH_4(SO_4)_2 \cdot 12H_2O$/0.1 *M* Na_2SO_4 at pH 1 under tungsten–halogen illumination of 100 mW cm^{-2}. From Frank (*25*).

with a thin (~15-Å) oxide layer to which polypyrrole binds. The nature of the oxide layer is likely to be affected by differences in the orientation of the planes and in the lattice spacings, which in turn would influence the adhesion of the polymer to the substrates.

Platinum (*69*) and gold (*22*) metallization of the single-crystal Si surface markedly improves adhesion of the polymer film. Adhesion is also improved by covalent attachment of a pyrrole derivative (*68*) to the surface prior to anodic polymerization of the pyrrole.

High photocurrent stability of single-crystal Si has been achieved with a 20-Å-thick layer of Pt beneath the polypyrrole film and I^-/I_3^- as the redox couple (*70*). The photocurrent density of the Si/Pt/polypyrrole electrode remained constant at about 11 mA cm^{-2} for 450 h. In addition to the proposed effect of improved adhesion, the enhanced photocurrent stability may also be related to more rapid charge transfer between the Si and the Pt layer and, possibly, between the Pt and the polypyrrole. However, the contribution of the I^-/I_3^- couple to the stability of the electrode must not be underestimated, because with the Fe^{3+}/Fe^{2+} couple at pH 1, the photocurrent density of a Si/Pt/polypyrrole electrode declined from 10 to 2 mA cm^{-2} after only 10 h (*71*). As Fig. 9 shows, relatively high photocurrent stability has been achieved with the Fe^{3+}/Fe^{2+} couple at the same pH without the platinization treatment of the polycrystalline Si electrode. In part, the ability of I_3^- to scavenge holes efficiently may be related to its strong adsorption to platinum. Thus the adhesive strength of the polymer to the electrode surface, the charge-transport properties at the interface, and the availability of an efficient hole scavenger are crucial factors in determining the overall protection of the semiconductor against photodegradation.

Fan *et al.* (*22*) combined the ideas of applying a metal underlayer to single-crystal Si photoelectrodes before polymerization of pyrrole with the use of a high concentration of an electrolyte to suppress anodic photodegradation. As has been previously demonstrated (*47*), concentrated LiCl solutions decrease the activity of water at the interface, which in turn lowers the solvation energy and thus decreases the driving force for photodecomposition. In 11 *M* LiCl solution containing Fe^{3+}/Fe^{2+}, the photocurrent density of a Si/Au/polypyrrole electrode remained constant at ~3.2 mA cm^{-2} for 48 h. Without the Au underlayer, the photocurrent of the polypyrrole-coated Si electrode decayed from 3 to 2 mA cm^{-2} after 6 h and thereafter remained constant. In a solution containing no LiCl, the photocurrent density of a Si–polypyrrole electrode declined nearly 90% in 12 min from an initial value of 4.5 mA cm^{-2}. A slower decay rate would be expected at a lower current density.

A different surface treatment approach (*68*) to improve the adhesion of polypyrrole to single-crystal *n*-type Si has involved covalent attachment of *N*-(3-trimethoxysilyl)propyl pyrrole (**III**), to the electrode by reaction of surface OH groups followed by anodic polymerization of pyrrole. Although the adhesion of the resulting polypyrrole film was substantially improved over the underivatized electrode surface, no detailed data on durability were given.

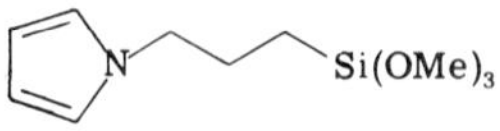

$Si(OMe)_3$

III

Photocorrosion studies have also been carried out on hydrogenated amorphous silicon (α-Si : H) (*42, 71, 72*). The stability and density of the photocurrent of the polypyrrole-coated electrodes are an improvement over the naked electrodes but are inferior to polypyrrole-coated single-crystal Si. A 10-Å-thick Pt layer on the α-Si : H electrode also improved the stability, but the magnitude of the photocurrent declined by nearly an order of magnitude, possibly because of pin holes in the Pt film (*72*).

VII. Surface States and Interface Energetics

Various sorts of imperfections on the surface of a semiconductor give rise to discrete energy levels that may lie within the band gap. These interface states may originate, for example, from dangling bonds owing to the abrupt termination of the lattice structure at the semiconductor surface, from physical disruption of the surface lattice as a result of abrasion, or from chemisorption of reactants or reaction intermediates. A large density of surface states can have a profound effect on the energetics and kinetics of the redox processes occurring at the interface. The energy levels of these surface states can equilibrate with the Fermi level of the semiconductor in the dark, leading to Fermi-level pinning (*3, 4*), or under illumination as charges accumulate (*52*). For the purpose of this discussion, these interface states are considered to produce junction effects that are independent of the redox species in the electrolyte and that limit the photovoltage of the device.

Several studies on the effects of polypyrrole coatings to the surface of various semiconductors have produced the following observations:

1. The open-circuit photovoltage V_{oc} depends nearly linearly on the redox potential $E°$ of species in the electrolyte (*56, 58*).
2. The flat-band potentials of the naked and polymer-coated electrodes are nearly identical (*57, 58*).
3. The redox electrolyte penetrates the polymer film (*7, 24*).

The linear dependence of V_{oc} on $E°$ indicates the absence or near absence of surface states and, hence, Fermi-level pinning of polypyrrole-coated electrodes. This contrasts with observations that thin platinum coatings on single-crystal *n*-type Si surfaces give rise to photovoltages lower than expected, and, hence, indicate Fermi-level pinning (*70*). In contrast, the effect of overlaying the Pt film with a coating of polypyrrole increases markedly the photovoltage, although not to a value comparable to that of a polymer-covered Si photoanode (*70*). The important implication of these results is that polymer coatings such as polypyrrole may be employed to control or passivate surface states and thus improve the photovoltage of the device. Polymeric coatings interacting with atomic species responsible for surface states have been shown to retard the recombination rate and thus improve the performance of *n*-type $MoSe_2$ and WSe_2 electrodes (*11, 80*).

Furthermore, the first observation, together with the flat-band potentials, suggest that the polypyrrole coatings do not alter appreciably the energy levels of the semiconductors. The third observation indicates that polypyrrole cannot be considered simply as a nonporous metallic-like film.

The preceding three observations are consistent with the energy-level diagrams depicted in Figs. 13 and 14. Figure 13 shows the energy relationships that arise at the contact interface between an *n*-type semiconductor and an electrolyte-permeable polymer film such as polypyrrole, with different redox couples in the electrolyte. Although the work function of the semiconductor is fixed with respect to vacuum, the work function of the polymer varies with the dopant redox species in the electrolyte. As dis-

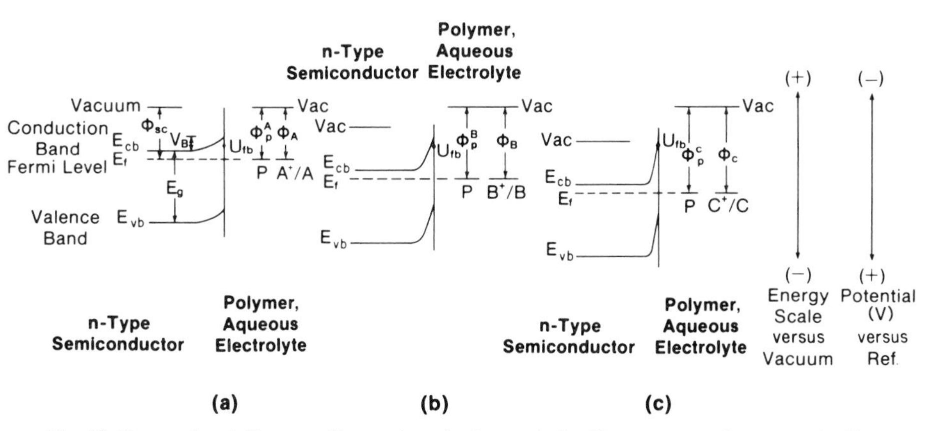

Fig. 13. Energy-level diagrams for semiconductor coated with a porous polymer contacting solutions of (a) A^+/A, (b) B^+/B, or (c) C^+/C. E_g, Semiconductor band gap; V_B, band bending; U_{fb}, flat-band potential; ϕ_{sc}, work function of the semiconductor; ϕ_P^A, ϕ_P^B, ϕ_P^C, work functions of the polymer.

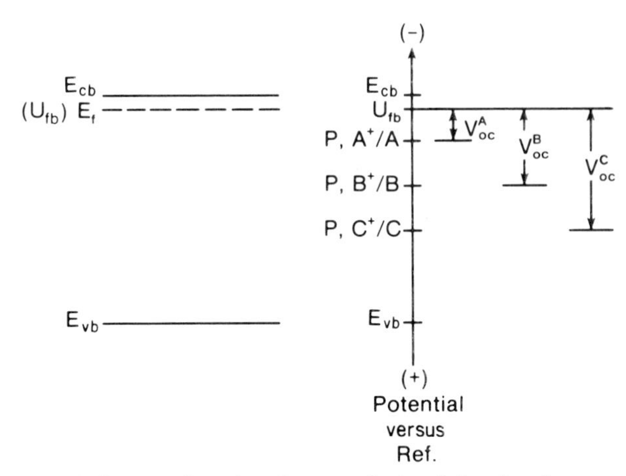

Fig. 14. Energy-level diagram of semiconductor at flat band showing the open-circuit photovoltage V_{oc} with different redox couples (A^+/A, B^+/B, and C^+/C) but with the same electrolyte-permeable polymer (P). The energy scale shows the redox potentials of the polymer and the redox couples are nearly the same.

cussed in Section VI, the high capacitance and large surface area of the polymer promote rapid electronic equilibrium with the redox species in the electrolyte. The high capacitance of the polymer is also presumed to be important in the reduction of surface-state effects because it permits a mechanism for rapid charge removal.

Figure 14 accounts for the observation that the maximum open-circuit photovoltage V_{oc} is determined by

$$V_{oc} = E^\circ - U_{fb}. \tag{7}$$

If the potential of the redox couple determines the Fermi level of the polymer, then under open-circuit conditions between an illuminated semiconductor electrode and a metal counterelectrode (not shown), the photovoltage produced would equal the difference between the Fermi level in the semiconductor and the redox potential of the electrolyte.

Figure 15 contrasts the interface energetics between a semiconductor coated with a highly porous conductive polymer film and a semiconductor coated with a nonporous metal film, in which Fermi-level pinning is exhibited. The amount of band bending V_B, or the Schottky barrier height, determines the maximum open-circuit photovoltage [see Eq. (7)]. In Fig. 15a the solution penetrates the polymer and contacts the semiconductor. Under these conditions the band bending depends on the potential of the redox couples, whereas the valence-band and conduction-band edges are fixed. In Fig. 15b the band bending of the nonporous metal-coated semi-

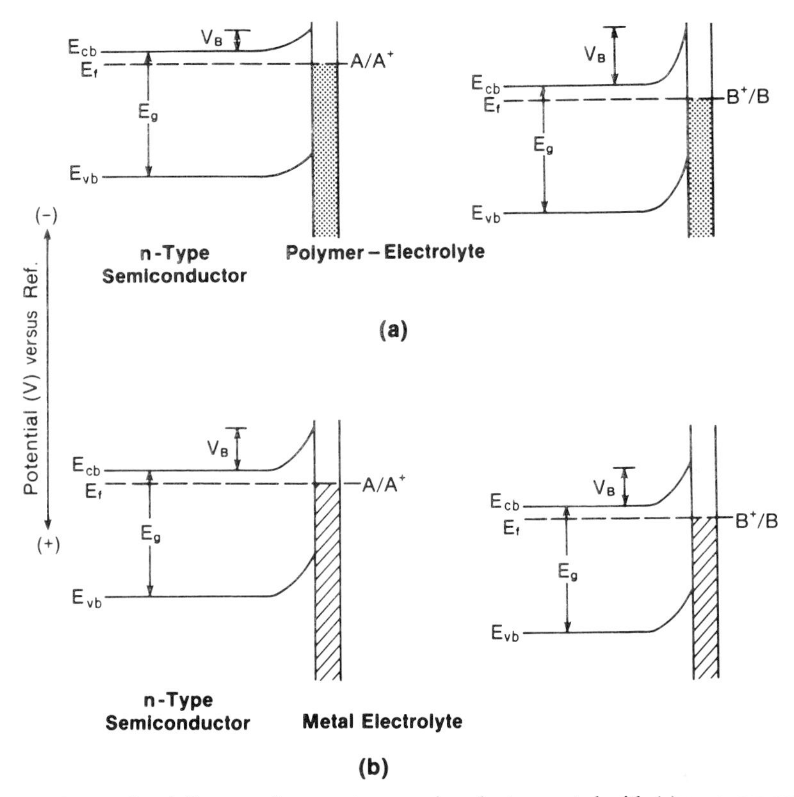

Fig. 15. Energy-level diagrams for an n-type semiconductor coated with (a) a porous polymer or (b) a nonporous metal film contacting a solution of A^+/A or B^+/B. (a) Band bending V_B varies with the redox couple, whereas E_{vb} and E_{cb} are fixed. (b) E_{vb} and E_{cb} shift with a change in the potential of the redox electrolyte, whereas V_B is constant.

conductor is fixed, and the band edges shift with a change in the redox potential of the electrolyte.

VIII. Photoelectrolysis of Water

A. Electrocatalytic Oxygen Evolution

In Section VI the focus of the discussion was on electrochemical photovoltaic cells, the net product of which is electrical power. Polymer films of polypyrrole were shown to provide greatly enhanced stability against the photodegradation of the semiconductor photoanodes in aqueous electrolytes. In those experiments, electroactive redox couples such as Fe^{3+}/

Fe^{2+}, I^-/I_3^-, $Fe(CN)_6^{3-}/Fe(CN)_6^{4-}$, Br^-/Br_3^-, and Ce^{4+}/Ce^{3+} were present in the electrolyte to scavenge efficiently the photogenerated holes before oxidation of the semiconductor occurred. Such single-charge-transfer reactions are simpler kinetically than the decomposition mechanism of the semiconductor.

A more difficult application of conducting polymer-coated electrodes is in photoelectrolysis cells in which the mechanism of the four-electron oxidation reaction of water may be as complex as the photodecomposi-

$$4h^+ + 2H_2O \longrightarrow O_2 + 4H^+ \tag{8}$$

tion reaction of the semiconductor. In this case, the polymer must protect the semiconductor against self-oxidation during oxygen evolution and at the same time undergo no irreversible chemical reaction itself. Because of the kinetic complexity of the oxygen-evolution reaction and the requirement of creating and utilizing holes at high positive potentials, all known nonoxide *n*-type semiconductors undergo photodecomposition preferentially or to the exclusion of water oxidation. The role of the conducting polymer must be, therefore, to direct the photogenerated holes to the O_2/H_2O redox couple at a rate much faster than the oxidation rate of the semiconductor.

The possibility of using polypyrrole under oxygen-evolving conditions raises certain important questions (*13*): (a) Is polypyrrole stable under O_2-evolving conditions? (b) Can the water oxidation reaction be confined to the outermost polypyrrole–liquid interface so as to avoid the physical destruction of the film by internal gas evolution? (c) Is polypyrrole sufficient as an electrocatalyst for O_2 evolution, and if not, can a catalytic agent be deposited on the film surface to achieve good O_2-evolution rates?

Affirmative answers to these questions were obtained from studies of polypyrrole films deposited on tantalum electrodes (*13*). Normally, a Ta electrode does not evolve O_2 because of the spontaneous development of an insulating oxide layer on the surface at potentials less positive than the O_2/H_2O redox potential.

Figure 16 shows the respective current–voltage curves for polypyrrole films, 2.5 and 5 μm thick, on Ta electrodes in the absence and in the presence of a 72-Å-thick platinum film formed by argon-ion-beam sputtering. It can be seen in the figure that the polypyrrole film itself is a poor catalyst for oxygen evolution. No oxygen bubbles could be observed on the polymer surface. The maximum current density obtained with the polypyrrole-coated electrode was less than 0.2 mA cm^{-2} at 1.4 V versus SCE. In marked contrast, the polypyrrole-covered Ta electrode with the catalytic platinum surface produced a current density of 1.3 mA cm^{-2} at 1.3 V. This particular current density is noteworthy because 1 mA cm^{-2} is

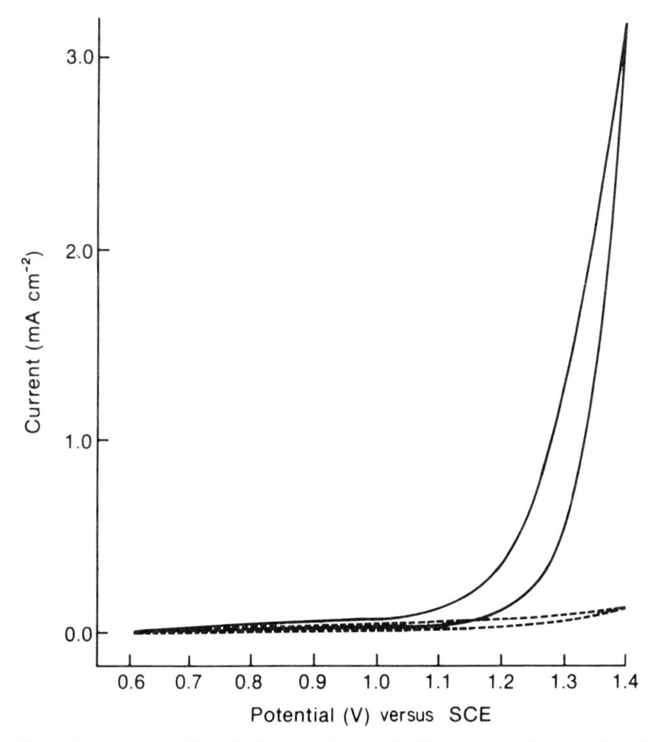

Fig. 16. Cyclic voltammogram for a 2.5-μm polypyrrole film on tantalum (---) and a 72-Å, Pt-coated, 5-μm polypyrrole film on tantalum (——) in 2 *M* phosphate buffer at pH 6.7. Data are shown for the samples after they were cycled 15 times from 0.6 to 1.0 V versus SCE. From Cooper *et al.* (*13*). Copyright 1982 Macmillan Journals Ltd.

about the threshold for visibly observing O_2 bubbles on the electrode surface. At 1.4 V the current density rose to over 3 mA cm^{-2}. Moreover, the electrochemical behavior of the Ta/polypyrrole/Pt electrode was indistinguishable from that of an ion-beam-sputtered Pt electrode with respect to O_2 production.

Oxygen evolution was observed on polypyrrole films of thicknesses ranging from 0.45 to 12.5 μm when a platinum layer was deposited on the surface of the polymer. Even with such thick films, however, it could not be ruled out conclusively that the Pt had penetrated the polypyrrole film, thereby short-circuiting the polymer and permitting O_2 evolution on a Pt-coated Ta surface. To investigate this possibility, a Ta electrode was coated with a 6.2-μm-thick polypyrrole film on which a nominal 45-Å-thick Pt layer was deposited. The surface of the electrode was then examined with Auger electron spectroscopy to determine the depth profile of the Pt perpendicular to the surface of the polymer. The Auger study showed that the Pt signal reached a maximum at ~100 Å and then declined to background noise level at about 900 Å beneath the polypyrrole surface. Thus the polymer transmits the holes nearly 62,000 Å before the platinum catalyst is encountered.

The results of the Ta/polypyrrole/Pt study indicate that (a) with a catalytic surface, polypyrrole protects the metal electrode from self-oxidation during O_2 evolution whereas normally the metal would passivate owing to an insulating oxide layer; (b) the polymers transmit charges over large distances (≥12.5 μm; (c) the overpotential of the polymer is too high for O_2 evolution; and (d) when O_2 evolution is confined to the outermost polymer–water interface, there is no sign of physical or chemical deterioration of the polymer after the passage of ~2.3 C cm^{-2}.

B. Visible-Light-Assisted Water Cleavage

Thermodynamically and kinetically a more demanding application of conductive-polymer-covered electrodes is in the actual photoelectrolysis cell. Unlike Ta and Si electrodes, which form insulating, normally water-insoluble oxide surface layers during photodegradation, the nonoxide semiconductors generally undergo light-activated dissolution in aqueous electrolytes, thus undermining the adhesion of the polymer to the surface of the substrate.

The crucial photoelectrolysis test was conducted on *n*-type CdS (*27, 28*). The band gap of CdS is 2.4 eV (λ = 517 nm). This material was chosen because of its photoinstability in aqueous electrolytes. In this medium, CdS photocorrodes to produce Cd^{2+} ions and a surface layer of sulfur:

$$\text{Photoanode:} \quad CdS + 2h^+ \longrightarrow Cd^{2+} + S \tag{9}$$

At the dark counterelectrode, hydrogen is evolved:

$$\text{Cathode:} \quad 2e^- + 2H_2O \longrightarrow H_2 + 2OH^- \tag{10}$$

If the photocorrosion reaction were inhibited, the kinetically more complex and thermodynamically less favorable water-oxidation reaction [Eq. (8)] would prevail. This premise has been verified in the case of catalyst-loaded CdS particles (*29, 43*).

Figure 17 is a conceptual representation of the approach. The conducting polymer channels the photogenerated holes to the RuO_2 catalyst before self-oxidation of the CdS photoanode occurs. The catalyst then mediates the chemical transformation of water into O_2. At the Pt counterelectrode, the cycle is completed with the evolution of H_2.

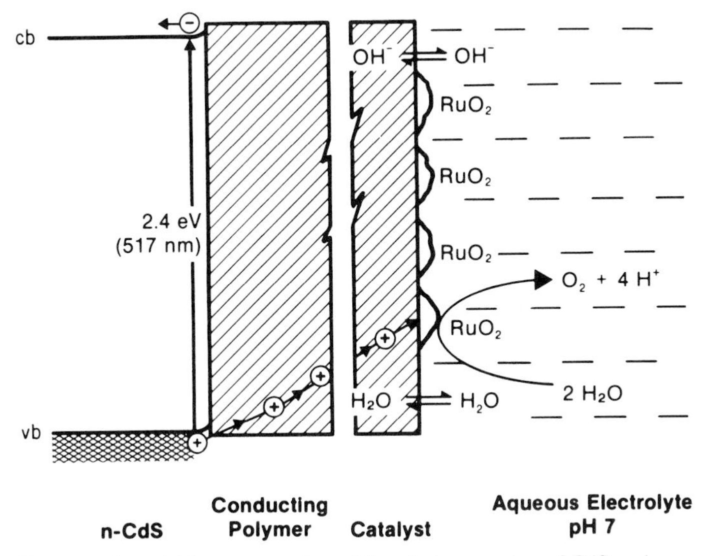

Fig. 17. Conceptual model for suppression of the photocorrosion of CdS and promotion of the water-oxidation reaction.

Studies were made on naked CdS electrodes, CdS coated with a catalyst, and CdS coated with polypyrrole containing a catalyst. The catalyst RuO_2 was immobilized with silver paint on the surface of either the bare electrode or the polymer-covered electrode. The average thickness of the polymer film was 0.8 μm. Oxygen production was either monitored directly with an electrochemical-type sensor or a gas chromatograph or indirectly with a pH electrode. The concentration of Cd^{2+} ions produced during photocorrosion was determined by atomic absorption spectrophotometry.

Because the photocorrosion rate of CdS increases with the photocurrent (*41*), the light intensity was adjusted to produce an initial photocurrent of ~150 μA cm^{-2}. The radiant powers at the bare, catalyst-coated, and polypyrrole-catalyst-covered CdS electrodes were 28, 110, and 80 mW cm^{-2}, respectively. Higher light intensities are required to produce similar initial photocurrents in the case of the surface-protected electrodes, because the coatings absorb light.

Figure 18 shows the time course of O_2 production by a naked electrode, a catalyst-covered electrode, and a polypyrrole-catalyst-coated CdS electrode. The rate of O_2 production by the naked electrode is nearly constant over 45 h of irradiation. During this period, ~2 μmol cm^{-2} O_2 are produced, and the rate constant is 6.0×10^{-10} cm s^{-1}. The O_2 production of the catalyst-coated CdS electrode exhibits an induction period of ~2 h,

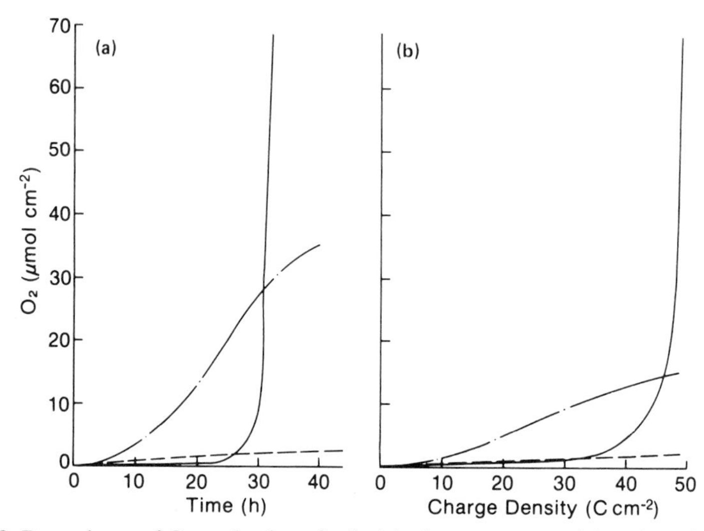

Fig. 18. Dependence of O_2 production of naked (---), catalyst-coated (-····-), and polypyrrole-catalyst-covered (——) CdS photoanodes in a 0.5 *M* Na_2SO_4 solution at pH 8.6 on (a) time and (b) charge density. From Frank and Honda (*27*). Copyright 1982 American Chemical Society.

increases linearly with time to ~40 h, and then tends to level off. The maximum rate constant is 1.5×10^{-8} cm s^{-1}, and 35 μmol cm^{-2} of O_2 are produced. An induction period for O_2 production of ~15 h occurs in the case of the polymer-catalyst-coated CdS electrode; thereafter, the rate of O_2 production rises sharply. The rate constant of maximum O_2 production is 3.5×10^{-7} cm s^{-1}, and 69 μmol cm^{-2} of O_2 are produced over 32 h. The induction period observed in the case of the coated electrodes involves, in part, the oxidation of the silver paint and configurational changes of the polymer (*28*). At 49 C cm^{-2}, the relative amounts of O_2 production from the naked, catalyst-covered, and polypyrrole-catalyst-coated CdS electrodes are 1, 8, and 35, respectively.

Figure 19 correlates the amounts of O_2 and Cd^{2+} ions produced after 49 C cm^{-2} are passed. With no coating, ~99% of the photogenerated holes that contribute to the photocurrent lead to the corrosion reaction [Eq. (9)]. The addition of the catalyst to the surface of CdS promotes the rapid removal and trapping of holes from the semiconductor for the subsequent water-oxidation reaction [Eq. (8)]. In this case, 21% of the photogenerated holes are directed to O_2 production, and 79% to the destruction of the semiconductor lattice. The most dramatic improvement in stabilizing CdS against photodegradation and allowing O_2 production results from the combination of polymer and catalyst in which 68% of the photogenerated holes are converted to O_2 and 32% produce Cd^{2+} ions. Moreover, depth-

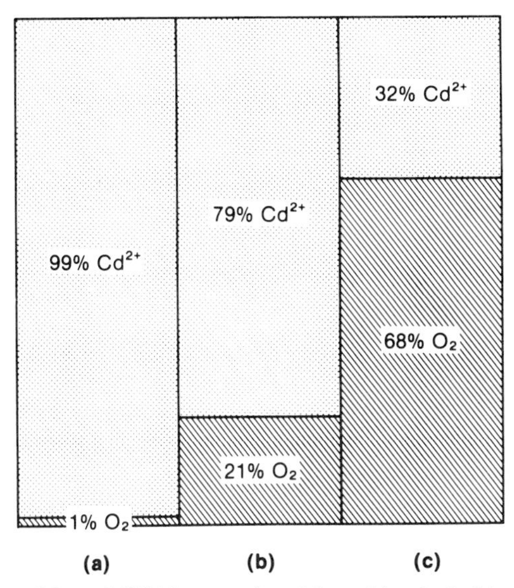

Fig. 19. Percentages of O_2 and Cd^{2+} ions produced from (a) naked, (b) catalyst-coated, and (c) polypyrrole-catalyst-covered CdS photoanodes in a 0.5 *M* Na_2SO_4 solution at pH 8.6 after 49 C cm^{-2}. From Frank and Honda (*27*). Copyright 1982 American Chemical Society.

profile results of Auger spectroscopy showed no Ru or Ag at the polymer–CdS interface; however, the Ru catalyst had penetrated deeper into the polymer than the silver. Finally, additional data show (*29*) that it is possible to suppress the photocorrosion reaction of CdS in aqueous solution almost completely and to direct nearly all the photogenerated holes to oxygen production.

IX. Guidelines for Control of Interface

In this section, the principal considerations are summarized for utilizing electrically conductive polymers to manipulate interfacial charge-transfer kinetics, thus suppressing photodecomposition and promoting desirable redox reactions (*25*).

The polymer should have good electronic transport properties at high solar light intensities (80–140 mW cm^{-2}). The polymer must be able to channel a high density of photogenerated minority carriers from the semiconductor to desirable redox species in the electrolyte at a rate greatly exceeding the rate of photodecomposition. The high capacitance of the polymer, because of its large surface area, can provide a driving force for rapid charge transport from the semiconductor. The saturation photocurrent will, in most cases, depend on the rate of charge removal from the polymer.

The specific interface energetics will depend on whether or not the redox electrolyte can penetrate the polymer film to the semiconductor. If the polymer film is permeable to the solvent, as in the case of polypyrrole films in water, rectification will be determined principally by the semiconductor–electrolyte junction. For this situation, protection of the semiconductor surface will hinge considerably on the good electronic transport properties of the polymer compared with the photodecomposition rate. Alternatively, a hydrophobic polymer may be desirable if it does not severely affect the desired redox kinetics. Hydrophobicity will reduce solvation effects and thus shift the decomposition potential of the electrode to positive values; however, it can also affect adversely the thermodynamics and kinetics of the desired redox processes. The use of high concentrations of electrolytes (*22, 47*) to lower the water activity while providing a conducting medium may be one strategy for balancing solvation effects and kinetic considerations. Another approach has been suggested in which a water-impermeable conducting polymer forms a Schottky barrier with the semiconductor (*25*).

To protect the semiconductor against photocorrosion, the polymer must be kinetically inert and/or more thermodynamically stable than both the semiconductor and the redox electrolyte. Inertness depends on the composition of the redox electrolyte (solvent, redox species, counterions, etc.). The redox electrolyte must efficiently scavenge the transmitted minority carriers from the polymer or from a catalytic agent associated with the polymer if chemical corrosion of the polymer itself is to be avoided. Disruption of the electronic unsaturation of the polymer through chemical reactions with the solvent or redox species can produce deterioration of the electrical conductivity and a diminished effectiveness in the stabilization of the semiconductor.

The polymer film must conduct either holes for *n*-type semiconductors or electrons for *p*-type materials; that is, the polymer must possess *p*- or *n*-type character, respectively, or transmit charges of both signs.

The extinction coefficient of the polymer should be small over the spectral region the semiconductor absorbs so as not to attenuate the excitation energy of the semiconductor. For example, because the oxidized state of polypyrrole tetrafluoroborate has strong, broad absorption bands in the near ir ($\varepsilon_{max} = 8 \times 10^4\ cm^{-1}$, $\lambda_{max} = 1000$ nm) and in the visible ($\varepsilon_{max} = 5 \times 10^4\ cm^{-1}$, $\lambda_{max} = 400$ nm), it is not suitable except as very thin coatings for semiconductors absorbing in these spectral regions. Also, the polymer should strongly adhere to the surface of the semiconductor.

In conclusion, the utility of electrically conductive polymers for practical photoelectrochemical solar-energy devices depends on the management of thermodynamic and kinetic factors affecting the semiconductor, polymer, and end-products (electricity or chemicals) desired.

Acknowledgment

The author gratefully acknowledges the helpful discussions with Arthur J. Nozik, Bruce Parkinson, and John A. Turner. This work was performed under the auspices of the Office of Basic Energy Sciences, Division of Chemical Sciences, United States Department of Energy, under Contract EG-77-C-01-4042.

References

1. Bard, A. J., and Wrighton, M. S., *J. Electrochem. Soc.* **124,** 1706 (1977).
2. Bard, A. J., *J. Photochem.* **10,** 50 (1979).
3. Bard, A. J., Bocarsly, A. B., Fan, F. F., Walton, E. G., and Wrighton, M. S., *J. Am. Chem. Soc.* **102,** 3671 (1980).
4. Bocarsly, A. B., Bookbinder, D. C., Dominey, R. N., Lewis, N. S., and Wrighton, M. S., *J. Am. Chem. Soc.* **102,** 3683 (1980).
5. Bookbinder, D. C., Bruce, J. A., Dominey, R. N., Lewis, N. S., and Wrighton, M. S., *Proc. Natl. Acad. Sci. USA* **77,** 6280 (1980).
6. Boudreaux, D. S., Williams, F., and Nozik, A. J., *J. Appl. Phys.* **51,** 2158 (1980).
7. Bull, R. A., Fan, F.-R. F., and Bard, A. J., *J. Electrochem. Soc.* **129,** 1009 (1982).
8. Butler, M. A., and Ginley, D. S., *J. Mater. Sci.* **15,** 1980 (1980).
9. Cahen, D., Hordes, G., and Manassen, J., *J. Electrochem. Soc.* **125,** 1623 (1978).
10. Cahen, D., Vainas, B., and Vandenberg, J. M., *J. Electrochem. Soc.* **128,** 1484 (1981).
11. Canfield, D., and Parkinson, B., *J. Am. Chem. Soc.* **103,** 1281 (1981).
12. Chen, S. N., Heeger, A. J., Kiss, Z., MacDiarmid, A. G., Gau, S. C., and Peebles, D. L., *Appl. Phys. Lett.* **36,** 96 (1980).
13. Cooper, G., Noufi, R., Frank, A. J., and Nozik, A. J., *Nature (London)* **295,** 578 (1982).
14. Diaz, A. F., Kanazawa, K. K., and Gardini, G. P., *J. Chem. Soc. Chem. Commun.*, p. 635 (1979).
15. Diaz, A., *Chem. Scr.* **17,** 145 (1981).
16. Diaz, A. F., Castillo, J. I., Logan, J. A., and Lee, W.-Y., *J. Electroanal. Chem.* **129,** 115 (1981).
17. Diaz, A. F., and Kanazawa, K. K., *in* "Extended Linear Chain Compunds" (J. S. Miller, ed.), Vol. 3, p. 417. Plenum, New York, 1982.
18. Dall'Olio, A., Dascola, Y., Varacca, V., and Bocchi, V., *C. Acad. Sci., Ser. C* **267,** 433 (1968).
19. Dominey, R. N., Lewis, N. S., Bruce, J. A., Bookbinder, D. C., and Wrighton, M. S., *J. Am. Chem. Soc.* **104,** 467 (1982).
20. Ehrenreich, H., "Solar Photovoltaic Energy Conversion, Report of the American Society Study Group." Amer. Physical Soc., New York, 1979.

21. Ellis, A. B., Kaiser, S. W., Bolts, J. M., and Wrighton, M. S., *J. Am. Chem. Soc.* **99,** 2839 (1977).

22. Fan, F.-R. F., Wheeler, B. L., Bard, A. J., and Noufi, R., *J. Electrochem. Soc.* **128,** 2042 (1981).

23. Ford, W. K., Duke, C. B., and Salaneck, W. R., *J. Chem. Phys.* **77,** 5030 (1982).

24. Frank, A. J., Turner, J. A., and Nozik, A. J., *Proc. Int. Conf. Photochem. Convers. Storage Sol. Energy 3rd,* p. 299 (1980).

25. Frank, A. J., *Mol. Cryst. Liq. Cryst.* **83,** 1373 (1982).

26. Frank, A. J., Cooper, J. Noufi, R., Turner, J. A., and Nozik, A. J., *in* "Photoelectrochemistry: Fundamental Processes and Measurement Techniques" (W. L. Wallace, A. J. Nozik, S. K. Deb, and R. H. Wilson, eds.), p. 248. Electrochem. Soc., Princeton, New Jersey, 1982.

27. Frank, A. J., and Honda, K., *J. Phys. Chem.* **86,** 1933 (1982).

28. Frank, A. J., and Honda, K., *J. Electroanal. Chem.* (1983).

29. Frank, A. J., and Honda, K., Unpublished results, 1982.

30. Frank, A. J., and Turner, J. A., Unpublished results, 1982.

31. Gardini, G. P., *Adv. Heterocycl. Chem.* **15,** 67 (1973).

32. Gerischer, H., *J. Electroanal. Chem.* **82,** 133 (1977).

33. Gerischer, H., *Top. Appl. Phys.* **31,** 115 (1979).

34. Ginley, D. S., and Butler, M. A., *J. Electrochem. Soc.* **125,** 1968 (1978).

35. Gissler, W., and McEvoy, A. J., *in* "Photoelectrochemistry: Fundamental Processes and Measurement Techniques" (W. L. Wallace, A. J. Nozik, S. K. Deb, and R. H. Wilson, eds.), Vol. 82–83, p. 212. Electrochem. Soc., Princeton, New Jersey, 1982.

36. Grant, P. M., and Batra, I. P., *Synth. Met.* **1,** 93 (1979–80).

37. Hardee, K. L., and Bard, A. J., *J. Electrochem. Soc.* **124,** 215 (1977).

38. Harris, L. A., Gerstner, M. E., and Wilson, R. H., *J. Electrochem. Soc.* **124,** 1511 (1977).

39. Harris, L. A., and Wilson, R. H., *Annu. Rev. Mater. Sci.* **8,** 99 (1978).

40. Harris, L. A., and Hugo, J. A., *J. Electrochem. Soc.* **128,** 1203 (1981).

41. Heller, A., Schwartz, G. P., Vadimsky, S., Menezes, S., and Miller, B., *J. Electrochem. Soc.* **125,** 1156 (1978).

42. Inganäs, O., Skotheim, T., and Lundström, I., *Phys. Scr.* **25,** 863 (1982).

43. Kalyanasundaram, K., Borgarello, E., and Grätzel, M., *Helv. Chim. Acta* **64,** 362 (1981).

44. Kanazawa, K. K., Diaz, A. F., Gill, W. D., Grant, P. M., and Street, B. G., *Synth. Met.* **1,** 329 (1979–80).

45. Kanazawa, K. K., Diaz, A. F., Krounbi, M. T., and Street, G. B., *Synth. Met.* **4,** 119 (1981).

46. Kohl, P. A., Frank, S. N., and Bard, A. J., *J. Electrochem. Soc.* **124,** 225 (1977).

47. Kubiak, C. P., Schneemeyer, L. F., and Wrighton, M. S., *J. Am. Chem. Soc.* **102,** 6898 (1980).

48. Levy-Clement, C., Heller, A., Bonner, W. A., and Parkinson, B., *J. Electrochem. Soc.* **129,** 1701 (1982).

49. Liou, F.-T., Yang, C. Y., and Levine, S. N., *in* "Photoelectrochemistry: Fundamental Processes and Measurement Techniques" (W. L. Wallace, A. J. Nozik, S. K. Deb, and R. H. Wilson, eds.), Vol. 82–3, p. 653. Electrochem. Soc., Princeton, New Jersey, 1982.

50. Memming, R., and Schwandt, G., *Surf. Sci.* **4,** 109 (1966).

51. Memming, R., *J. Electrochem. Soc.* **125,** 117 (1978).

52. Memming, R., and Kelly, J. J., *in* "Photochemical Conversion and Storage of Solar Energy" (J. S. Connolly, ed.), p. 243. Academic Press, New York, 1981.
53. Morisaki, H., Ono, H., Donkoski, H., and Yazawa, K., *Jpn. J. Appl. Phys.* **19,** L148 (1980).
54. Nakato, Y., Abe, K., and Tsubomura, H., *Ber. Bunsenges. Phys. Chem.* **80,** 1002 (1976).
55. Nakato, Y., Tonomura, S., and Tsubomura, H., *Ber. Bunsenges. Phys. Chem.* **80,** 1289 (1976).
56. Noufi, R., Tench, D., and Warren, L. F., *J. Electrochem. Soc.* **127,** 2310 (1980).
57. Noufi, R., Frank, A. J., and Nozik, A. J., *J. Am. Chem. Soc.* **103,** 1849 (1981).
58. Noufi, R., Tench, D., and Warren, L. F., *J. Electrochem. Soc.* **128,** 2596 (1981).
59. Nozik, A. J., *Appl. Phys. Lett.* **29,** 150 (1976).
60. Nozik, A. J., *Annu. Rev. Phys. Chem.* **39,** 189 (1978).
61. Nozik, A. J., *Proc. Int. Conf. Photochem. Convers. Storage Sol. Energy 2nd,* p. 70 (1978).
62. Nozik, A. J., Boudreaux, D. S., Chance, R. R., and Williams, F., *Adv. Chem. Ser.* No. 184, 155 (1980).
63. Parkinson, B., Unpublished results, 1982.
64. Prejza, J., Lundstrüm, I., and Skotheim, T., *J. Electrochem. Soc.* **129,** 1685 (1982).
65. Salaneck, W. R., Erlandsson, R., Prejza, J., Lundstrom, L., Duke, C. B., and Ford, W. K., *Polym. Prepr.* **23,** 120 (1982).
66. Salmon, M., Diaz, A. F., Logan, A. J., Krounbi, M., and Barbon, J., *Mol. Cryst. Liq. Cryst.* **83,** 1297 (1982).
67. Schroeder, A. J., and Kaufman, F. B., *J. Electroanal. Chem.* **113,** 209 (1980).
68. Simon, R. A., Ricco, R. A., and Wrighton, M. S., *J. Am. Chem. Soc.* **104,** 2031 (1982).
69. Skotheim, T., Lundström, I., and Prejza, J., *J. Electrochem. Soc.* **128,** 1625 (1981).
70. Skotheim, T., Petersson, L.-G., Inganäs, O., and Lundström, I., *J. Electrochem. Soc.* **129,** 1737 (1982).
71. Skotheim, T., Inganäs, O., Prejza, J., and Lundström, I., *Mol. Cryst. Liq. Cryst.* **83,** 1361 (1982).
72. Skotheim, T., Lundström, I., Delahoy, A. E., Kampas, F. J., and Vanier, P. E., *Appl. Phys. Lett.* **40,** 281 (1982).
73. Street, G. B., Clarke, T. C., Krounbi, M., Kanazawa, K., Lee, V., Pfluger, P., Scott, J. C., and Weiser, G., *Mol. Cryst. Liq. Cryst.* **83,** 1285 (1982).
74. Tanaka, M. Watanabe, A., Fujimoto, H., and Tanaka, J., *Mol. Cryst. Liq. Cryst.* **83,** 1309 (1982).
75. Tomkiewicz, M., and Woodall, J. M., *J. Electrochem. Soc.* **124,** 1437 (1977).
76. Tributsch, H., *Z. Naturforsch. A* **32,** 972 (1977).
77. Tributsch, H., and Bennet, J. C., *J. Electroanal. Chem.* **81,** 97 (1977).
78. Turner, J. A., and Nozik, A. J., *Appl. Phys. Lett.* **41,** 101 (1981).
79. Wagner, S., and Shay, J., *Appl. Phys. Lett.* **31,** 446 (1977).
80. White, H. W., Abruna, H. D., and Bard, A. J., *J. Electrochem. Soc.* **129,** 265 (1982).
81. Williams, R., *J. Chem. Phys.* **32,** 1505 (1960).
82. Wrighton, M. S., Austin, R. G., Bocarsly, A. B., Bolts, J., Haas, O., Legg, K. D., Nadjo, L., and Palazzotto, M., *J. Am. Chem. Soc.* **100,** 1602 (1978).
83. Wrighton, M., *Acc. Chem. Res.* **12,** 303 (1979).
84. Wrighton, M. S., Bocarsly, A. B., Bolts, J. M., Bradley, M. G., Fischer, A. B., Lewis, N. S., Palazzotto, M. C., and Walton, E. G., *Adv. Chem. Ser.* No. 184, 269 (1980).
85. Yoneyama, H. Sakamoto, H., and Tamura, H., *Electrochim. Acta* **20,** 341 (1975).

15 Photochemical Fixation of Carbon Dioxide

M. Halmann

Isotope Department
The Weizmann Institute of Science
Rehovot, Israel

I. Introduction

A. Natural Photosynthesis, Geochemical Carbon Cycle, and Climatic Effects of Increase in Atmospheric Carbon Dioxide

Photosynthesis, which brings about the fixation of carbon dioxide, water, and other essential elements such as nitrogen, phosphorus, and various metal ions, is the basis of life on Earth. Photosynthesis occurs on a vast scale in the oceans, by algae and photosynthetic bacteria consuming carbon dioxide according to the general stoichiometry

$$106CO_2 + 16NO_3^- + HPO_4^{2-} + 122H_2O + 18H^+ \longrightarrow C_{106}H_{263}O_{110}N_{16}P + 138O_2 \quad (1)$$

which results in a stable C : N : P atomic ratio of 106 : 16 : 1 in the plankton of the open ocean (*100*). On the continents, carbon dioxide is similarly

ENERGY RESOURCES THROUGH
PHOTOCHEMISTRY AND CATALYSIS

ISBN 0-12-295720-2

consumed by higher plants as well as by the phytoplankton of lakes and rivers. However, in unpolluted or only moderately polluted freshwater, the plankton contain a much higher carbon content, indicating a larger consumption of carbon dioxide. Thus, in samples of lake-water plankton, an atomic ratio C : N : P of 500 : 33 : 1 was obtained (*56*).

The origin of the Earth's atmosphere is presumably the result of outgassing of volcanic rocks, releasing carbon dioxide, nitrogen, water vapor, and other gases (*104*). It is generally believed that the contemporary oxygen-containing atmosphere is predominantly owing to plant photosynthesis. At the dawn of life, possibly more than 3.5×10^9 yr ago, organisms similar to photosynthetic blue-green algae appeared and left their traces in microfossils.

Our present civilization is extensively adapted to the use of organic fossil materials for energy sources, chemical intermediates, and raw materials. The use of these fossil resources, petroleum, gas, and coal results in their inevitable depletion, their combustion to oxidized products ending in carbon dioxide.

Although part of the carbon dioxide is recycled to organic materials by natural photosynthesis in green plants or is precipitated in oceanic sediments, the rate of industrial consumption of fossil fuel has become so large that the natural sinks for carbon dioxide are unable to cope with the increasing supply. During the last 150 yr, the amount of fuel burned has increased by about 4.3% each year (*134*). Hence the global concentration of the Earth's atmospheric concentration has been rising steadily, altogether by ~10% since the industrial revolution. The "greenhouse effect" of this increase in atmospheric carbon dioxide on the future climate of the Earth is the subject of considerable discussion (*62, 108*). The geochemical cycle of carbon is complex and involves essentially two different time scales. On a geological time scale, the deposition and oxidation of organic matter in sediments is measured on the scale of 10^8 yr, whereas on the scale of human generations, the residence time of atmospheric carbon dioxide is estimated to be ~22 yr (*134*).

The photochemical reactions of carbon dioxide are of interest also in the context of the evolution of the atmospheres of the inner planets. Thus, on the surface of Venus, the atmosphere consists mainly (97%) of CO_2, at 95 bar pressure (*128*).

B. Methanol versus Hydrogen Economy

There has been considerable effort to propagate the use of hydrogen as the ideal fuel of the future. Although hydrogen has definite advantages because of its high heat of combustion (143 kJ g^{-1}), it also has severe drawbacks such as the difficulty of storage, its low boiling point (−253°C), and its safety hazard. Our present economy is primarily geared to the use of liquid hydrocarbon fuel, particularly for transportation. A more likely substitute for liquid hydrocarbon fuel could be methanol. Although its heat of combustion is much lower (22.7 kJ g^{-1}) than that of hydrogen, its convenient boiling point (64.8°C) makes it a useful fuel, which could be used for automobiles. Also, methanol can be converted with high efficiency to a mixture of hydrocarbons (*28*). Methanol is at present primarily produced from synthesis gas,

$$CO + 4H_2 \longrightarrow CH_3OH \tag{2}$$

and most of it is further converted by oxidation to formaldehyde, which is one of the most important feedstocks of the chemical industry. Although much of the interest in synthetic fuel has been directed toward methanol, there also exists a definite potential for formic acid as an energy resource. Thus formic acid has been successfully used as the fuel in alkaline fuel cells (*61*). Formic acid has also been proposed as a convenient storage material for hydrogen (*126, 127*).

C. *Hydrogenation of Carbon Dioxide*

Carbon dioxide by itself, or together with carbon monoxide, can be readily reduced by molecular hydrogen, producing methanol or methane, depending on the catalyst or reaction conditions (*32*). Thus on metal catalysts (e.g., Ni), the reaction

$$CO_2 + 4H_2 \longrightarrow CH_4 + 2H_2O \tag{3}$$

occurs at lower temperatures and pressures and with higher selectivity than the corresponding reduction of carbon monoxide (*24*). Hydrogenation of carbon dioxide has also been achieved in the gas phase by a glow discharge (*34, 35*) or in aqueous solutions with rhodium hydride complexes (*133*).

Some reduction of carbon dioxide in aqueous solutions can be obtained with certain transition-metal compounds even in the absence of irradiation. With molybdates such as Na_2MoO_4 or H_2MoO_4 in the presence of Zn amalgam in acid media, the primary product was malic acid, $HOOCCH_2$-$CH(OH)COOH$ (*85*). Obviously, such dark reactions cannot be sustained beyond the consumption of the reducing agent available, in this case the Zn amalgam. With $TiCl_4$ in acid solutions in the presence of Zn amalgam, these authors obtained tartaric and malic acids, whereas with $FeCl_3$ and

Zn amalgam the products were malic, citric, and traces of oxalic and lactic acids. In the absence of transition-metal compounds, but in the presence of Zn amalgam in a strongly acid medium (pH 2), traces of formic acid were produced. Analogous results to the preceding experiments were obtained in the electroreduction of carbon dioxide in the presence of Na_2MoO_4 in acid solutions on a stainless steel cathode. The products identified included malic and lactic acids, using a reduction potential of 0.6–0.8 V.

Obviously, the reduction of carbon dioxide with molecular hydrogen or hydrides requires that one first produce these forms of reduced hydrogen. It would be, therefore, extremely attractive to find a means of using water as the source of hydrogen and sunlight as the source of energy. Several reviews have stressed the importance of solar-energy conversion and storage by fuel-forming reactions, such as the reduction of carbon dioxide to organic compounds (*1, 21*).

Carbon dioxide is an extremely abundant and inexpensive resource because it is a waste product of many industries, such as in the production of limestone and cement, fermentation of carbohydrates, and on a vast scale in fossil-burning power plants. It may also be produced by concentrated solar energy, by calcination of limestone at 800–900°C (*37, 38*).

II. Energetics of Carbon Dioxide Reduction

The study of energy-storing chemical reactions can be directed to several fuel-forming reactions, which all involve only gaseous and liquid reactants and products—thus avoiding the difficulties of handling solids.

For the reaction of gaseous water at 1 atm pressure

$$H_2O(l) = H_2(g) + \tfrac{1}{2}O_2(g) \tag{4}$$

which involves a two-electron transfer ($n = 2$), the Gibbs free energy of the reaction at 25°C is $\Delta G° = 56.69$ kcal mol^{-1} (237 kJ mol^{-1}). Thus the energy required per electron transferred is (*82*)

$$\Delta G/n = 56.69/2 = 28.34 \quad \text{kcal} \quad \text{mol}^{-1} \quad \text{or} \quad 1.23 \quad \text{eV}, \tag{5}$$

which is the equilibrium electrochemical potential of this reaction. In reality, because of overvoltages, a much higher potential is required to effect water electrolysis. In photoelectrolysis, part or all of this potential may be derived from the energy of sunlight.

A variant of the photoelectrolysis of water is to introduce into the electrolysis cell not only water but also carbon dioxide. In this case, in addition to water splitting there will be on the cathode a competing reaction of carbon dioxide reduction, producing formic acid, formaldehyde, methanol, and methane. The anodic reaction will include in addition to water oxidation also the oxidation of the preceding organic products. Using a sufficiently fast stream of carbon dioxide, it may be possible to sweep out the bulk of the organic products out of the reaction zone into a cold trap, thus decreasing the extent of such a back reaction.

There seems to be a considerable advantage in carrying out the photoelectrolysis of aqueous carbon dioxide rather than that of water alone. Although both reactions are thermodynamically uphill, the reactions leading from carbon dioxide to organic compounds such as methanol or methane require a smaller energy input per electron transferred than that required for water splitting (*69*).

The reactions leading from carbon dioxide and water directly to formic acid, formaldehyde, methanol, and methane involve 2, 4, 6, and 8 electron transfers, respectively. They are represented by Eqs. (6)–(9), with the free-energy change per electron transfer given by $E°$.

$$CO_2(g) + 2H_2O(l) = HCOOH(aq) + \tfrac{1}{2}O_2 \qquad E° = 1.428 \quad eV \tag{6}$$

$$CO_2(g) + H_2O(l) = HCHO(aq) + O_2(g) \qquad E° = 1.350 \quad eV \tag{7}$$

$$CO_2(g) + 2H_2O(l) = CH_3OH(aq) + \tfrac{3}{2}O_2(g) \qquad E° = 1.119 \quad eV \tag{8}$$

$$CO_2(g) + 2H_2O(l) = CH_4(g) + 2O_2(g) \qquad E° = 1.037 \quad eV \tag{9}$$

For the water-dissociation reaction, it had been shown that the value of the free-energy change, hence the electrochemical potential, decreases as the temperature is raised (*31*). It is interesting to test to what extent the values of the electrochemical potentials of reactions (6)–(9) of CO_2 are affected by changes in temperature. We calculated these, taking the known data for the enthalpy $\Delta H°$, the free energy $\Delta G°$, the entropy $S°$, and the heat capacity $c°_p$ (*102*), and using Kirchoff's equation $[d(\Delta H)/dT]_p = c_p$ (*31*). For the reduction of carbon dioxide and water to formic acid [Eq. (6)], such a calculation indicates that an increase in temperature causes an enhancement in the value of the electrochemical potential and is thus unfavorable for carrying out this reaction. In contrast, for the carbon dioxide reduction directly to formaldehyde, methanol, or methane, such calculations indicate a slightly favorable temperature effect. Thus the predicted temperature effect on the energy requirement of the

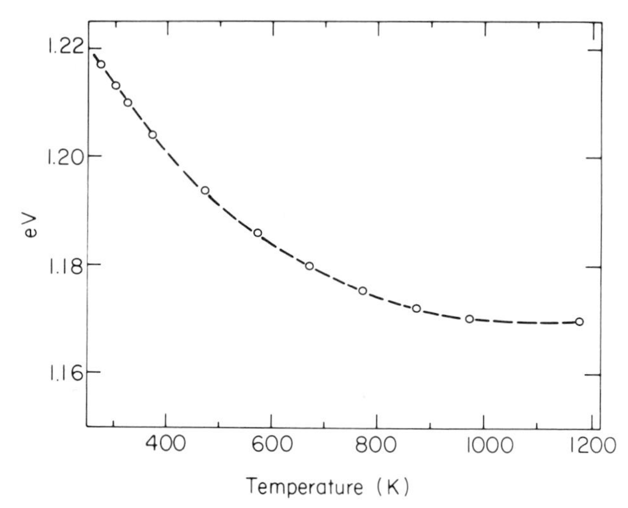

Fig. 1. Temperature dependence of the energy requirement per electron transfer (in electron volts) for the reaction $CO_2(g) + 2H_2O(l) = CH_3OH(l) + \frac{3}{2}O_2(g)$.

reduction of carbon dioxide to methanol (a six-electron transfer) by the reaction

$$CO_2(g) + 2H_2O(l) = CH_3OH(l) + \tfrac{3}{2}O_2(g) \tag{10}$$

is shown in Fig. 1 as a plot of energy per electron transferred (in electron volts) versus temperature. This calculation predicts a slight advantage of decreasing potential requirement with rising temperature, up to about 400–500 K.

III. Photochemical Fixation of Carbon Dioxide

Interest in the direct photolysis of gaseous carbon dioxide has been stimulated by the detection of many simple one-, two-, and three-carbon compounds in interstellar and intergalactic space (*25, 52, 72, 99, 124*). Carbon dioxide is an important constituent of the atmospheres of the inner planets of our solar system (*84*). In an attempt to simulate photosynthesis in outer space by high-energy radiation, aqueous carbon dioxide was bombarded with 40-MeV helium ions from a cyclotron. The products identified were HCOOH and HCHO (*45*). A careful and detailed review of prebiotic photosynthetic reactions has been given by Chittenden and Schwartz (*29*).

Because neither carbon dioxide nor water absorb light in the visible region, the direct photolysis of carbon dioxide requires vacuum-uv irradiation. By illumination of carbon dioxide in the 147- and 130.2- to 130.6-nm regions (from a xenon discharge), the quantum yield for the reaction

$$CO_2 \longrightarrow CO + O(D) \qquad (11)$$

was found to be unity (*109*). Vacuum-uv irradiation in the same wavelength regions of a mixture of gaseous carbon dioxide and water also caused water dissociation (*54*)

$$H_2O \longrightarrow OH + H \qquad (12)$$

and production of the formyl radical HCO

$$H + CO \longrightarrow HCO \qquad (13)$$

The stable products identified were formaldehyde and glyoxal:

$$2HCO \longrightarrow HCHO + CO + CHOCHO \qquad (14)$$

The cross sections for the photodissociation of CO_2 in the vacuum-uv region to electronically excited carbon atoms have been applied to a model of chemical reactions in the atmospheres of Mars and Venus (*132*).

The realization that all life on Earth depends directly or indirectly on the photosynthetic fixation of carbon dioxide in green plants and photosynthetic bacteria has stimulated many attempts to duplicate carbon dioxide fixation in artificial systems. It was Baeyer (*8*) who first proposed that the initial stage in the synthesis of organic matter from inorganic constituents by green plants involves the reaction of carbon dioxide and water to produce formaldehyde and oxygen, the energy for the endothermic reaction being supplied by light.

Such experiments on nonbiological photosynthesis have usually been carried out by irradiating aqueous solutions with either uv and visible light from a mercury arc or with sunlight. The particular products sought were formaldehyde and sugars. Homogeneous and heterogeneous catalysis have been described in which dissolved metal ions or suspended solids caused an increase in the rates of yields of formation of the organic products. An extensive literature has appeared since the turn of the last century describing both such experiments as well as the controversy caused by the many unsuccessful attempts to repeat these observations (*6, 7, 10, 11, 14, 15, 36, 44, 50, 91, 111, 120*).

One example of this early work was the photosynthesis of carbohydrates from hydrated carbon dioxide in the presence of a $NiO–ThO_2$ mixture (molar ratio 24 : 1) adsorbed on kieselguhr (*11*). The product could be hydrolyzed by takadiastase to reducing sugars.

Getoff *et al.* (*47*) and Getoff (*48, 49*) studied the uv irradiation of aqueous carbon dioxide, using a low-pressure Hg discharge as the light source. The products identified were carbon monoxide, aldehydes, and formic acid. An enhancement in the yields of aldehydes and formic acid was observed in the presence of ferrous ions. The initial step was proposed to be the formation of CO_2^- radicals:

$$Fe^{2+} + CO_2 \longrightarrow Fe^{3+} + CO_2^- \tag{15}$$

$$CO_2^- + H \longrightarrow CO + OH^- \tag{16}$$

Hydrogen peroxide was also produced, presumably by dimerization of OH radicals.

A study of the photochemical metal-ion-promoted reduction of carbon dioxide and of formaldehyde in aqueous solutions was reported by Åkermark *et al.* (*2*). Transition-metal salts were tested for their effectiveness in achieving the photoreduction of CO_2. Best results were obtained with Fe^{2+}, $Fe(bipyridyl)_3{}^{2+}$, and Cr^{2+}. Based on the amount of ferrous ions that had been consumed, the chemical yield of the formaldehyde produced was ~0.06%. After irradiation of aqueous formaldehyde or methanol, the reduction products were methanol or methane, respectively. The conversion of formaldehyde to methanol in the presence of Cr^{2+} or Fe^{2+} ions occurred with relatively high yields, (~6–8%), whereas the conversion of methanol to methane was found to proceed only with extremely low yields.

In the search for models of prebiotic chemical evolution, photosynthetic carbon dioxide fixation experiments were made using model protocells (prepared by quenched spark discharges through gas mixtures of N_2, CH_4, and CO_2 over a distilled water surface) (*39*). Ultraviolet irradiation was carried out at 254 nm, in the presence of 710 torr N_2, 50 torr H_2 and $NaH^{14}CO_3$. The most abundant product was formaldehyde. Nitrogen-containing products identified included glycine, alanine, β-alanine, and valine. Quenched spark discharges through molecular nitrogen over aqueous suspensions of calcium carbonate were also carried out in the simulated interphase of a primitive Earth atmosphere–hydrosphere. In addition to hydrazine, the product mixture contained in low yield an

$$CO_2 + H_2NNH_2 \rightarrow H_2NNHCOOH \rightarrow H_2NNHCONHNH_2 + H_2O \tag{17}$$

interesting compound, carbohydrazide (*40*). Carbazic acid (H_2N-NHCOOH) is the postulated intermediate in this reaction.

In a related study, carbon dioxide photoreduction was reported by illumination with visible or uv light of organomolybdenum microstructures (prepared by sunlight irradiation of ammonium molybdate, containing a mixture of diammonium phosphate, formaldehyde, calcium acetate, NaCl, KH_2PO_4, $MgSO_4$, $MnSO_4$, K_2SO_4, and $FeSO_4$) (*110*).

In an effort to mimic more closely the photosynthesis of carbon compounds in green plants, platinized chlorophyll *a* dihydrate polycrystals were illuminated in the presence of carbon dioxide and water (*42*). Using isotope labeling and mass spectrometry for product analysis, they detected formic acid as the only product. The oxygen produced caused a

$$2CO_2 + 2H_2O \rightarrow 2HCOOH + O_2 \tag{18}$$

poisoning of the Pt–chlorophyll photoactivity and thus inhibited further reaction.

In a photogalvanic system, tetrasulfonated metal phthalocyanins in aqueous solutions were used as photosynthesizers for visible light (*30*). With Mg or Zn as the metal in the phthalocyanin, and using platinum electrodes, an increase in photocurrents was observed in the presence of carbon dioxide and small amounts of formaldehyde were detected as the reaction products.

A model for photochemical reactions in a hypothetical primitive Earth atmosphere was proposed in which a mixture of molecular nitrogen, carbon dioxide, water vapor, and traces of carbon monoxide were assumed to react under a given flux of uv and visible light (*97*). Based on known rate constants, a model was created for the rates of production and loss of HCO radicals and formaldehyde.

In homogeneous polar media, such as aqueous acetonitrile, photofixation of carbon dioxide to formic acid could be achieved, using aromatic hydrocarbons both as photoactive agents and as electron donors (e.g., pyrene or perylene) (*113*). As electron acceptors, 1,4-dicyanobenzene or 9,10-dicyanoanthracene were used, which transferred electrons to carbon dioxide. Presumably, the radical anion CO_2^- was produced as an intermediate, which was further reduced to formic acid or dimerized to oxalic acid.

Some attempts have been made to apply carbon dioxide fixation by organometallic complexes to photochemical reactions. Such a reversible binding was achieved with the Cu(I)phenylacetylide–phosphine complex, which acts as reversible CO_2 carrier at ambient temperature and ordinary pressure by CO_2 insertion into the Cu—C bond of the complex (*117*).

Ruthenium bipyridyl complexes [$Ru(bipy)_3^{2+}$] have been intensively tested as sensitizers for photochemical water splitting using visible light. Two reports describe applications of such complexes to carbon dioxide fixation. Using $Ru(bipy)_3^{2+}$ as a photosensitizer to visible light, EDTA as a sacrificial electron donor, a strongly reducing electron relay agent (1,1′-trimethylene-4,4′-dimethyl-2,2′-dipyridylium dibromide), and colloidal TiO_2 as the catalyst, formic acid was detected as the organic product (*79*). The very interesting conversion of carbon dioxide and water under illumination with visible light to carbon monoxide and hydrogen ("synthesis gas") was achieved using solutions of $Ru(2,2'\text{-bipy})_3^{2+}$, $CoCl_2$, and CO_2 in aqueous acetonitrile as solvent, with triethylamine as electron donor (*83*). The endothermic reaction leading to carbon monoxide involves the stan-

$$CO_2(g) = CO(g) + \tfrac{1}{2}O_2 \quad (19)$$

dard free-energy change $\Delta G° = 61.4$ kJ mol^{-1} and the standard potential per electron transfer $E° = 1.33$ V (two-electron transfer).

IV. Electrochemical Reduction of Carbon Dioxide

The earliest report on the electrochemical reduction of carbon dioxide was more than a century ago (*103*), by electrolyzing aqueous sodium bicarbonate with a zinc or zinc amalgam electrode. The product was found to be formic acid. This work has since been repeated many times, using a large variety of electrode materials and electrolyte media. Thus, on mercury electrodes, in an 0.1 *M* solution of quarternary ammonium salts through which carbon dioxide was bubbled, a polarographic wave was obtained having a half-wave potential of 2.1 V versus SCE (*114, 115*).

In a study of the electrochemical reduction of CO_2 to formic acid, optimal conditions were found using a mercury cathode, a neutral electrolyte (0.05 *M* phosphate buffer, pH 6.8), and at a cell potential of 3.5 V, a current density of 20 mA cm^{-2}, resulting in a coulombic efficiency of 81.5% (*105*). The further reduction of formic acid to methanol was found to require an acid medium. Optimal results were on an electroetched tin cathode, in a 0.25 *M* HCOOH and 0.1 *M* $NaHCO_3$ buffer solution (pH 3.8). Almost 100% current efficiency in the reduction to methanol was achieved, but only in a narrow electrode potential region of −0.72 to −0.68 V and at the very low current densities of 2–3 μA cm^{-2}. No formaldehyde was detected in the electrolysis products from formic acid. For the electrochemical reduction of formaldehyde to methanol, a basic medium was required (about pH 11). By electrolyzing 0.1 *M* CH_2O in 0.1 *M* Na_2CO_3 on a mercury pool at a current density of 10 mA cm^{-2}, the efficiency for methanol production was 100%. Owing to the high overpotential required (~1.2 V), the energy efficiency was low.

In a comparison of various electrode materials for the cathodic reduction of carbon dioxide to formic acid, the lowest overpotential and the highest current efficiency (92%) were obtained at an indium electrode in an aqueous solution of lithium carbonate (*75*). However, such high efficiency was achieved at the rather low current density of ~5 mA cm^{-2}, whereas at higher current densities the efficiency dropped markedly. A cation-exchange membrane was used to separate the cathodic and anodic compartments, and platinum was used as the anode.

The earlier work on the electrochemical reduction of carbon dioxide on metal electrodes has been carefully reviewed by Russel *et al.* (*105*).

V. Dynamics of Carbon Dioxide Reduction

Evidence on the rates and mechanisms of the elementary reactions involved in the reduction of carbon dioxide has come from pulse radiolysis, flash photolysis, electron spin resonance, and electrochemical studies.

In the electrochemical reduction of carbon dioxide on mercury electrodes, the species that undergoes electron capture was proven to be the neutral CO_2 molecule, not the bicarbonate or carbonate ion (*114, 115*). The postulated mechanism involves electron capture by CO_2, producing the CO_2^- radical-anion, which then undergoes rapid protonation to form the neutral HCO_2 radical. In a subsequent step, this species undergoes a second electron capture, producing the formate anion HCO_2^- (*77*).

Carbon dioxide in aqueous solutions reacts with the hydrated electron $e^-(aq)$ at a rate that is close to the diffusion-controlled limit, with a second-order rate constant $k = 7.7 \times 10^9\ M^{-1}\ s^{-1}$ (*66*). In contrast, the

$$e^-(aq) + CO_2 \longrightarrow CO_2^- \qquad (20)$$

$$e^-(aq) + HCO_3^- \longrightarrow HCO_3^{2-} \qquad (21)$$

reaction with the bicarbonate ion is very much slower, with an estimated rate constant $k = 6 \times 10^5\ M^{-1}\ s^{-1}$.

By pulse radiolysis of deaerated, CO_2-saturated aqueous solutions, a strong transient uv-absorption maximum at 250 nm was produced (*78*). This absorption maximum was assigned to the CO_2^- radical-anion, which is in equilibrium with its conjugate acid HCOO. Both the anion and the

$$CO_2^- + H_2O \rightleftharpoons HCOO + OH^- \qquad (22)$$

acid form seem to have the same absorption maximum and also the same extinction coefficient, which in aqueous solution is 2400 $M^{-1}\ cm^{-1}$. The pK of the carboxyl radical is estimated to be in the region 4–5. The same absorption spectrum was found with aqueous solutions of carbon monoxide, sodium bicarbonate, formic acid, and sodium formate and in all cases was ascribed to formation of the CO_2^- species. The decay of this transient absorption maximum obeyed second-order kinetics and depended on the ionic strength of the solution. At zero ionic strength, the second-order rate constant for the reaction

$$CO_2^- + CO_2^- \longrightarrow CO_2^-{-}CO_2^- \quad (23)$$

is $k_o = 0.5 \times 10^9\ M^{-1}\ s^{-1}$. The same molar extinction coefficient at 250 nm and the same second-order rate constant k_o were found both at pH 5 (in a CO_2-saturated sodium formate solution) and at pH 2.4 (in CO_2-saturated formic acid solution).

There was no measureable reaction (on the time scale of the pulse radiolysis experiments, 0.2–2 μs) between CO_2 and H or OH radicals. However, OH was found to react with the bicarbonate ion, producing the CO_3^- radical-ion, which is in equilibrium with its conjugate acid HCO_3.

$$HCO_3^- + OH \longrightarrow CO_3^- + H_2O \rightleftharpoons HCO_3 + OH^- \quad (24)$$

This species absorbed strongly at 600 nm. Thus pulse radiolysis of 0.5 M sodium bicarbonate saturated with CO_2 (pH 7.6) caused absorption peaks both at 250 and 600 nm, owing to the CO_2^- and CO_3^- species, respectively. The molar extinction coefficient of CO_3^- in aqueous solution at 600 nm was 2900 $M^{-1}\ cm^{-1}$. The second-order rate constant for the formation of the CO_3^- radical-ion [reaction (24)] was found to be $k = 1 \times 10^7\ M^{-1}\ s^{-1}$ (*78*).

Although neutral carbon dioxide or the bicarbonate ion can undergo reduction reactions, the carbonate ion CO_3^{2-} may also undergo oxidation reactions. This was observed by pulse radiolysis with 15-MeV electrons from a linear accelerator on aqueous sodium carbonate, which contained nitrous oxide to convert e^-(aq) to OH radicals (*125*). The CO_3^- radical was detected by its intense absorption maximum at 600 nm, with an extinction coefficient of 1860 $M^{-1}\ cm^{-1}$. The reactions observed and second-order rate constants determined in this study were

$$OH + CO_3^{2-} \longrightarrow OH^- + CO_3^- \quad (25)$$

$$k = 4.2 \times 10^8\ M^{-1}\ s^{-1}$$

$$CO_3^- + CO_3^- \longrightarrow 2CO_2 + HO_2 + OH^- \quad (26)$$

$$2k = 1.25 \times 10^7\ M^{-1}\ s^{-1}$$

$$OH + HCO_3^- \longrightarrow CO_3^- + H_2O \quad (27)$$

$$k = 1.5 \times 10^7\ M^{-1}\ s^{-1}$$

Formation of the CO_3^- radical was also observed by interaction of the carbonate ion with the triplet state of duroquinone (DQ_T) in a water–ethanol solution, according to

$$DQ_T + CO_3^{2-} \longrightarrow DQ^- + CO_3^- \quad (28)$$

with a rate constant of $7 \times 10^7\ M^{-1}\ s^{-1}$ (*106*). The progress of the reaction could be followed by the appearance of the CO_3^- absorption maximum at

600 nm, with an extinction coefficient in this solvent of 1860 M^{-1} cm^{-1} (*125*). In an aqueous micellar solution of dodecyltrimethylammonium chloride, the reaction (28) was much faster than in water–ethanol and was followed by a subsequent reaction step in which the CO_3^- radical reacted with a second triplet duroquinone molecule, forming carbon superoxide

$$DQ_T + CO_3^- \longrightarrow DQ^- + CO_3 \quad (29)$$

CO_3 (*107*), with a rate constant 2×10^9 M^{-1} s^{-1}. The carbon superoxide eventually decomposed to oxygen and carbon dioxide:

$$CO_3 \longrightarrow CO_2 + \tfrac{1}{2}O_2 \quad (30)$$

A fundamental series of studies on the reduction of carbon dioxide in aprotic solvents was performed with the purpose of determining the kinetic parameters for the individual reaction steps (*81*). Using dimethylformamide (DMF) as solvent, in the presence of suspended active alumina to decrease the residual water concentration, Lamy *et al.* achieved less solvation of the CO_2^- radical-ion intermediate, and hence more coulombic repulsion against dimerization. The supporting electrolyte was 0.1 M tetraethylammonium chloride and the CO_2 concentration was 1–8 $\times$ 10^{-3} M. The standard potential of the CO_2/CO_2^- redox couple thus determined in dry DMF was −2.21 V versus SCE. The standard rate constant of electron transfer was 6×10^{-3} cm s^{-1}, and the transfer coefficient was 0.4. The second-order rate constant of CO_2^- deactivation (by the dimerization reaction) was 10^{-7} M^{-1} s^{-1}. For the overall carbon dioxide reduction in DMF, the rate-determining step is the charge transfer itself. The reaction path leading to CO and carbonate may possibly involve an intermediate with a C—O—C linkage. In a continuation of this study—using either Hg or Pb

$$O{=}C^-{-}O{-}C({=}O){-}O^- \longrightarrow CO + CO_3^{2-} \quad (31)$$

cathodes and DMF or dimethyl sulfoxide (DMSO) as solvents, the main products were found to be oxalic, formic, and glycolic acids and carbon monoxide (*53*). Formic and glycolic acids were produced if the electrolyte also contained water. Lead was found to be the better electrode for oxalate production, which involves the reaction

$$CO_2^- + CO_2 + e^- \longrightarrow C_2O_4^{2-} \quad (32)$$

On Hg cathodes, carbon monoxide production is the predominant reaction.

Using fast microelectrolytic techniques, a further study by the same research group was made on the dynamics of the electroreduction of carbon dioxide (*3*). Although formic acid is the predominant product in water as the solvent, in solvents with low proton availability such as DMF, the products are oxalate and carbon monoxide, along with formate.

The authors distinguished three competing pathways for reactions of the intermediate CO_2^- radical:

1. Oxalate formation through self-coupling of the radical

$$2CO_2^- \xrightarrow{k_1} C_2O_4^{2-} \tag{33}$$

2. Carbon monoxide formation via oxygen–carbon coupling of the radical with neutral carbon dioxide

$$CO_2^- + CO_2 \xrightarrow{k_5} C(O)—O—C(O)—O^- \tag{34}$$

3. Formate production through protonation of CO_2^- by residual or added water, followed by an additional electron transfer

$$CO_2^- + H_2O \xrightarrow{k_2} HCO_2 \tag{35}$$

$$HCO_2 + e^- \rightleftharpoons HCOO^- \tag{36}$$

In DMF as solvent, at a mercury electrode, the rate constants obtained were $k_1 = 10^7\ M^{-1}\ s^{-1}$, $k_2 = 7.7 \times 10^2\ M^{-1}\ s^{-1}$, and $k_5 = 3.2 \times 10^3\ M^{-1}\ s^{-1}$.

An important tool in the detection and identification of free radicals is electron spin resonance (esr). Many radical intermediates in photochemical or radiation induced reactions are short lived but can be detected directly by esr spectroscopy at low temperatures, such as in a frozen or glassy state. In a model of a simulated Martian surface, photoinduced free radicals were detected by esr spectroscopy on mixtures of CO, CO_2, and H_2O adsorbed on the surface of silica gel at a temperature of −170°C, while irradiating with light of 253.7 nm (*116*). When irradiating frozen water alone on silica gel, the only radical species identified were OH radicals, whereas in the presence of oxygen the additional signals were owing to the O_2H radicals. Irradiation of CO and CO_2 in the presence of frozen water gave signals identified as CO_2H radicals. On letting the samples warm up to room temperature, a new signal appeared, assigned to the CO_2^- radical-ion. The stable products detected after the uv irradiation of CO_2, CO, and water on silica gel were found to be formaldehyde and glycolic acid.

An alternative method for detection of radicals by esr is to stabilize them by letting them react with specific "spin traps," producing stable or long-lived radicals that can be identified by esr at room temperature. This tool was used to detect radical intermediates in the photoassisted decomposition of water in the presence of TiO_2, proving the formation of OH, O_2H, and other radicals (*55, 63, 64, 76, 80*). This method was applied to elucidate the effects of illumination on various one-carbon compounds in the presence of suspended tungsten oxide (*5*). The spin traps used were α-

phenyl-*N*-*tert*-butylnitrone (PBN) and α-(4-pyridyl 1-oxide)-*N*-*tert*-butylnitrone (POBN).

Irradiation of WO_3 in aqueous suspension in the presence of PBN revealed the esr signal of the spin trap of the OH radical, revealed as a triplet of doublets, with the hyperfine splitting constants of $a_N = 15.4$ G and $a_H = 2.7$ G, similar to those previously reported in various aqueous systems [e.g., see Jaeger and Bard (*76*).] The OH radical is probably formed by interaction of holes from the irradiated tungsten oxide with water:

$$h^+ + H_2O \longrightarrow H^+ + OH \quad (37)$$

On adding sodium formate to the preceding mixture, an additional spin-trap signal appeared, which was assigned to the CO_2^- radical, presumably produced by reaction of OH radicals with formate ion:

$$OH + HCO_2^- \longrightarrow H_2O + CO_2^- \quad (38)$$

The same radical had previously been obtained by reaction (38) from formate ions and OH radicals produced with Ti^{3+}/H_2O_2 (*64*) or by irradiation of aqueous formate in the presence of suspended zinc oxide (*55*). Irradiation of an aqueous suspension of the spin trap, tungsten oxide, and methanol also gave in addition to the signal owing to the OH radical adduct a signal assigned to the CH_2OH radical, probably formed by

$$OH + CH_3OH \longrightarrow H_2O + CH_2OH \quad (39)$$

With suspensions of WO_3 in aqueous sodium hydrogen carbonate (pH 6.7) in the presence of a spin-trap reagent, a complicated spectrum was obtained, which on short-time illumination revealed the presence of OH and CO_3^- radicals, whereas prolonged irradiation (after 4.5 h) produced additional spin adducts that could tentatively be assigned to CO_2^- and CH_2OH radicals (*5*). No signals were obtained with the spin traps and aqueous suspensions of WO_3 containing formaldehyde.

VI. Photoelectrochemical Reduction of Carbon Dioxide

Since the 1970s there has been a considerable research effort directed toward harnessing photochemical reactions for solar energy conversion and storage. Energy conversion in photochemical reactions usually depends on charge transfer, with production of reduced and oxidized species. In homogeneous solutions a major loss mechanism is the recombination of these charged intermediate species. This difficulty was partly overcome with the discovery in 1972 that such a charge separation can be achieved indirectly by band-gap illumination of semiconductors in liquid-junction contact with electrolyte solutions. Thus Fujishima and Honda (*43*) showed that *n*-type titanium oxide under illumination with near-uv light gives an anodic response. It can therefore serve as a photoanode in a photoelectrochemical cell, releasing oxygen from water. Also, by combining *n*- and *p*-type semiconductors, photochemical diodes were constructed that were able to split water both to oxygen and hydrogen without requiring an external bias (*95*). The theory, scope, and limitations of liquid-junction solar cells for the conversion of optical to electrical and chemical energy have been the subject of many excellent reviews (*12, 13, 26, 46, 65, 71, 86, 87, 96, 131*). Essentially, the semiconductor–electrolyte interface is considered to operate analogously to the Schottky barrier at a semiconductor metal junction.

For solar-to-electrical energy conversion, it has been possible to achieve remarkably high conversion efficiencies, up to 12% with single-crystal *n*-GaAs and *p*-InP (*68*) and up to 12.4% with single-crystal *n*-CdSe (*41*). In these cases the semiconductors chosen had been optimally matched to the solar spectrum, that is, with band gaps in the range 1.2–1.6 eV. In contrast, there has been less success in photoelectrolysis for water splitting (*16*). This requires supplying not only the energy needed on the basis of the free energy of formation of the water molecule, but also additional energy for the overvoltages in the anodic and cathodic parts of the reaction. With semiconductor electrodes under band-gap irradiation, a partial gain may be obtained because of the photovoltaic effect—which permits carrying out electrolysis with a lower bias potential than with a metal electrode in the dark.

The first stage in the reduction of carbon dioxide to stable organic compounds is the two-electron-transfer reaction producing formic acid:

$$CO_2 + H_2O \longrightarrow HCOOH + \tfrac{1}{2}O_2 \tag{40}$$

For the redox couple $CO_2/HCOOH$, the standard reduction potential is −0.2 V, which is thus only slightly more negative than the standard hydrogen-electrode potential of 0 V. Therefore, the conditions optimal for carbon dioxide reduction can be expected to be very similar to those for water electrolysis. In common with water splitting, process (40) involves the oxidation of water to oxygen as the anodic reaction, with the standard redox potential of the O_2/H_2O couple at +1.23 V. For water reduction, it was found that *p*-type semiconductors, under band-gap illumination, are able to serve as photocathodes, in which part of the potential required for electrolysis is supplied by light energy (*17*). Electrons from the conduction band of semiconductor electrodes should be able to reduce carbon dioxide, just as they are able to reduce water or H_3O^+. One of the relatively few known semiconductors that fulfill the conflicting requirements of relatively good stability in aqueous media, a band gap large enough to supply the potential requirements of electrolysis (both the thermodynamic potential and the electrodic overpotentials as well as ohmic losses), yet small enough to at least partially match the solar spectrum, is *p*-gallium phosphide, which has a band gap of 2.25 eV. Using single-crystal *p*-GaP as a photocathode, a CO_2-saturated phosphate buffer (pH 6.8) as the electrolyte, and a graphite rod as the anode, a photocurrent response was achieved on the semiconductor electrode when a cathodic bias of at least −0.4 V was applied to it (*57*). The *IV* curve for such an experiment is presented in Fig. 2a. The predominant product of the photoelectrolysis was formic acid, as during electrolysis of carbon dioxide on metal electrodes. However, in addition there was also production of smaller amounts of formaldehyde and of methanol. Thus, after 18 h of illumination with a 75-W high-pressure mercury lamp (light flux 210 mW cm^{-2}), with a cathodic bias of −1.0 V, the concentrations of formic acid, formaldehyde, and methanol were 1.2×10^{-2}, 3.2×10^{-4}, and 1.1×10^{-4} *M*, respectively. For estimation of the solar-to-chemical energy-conversion efficiency, illu-

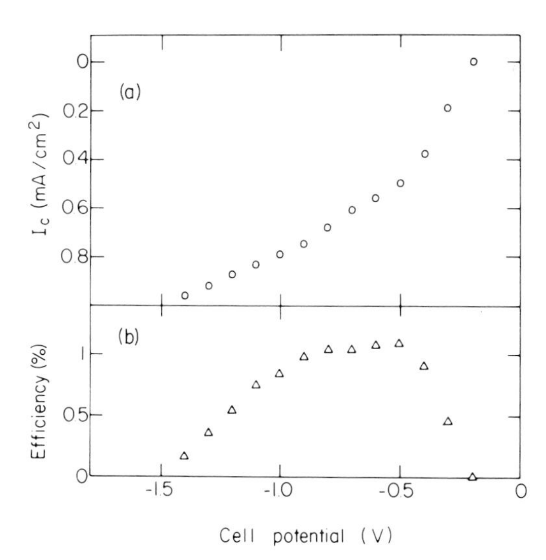

Fig. 2. (a) Current versus cell potential for the reduction of CO_2 on *p*-GaP photocathode. Counterelectrode: carbon rod. Electrolyte: 0.05 *M* potassium phosphate buffer, pH 6.17. Light source: monochromatic 365 nm; flux I_o = 6.85 mW cm^{-2}. Temperature: 82.5°C. (b) Efficiency of solar-energy conversion as a function of the cell potential.

mination was carried out with 365-nm monochromatic light. The efficiency (%) of solar-to-chemical energy conversion was calculated by

$$\text{Efficiency} = 100 I_c S F[(\Delta G^\circ/z) - V_b]/I_o, \tag{41}$$

where S is the fraction of the solar spectrum that is effective in exciting the semiconductor used (for GaP, S is taken as 0.17), F is the Faradaic current efficiency, I_c the current density (in mA cm^{-2}), and I_o the incident light energy (in mW cm^{-2}), and ΔG° the standard free-energy change of formation of the reduction products. Thus, for liquid formic acid at 25°C, ΔG° is 68.35 kcal mol^{-1}. Let z be the number of electrons involved in the reduction of one molecule of carbon dioxide to one molecule of product. Hence, for HCOOH(l) production, $\Delta G^\circ/z$ is 1.48 V (V_b is the electrical bias measured in volts). The results for such an experiment, taking F = 0.9, are shown in Fig. 2b, which shows a maximum solar-to-chemical conversion efficiency of 1.1% at a cell potential of 0.5 V. The values of $\Delta G^\circ/z$ for the production of HCHO(g) and of CH_3OH(l) are 1.35 and 1.21 V, respectively. The evolution of organic-product concentration in a similar but preparative experiment as the function of photoelectrolysis time is presented in Fig. 3.

Similar carbon dioxide reduction experiments with an illuminated *p*-GaP electrode were carried out by Inoue *et al.* (*74*), but using 0.5 *M* sulfuric acid as the electrolyte. At a cathodic bias of −1.5 V versus SCE, illuminating for 2 h with either a 500-W xenon lamp or a 500-W high-pressure mercury arc, and filtering out the light of wavelengths above 500 nm, the concentrations of formaldehyde and of methanol were 3.6 and 7.3 m*M*, respectively. Almost the same results were obtained by the previous authors in the absence of illumination, using single-crystal *n*-TiO_2 electrodes at the same cathodic bias of −1.5 V versus SCE. Concentrations of formaldehyde and methanol were 2.8 and 5.7 m*M*, respectively. The production of formic acid and methane was also reported.

The reduction of carbon dioxide at *p*-GaP photocathodes in aqueous lithium carbonate electrolytes was the subject of a careful study by Taniguchi *et al.* (*135*). They observed that the reduction of carbon dioxide occurs in competition with hydrogen evolution, and that the current efficiency is markedly enhanced in the presence of crown ethers such as 15-crown-5 ether in the electrolyte. A chemical reaction mechanism was proposed, which involves the reduction of carbon dioxide by cathodically deposited lithium metal, with formation of the COO^- radical-anion:

$$Li + CO_2 = Li^+ + COO^- \tag{42}$$

Such a reaction could be in parallel with the direct cathodic reduction, by electron transfer from the conduction band of *p*-GaP to the carbon

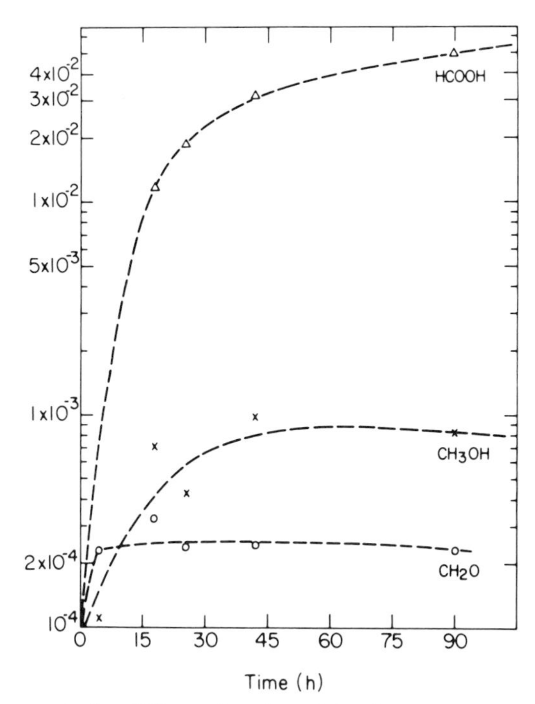

Fig. 3. Product concentration (M) as a function of time in the photoelectrolysis of CO_2 by p-GaP (bias -1.1 V versus SCE). Electrolyte: potassium phosphate buffer 0.05 M, pH 6.8. Counterelectrode: carbon rod. Light source: 45-W high-pressure Hg lamp, unfiltered.

dioxide. The subsequent reactions would then be as shown above in Eqs. (35) and (36). Evidence for reaction (42) was obtained by carrying out the reaction of lithium metal with carbon dioxide in propylene carbonate solution. The COO^- radical-anion was detected by its strong absorption maximum at 265 nm (*135*).

Polycrystalline TiO_2 electrodes (prepared by thermal decomposition in air of $TiCl_4$ on titanium plate) were also effective for the reduction of carbon dioxide, even in the dark (*89, 90*). Cyclic voltammetric measurements in CO_2-saturated 0.5 M KCl solutions indicated that the reduction of CO_2 occurred at potentials positive with respect to the hydrogen electrode potential and before the start of the evolution of hydrogen. Thus there exists even in the dark a relatively small overvoltage for CO_2 reduction on TiO_2, which seems to have remarkably good electrocatalytic properties for carbon dioxide reduction. With a polycrystalline TiO_2 cathode and a graphite anode, long-term electrolysis provided reduction of CO_2 up to methanol.

VII. Photoreduction of Carbon Dioxide with Semiconductors

Heterogenous catalysis on solid semiconductors may be considered to involve charge transfer across a solid–liquid or solid–gas junction. Illumination of suspended solid semiconductors in liquid media by band-gap irradiation will cause electron excitation from the valence band to the conductance band. Various sites on the semiconductor particles will act as cathodic or anodic centers, which will provide or accept electrons from suitable redox reagents in solution.

In photocatalytic reactions, powdered semiconductors are used, either in suspensions in water or some other solvent, as fluidized beds suspended in a gas stream, or as coated surfaces in liquid–solid or gas–solid contact. Under band-gap illumination, each semiconductor particle acts as a minute photoelectrolysis cell, with separate anodic and cathodic sites on the surface providing positive and negative charges to surface-adsorbed molecules and ions. The cell is short-circuited by electronic conduction through the semiconductor particle and by electrolytic conduction through the surface layer (*13, 23, 51, 98, 121–123*).

The effects of light excitation of semiconductors in contact with solutions may be very complex and may result in changes both in the absorptivity of the semiconductor toward various solutes as well as in its catalytic activity, owing to changes in the surface concentrations of electrons and holes (*129, 130*).

Because of the importance of titanium dioxide as a pigment, there have been many studies on the adsorption of gases and of the sunlight-induced decomposition of such adsorbed species on titanium dioxide. Carbon dioxide adsorption on the anatase form of TiO_2 generally occurs in the form of carbonate-like species, which reveal surface basicity (*93, 94*). Using infrared spectroscopy and microgravimetry, the chemisorption of carbon dioxide on anatase at ambient temperature showed that there exist two forms of adsorbed carbon dioxide:

1. A CO_2 species that is linearly held at surface Lewis-acidic sites
2. Carbonate-like species apparently attached to surface basic sites

Even without illumination, carbon dioxide and water were found to undergo decomposition on prereduced anatase (*101*). This occurred at 773 K and resulted in the production of oxygen that remained adsorbed on the anatase, while hydrogen and carbon monoxide were released.

The amphoteric nature of metal-oxide surfaces, such as those of anatase, rutile, α-Fe_2O_3, and SnO_2, is the result of the structure of the hydroxylated surfaces, one-half of which are acidic, whereas the other half are basic and may be exchanged for other anions (*18*).

Suspended particles of semiconductors such as titanium oxide can be regarded as "microphotocells." By measurements of the dark- and photoconductivity of compressed TiO_2 powders at various temperatures and in controlled atmospheres, TiO_2(rutile) was found to be *n*-type conducting at room temperature and *p*-type conducting at elevated temperatures (about 400–600°C) (*73*). Similar results were obtained by conductivity measurements on single crystals of rutile at various O_2 pressures at 200–500°C (*67*).

A photosynthetic reaction was found to take place between water vapor and carbon dioxide in the presence of a strontium titanate surface in contact with platinum (*69, 70*). Using a clean, single-crystal surface of $SrTiO_3$ (111 orientation, which had been reduced under hydrogen for 2 h at 1000 K) in contact with platinum foil, in about 15 torr CO_2 and 17 torr H_2O, and illuminating with a 500-W high-pressure Hg lamp, there was production of methane, as determined by gas chromatography. The initial production rate of methane was about one molecule for every 10^4 incident photons. The methane production decreased rapidly within ~10 min. The $SrTiO_3$ could be reactivated by recleaning (argon bombardment and brief heating in oxygen). Because the number of methane molecules produced was at least five times larger than the number of surface sites on the crystal, the CO_2 reduction may possibly be a true catalysis. However, the same CO_2 reduction to methane was also observed with the $SrTiO_3$–Pt material heated in the dark to 420 K. Therefore, the reduction could be owing to Ti^{3+} surface ions, which stoichiometrically react with carbon dioxide. The action of the band-gap illumination could then be to regenerate Ti^{3+} sites. uv-Photoelectron spectroscopy (ups) measurements on the cleaned strontium titanate surface showed the presence of Ti^{3+} ions (*70*). The signal owing to these ions decreased markedly after adsorption of water. This was explained as a result of water dissociation, resulting in oxidation of Ti^{3+} to Ti^{4+}, with formation of molecular hydrogen. The reduction of carbon dioxide was proposed by these authors to occur by a multistep process similar to the methanation reaction

$$CO_2 + 4H_2 \rightleftharpoons CH_4 + 2H_2O$$

in which $\Delta G = -1.4$ eV, which is thus thermodynamically downhill.

Carbon dioxide at a gas pressure of 10–40 kg cm^{-2} over an aqueous suspension of Fe_2O_3 was converted by illumination with uv light, radiation from a radioactive source, or electrolysis into formaldehyde and oxygen—as reported in a patent claim (*112*).

A systematic study of a variety of semiconductor powders in aqueous suspensions was carried out by Inoue *et al.* (*74*). Product yields were reported for formaldehyde and for methanol and were found to decrease in the following order of semiconductor materials: SiC, CdS, GaP, ZnO, TiO_2, and WO_3. Experiments were carried out by suspending 1 g of the material in 100 ml of water and bubbling through carbon dioxide, while illuminating with a 500-W lamp (either xenon or high-pressure Hg lamp). The estimated quantum yields for the photosynthetic reduction of carbon dioxide with SiC as the catalyst, measured relative to the absorbable incident photons, were 5.0×10^{-4} for formaldehyde and 4.5×10^{-3} for methanol. The authors found a positive correlation between the yield of methanol produced on each of the preceding semiconductors and the potential of the conduction band of the semiconductor. Silicon carbide with the most negative conduction-band level (on the hydrogen electrode energy scale) thus was the best photocatalyst. The charge transfer between the photogenerated electrons of the conduction band and the carbon dioxide molecule was more effective with the more positive redox potential relative to the conduction-band level.

In a further study of the photoassisted reduction of carbon dioxide on powdered semiconductors, a variety of experimental modes was applied, using either aqueous suspensions or coated surfaces of these materials (*4*, *58*). Light sources were high-pressure Hg lamps (70 or 500 W) or sunlight. Irradiations were usually carried out in open systems, with carbon dioxide flowing continuously through the irradiation chamber and a series of cold traps (for collection of the volatile products) and then released to the atmosphere. However, in one series of experiments, a closed-loop configuration was used, in which the outgoing carbon dioxide after the cold traps was recirculated with a peristaltic pump to the irradiation chamber. Using a 70-W high-pressure Hg lamp (light intensity in the irradiation zone was ~600 W m^{-2}), with strontium titanate suspended in water (~1 g per 160 ml, material was pretreated at 600°C for 6 h under vacuum), at a CO_2 flow rate of 164 ml min^{-1}, the formaldehyde and methanol production rates were 0.04 and 7.0 μmol h^{-1}, respectively. With other semiconductor materials, lower conversion rates were found.

Because some of the photoactive agents exist as natural minerals, such as titanite, hematite, wolframite ilmenite, and nontronite, which probably already existed on the surface of the Earth before the emergence of life, it is reasonable to consider that such minerals may have acted as prebiotic photosynthetic agents, enabling the conversion of inorganic gases such as carbon dioxide, nitrogen, and water under sunlight to organic compounds (*59*).

To test the effects of sample treatments and of rare-earth dopants on the photosynthetic reduction of carbon dioxide, experiments with the large-band-gap semiconductors barium titanate and lithium niobate were carried out (*118, 119*). Using a high-pressure Hg lamp (75 W) with aqueous suspensions of the photoactive materials, the production rates of formic acid and of the other organic products were found to depend markedly on the nature of the semiconductor materials. Some results of these experiments are shown in Table I. With both lithium niobate and barium titanate, the carbon dioxide reduction was markedly improved by the addition of neodymium oxide. Undoped lithium niobate had an onset of light absorption only below 350 nm (thus its band gap was ~3.5 eV), whereas barium titanate absorbed below 400 nm. The rare-earth-doped samples had their absorption onset shifted to the visible region, with a shallow, broad absorption starting below 600 nm.

Photoassisted carbon dioxide reduction was also successful when the semiconductor powder was painted on a flat surface, in a photochemical solar reactor, designed somewhat similarly to solar hot-water collectors (*60*). The photoactive material was ground with water to a paste, which was painted on the inside back plate of this reactor. The front plate was of transparent methylmethacrylate glass, and the space between these two plates (~1 cm apart) was filled with water, through which carbon dioxide

TABLE I

Carbon Dioxide Reduction During Illumination of Powdered Semiconductors in Aqueous Suspensions[a]

Photoactive material[b]	Production rate (μmol h^{-1})			Efficiency (%)
	HCOOH	HCHO	CH_3OH	
(a) $SrTiO_3$	1.2	0.05	0.02	0.022
(b) TiO_2	0.71	0.014	—	0.011
(c) TiO_2	2.3	0.066	0.022	0.039
(d) $BaTiO_3$	0.11	0.012	0.001	0.003
(e) $BaTiO_3$	—	—	—	—
(f) $BaTiO_3$	0.23	0.027	—	0.005
(g) $LiNbO_3$	—	0.029	—	0.001
(h) $LiNbO_3$	1.56	0.15	—	0.030

[a] Aqueous suspension: 0.8–1.0 g 160 ml^{-1}. Flow rate of CO_2 : 100 ml min^{-1}. Illumination: 75-W high-pressure Hg lamp. Reaction temperature: 60°C.

[b] Material pretreatments: (a) Heated 6.5 h at 1100°C under vacuum, (b) 7 h at 1000°C under vacuum, (c) 2 h at 500°C under vacuum, (d) 22 h at 1000°C under H_2 + Ar (1 : 1), (e) 4 h at 700°C in air, (f) 4 h at 700°C in air, with 5×10^{-3} mol % Nd_2O_3, (g) 2 h at 1150°C in air, (h) 2.5 h at 1150°C in air, with 0.5 mol % Nd_2O_3.

was bubbled. During operation in sunlight the water temperature inside the reactor rose to 65°C. Volatile organic products were swept out with the carbon dioxide and were collected in a series of three cold traps at 0°C. The total dose of incident sunlight was measured with a pyranometer connected to an integrator. Thus, by measuring the yields of the organic reduction products together with the total light dose, the efficiency of the solar-to-chemical energy conversion could be directly calculated. As an example of the use of this photochemical reactor, strontium titanate (which had been pretreated by heating for 9 h at 1100°C under vacuum) was covered with CO_2-saturated water. The CO_2 flow rate was 500 ml min^{-1}. Maximal temperature (at noon) was 50°C. Production rates ($\mu mol\ h^{-1}$) were: HCOOH, 6.4; HCHO, 0.7; CH_3OH, 0.4; and CH_3CHO, 0.3. The solar-to-chemical energy-conversion efficiency was $8.5 \times 10^{-3}\%$.

A major problem with the photoassisted reduction of carbon dioxide on semiconductor materials is the dissociation of the organic products on these materials. Thus formic acid was found to dissociate on anatase in a reaction in which anatase essentially acted as a dehydration catalyst for formic acid, producing carbon monoxide at a rate of 17.9×10^{-9} mol $m^{-2}\ s^{-1}$ (*33*). Photocatalytic oxidation of formic acid and of methanol on illuminated oxide semiconductors has been the subject of many studies (*88, 92*).

VIII. Conclusions

Natural photosynthesis in plants exists. It thus provides a great challenge to equal or surpass this photosynthesis in an artificial system (*27*). From the start of such efforts (in the 1890s) there has been considerable progress in achieving photoassisted reduction of carbon dioxide to organic compounds, using sunlight as the energy source. There is no clear evidence yet on what should be the upper limit of efficiency for such a process. For a photochemical solar energy storage system, in general, an estimate of a realistic maximum of energy conversion efficiency has led to a value of 15–16% (*19–22*). It will be of considerable theoretical and practical importance to determine the actual upper limit to the photoassisted reduction of carbon dioxide with water under optimal conditions.

Acknowledgment

I wish to thank Professor I. Dostrovsky for advice on the thermochemical calculations. This study was supported in part by a grant from the National Council of Research and Development, Israel, and the K.F.A. Jülich, Germany.

References

1. Åkermark, B., *in* "Solar Energy—Photochemical Conversion and Storage" (S. Claesson, and L. Engstrom, eds.), p. 1. Natl. Swedish Board Energy Source Development, Stockholm, 1977.
2. Åkermark, B., Eklund-Westlin, U., Baeckstrom, P., and Lof, R., *Acta Chem. Scand. Ser. B* **34,** 27 (1980).
3. Amatore, C., and Saveant, J.-M., *J. Am. Chem. Soc.* **103,** 5021 (1981).
4. Aurian-Blajeni, B., Halmann, M., and Manassen, J., *Sol. Energy* **25,** 165 (1980).
5. Aurian-Blajeni, B., Halmann, M., and Manassen, J., *Photochem. Photobiol.* **35,** 157 (1982).
6. Bach, A., *C. R. Acad. Sci., Paris* **116,** 1145 (1893).
7. Bach, A., *Chem. Ber.* **39,** 1672 (1906).
8. Baeyer, A., *Chem. Ber.* **3,** 63 (1870).
9. Baly, E. C. C., Davies, J. B., Johnson, M. R., and Shanassy, H., *Proc. R. Soc. London Ser. A* **116,** 197 (1927).
10. Baly, E. C. C., Stephen, W. E., and Hood, N. R., *Proc. R. Soc. London Ser. A* **116,** 212 (1927).
11. Baly, E. C. C., *Proc. R. Soc. London Ser. A* **172,** 445 (1939).
12. Bard, A. J., *J. Photochem.* **10,** 59 (1979).
13. Bard, A. J., *Science (Washington, D.C.)* **207,** 139 (1980).
14. Baur, E., *Z. Phys. Chem.* **80,** 668 (1912).
15. Baur, E., and Buchi, P., *Helv. Chim. Acta* **6,** 960 (1923).
16. Bockris, J. O'M., and Uosaki, K., *Adv. Chem. Ser.* No. 163, 33 (1977).
17. Bockris, J. O'M., and Uosaki, K., *Int. J. Hydrogen Energy* **2,** 123 (1977).
18. Boehm, H. P., *Discuss. Faraday Soc.* **52,** 264 (1971).
19. Bolton, J. R., *Science (Washington, D.C.)* **202,** 705 (1978).
20. Bolton, J. R., *Sol. Energy* **20,** 181 (1978).
21. Bolton, J. R., and Hall, D. O., *Annu. Rev. Energy* **4,** 353 (1979).
22. Bolton, J. R., Haught, A. F., and Ross, R. T., *in* "Photochemical Conversion and Storage of Solar Energy" (J. S. Connolly, ed.), p. 297. Academic Press, New York, 1981.
23. Borgarello, E., Kiwi, J., Pelizzetti, E., Visca, M., and Grätzel, M., *J. Am. Chem. Soc.* **103,** 6324 (1981).
24. Braunstein, P., *Nachr. Chem. Tech. Lab.* **29,** 695 (1981).
25. Brown, R. D., *in* "Origin of Life" (Y. Wolman, ed.), p. 1. Reidel, Dordrecht, 1981.
26. Butler, M. A., and Ginley, D. S., *J. Mater. Sci.* **15,** 1 (1980).
27. Calvin, M., *in* "Photochemical Conversion and Storage of Solar Energy" (J. S. Connolly, ed.), p. 1. Academic Press, New York, 1981.
28. Chang, C. D., and Silvestri, A. J., *J. Catal.* **47,** 249 (1977).
29. Chittenden, G. J. F., and Schwartz, A. W., *BioSystems* **14,** 15 (1981).
30. DeBacker, M. G., Richoux, M. C., Leclercq, F., and Lepoutre, G., *Proc. Int. Conf. Photochem. Convers. Storage Sol. Energy 2nd,* p. 51 (1978).
31. Denbigh, K., "The Principles of Chemical Equilibrium," 3rd ed. Cambridge Univ. Press, Cambridge, 1971.
32. Denise, B., and Sneeden, R. P. A., *CHEMTECH* **12,** 108 (1982).
33. Enriquez, M. A., and Fraissard, J. P., *J. Chim. Phys.* **78,** 457 (1981).
34. Eremin, E. N., Maltsev, A. N., and Ivanter, V. L., *Zh. Fiz. Khim.* **54,** 150 (1980).
35. Eremin, E. N., Maltsev, A. N., Ivanter, V. L., and Belova, V. M., *Zh. Fiz. Khim.* **55,** 1365 (1981).

36. Euler, H., *Chem. Ber.* **37,** 3411 (1904).
37. Flamant, G., Hernandez, D., Bonet, C., and Traverse, J.-P., *Sol. Energy* **24,** 385 (1980).
38. Flamant, G., *Rev. Phys. Appl.* **15,** 503 (1980).
39. Folsome, C., and Brittain, A., *Nature (London)* **291,** 482 (1981).
40. Folsome, C. E., Brittain, A., Smith, A., and Chang, S., *Nature (London)* **294,** 64 (1981).
41. Frese, K. W., Jr., *Appl. Phys. Lett.* **40,** 275 (1982).
42. Fruge, D. R., Fong, G. D., and Fong, F. K., *J. Am. Chem. Soc.* **101,** 3694 (1979).
43. Fujishima, A., and Honda, K., *Nature (London)* **238,** 37 (1972).
44. Galwialo, M. J., *Biochem. Z.* **158,** 65 (1925).
45. Garrison, W. M., Morrison, D. C., Hamilton, J. G., Benson, A. A., and Calvin, M., *Science (Washington, D.C.)* **114,** 416 (1951).
46. Gerischer, H., *J. Electroanal. Chem.* **58,** 263 (1975).
47. Getoff, N., Scholes, G., and Weiss, J., *Tetrahedron Lett.* **18,** 17 (1960).
48. Getoff, N., *Z. Naturforsch. B* **17,** 87 (1962).
49. Getoff, N., *Z. Naturforsch B* **18,** 169 (1963).
50. Gore, V., *J. Phys. Chem.* **39,** 399 (1935).
51. Grätzel, M., *Acc. Chem. Res.* **14,** 376 (1981).
52. Green, S., *Annu. Rev. Phys. Chem.* **32,** 103 (1981).
53. Gressin, J. C., Michelet, D., Nadjo, L., and Saveant, J. M., *Nouv. J. Chim.* **3,** 545 (1979).
54. Groth, W., and Suess, H., *Naturwissenschaften* **26,** 77 (1938).
55. Hair, M. L., and Harbour, J. R., *Adv. Chem. Ser.* No. 184, p. 173 (1980).
56. Halmann, M., *Isr. J. Chem.* **10,** 841 (1972).
57. Halmann, M., *Nature (London)* **275,** 115 (1978).
58. Halmann, M., and Aurian-Blajeni, B., *Proc. Eur. Commun. Photovolt. Sol. Energy Conf. 2nd,* p. 682 (1979).
59. Halmann, M., Aurian-Blajeni, B., and Bloch, S., *in* "Origin of Life" (Y. Wolman, ed.), p. 143. Reidel, Dordrecht, 1981.
60. Halmann, M., Ulman, M., and Aurian-Blajeni, B., *Solar Energy,* in press (1983).
61. Hamann, C., and Schmode, P., *J. Power Sources* **1,** 141 (1976–77).
62. Hansen, J., Johnson, D., Lacis, A., Lebedeff, S., Lee, P., Rind, D., and Russel, G., *Science (Washington, D.C.)* **213,** 957 (1981).
63. Harbour, J. R., and Hair, M. L., *J. Phys. Chem.* **81,** 1791 (1977).
64. Harbour, J. R., and Hair, M. L., *Can. J. Chem.* **57,** 1150 (1979).
65. Harris, L. A., and Wilson, R. H., *Annu. Rev. Mater. Sci.* **8,** 99 (1978).
66. Hart, E. J., and Anbar, M., "The Hydrated Electron." Wiley (Interscience), New York, 1970.
67. Hauffe, K., Hupfeld, J., and Wetterling, T., *Z. Phys. Chem.* (*Wiesbaden*) **103,** 115 (1976).
68. Heller, A., *Acc. Chem. Res.* **14,** 154 (1981).
69. Hemminger, J. C., Carr, R., and Somorjai, G. A., *Chem. Phys. Lett.* **57,** 100 (1978).
70. Hemminger, J. C., Carr, R., Lo, W. J., and Somorjai, G. A., *Adv. Chem. Ser.* No. 184, 233 (1980).
71. Hodes, G., Manassen, J., and Cahen, D., *Nature (London)* **261,** 403 (1976).
72. Hoffmann, M. R., and Schaefer, H. F., *Astrophys. J.* **249,** 563 (1981).
73. Hupfeld, J., Ph.D. Thesis, Univ. of Gottingen, 1975.
74. Inoue, T., Fujishima, A., Konishi, S., and Honda, K., *Nature (London)* **277,** 637 (1979).

75. Ito, K., Murata, T., and Ikeda, S., *Bull. Nagoya Inst. Tech.* **27,** 209 (1975).
76. Jaeger, C. D., and Bard, A. J., *J. Phys. Chem.* **83,** 3146 (1979).
77. Jordan, J., and Smith, P. T., *Proc. Chem. Soc.* 246 (1960).
78. Keene, J. P., Raef, Y., and Swallow, A. J., *in* "Pulse Radiolysis" (M. Ebert, J. P. Keene, and A. J. Swallow, eds.), p. 100. Academic Press, London, 1965.
79. Kiwi, J., Kalyanasundaram, K., and Grätzel, M., *Struct. Bonding (Berlin)* **49,** 37 (1982).
80. Kraeutler, B., Jaeger, C. D., and Bard, A. J., *J. Am. Chem. Soc.* **100,** 4903 (1978).
81. Lamy, E., Nadjo, L., and Saveant, J. M., *J. Electroanal. Chem.* **78,** 403 (1977).
82. Latimer, W. M., "Oxidation Potentials," 2nd ed. Prentice-Hall, New York, 1952.
83. Lehn, J.-M., and Ziessel, R., *Proc. Natl. Acad. Sci. USA* **79,** 701 (1982).
84. Levine, J. S., *in* "Comparative Planetology" (C. Ponnamperuma, ed.), p. 165. Academic Press, New York, 1978.
85. Lysyak, T. V., Kharitinov, Yu. Ya., and Kolomnikov, I. S., *Zh. Neorg. Khim.* **25,** 2562 (1980).
86. Manassen, J., Cahen, D., Hodes, G., and Sofer, A., *Nature* (*London*) **263,** 97 (1976).
87. Maruska, H. P., and Ghosh, A. K., *Sol. Energy* **20,** 443 (1978).
88. Miyake, M., Yoneyama, H., and Tamura, H., *J. Catal.* **58,** 22 (1979).
89. Monnier, A., Augustynski, J., and Stalder, C., *J. Electroanal. Chem.* **112,** 383 (1980).
90. Monnier, A., Augustynski, J., and Stalder, C., *Proc. Int. Conf. Photochem. Convers. Storage Sol. Energy 3rd,* p. 423 (1980).
91. Moore, B., and Webster, T. A., *Proc. R. Soc. London Ser. B* **87,** 163 (1913).
92. Morrison, S. R., and Freund, T., *J. Chem. Phys.* **47,** 1543 (1967).
93. Morterra, C., Ghiotti, G., and Garrone, E., *J. Chem. Soc. Faraday Trans. 1* **76,** 2102 (1980).
94. Morterra, C., Chiorino, A., and Boccuzzi, F., *Z. Phys. Chem. (Neue Folge)* **124,** 211 (1981).
95. Nozik, A. J., *Appl. Phys. Lett.* **30,** 567 (1977).
96. Nozik, A. J., *Annu. Rev. Phys. Chem.* **29,** 189 (1978).
97. Pinto, J. P., Gladstone, G. R., and Yung, Y. L., *Science (Washington, D.C.)* **210,** 183 (1980).
98. Porter, G., *J. Photochem.* **17,** 193 (1981).
99. Prasad, S. S., and Huntress, W. T., *Astrophys. J.* **228,** 123 (1979).
100. Redfield, A. C., *Am. Sci.* **46,** 205 (1958).
101. Reymond, J.-P., and Gravelle, P. C., *J. Chim. Phys.* **76,** 879 (1979).
102. Rossini, F. D., Wagman, D. D., Evans, W. H., Levine, S., and Jaffe, I., "Selected Values of Chemical Thermodynamic Properties," Natl. Bur. Stand. Circular 500. U.S. Gov. Printing Office, Washington, D.C., 1952.
103. Royer, E., *C. R. Acad. Sci., Paris* **70,** 731 (1870).
104. Rubey, W. W., *Bull. Geol. Soc. Am.* **62,** 1111 (1951).
105. Russel, P. G., Kovac, N., Srinivasan, S., and Steinberg, M., *J. Electrochem. Soc.* **124,** 1329 (1977).
106. Scheerer, R., and Grätzel, M., *Ber. Bunsenges. Phys. Chem.* **80,** 979 (1976).
107. Scheerer, R., and Grätzel, M., *J. Am. Chem. Soc.* **99,** 865 (1977).
108. Schiff, H. I., *Planet. Space Sci.* **29,** 935 (1981).
109. Slanger, T. G., and Black, G., *J. Chem. Phys.* **68,** 1844 (1978).
110. Smith, A., Folsome, C., and Bahadur, K., *Experientia* **37,** 357 (1981).
111. Stoklasa, J., and Zdobnicky, W., *Biochem. Z.* **30,** 433 (1911).
112. Tadami, E., Japanese Patent No. 76 06 895, 20. 1 (1976). (Chem. Abstr. 85, 7888j).
113. Tazuke, S., and Kitamura, N., *Nature (London)* **275,** 301 (1978).

114. Teeter, T. E., and Van Rysselberghe, P., *J. Chem. Phys.* **22,** 759 (1954).
115. Teeter, T. E., and Van Rysselberghe, P., *Proc. Meet. Int. Cmt. Electrochem. Thermodyn. Kinet. 6th,* p. 538 (1955).
116. Tseng, S. S., and Chang, S., *Nature (London)* **248,** 575 (1974).
117. Tsuda, T., Chujo, Y., and Saegusa, T., *J. Chem. Soc. Chem. Commun.* p. 963 (1975).
118. Ulman, M., Aurian-Blajeni, B., and Halmann, M., *Isr. J. Chem.* **22,** 177 (1982).
119. Ulman, M., Halmann, M., and Aurian-Blajeni, B., *Proc. Int. Conf. Photochem. Convers. Storage Sol. Energy 4th,* p. 111 (1982).
120. Usher, F. L., and Priestley, J. H., *Proc. R. Soc. London Ser. B* **78,** 318 (1906).
121. Van Damme, H., and Hall, W. K., *J. Am. Chem. Soc.* **101,** 4373 (1979).
122. Vonach, W., and Getoff, N., *Z. Naturforsch. A* **36,** 876 (1981).
123. Watanabe, T., and Fujishima, A., *in* "Solar–Hydrogen Energy Systems" (T. Ohta, ed.), p. 162. Pergamon, Oxford, 1979.
124. Watson, W. D., *Acc. Chem. Res.* **10,** 221 (1977).
125. Weeks, J. L., and Rabani, J., *J. Phys. Chem.* **70,** 2100 (1966).
126. Williams, R., Crandall, R. S., and Bloom, A., *Appl. Phys. Lett.* **33,** 381 (1978).
127. Williams, R., and Bloom, A., U.S. Patent No. 4,160,816 (1979).
128. Wisseroth, K., *Chem. Z.* **105,** 1 (1981).
129. Wolkenstein, Th., *Adv. Catal.* **12,** 189 (1960).
130. Wolkenstein, Th., *Adv. Catal.* **23,** 157 (1973).
131. Wrighton, M. S., *Chem. Eng. News* **57,** 29 (1979).
132. Wu, C. Y. R., Phillips, E., Lee, L. C., and Judge, D. L., *J. Geophys. Res.* **83,** 4869 (1978).
133. Yoshioda, T., Thorn, D. L., Okano, T., Ibers, J. A., and Otsuka, S., *J. Am. Chem. Soc.* **101,** 4212 (1979).
134. Lerman, A., "Geochemical Processes: Water and Sediment Environments," pp. 23–42. Wiley-Interscience, New York, 1979.
135. Taniguchi, Y., Yoneyama, H., and Tamura, H., *Bull. Chem. Soc. Jpn.* **55,** 2034 (1982).

16 Catalytic Nitrogen Fixation in Solution

A. E. Shilov

Institute of Chemical Physics of the U.S.S.R. Academy of Sciences
Chernogolovka, U.S.S.R.

I. Introduction

Since the publication by Vol'pin and Shur (*8*), a new branch of coordination catalysis has been developing—that of molecular nitrogen reduction in solutions in the presence of transition-metal complexes. Almost at the same time, complexes of molecular nitrogen with metal compounds were discovered, and this led to the formation of a new field in coordination chemistry (*5, 32*). These two undoubtedly related regions were linked in 1969 when intermediate nitrogen complexes were found to be able of further converting N_2 into hydrazine and ammonia derivatives (*33*).

In 1970, the first systems reducing nitrogen to hydrazine and ammonia in protic media were discovered (*16–18*). Thus a bridge was created be-

ENERGY RESOURCES THROUGH
PHOTOCHEMISTRY AND CATALYSIS

ISBN 0-12-295720-2

tween chemical reduction of nitrogen and its biological fixation with the participation of the enzyme nitrogenase, in which the first step is a reduction of N_2 to NH_3, and an important role belongs to the transition-metal complexes, those of iron and molybdenum.

Now the chemistry of nitrogen reactions in solutions comprises a very large number of systems. The two lone processes that were known, namely, heterogeneous catalytic synthesis of ammonia from nitrogen and hydrogen at high temperatures and pressures and enzymatic nitrogen fixation, are but extreme cases in this field, containing numerous N_2 reactions that can proceed under a variety of conditions and that include free transition metals and many of their compounds (see reviews in *2–4, 9*).

A joint consideration of these reactions will help to better understand both general features and differences existing between them. The aim of this chapter is to summarize the results concerning nitrogen fixation in solution, especially nitrogen fixation in protic media, which is close in its mechanism to the enzymatic nitrogen fixation.

II. Peculiarities of the Thermodynamics of Molecular Nitrogen Reduction

The chemical inertness of the N_2 molecule is a result of not only the great amount of energy (225 kcal/mol) required to break the triple $N{\equiv}N$ bond, but also mainly because more than half of the required energy is used to break the first bond. Therefore, the addition, for example, of one H_2 molecule to N_2 to produce N_2H_2 is an endothermic process.

This distinguishes nitrogen from, for example, the acetylene molecule, which is isoelectronic to nitrogen (Fig. 1). In C_2H_2, on the contrary, the first bond broken is the least stable. Hence, for their reaction with nitrogen, one- and two-electron reducing agents should possess negative redox potential with high absolute values (Fig. 2).

At the same time, to break two or even three bonds simultaneously, if such a reaction could be possible, we may use a comparatively weak reducing agent because the cleavage of the second and particularly third bonds in N_2 requires much less energy. This leads to the suggestion about a possible mechanism of catalysis in nitrogen reduction—if a catalyst can switch from the one-electron mechanism over to the four- or even six-electron one with a simultaneous rupture of two or three bonds, then a much weaker reducing agent can be used than for the one- or two-electron reduction. It is natural to assume that such a catalyst should be a cluster

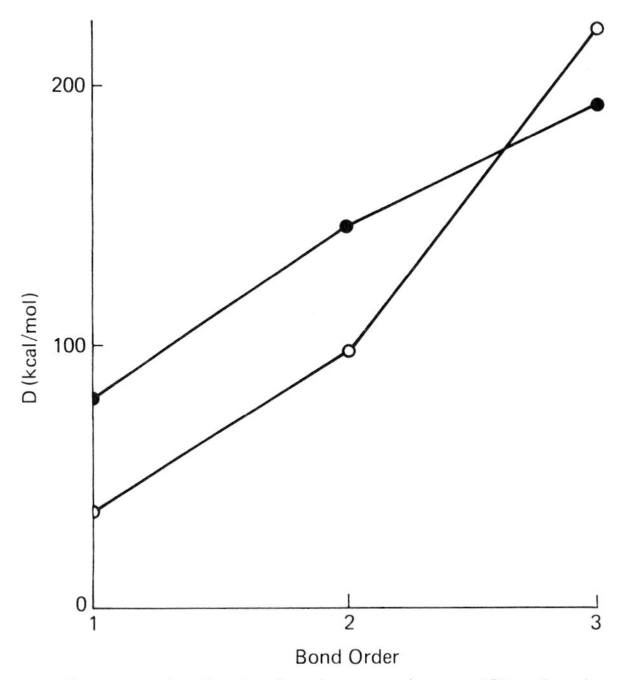

Fig. 1. Bond energies versus bond order for nitrogen–nitrogen (○) and carbon–carbon (●) bonds.

involving a number of ions of transition metals bound together. In this case, one can visualize a one-by-one electron transfer from the external reducing agent to the catalyst followed by a collective multielectron process involving the N_2 molecule activated by the catalyst (*2–4*).

III. Nitrogen Reduction in Aprotic Media

Nitrogen reduction in hydrocarbon and ether media proceeds under the action of complexes of low-valent transition metals formed using such reducing agents as organometallic compounds (RMgX, RLi, or R_3Al), metal hydrides (LiAlH or LiH), metals (Li, Na, Mg, etc.), and various transition-metal compounds (*9*). The most active reagents toward nitrogen are the metals of the IV, V, and VI groups of the periodic table, such as Ti, V, Cr, Mo, W, as well as Zr and Fe; the less reactive ones are Co and Ni. The reaction products are usually nitride derivatives of transition

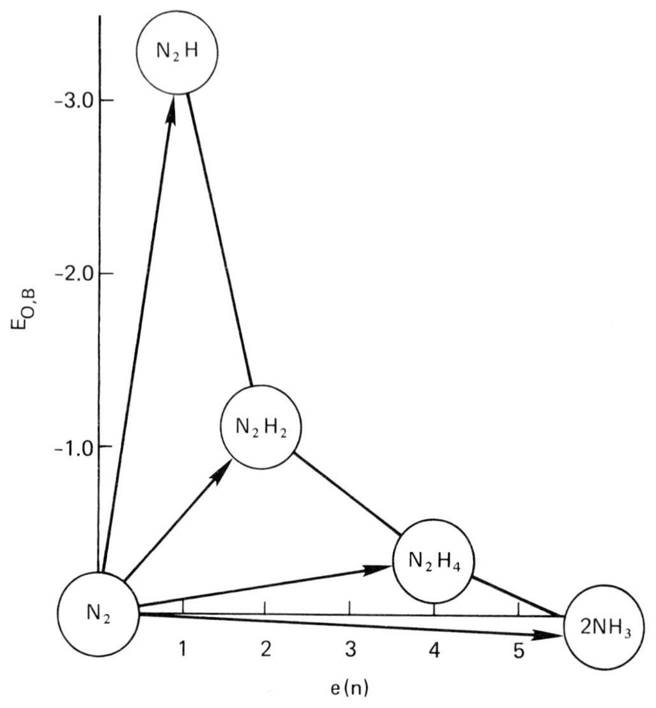

Fig. 2. Redox potential dependence on the number of electrons transferred to a N_2 molecule.

metals, forming ammonia when hydrolyzed. The reaction proceeds usually stoichiometrically producing up to one (or two) mole(s) of ammonia per mole of the transition metal.

The reduction may also proceed in a catalytic way in some rare systems, for example, with the participation of compounds of titanium and aluminum bromide to form an aluminum nitride compound:

$$N_2 + 2Al + 4AlX_3 \xrightarrow{Ti(II)} 2[N(AlX_2)_3]$$

For example, in the $TiCl_4 + Al + AlBr_3$ system with the reagents ratio 1 : 600 : 1000 at 130°C 268 mol of aluminum nitride per mole of the titanium compound are formed. The role of $AlBr_3$ apparently lies in the rupture of the nitride Ti—N bond formed in the dinitrogen reduction. N_2 reduction by transition-metal compounds involves the primary N_2 coordination to give molecular complexes.

The consideration of the intermediate complexes structure allows to understand the mechanism of the subsequent dinitrogen reduction.

IV. Molecular Nitrogen Complexes with Transition-Metal Compounds and the Mechanism of Nitrogen Reduction in the Coordination Sphere of the Complex

Nitrogen complexes with transition-metal compounds contain a nitrogen molecule usually coordinated with one or two atoms of the metal M. The vast majority of the presently known complexes have fragments in a linear configuration (*1, 15*):

$$M—N{\equiv}N \quad \text{or} \quad M—N{\equiv}N—M$$

For many stable N_2 complexes, both mononuclear and binuclear, all attempts to involve the N_2 molecule in any chemical reaction (besides formation of free N_2 at the decomposition) have failed. Therefore, it is of interest to consider the properties of N_2 complexes from the viewpoint of requirements for their participation in the catalytic N_2 reduction.

Figures 3 and 4 visualize molecular orbitals of mono- and bimolecular complexes. The d_π–p_π interaction is essential for the lowering of the NN bond order and increasing of the negative charge on nitrogen atoms, both effects play a significant role in the subsequent reactions of the coordinated N_2 molecule. At the d_π–p_π interaction, *d* electrons of the metal pass partially onto the doubly degenerate antibonding $1\pi_g^*$ orbital of the nitrogen molecule. In the case of d^1–d^3 and, possibly, d^4 electronic configura-

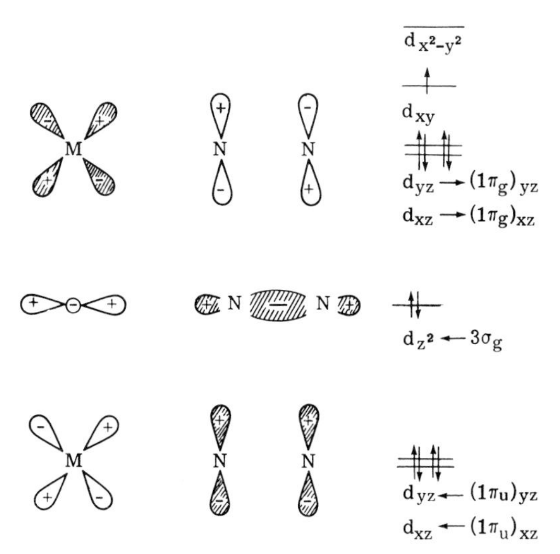

Fig. 3. Molecular orbitals for the mononuclear M—N≡N complex.

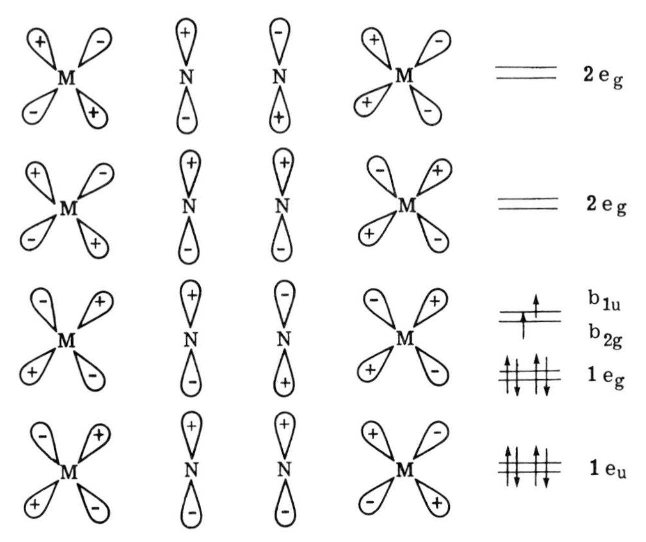

Fig. 4. Molecular orbitals for the binuclear M—N≡N—M complex for octahedral surrounding for Ms (D_{4n} symmetry).

tions of M, the maximum overlapping of the occupied metal d orbitals and the $1\pi_g^*$ nitrogen orbital will occur for the M—N≡N—M binuclear complex, whereas the increase of the number of d electrons at M to 5, 6, etc. will decrease the $d_\pi \rightarrow p_\pi$ interaction for a given value of redox potential, thus producing an undesirable effect for the M—N≡N—M complex as a potential catalyst. The optimum number of electrons at each M for such a complex, at least for an octahedral ligand surrounding of M appears to be 3 (*2–4*). Thus d^3 electronic configuration combined with definite reducing properties of M is required for the best catalyst of N_2 reduction via M—N≡N—M complex. Mo(III) and V(II) are obvious candidates for participation in such complexes, particularly in protic media, in which these ions can be sufficiently stable.

For the d^5–d^{10} electronic configuration, a mononuclear M—N≡N complex will be more suitable for further N_2 reduction than a binuclear one, optimum electronic configuration being perhaps d^5 for a given redox po-

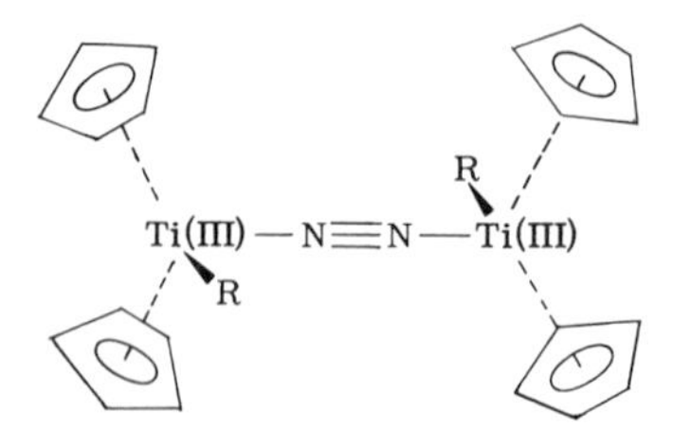

Fig. 5. Structure of the $[(\pi\text{-}C_5H_5)_2TiR]_2N_2$ complex.

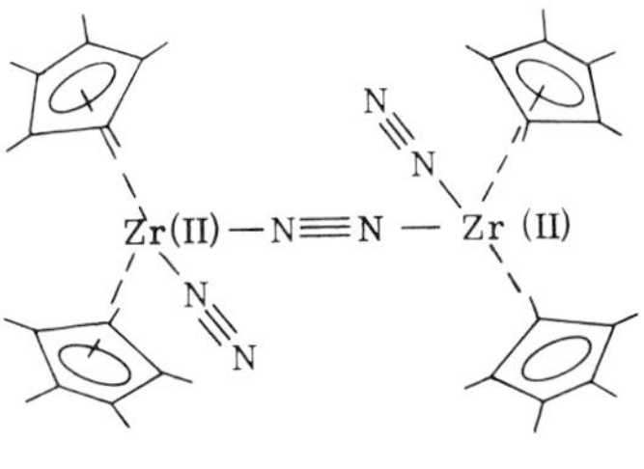

Fig. 6. Structure of the $\{[\pi\text{-}C_5(CH_3)_5]_2Zr(N_2)\}_2N_2$ complex.

tential of M. Other *d* electronic configurations are also acceptable, particularly when M is a very strong reducing agent reacting with N_2 in aprotic media. The protonation of the nitrogen molecule in a complex, as well as the addition of aprotic acid, should naturally enhance the electron transfer to the coordinated nitrogen molecule for both mono- and binuclear complexes.

The experimental results obtained for the intermediate N_2 complexes are in agreement with these conclusions. In the case of d^1–d^2 electronic configurations [Ti(III) (*33, 37*), Ti(II), Zr(II) (*10, 28*)], and Ta(III) and Nb(III) (*38*), binuclear complexes of the M—N≡N—M type have been detected and in some cases isolated with N_2 capable of further reduction to a hydrazine level (Figs. 5 and 6). For d^6–d^8 electronic configurations, the nitrogen reduction takes place in mononuclear complexes of the M—N≡N type (*15*) (Fig. 7).

The addition of acid ensures protonation and facilitates the reduction of the nitrogen molecule. When nitrogen is reduced in aprotic media, other positive ions, such as Li^+ in a Fe(0) complex, may behave as electron acceptors (*12*) (Fig. 8).

Some data available concerning the structure of intermediate forms arising from nitrogen reduction in a complex testify for the formation of hydrazine derivatives in the presence of four accessible electrons, for example, W=N—NH_2 (*15*).

For binuclear complexes, the structure of intermediate species formed after protonation is less clear. We may assume that a binuclear hydrazine

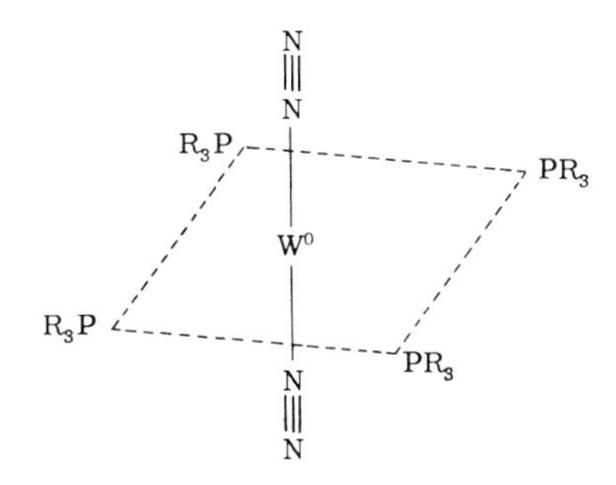

Fig. 7. Structure of the $W(N_2)_2(PR_3)_4$ complex.

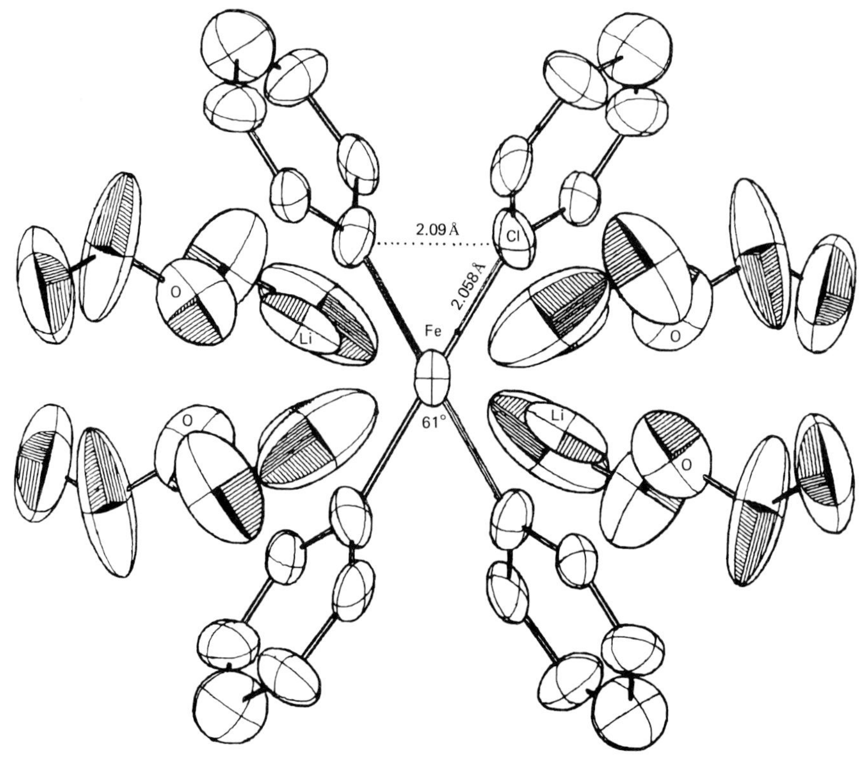

Fig. 8. Structure of the $[FePh_4(Et_2O)_4]Li_4$ complex.

structure formed, is primarily of the type

$$\begin{array}{c} \quad\quad \mathrm{H} \\ \mathrm{N{-}N}^{\oplus} \\ \mathrm{M} \!\!\!\!\!\!\quad \quad \mathrm{M} \end{array}$$

The intermediate binuclear complex can be expected to form in the process of N_2 reduction in some cases when initially mononuclear complex reacts with acid, for example,

$$\mathrm{Mo(II){=}N{-}NH_2 + Mo(0)} \xrightarrow{H^+} \mathrm{Mo{=}\overset{H}{N}{-}\overset{H}{N}{=}Mo} \longrightarrow \mathrm{2Mo(III) + 2NH_3}$$

V. Nitrogen Reduction in Protic Media

Nitrogen reduction in protic media has some advantages for catalytic reduction; at the same time, it has some disadvantages. The molecules formed from nitrogen are more basic than the starting N_2 molecule. The intermediate complex formation also enhances the basic properties of the coordination N_2 molecule. Hence, in protic media thermodynamics of the reduction owing to the hydrogen bond formation is more favorable than in aprotic media, and the protonation of the intermediate complex in protic media facilitates its subsequent reduction. Besides, a catalytic reaction may be expected to proceed via hydrolytic cleavage, by an AH solvent, of the M—N bond formed in the interaction with a metal-catalyst

$$N_2 + 4e + 4AH \xrightarrow{M} N_2H_4 + 4A^-$$

$$N_2 + 6e + 6AH \xrightarrow{M} 2NH_3 + 6A^-$$

ion with a constant renewal of the initial catalyst M, whereas in aprotic media nitrides are usually the final products.

At the same time, because in protic media a comparatively strong reducing agent (stronger than H_2) is required to reduce nitrogen, even for a four-electron mechanism, there may arise a side reaction of H_2 formation that is simpler than N_2 reduction:

$$2e + 2AH \xrightarrow{M} H_2 + 2A^-$$

This naturally hinders the search for an appropriate catalyst, and is evidently the reason why until 1970 the biological nitrogen reduction remained the only process of N_2 reduction in protic media and numerous attempts to observe reduction in chemical systems failed. At present, several systems are known that reduce nitrogen in aqueous and alcoholic media (*2, 4*). They are listed in Table I.

Their consideration should begin with biological nitrogen fixation, the study of which helped very much when searching for nitrogen reducing chemical systems.

A. Nitrogenase

In enzymatic nitrogen reduction, water protons take part in the formation of ammonia, which is the primary product. Taking into account that hydrophobic areas exist in enzymes, it is not clear to what extent one can speak of a "protic medium," in the strict sense of the word, as a solvent in which enzymatic N_2 fixation develops. However, there is no doubt that the enzyme active center is accessible to the action of protons that are reduced to molecular hydrogen in the absence of N_2. Therefore, when considering enzymatic nitrogen fixation, it is necessary to take into account the advantages and limitations characterizing nitrogen reduction

TABLE I

Systems Reducing Nitrogen in Protic Media[a]

Reducing agent	Products	Yield (%)	References
M=Mo(III)			
$Ti(OH)_3$	N_2H_4, NH_3	~100 (60°C)	*21*
$Ti(OH)_3 + Mg(OH)_2$	N_2H_4,NH_3	17,000 (110°C)	*34*
$Cr(OH)_2$	N_2H_4, NH_3	80 (90°C)	*19*
Na(Hg) + Mg^{2+}	N_2H_4	250 (20°C)	*22*
Na(Hg) + Mg^{2+} + phospholipid	N_2H_4, NH_3	3,000 (20°C)	*23*
Cathode ana $Ti(OH)_3$	N_2H_4, NH_3	50 (20°C)	*24*
Cathode, Ti(III), and guanidine	N_2H_4	~6 (20°C)	*26*
M=V(II)			
$V(OH)_2 + Mg(OH)_2$ in aqueous methanol	N_2H_4, NH_3 (pH > 12.5)	70 (20°C)	*20*
	NH_3 (no N_2H_4) (pH < 12)	35 (20°C)	*20*
V(II) + catechol	NH_3 (no N_2H_4) (pH 10.5)	75 (20°C)	*30*

[a] For the first four systems based on Mo(III) the yield is calculated per Mo atom. For electrochemical reduction (the fifth and sixth), the yield on current is given. For the systems based on V(II), the yield is related to the reaction stoichiometry taking into consideration that V(II) is a one-electron reducing agent when converted into V(III).

with the participation of protons. All this justifies our consideration of this process together with chemical reactions of nitrogen reduction in protic media.

The structure and mechanism of nitrogenase action have been studied with special intensity from the early 1960s when a cell-free N_2 reduction was observed for the first time. Later, the enzyme in pure form was isolated. Different aspects of the functioning of nitrogenase isolated from nitrogen-fixing bacteria and algae have been reported elsewhere (*25*).

Here we formulate only the main chemical features of the enzymatic nitrogen reduction. The enzyme isolated, for example, from *Azotobacter vinelandii,* is a combination of two proteins, one of them (MoFe protein) with MW ~230,000, the other (Fe protein) with MW ~60,000. The first contains 2 Mo, 30 Fe, and 30 S atoms per molecule and includes 4 Fe_4S_4 clusters as well as two so-called FeMo cofactors (see the following text), each containing 6–8 atoms of iron and one atom of molybdenum. Fe protein includes 4 Fe and 4 S per molecule in the form of a Fe_4S_4 cluster and consists of two subunits. Nitrogen fixation takes place only when both the proteins are present.

Nitrogen reduction *in vitro* can be observed when sodium dithionite is added as a reducing agent. In the N_2 reduction, no intermediate (e.g., N_2H_4) is recorded. However, when adding acid or alkali to the solution in which N_2 reduction is proceeding, the formation of some amount of hydrazine is observed, indicating a substance with a single N—N bond formed intermediately in the course of the reduction (*35*). The intermediate is apparently a metal-bound hydrazine derivative giving N_2H_4 at the decomposition of the functioning enzyme. Free hydrazine is slowly reduced by nitrogenase, and the very small concentrations of the observed intermediate, if it were free hydrazine, could not ensure the observed rates of N_2 fixation.

Nitrogen reduction is performed with the participation of ATP, the hydrolysis of which evidently provides a sufficiently low negative potential of the reducing agent. The reducing agent initially transfers electrons onto the Fe protein, which then passes them to the MoFe protein, this latter process coupled with ATP hydrolysis is the limiting stage of N_2 reduction. Besides nitrogen, other molecules having a triple bond, such as acetylene, CN^-, N_3^-, N_2O, CH_3CN, and CH_3NC as well as cyclopropane, can be reduced by nitrogenase.

As already mentioned, proton reduction with the formation of H_2 takes place in the absence of nitrogen, the yield of H_2 decreasing with the increase of N_2 pressure. Nevertheless, even at atmospheric and higher nitrogen pressures, the yield of H_2 does not drop below 25% of the number of electrons transferred by the reducing agent, and the yield of NH_3 does not exceed 75%. Thus, in the presence of nitrogen, H_2 is produced in the reaction coupled with nitrogen reduction, and the stoichiometry of the reaction may be presented as

$$N_2 + 8e^- + 8H^+ \longrightarrow 2NH_3 + H_2$$

In the presence of D_2, nitrogenase catalyzes HD formation. Here, along with the reaction proceeding independently of N_2, main HD formation occurs coupled with N_2 reduction (*11, 36*), and the overall stoichiometry appears to be

$$N_2 + 8e^- + 8H^+ + D_2 \longrightarrow 2NH_3 + 2HD$$

Carbon monoxide inhibits N_2 reduction as well as coupled H_2 evolution and HD formation, but it does not inhibit the N_2 independent evolution of H_2.

Acetylene is reduced only to ethylene, which does not undergo a subsequent reduction to ethane, and it does not inhibit the N_2 reduction. When C_2D_2 is reduced, *cis*-dideuteroethylene is selectively produced.

The rate of nitrogen reduction reaches about 0.1–1 mol per active center per second. The activation energy is 14 kcal/mol at temperatures higher than 20°C.

Molybdenum present in nitrogenase apparently has an important role in activation and reduction of molecular nitrogen. Hence, it is essential that a substance containing iron and molybdenum [a so-called iron–molybdenum cofactor, (FeMo-co)] (*7, 31*) can be isolated from MoFe protein. It can reactivate extracts from bacteria mutant forms, for example, from *A. vinelandii,* which does not reduce nitrogen in the absence of FeMo-co. FeMo-co contains molybdenum, iron, and sulfur in the ratio 1 : 6–8 : 4–6 but does not contain any aminoacids. The iron–molybdenum cofactor is, by all evidence, a cluster in which molybdenum is surrounded by sulfide ligands binding Mo with iron atoms; MoFe protein apparently contains two cofactors.

Thus the following most general conclusion can be made based on the results of investigations of biological N_2 fixation: *nitrogenase is a multicentral reducing agent capable of accepting several (at least six) electrons from an external one- or two-electron reducing agent, whereas the* N_2 *molecule in the process of activation appears to be bound to one or two molybdenum atoms.*

It should be noted that nitrogenase contains phospholipids and is a phospholipid-dependent enzyme: when phospholipids are removed, the rate of the enzymatic process drops, and the activation energy is changed (*14*).

B. Reduction of Nitrogen by Hydroxides

Hydroxides of certain metals provide the opportunity to fulfill in the simplest possible way the main condition necessary to model nitrogenase functioning, that is, binding together several ions that are able to donate electrons to the activated N_2 molecule, such as hydroxides, which are sufficiently strong reducing agents and can evolve hydrogen from water in alkaline media. Hydroxides of titanium(III), chromium(II), and vanadium(II) belong to this group (*19–21, 34*). Hydroxides that are not sufficiently strong reducing agents, such as $Fe(OH)_2$ or $Mn(OH)_2$, do not reduce nitrogen.

It is essential that in the case of $Ti(OH)_3$ and $Cr(OH)_2$ the reduction occur at a considerable rate only in the presence of added molybdenum compounds. Although it has been demonstrated (*6, 13*) that small amounts of hydrazine and ammonia are formed from nitrogen under the action of $Ti(OH)_3$ and $Cr(OH)_2$ without additions of Mo, reaction rates and yields in these cases are much less than in the presence of molybdenum. As for the case of $V(OH)_2$, the N_2 reduction proceeds readily without molybdenum. Taking into account that molybdenum has been found to be reduced to Mo(III) (electronic configuration d^3) by functioning hydroxide systems, we may conclude that the next condition for the effective nitrogen reduction in protic media, which could be fulfilled in the case of nitrogenase as well, is the d^3 electronic configuration of the nitrogen-coordinating active center, which is the case for both V(II) and Mo(III).

The reaction products in the nitrogen reduction are hydrazine and ammonia. It is noteworthy that although hydrazine can be reduced to ammonia in the conditions of N_2 reduction, the main portion of ammonia formed in the reaction is not a product of the hydrazine reduction. Kinetic experiments show that hydrazine and ammonia are produced in parallel, independent reactions. The ratio of reaction paths in N_2 reduction to NH_3 and N_2H_4 depends strongly on pH. Although hydrazine (further partially reduced to ammonia) is the reaction product in strongly alkaline media, at pH $\leq$ 12 the reaction goes directly to ammonia, omitting the stage of hydrazine formation.

Let us now consider separately some of the hydroxide N_2-fixing systems.

1. Stoichiometric Reduction of N_2 in $Ti(OH)_3$–Mo(III)

This system was the first to reduce nitrogen in protic media. The nitrogen reduction takes place only in the case when both Ti(III) and Mo(III) salts are present in solution before the addition of alkali. If the hydroxides are prepared separately and then mixed, there is practically no reduction.

Both in pure aqueous and methanol solutions, the reaction proceeds with very low yields. The best medium to carry out the reaction is methanol with water additions (5 vol%) (*21*). The presence of Mo(III) in the hydroxide active toward N_2 has been confirmed by an esca technique. If the initial solution contains the Mo(V) or Mo(VI) compound ($MoOCl_3$ or MoO_4^{2-}), then molybdenum is first reduced to Mo(III) under the action of titanium(III).

The reaction with N_2 readily proceeds at 60°C, and slowly at room temperature. In the absence of special additions, the yield of products from nitrogen reduction does not exceed the amount of molybdenum present.

Similar results have been obtained for the $Cr(OH)_2$–Mo(III) system (*34*).

2. Catalytic Nitrogen Reduction in the $Ti(OH)_3$–$Mg(OH)_2$–Mo(III) System

The additions of salts of some metals (Ca^{2+}, Ba^{2+}, and Mg^{2+}) to the solution of Ti^{3+} and Mo before hydroxide formation by alkali action considerably increases the hydrazine and ammonia yields in the nitrogen reduction. The most pronounced effect has been recorded for elevated temperatures for magnesium salt additions (optimum ratio Ti : Mg is 2 : 1).

In this case, the reduction of hundreds of N_2 molecules is observed per Mo atom; hence molybdenum becomes a catalyst of nitrogen reduction (*19*).

An important feature of this system is the formation of a crystalline mixed hydroxide (Fig. 9) consisting of $Ti(OH)_3$ and $Mg(OH)_2$ that possesses semiconductor properties, and the Mo(III)-containing catalytic centers are situated on the surface of crystals formed (*39*).

In agreement with this conclusion, N_2 reduction can effectively proceed if, *first,* crystals of mixed hydroxide are formed, *then* molybdenum compound is added. The catalytic effect of this system is owing to electron transfer from Ti ions surrounding the Mo(III) catalyst. Molybdenum ions are adsorbed at the hydroxide surface; they coordinate the N_2 molecule and ensure the electron transfer from Ti^{3+} ions to the activated N_2 molecule.

3. *Vanadium(II)–Magnesium Hydroxide*

One of the simplest systems capable of reducing dinitrogen in protic media is the mixed vanadium–magnesium hydroxide that is produced at the addition of alkali to both V^{2+} and Mg^{2+} ions, the latter being in excess. The system is also one of the most active toward N_2. The formation of hydrazine and ammonia proceeds with an appreciable rate even at room temperature and at atmospheric pressure of N_2 (*20*).

In the absence of Mg^{2+}, the hydroxide produced almost immediately loses its activity. In the presence of excess magnesium, V^{2+} ions trapped inside a solid mixed Mg(II)–V(II) hydroxide first diffuse to the surface of the particle and form intermediate V(II) clusters active toward N_2. These clusters form the stable form $V(OH)_2$ with time, and its activity toward N_2 disappears (*40*).

C. *Electrochemical Reduction of N_2 with the Participation of Molybdenum Complexes on the Cathode Surface: The Reduction by Sodium Amalgam*

Taking into consideration that a multielectron reducing agent with a sufficiently low negative redox potential is required to reduce nitrogen in protic media, it is natural to assume that this condition may be reached at the electrochemical nitrogen reduction at the cathode. To prevent hydro-

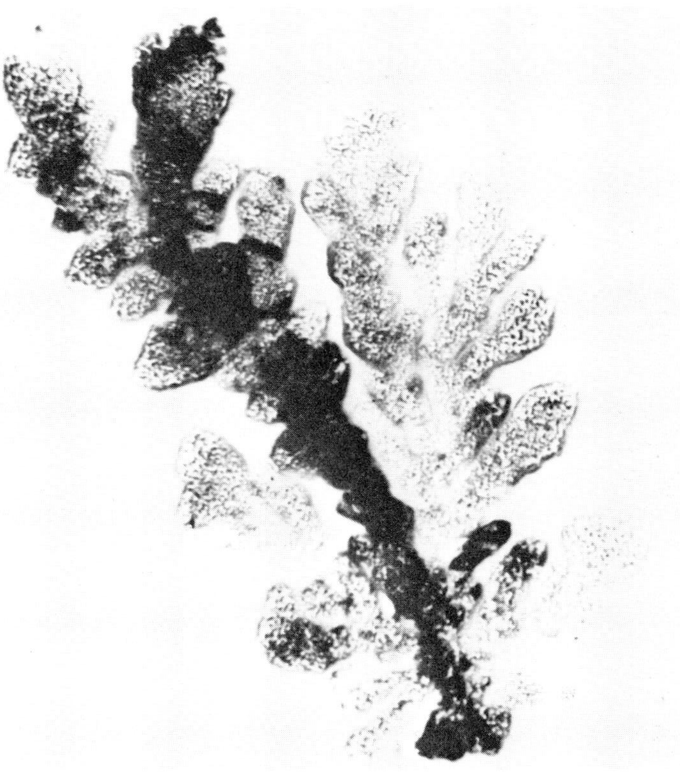

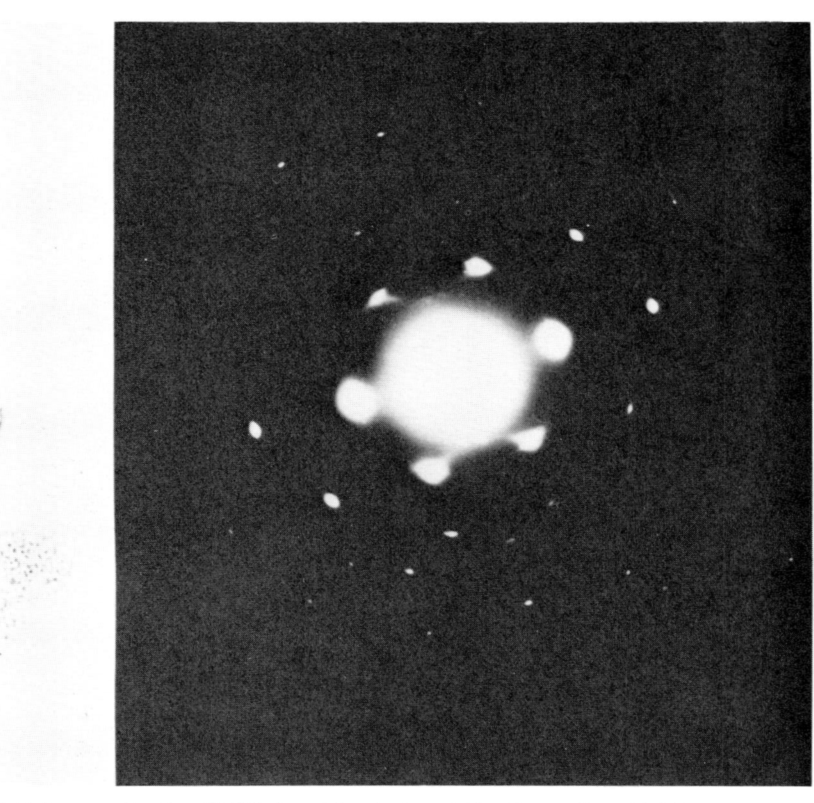

Fig. 9. Crystal structure of $Ti(OH)_3 \cdot 0.5Mg(OH)_2$ (left, magnification 50,000×) and its electron-diffraction pattern (right).

gen evolution at negative cathode potentials, it is necessary to use materials for the cathode that have a considerable overvoltage, for example, mercury and lead. At a negative potential of the mercury cathode, ions of alkali metals can be discharged to produce amalgam, which may serve as a reducing agent of N_2 in the presence of the catalyst. Thus the electrochemical reduction of N_2 on the cathode may include, as a first stage, the formation of the amalgam, which becomes a real reducing agent of nitrogen.

1. Sodium Amalgam as a Nitrogen Reducing Agent

If sodium amalgam is added to the N_2-fixing system $Ti(OH)_3$–Mo(III), the activity toward nitrogen increases, and the yields of the reduction products, those of hydrazine and ammonia, rise significantly (*29*). Here, while the amalgam is being spent, the titanium in the system remains trivalent.

Hence, Ti(III) ions in the hydroxide transfer the electrons from the amalgam to the Mo(III)-containing center.

The Mo(III) complex containing Mg^{2+} ions has been shown to be the catalyst of N_2 reduction by sodium amalgam in methanol solution without Ti(III). So far, this reaction is the only nonbiological catalytic reaction of molecular nitrogen reduction at room temperature (*22*).

As usual, nitrogen reduction proceeds in parallel with hydrogen evolution, which is particularly fast at comparatively high molybdenum concentrations. Thus it is necessary to work at low concentrations of the molybdenum complex ($1–3 \times 10^{-5}$ mol/liter). Because the catalytic effect is small (2–3 cycles per atom of Mo), the yield of hydrazine, which is the product of this reaction, is about $3–5 \times 10^{-5}$ mol/liter.

To create an effective catalyst, it is necessary to ensure its stable contact with the amalgam surface where nitrogen reduction takes place. This effect has apparently been achieved by the addition of a phospholipid to the system, namely, phosphatidylcholine (*23*). The reaction rate and yield of products in certain conditions are increased a factor of more than 10 (Fig. 10).

The reaction products in this case are hydrazine and ammonia (with ratio of about 2 : 1), which are formed in parallel reactions. The ratio of products does not alter at the change of nitrogen pressure or temperature. The effect of phosphatidylcholine action is very specific. The addition of many other substances, among them various phospholipids and surfactants, does not produce any positive result.

There is no doubt that the phosphatidylcholine effect is connected with film formation on the amalgam surface in which catalysts particles are

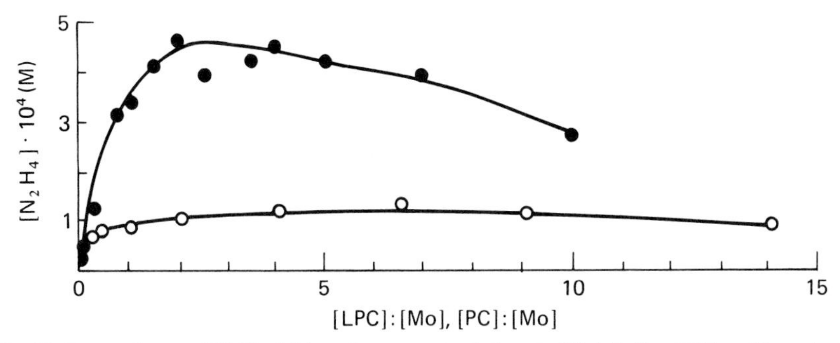

Fig. 10. Dependence of N_2H_4 yield on the amount of phosphatidylcholine (PC) and lysophosphatidylcholine (LPC) in the system Mo(III)–Mg(II)–Na(Hg). ○, LPC; ●, PC.

incorporated. Here, the action of the phospholipid is analogous to the functioning of the phospholipid-dependent enzymatic system in which phospholipids form membranes with incorporated enzyme molecules.

Apparently, the solid-like phospholipid film on the surface of the amalgam is formed because of the attraction of the positively charged heads (choline groups) of the phosphatidylcholine to the negatively charged amalgam surface. The specificity of the phosphatidylcholine action may be explained by the catalyst complex formation with a phospholipid molecule, which occurs presumably because of the oxygen atoms of phosphatide acid or carbonyl groups that are present in the polar head. The preceding result is an encouraging example of employing the principles of enzyme action in creating nonbiological catalytic systems.

2. *Electrocatalytic Nitrogen Reduction*

Nitrogen reduction has been detected under the conditions of electrolysis for the heterogeneous system containing a mercury cathode plus $Ti(OH)_3$–Mo(III) as well as for the analogous homogeneous system with additions of guanidine. The first system showed a very pronounced dependence of the products (hydrazine and ammonia) yield on current density. The maximum for ammonia yield is observed at 0.5 A/cm^2 current density. The authors (*24*) explained it by the change of the mercury electrode interface charge. At a current density of 0.5 A/cm^2, the metal amalgam having a zero surface charge is formed, which brings about the optimum adsorption of the negatively charged hydroxide on the amalgam surface, ensuring a potential sufficient to reduce nitrogen. At lesser current densities, the potential appears to be insufficient; at larger ones the negative charge increases, impeding the hydroxide adsorption on the electrode. Under optimum conditions, the current yield reaches 43% with respect to ammonia and 4% with respect to hydrazine.

In the $Ti(OH)_3$–Mo(III)–guanidine system, the rate of hydrazine formation v passes its maximum at the potential $\phi = -1.9$ V (*26*). In the range of -1.6 to -1.9 V, a linear dependence of log $v_{N_2H_4}$ on ϕ is reported. Thus the system behaves according to the Tafel equation

$$\phi = a + b \log i,$$

where i is the partial current used to reduce nitrogen to hydrazine. Hence, the rate-determining stage of the electrode process is thought to be electron transfer from the cathode to the catalyst active center containing the coordinated nitrogen molecule.

Electrolysis in this system was carried out under H_2 depolarization of the anode. Hydrogen oxidation occurs on a nickel catalyst in the anode region that is separated from the cathode region by a ceramic diaphragm. Thus the following electrode processes take place in the cell:

$$\begin{array}{rl} \text{on the cathode:} & N_2 + 4e^- + 4AH \longrightarrow N_2H_4 + 4A^- \\ \text{on the anode:} & 2H_2 - 4e^- + 4A^- \longrightarrow 4HA \\ \hline & N_2 + 2H_2 \longrightarrow N_2H_4 \end{array}$$

D. *Nitrogen Reduction in Homogeneous Aqueous and Alcoholic Solutions of V(II) Complexes with Catechol*

Complexes of V(II) with catechol and its analogs form homogeneous aqueous and alcoholic solutions that possess the unique ability to reduce nitrogen to ammonia, even at room temperature and atmospheric nitrogen pressure (*30*). At elevated pressures (P_{N_2} = 100 atm, T = 25°C), the reaction is completed within a few minutes. The reaction rate depends strongly on the concentration of the base present. Maximum ammonia yields are detected in aqueous solutions at pH 10.4. At pH < 8 and pH $>$ 13.5, nitrogen reduction practically does not occur. Together with the nitrogen reduction, H_2 evolution occurs both in the absence and in the presence of N_2, the former evolving even at high N_2 pressures. The study has shown that hydrogen evolution proceeds in two parallel reactions. One is an independent reduction of solvent protons by V(II) complexes, and the other is coupled with N_2 reduction and produces one H_2 per each N_2 reduced. Therefore, reduction of nitrogen corresponds to an eight-electron stoichiometry. When the nitrogen reduction is stopped in its initial stages by adding excess acid or oxidant ($VOSO_4$), a small quantity of hydrazine is found, which may confirm its intermediate formation. A kinetic study of hydrazine reduction by the same system has shown, however, that its rate is at least two orders of magnitude higher than it would have been had hydrazine, found in the system when nitrogen is being reduced, been the intermediate during ammonia formation. From this it can be inferred that hydrazine, which is detected at addition of acid, is produced from the hydrazine derivative of vanadium. This derivative, which is a real intermediate, is reduced faster than free N_2H_4.

Both the eight-electron stoichiometry and intermediate hydrazine derivative formation in the V(II)–catechol system are similar to these phenomena in enzymatic N_2 reduction. Similar to nitrogenase, only *cis*-deuteroethylene is formed when C_2D_2 is reduced by V(II) catechol system in D_2O. Therefore, investigation of the mechanism of N_2 reduction in this homogeneous system may provide a clue to some of the mechanistic aspects of the nitrogenase reaction.

The study of N_2 reduction kinetics has shown this reaction to be of the second order with respect to the total vanadium concentration and of the first order with respect to the N_2 concentration:

$$-d[\mathrm{V(II)}]/dt = k_1[\mathrm{V(II)}]^2[N_2] + k_2[\mathrm{V(II)}]^{1/2}.$$

The second term of the equation corresponds to the parallel and independent reaction with a solvent producing H_2.

The N_2-reduction rate dependence on the base concentration passes through a sharp maximum (Fig. 11). This indicates the existence of a vanadium(II) complex that is active toward N_2, apparently in equilibrium with other inactive complexes, the basicity of the solution being responsible for the shift in the equilibrium.

Two complex species are believed to participate in the reaction with one N_2 molecule. This leads to the second-order kinetics of the reaction with respect to V(II) concentration and a pronounced dependence of the reaction rate on the base concentration.

Electron spin resonance studies have made possible certain conclusions as to the nature of the complexes present in the system (*27*). At least three V(II) complexes with catechol have turned out to be present in alkaline media, two of these are trimeric and one is monomeric. With an

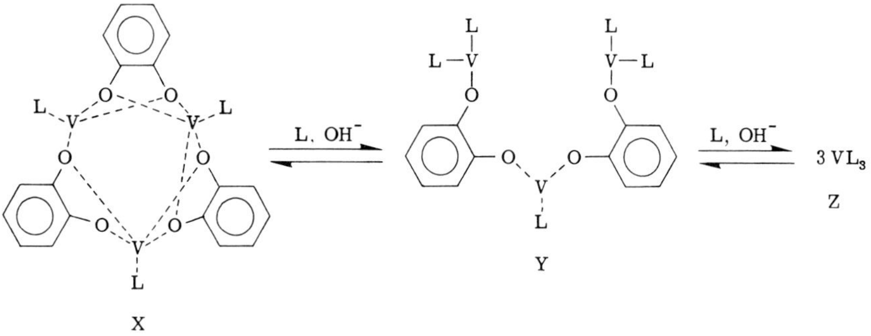

X $\underset{}{\overset{\mathrm{L,\ OH^-}}{\rightleftharpoons}}$ Y $\underset{}{\overset{\mathrm{L,\ OH^-}}{\rightleftharpoons}}$ $3\,\mathrm{VL_3}$ Z

L = catechol

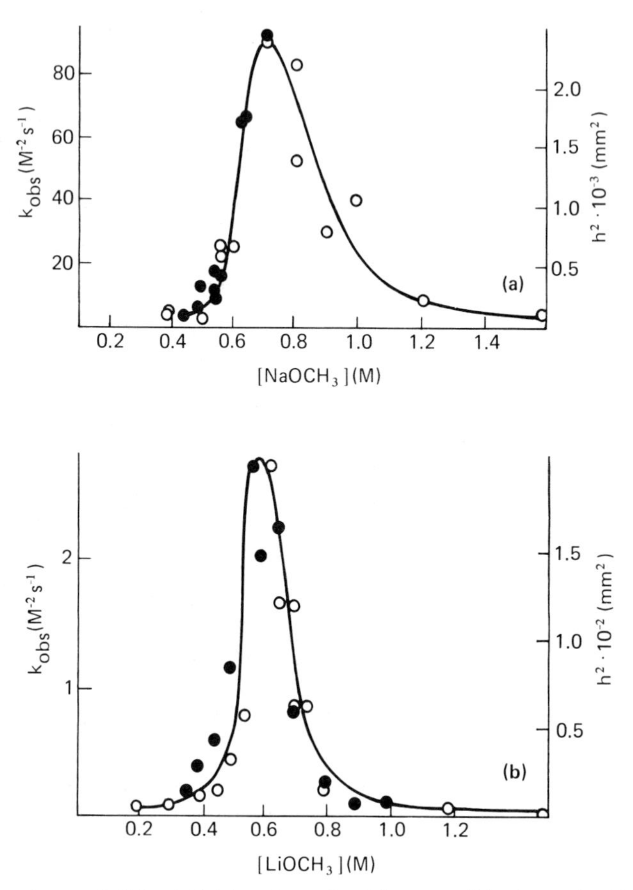

Fig. 11. Dependence of NH_3 yield (●) and esr signal intensity (h) of complex Y (○) on the base concentration.

increase of catechol and concentration basicity, the equilibrium is shifted from left to right.

The trimeric complex Y has turned out to be active toward nitrogen, which follows from a comparison of the reaction rate and its concentration (Fig. 11). Its flexible nature apparently allows nitrogen coordination and its further reduction. The reduction mechanism may be presented as follows:

$$\underset{\text{Y}}{\text{V--L}\cdots\text{V}\cdots\text{L--V}} + N_2 \rightleftharpoons \underset{\text{YN}_2}{\text{V--N}\equiv\text{N--V},\ \text{V--L}\cdots\text{V}\cdots\text{L--V}}$$

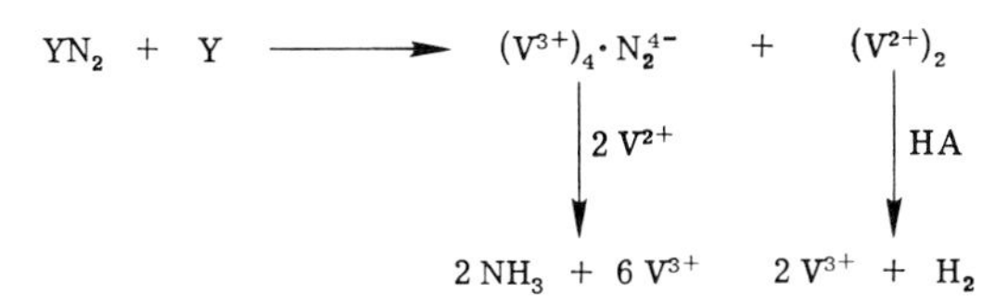

The reaction rate is of the second order with respect to the concentration of the trimeric complex Y. Thus a polynuclear complex V(II) is proved to be active in nitrogen reduction. It contains six ions of V^{2+}, four of which reduce N_2 to a hydrazine derivative, the other two ensuring coupled hydrogen formation.

VI. Conclusion

The experimental evidence on the molecular nitrogen reduction in solutions of transition-metal complexes confirm, on the whole, the initial assumptions and general conclusions that could be made by the analysis of the thermodynamics of N_2 consecutive reduction and the structure of N_2 complexes with transition-metal complexes.

In the presence of strong reducing agents in aprotic media, the reduction may proceed with the participation of binuclear complexes such as Ti(III), Ti(II), Zr(II) (electronic configuration d^1 and d^2), or mononuclear complexes: Mo(0), W(0), Fe(0) (electronic configuration d^6 and d^8). The reduction of coordinated nitrogen may proceed by different mechanisms, among them two-electron reduction with a diimide derivative formation. The addition of proton donors facilitates the electron transfer and nitrogen reduction in the coordination sphere of the complex.

In protic media there arise some additional limitations connected with difficulties for employing reducing agents that are too strong. In this case, an effective process is carried out using a polynuclear center capable of four-electron N_2 reduction[1] to form a hydrazine derivative. The process may proceed as a rearrangement of bonds rather than a simple electron transfer:

$$M\cdots N\equiv N\cdots M \xrightarrow{HA} M{=}N{-}\overset{+}{N}H{=}M \quad A^-$$

As can be expected, the V(II) and Mo(III) derivatives in which sufficiently strong reducing properties of M are combined with an appropriate electronic configuration (d^3) react most effectively. In the case of V(II),

[1] We define this term as the reaction without intermediate formation of free diimide, N_2H_2. It is not necessarily a one-step process.

no additional reducing agent is required, and the system is stoichiometric. In the case of Mo(III), an additional, stronger reducing agent [Ti(III), Cr(II), Na(Hg), or a cathode] is necessary, and the system can be turned into a catalytic one with Mo(III) as a catalyst.

If an active center is a strong reducing agent (hydroxides at high pH or Na amalgam), the hydrazine derivative formed may evolve free N_2H_4 under the action of the solvent.

In the case of weaker reducing agents [hydroxides at low pH or homogeneous V(II)–catechol system], the hydrazine derivative prefers to await two additional electrons to split the N—N bond to form ammonia. The hydrazine derivative can evolve N_2H_4 on the addition of acid as a proton donor (or alkali, strengthening the metal donor properties).

The accumulation of electrons in the polynuclear active center by numbers larger than those required to reduce N_2 may lead to a coupled hydrogen formation.

All of these properties belong both to nonbiological systems and to the enzymatic nitrogenase process. Evidently, this cannot be just an accidental coincidence.

The study of the enzymatic process encouraged the search for chemical nitrogen-fixing systems in protic media.

Now the study of the mechanism of nonbiological reduction allows one to make some conclusions as to the mechanism of enzymatic reaction. Certain peculiarities of the enzymatic process were first found for nonbiological models (e.g., the possibility of hydrazine formation at the decomposition of an intermediate hydrazine derivative). Molybdenum in the active state of MoFe protein might be in the form of Mo(III), which is also active in nonbiological model systems, as was already clear in 1970.

It might be suggested from what has just been described that a binuclear Mo(III)—N≡N—Mo(III) complex with optimum d^3 configuration of molybdenum is an intermediate state in both chemical and enzymatic N_2 reduction. For the latter case two molybdenum atoms must be provided with two FeMo cofactors. So far there is no evidence that molybdenum atoms are situated close enough in the enzyme to form the complex. It should be mentioned that the distance between the two molybdenum atoms should not be too short (perhaps longer than 5 Å) because the structure of the initial complex must allow the insertion of an N_2 molecule between the two molybdenum atoms to form a linear binuclear fragment.

It is not excluded, however, that in binuclear complex with an N_2 molecule, iron takes part together with molybdenum [e.g., Mo(IV)NN · Fe(II) or Mo(III)NN · Fe(III)], with the number of *d* electrons (8) that is still close to optimum value for N_2 reduction (Section V,D).

From the model studies we may suggest the mechanism for coupled H_2

formation as well as HD formation catalyzed by N_2. If six electrons are available for four-electron reduction to hydrazine, then two extra electrons can be used to form H_2 [as in the case of the V(II)–catechol complex, in which two trimers react with one N_2 molecule]. The mechanism of this reaction might include "opening" to the solvent protons of a low-valent metal ion that would react with two H^+ ions to form H_2. The same low-valent metal ion (possibly Mo) may react with D_2 to form dideuteride D—M—D in the oxidation addition reaction, and the latter then will produce two HD molecules in the usual hydrolytic reaction with solvent protons.

No HD was detected for the V(II) catechol system in the presence of D_2, which could be explained by a lesser tendency of V to form hydrides as compared with molybdenum.

The author recognizes that the ideas and suggestions presented here are far from being universally accepted. The investigators working in the field of biological and chemical nitrogen fixation usually have somewhat different views. For example, some suggest a mononuclear complex of Mo with N_2 as an intermediate for nitrogenase reaction.

The future will show to what degree the considerations developed here are correct. At any rate, these considerations have helped to discover the effective chemical nitrogen-fixing systems that have as yet no analoges in other works.

References

1. Bottomley, F., *in* "A Treatise on Dinitrogen Fixation" (R. Hardy, F. Bottomley, and R. Burns, eds.), Sect. I, p. 109. Wiley, New York, 1979.
2. Shilov, A. E., *in* "Biological Aspects of Inorganic Chemistry" (A. W. Addison, W. R. Cullen, D. Dolphin, and B. R. James, eds.), p. 197. Wiley, New York, 1977.
3. Shilov, A. E., *in* "A Treatise on Dinitrogen Fixation" (R. Hardy, F. Bottomley, and R. Burns, eds.), Sect. I, p. 31. Wiley, New York, 1979.
4. Shilov, A. E., *in* "New Trends in the Chemistry of Nitrogen Fixation" (J. Chatt, L. M. da Camara Pina, and R. L. Richards, eds.), p. 129. Academic Press, London, 1980.
5. Allen, A. D., and Senoff, C. V., *J. Chem. Soc. Chem. Commun.* No. 23, p. 621 (1965).
6. Kobeleva, S. I., and Denisov, N. T., *Kinet. Katal.* **18,** 794 (1977).
7. Shah, V. K., and Brill, W. J., *Proc. Natl. Acad. Sci. USA* **74,** 3249 (1977).
8. Vol'pin, M. E., and Shur, V. B., *Dokl. Akad. Nauk SSSR* **156,** 1103 (1964).
9. Vol'pin, M. E., and Shur, V. B., *J. Organomet. Chem.* **200,** 319 (1980).
10. Bercaw, J. E., Rosenberg, E., and Roberts, J. D., *J. Am. Chem. Soc.* **96,** 612 (1974).
11. Burgess, B. K., Wherland, S., Newton, W. E., and Stiefel, E. I., *Biochemistry* **20,** 5140 (1981).
12. Bazhenova, T. A., Lobkovskaya, R. M., Shibaeva, R. P., Shilov, A. E., Gruselle, M., Leny, G., and Tchoubar, B., *Kinet. Katal.* **23,** 246 (1982).

13. Burbo, E. M., Denisov, N. T., Shestakov, A. F., and Shilov, A. E., *Kinet. Katal.* **23,** 226 (1982).
14. Centerick, F., Peeters, J., Heremans, K., DeSmedt, H., and Olbrichts, H., *Eur. J. Biochem.* **87,** 401 (1978).
15. Chatt, J., Dilworth, J. R., and Richards, R. L., *Chem. Rev.* **78,** 589 (1978).
16. Denisov, N. T., Efimov, O. N., Shuvalova, N. I., Shilova, A. K., and Shilov, A. E., *Zh. Fiz. Khim.* **44,** 2294 (1970).
17. Denisov, N. T., Shuvalov, V. F., Shuvalova, N. I., Shilova, A. K., and Shilov, A. E., *Dokl. Akad. Nauk SSSR* **195,** 879 (1970).
18. Denisov, N. T., Shuvalov, V. F., Shuvalova, N. I., Shilova, A. K., and Shilov, A. E., *Kinet. Katal.* **11,** 813 (1970).
19. Denisov, N. T., Terekhina, G. G., Shuvalova, N. I., and Shilov, A. E., *Kinet. Katal.* **14,** 939 (1975).
20. Denisov, N. T., Kobeleva, S. I., Shilov, A. E., and Shuvalova, N. I., *Kinet. Katal.* **21,** 1257 (1980).
21. Denisov, N. T., Kobeleva, S. I., and Shuvalova, N. I., *Kinet. Katal.* **21,** 1251 (1980).
22. Didenko, L. P., Shilov, A. E., and Shilova, A. K., *Kinet. Katal.* **20,** 1488 (1979).
23. Didenko, L. P., Shilova, A. K., and Shilov, A. E., *Dokl. Akad. Nauk SSSR* **254,** 643 (1980).
24. Gorodyskii, A. V., Danilin, V. V., Efimov, O. N., Nechaeva, N. E., and Tsarev, V. N., *React. Kinet. Catal. Lett.* **11,** 337 (1979).
25. Hardy, R., Bottomley, F., and Burns, R. (Eds.) "A Treatise on Dinitrogen Fixation" Sect. II, p. 383. Wiley, New York, 1979.
26. Kulakovskaya, S. I., Tsarev, V. N., Efimov, O. N., and Shilov, A. E., *Kinet. Katal.* **18,** 1045 (1977); Kulakovskaya, S. I., Vasilieva, L. P., and Efsimov, O. N., *React. Kinet. Catal. Lett.* **14,** 181 (1980).
27. Luneva, N. P., Moravsky, A. P., and Shilov, A. E., *Nouv. J. Chim.* **6,** 245 (1982).
28. Manriques, J. M., Sanner, R. D., Marsh, R. E., and Bercaw, J. E., *J. Am. Chem. Soc.* **98,** 3042 (1976).
29. Nikolaeva, G. V., Efimov, O. N., Brikenshtein, A. A., and Denisov, N. T., *Zh. Fiz. Khim.* **50,** 3030 (1976).
30. Nikonova, L. A., Isaeva, S. A., Pershikova, N. I., and Shilov, A. E., *J. Mol. Catal.* **1,** 367 (1975–76).
31. Pienkos, P. I., Shah, V. K., and Brill, W. J., *Proc. Natl. Acad. Sci. USA* **74,** 5468 (1977).
32. Shilov, A. E,. Shilova, A. K., and Borod'ko, Yu. G., *Kinet. Katal.* **7,** 768 (1966).
33. Shilov, A. E., Shilova, A. K., and Kvashina, E. F., *Kinet. Katal.* **10,** 1402 (1969).
34. Shilov, A. E., Shilova, A. K., and Vorontsova, T. A., *React. Kinet. Catal. Lett.* **3,** 143 (1975).
35. Thorneley, R. N. F., Eady, R. R., and Lowe, D. J., *Nature (London)* **272,** 557 (1978).
36. Wherland, S., Burgess, B. K., Stiefel, E. I., and Newton, W. E., *Biochemistry* **20,** 5132 (1981).
37. Zeinstra, J. B., Teuben, J. H., and Jellinek, F., *J. Organomet. Chem.* **170,** 39 (1979).
38. Pocklage, S. M., and Schrock, R. R., *J. Am. Chem. Soc.* **104,** 3077 (1982).
39. Denisov, N. T., Kobeleva, S. I., Korostelyova, A. I., Shevchenko, V. V., and Shuvalov, V. F., *Kinet. Katal.* **24,** 168 (1983).
40. Denisov, N. T., Ovcharenko, A. G., Svirin, V. G., Shilov, A. E., and Shuvalova, N. I., *Nouv. J. Chim.* **3,** 403 (1979).

Index

A

B

C

D

F

G

H

I

K

L

Q

R

S

T

U

V

W

Z